Günter Obe • Vijayalaxmi (Editors)

Chromosomal Alterations

Methods, Results and Importance in Human Health

Günter Obe

Vijayalaxmi (Editors)

Chromosomal Alterations

Methods, Results and Importance in Human Health

With 99 Figures, 11 in Color, and 47 Tables

Springer

Professor Dr. Günter Obe
Gershwinstrasse 33
14513 Teltow
Germany

Professor Dr. Vijayalaxmi
Department of Radiation Oncology
The University of Texas
Health Science Center
7703 Floyd Curl Drive
San Antonio, TX 78229
USA

Cover:

In Remembrance of Sarah

Chromosomal aberrations in a metaphase of cultured fibroblasts exposed to mitomycin C (50 ng/ml for 24 hrs). Fibroblasts are from a girl with Fanconi anemia, complementation group FA-N, caused by mutations in the PALB2 gene (Reid S et al., Nat Genet 39, 162-164, 2007). The checkpoint defect in FA-N also leads to aneuploidy due to non-disjunction.
Sarah died from childhood cancer at the age of 12 months due to severe chromosome instability.

Library of Congress Control Number: 2007923871

ISBN: 978-3-540-71413-2 Springer-Verlag Berlin Heidelberg New York

eISBN: 978-3-540-71414-9 Springer-Verlag Berlin Heidelberg New York

Springer-Verlag is a part of Springer Science+Business Media

springer.com

Editor: Dr. Dieter Czeschlik, Heidelberg
Desk Editor: Anette Lindqvist, Heidelberg
Production: SPi
Typesetting: SPi
Cover Design: Design & Production, Heidelberg

Printed on acid-free paper 31/3152-HM 5 4 3 2 1 0

Preface

This book deals mainly with chromosomal alterations which represent changes in the structure (chromosome mutations) or number (genome mutations) of chromosomes and with sister chromatid exchanges (SCE).

Double-strand breaks (DSB) in DNA are the ultimate lesions for the formation of chromosome mutation and SCE evaluated in the light microscope. Typical chromosomal alterations are chromosomal aberrations (CA) and micronuclei (MN). SCE result from repair of lesions during S phase ("true" SCE), or from chromosome-type CA induced during G1 phase ("false" SCE). Numerical alterations of chromosomes are discussed with respect to their importance in clinical cytogenetics, to their ability to give rise to MN containing whole chromosomes and to their origin during divisions of CA containing cells.

Evaluation of chromosomal alterations, especially CA, MN and SCE, has long been used in basic and applied research. Investigations into the mechanism(s) of the origin of chromosomal alterations lead to insights into the structure and function of chromosomes. Increased indices of chromosomal alterations in cells, in vitro and in vivo, are used as indicators of exposure to biological, chemical and physical genotoxins in our environment. Furthermore, elevated genotoxicity has been correlated with increased carcinogenicity. This fits with the observation of specific types of CA in cancer cells and with a positive correlation between elevated frequencies of CA and cancer risk in human populations.

DSB are induced directly by powerful DNA-damaging agents such as ionizing radiation and restriction endonucleases which lead to the formation of CA in the same cell cycle stage in which the DSB are induced: in G0/G1 phase as chromosome-type CA, in S phase as both chromosome-type and chromatid-type CA (depending on whether DSB occur in unreplicated or in replicated DNA, respectively) and in G2 phase as chromatid-type CA. Most genotoxic agents induce DNA lesions other than DSB which during S phase lead to chromatid-type CA and to "true" SCE.

CA, MN and SCE are analyzed in the light microscope after appropriate staining (generally Giemsa stain). Special pretreatment of fixed metaphase chromosomes on slides followed by staining with Giemsa stain leads to specific banding patterns that allow in-depth analyses of specific types of CA such as translocations and intra chromosomal and interchromosomal distributions of CA. Application of fluorescence in situ hybridization (FISH) opened new insights into unexpected complexities of CA. Following differential substitution of chromosomal DNA with specific agents such as bromodeoxyuridine, SCE can be made visible by staining with Giemsa stain. MN result from CA or from whole chromosomes not distributed to the cell poles. Analyses of MN are usually carried out in binucleate second-division cells after exposure to a mutagen. FISH techniques are extremely useful to investigate mechanistic aspects of the formation of MN.

The book starts with a chapter on structural and functional aspects of human chromosomes. Chapters 2–8 describe DNA lesions leading to chromosomal alterations (including gene mutations), and the ability of the cells to repair such lesions. Chapters 9 and 10 explain how DNA damage can be measured by means of γ-H2AX foci and by the comet assay. Chapters 11–14 explore mechanisms of the origin of CA (Chap. 11), the influence of nuclear and chromatin structure on CA formation (Chaps. 12, 13) and CA and SCE in telomeric regions (Chap. 14). Chapters 15–17 are devoted to MN and SCE. FISH methods and their applications for the analysis of chromosomal alterations are presented in Chaps. 18 and 19. In the following chapters topics such as changing patterns of chromosomal alterations in ongoing cell cycles (Chap. 20), human-population monitoring (Chap. 21), biological dosimetry (Chaps. 22–24) and CA in peripheral lymphocytes of astronauts (Chap. 25) are included. Chapters 26–29 are devoted to the question of possible carcinogenic and chromosome-damaging effects of low- and high-frequency electromagnetic fields. The importance of chromosomal alterations in cancer cells and as indicators of cancer risk in human populations is highlighted in Chaps. 30 and 31.

We would like to thank our colleagues who generously and willingly contributed to this book.

The editors hope that the book will serve as a textbook for graduate students in biological sciences, residents in radiology and radiation oncology, as well as researchers interested in occupational exposures, ionizing and non-ionizing radiation, environmental mutagenesis and carcinogenesis.

January 2007
Günter Obe
Vijayalaxmi

Contents

List of Contributors

SIDDHARTH G. ADHVARYU
Department of Pathology, University of Texas Health Science Center,
7703 Floyd Curl Drive, San Antonio, TX 78229-3900, USA

PASCALE V. BELLIER
Consumer and Clinical Radiation Protection Bureau, Health Canada,
775 Brookfield Road, Ottawa, ON K1A 1C1, Canada

FERDINANDO BERSANI
Department of Physics, University of Bologna, Viale Berti Pichat 6/2, 40127
Bologna, Italy

STEFANO BONASSI
Unit of Molecular Epidemiology, National Cancer Research Institute, Largo
Rosanna Benzi 10, 16132 Genoa, Italy

PETER E. BRYANT
Bute Medical School, University of St. Andrews, St. Andrews,
Fife KY16 9TS, UK

MARKUS CHRISTMANN
Department of Toxicology, University of Mainz, Obere Zahlbacher Str.
67, 55131 Mainz, Germany

ILSE CHUDOBA
MetaSystems, Robert-Bosch-Str. 6, 68804 Altlussheim, Germany

CHRISTOPH CREMER
Kirchhoff Institute of Physics, INF 227, 69120 Heidelberg, Germany

RINNE DE BONT
Study Centre Carcinogenesis and Primary Prevention of Cancer,
Department of Radiotherapy, Nuclear Medicine and Experimental
Cancerology, Ghent University, University Hospital 3K3,
De Pintelaan 185, 9000 Ghent, Belgium

María Vittoria Di Tomaso
Department of Genetic Toxicology and Chromosome Pathology, Instituto de Investigaciones Biológicas Clemente Estable, Ministry of Education and Culture, Avenida Italia 3318, CP 11600, Montevideo, Uruguay

Máximo E. Drets
Laboratory of Quantitative Microscopy, Instituto de Investigaciones Biológicas Clemente Estable, Av. Italia 3318, 11600 Montevideo, Uruguay

Sarah Driessen
femu – Research Center for Bioelectromagnetic Interaction, RWTH Aachen University, 30 Pauwelsstrasse, 52074 Aachen, Germany

Marco Durante
Department of Physics, University Federico II, Monte S. Angelo, Via Cintia, 80126 Naples, Italy

Alan A. Edwards
Radiation Protection Division, Health Protection Agency, Chilton, Didcot OX11 0RQ, UK

Michael Fenech
CSIRO Human Nutrition, Genome Health Nutrigenomics Project, PO Box 10041, Adelaide BC, Adelaide, SA 5000, Australia

Kerry George
Wyle Laboratories, 1290 Hercules Drive, Houston, TX 77058-2759, USA

Wolfgang Goedecke
Department Biology and Geography, University of Duisburg-Essen, Universitätsstraße 5, 45117 Essen, Germany

Makoto Hayashi
Division of Genetics and Mutagenesis, National Institute of Health Sciences, 1-18-1 Kamiyoga, Setagaya-ku, Tokyo 158-8501, Japan

George Iliakis
Institute of Medical Radiation Biology, University of Duisburg-Essen Medical School, 45122 Essen, Germany

Christian Johannes
Department of Genetics, Universität Duisburg-Essen, Universitätsstraße 5, 45117 Essen, Germany

Bernd Kaina
Department of Toxicology, University of Mainz, Obere Zahlbacher Str. 67, 55131 Mainz, Germany

Jürgen Kiefer
Am Dornacker 4, 35435 Wettenberg, Germany

Frank Klubertz
femu – Research Center for Bioelectromagnetic Interaction, RWTH Aachen University, 30 Pauwelsstrasse, 52074 Aachen, Germany

Gregor Kreth
Kirchhoff Institute of Physics, INF 227, 69120 Heidelberg, Germany

Tracy Lightfoot
Epidemiology and Genetics Unit, Department of Health Sciences, University of York, York, UK

David C. Lloyd
Radiation Protection Division, Health Protection Agency, Chilton, Didcot OX11 0RQ, UK

Wilner Martínez-López
Department of Genetic Toxicology and Chromosome Pathology, Instituto de Investigaciones Biológicas Clemente Estable, Ministry of Education and Culture, Avenida Italia 3318, CP 11600, Montevideo, Uruguay, and Institute of Biology, Faculty of Sciences, Montevideo, Uruguay

James P. McNamee
Consumer and Clinical Radiation Protection Bureau, Health Canada, 775 Brookfield Road, Ottawa, ON K1A 1C1, Canada

Leticia Méndez-Acuña
Department of Genetic Toxicology and Chromosome Pathology, Instituto de Investigaciones Biológicas Clemente Estable, Ministry of Education and Culture, Avenida Italia 3318, CP 11.600, Montevideo, Uruguay

William F. Morgan
Radiation Oncology Research Laboratory, and Marlene and Stewart Greenebaum Cancer Center, BRB 7-011, University of Maryland, 655W. Baltimore St., Baltimore, MD 21201-1551, USA

María Mühlmann
Molecular Cytogenetic Laboratory, Radiobiology Department, National Commission of Atomic Energy, Buenos Aires, Argentina

Wolfgang-Ulrich Müller
Institut für Medizinische Strahlenbiologie, Universität Duisburg-Essen, Universitätsklinikum Essen, Hufelandstraße 55, 45122 Essen, Germany

HEIDEMARIE NEITZEL
Institut für Humangenetik, Charité – Universitätsmedizin Berlin, Campus Virchow-Klinikum, Augustenburger Platz 1, 13353 Berlin, Germany

GÜNTER OBE
Gershwinstrasse 33, 14513 Teltow, Germany

URSULA OESTREICHER
Division for Radiation Protection and Health, Federal Office for Radiation Protection, Ingolstaedter Landstrasse 1, 85764 Oberschleissheim, Germany

GABRIEL E. PANTELIAS
Institute of Nuclear Technology and Radiation Protection, National Centre for Scientific Research "Demokritos", 15310 Ag. Paraskevi Attikis, Athens, Greece

LOTHAR RIEDEL
Bundesamt für Verbraucherschutz und Lebensmittelsicherheit, Mauerstraße 39-42, 10117 Berlin, Germany

HORST ROMM
Division for Radiation Protection and Health, Federal Office for Radiation Protection, Ingolstaedter Landstrasse 1, 85764 Oberschleissheim, Germany

WYNAND P. ROOS
Department of Toxicology, University of Mainz, Obere Zahlbacher Str. 67, 55131 Mainz, Germany

PAVEL RÖSSNER
Laboratory of Genetic Ecotoxicology, Institute of Experimental Medicine AS CR, Vídeňská 1083, 142 20 Prague 4, Czech Republic

FEDERICO F. SANTIÑAQUE
Laboratory of Quantitative Microscopy, Instituto de Investigaciones Biológicas Clemente Estable, Av. Italia 3318, 11600 Montevideo, Uruguay

JOHN R.K. SAVAGE
34 City Road, Tilehurst, Reading RG31 5HB, UK

MARIA ROSARIA SCARFÌ
CNR Institute for Electromagnetic Sensing of Environment, Via Diocleziano 328, 80124 Naples, Italy

JOACHIM SCHÜZ
Institute of Cancer Epidemiology, Danish Cancer Society, Copenhagen, Denmark

JIRI SILNY
Femu – Research Center for Bioelectromagnetic Interaction, RWTH Aachen University, 30 Pauwelsstrasse, 52074 Aachen, Germany

MARIANNE B. SOWA
Chemical Structure & Dynamics, Pacific Northwest National Laboratory, Richland, WA 99354, USA

RADIM J. SRAM
Laboratory of Genetic Ecotoxicology, Institute of Experimental Medicine AS CR, Vídeňská 1083, 142 20 Prague 4, Czech Republic

GÜNTER STEPHAN
Division for Radiation Protection and Health, Federal Office for Radiation Protection, Ingolstaedter Landstrasse 1, 85764 Oberschleissheim, Germany

MARTA SZŁUIŃSKA
Radiation Protection Division, Health Protection Agency, Chilton, Didcot OX11 0RQ, UK

IRENA SZUMIEL
Department of Radiobiology and Health Protection, Institute of Nuclear Chemistry & Technology, ul Dorodna 16, 03-195 Warsaw, Poland

GEORGIA I. TERZOUDI
Institute of Nuclear Technology and Radiation Protection, National Centre for Scientific Research "Demokritos", 15310 Ag. Paraskevi Attikis, Athens, Greece

MARC TRIMBORN
Institut für Medizinische Genetik, Charité – Universitätsmedizin Berlin, Campus Virchow-Klinikum, Augustenburger Platz 1, 13353 Berlin, Germany

NIK VAN LAREBEKE
Study Centre Carcinogenesis and Primary Prevention of Cancer, Department of Radiotherapy, Nuclear Medicine and Experimental Cancerology, Ghent University, University Hospital 3K3, De Pintelaan 185, 9000 Ghent, Belgium

URSULA VON RECKLINGHAUSEN
Institute of Hygiene and Occupational Health, Universitätsklinikum Essen, Hufelandstraße 55, 45122 Essen, Germany

CLEMENS VON SONNTAG
Max-Planck-Institut für Bioanorganische Chemie, Stiftstr. 34-36, 45413 Mülheim an der Ruhr, Germany, and Leibniz-Institut für Oberflächenmodifizierung (IOM), Permoserstr. 15, 04303 Leipzig, Germany

Minli Wang
Institute of Medical Radiation Biology, University of Duisburg-Essen Medical School, 45122 Essen, Germany

Roman Wienert
femu – Research Center for Bioelectromagnetic Interaction, RWTH Aachen University, 30 Pauwelsstrasse, 52074 Aachen, Germany

Andrzej Wojcik
Department of Radiation Biology and Health Protection, Institute of Nuclear Chemistry and Technology, ul. Dorodna 16, 03-195 Warsaw, Poland, and Department of Radiobiology and Immunology, Institute of Biology, Swietokrzyska Academy, ul. Swietokrzyska 15, 25-406 Kielce, Poland

Wenqi Wu
Institute of Medical Radiation Biology, University of Duisburg-Essen Medical School, 45122 Essen, Germany

1 Human Chromosomes: Structural and Functional Aspects

HEIDEMARIE NEITZEL AND MARC TRIMBORN

Abstract

Understanding the structural and molecular basis of the mitotic chromosome remains a basic challenge in cell biology and cytogenetics. The chromosomal behavior during cell division was first described in 1882. At the beginning of the last century, the chromosome theory of inheritance combined the cytological observations with the principles of Mendelian inheritance deduced from breeding experiments. This fundamental theory explained for the first time that the chromosomes are the units of heredity which are arrayed linearly on chromosomes as well as genetic linkage, chromosomal recombination, and the independent assortment of alleles localized on different chromosomes. The chromosome theory of inheritance was the prerequisite for the important improvements in the fields of experimental and clinical cytogenetics. After the definition of the exact number of human chromosomes $2n = 46$ in 1955, it was demonstrated that different human aneuploidies are the leading cause of fetal loss, birth defects, and mental retardation and that duplications or deletions of even smaller chromosomal segments have profound consequences on normal gene expression, resulting in severe developmental and physiologic abnormalities. In the 1980s and 1990s, the field of conventional cytogenetics was revolutionized by the introduction of molecular cytogenetic techniques resulting in the recognition of the importance of subtle cytogenetic aberrations, such as microdeletions and microduplications. Our insights into cell cycle progression, which is coordinated by a complex network of checkpoints to monitor chromosome structure, with DNA repair and spindle formation led to the identification of Mendelian disorders affecting chromosome integrity, some of them associated with high genomic instability and a markedly increased cancer risk. It was the thorough observation of chromosomal changes by many researchers during the last four decades which paved the way for our understanding of the underlying mechanisms of many congenital disorders, as well as of chromosome surveillance, DNA repair, and cancer susceptibility.

Günter Obe and Vijayalaxmi (Eds.)
Chromosomal Alterations: Methods, Results and Importance in Human Health

1.1 History of Chromosome Research

"Chromosomes have attracted many microscopists not only because these sausage-like bodies represent vehicles of genetic material (and hence, are biologically important) but also because they are hypnotically beautiful objects" (Hsu 1979).

The first cytologist who described chromosome behavior during cell division and how chromosomes move during mitosis was Walter Flemming (1882) in 1882. His terms "prophase," "metaphase," and "anaphase" are still used to describe the different steps of mitosis. In 1888 the structures were termed "chromosomes" (Greek *chroma* meaning "color" and *soma* meaning "body") by the German anatomist Heinrich Waldeyer, because they were particularly well stained with a certain nuclear dye. Flemming's work and the rediscovery of Mendel's laws were the basis of the chromosome theory of inheritance. Using the fruit fly *Drosophila melanogaster* as a model organism, Thomas Hunt Morgan and his students at Columbia University, who included such important geneticists as Alfred Sturtevant, Hermann Muller, and Calvin Bridges, made many important contributions to genetics. They showed that genes, strung on chromosomes, are the units of heredity which are arrayed linearly on chromosomes. They described the independent assortment of alleles localized on different chromosomes, X-linked inheritance, genetic linkage, and chromosomal recombination.

The chromosome theory of inheritance combined the cytological observations with the principles of Mendelian inheritance deduced from breeding experiments (Fig. 1.1). The two homologous chromosomes in somatic cells correspond to two alleles, one of each inherited from the mother and the father. The chiasmata observed in meiosis I corresponds to the recombination or crossing over events. The segregation of homologous chromosomes during meiosis correlates with the segregation of alleles into separate gametes.

In 1953 Francis Crick and James Watson described the double-helix structure of DNA and concluded in their paper, published in *Nature*, "It has not escaped our notice that the specific pairing we have postulated immediately suggests a possible copying mechanism for the genetic material" (Watson and Crick 1953). Subsequently, Matthew Meselson and Franklin Stahl (1958) demonstrated that DNA replicates semiconservatively, with each strand in a DNA molecule from the parent generation pairing with a new strand in the daughter generation.

In 1955 Tjio and Levan (1956) defined the exact number of human chromosomes $2n = 46$ and in 1959 Lejeune et al. (1959) described the first aneuploidy in humans, demonstrating that children with Down syndrome have three instead of two copies of chromosome 21. Although more than 100 years has passed since the first observation of the chromosome (Flemming 1882) a number of researchers have continued to study chromosomes and their behavior during the cell cycle. Important improvements were made during

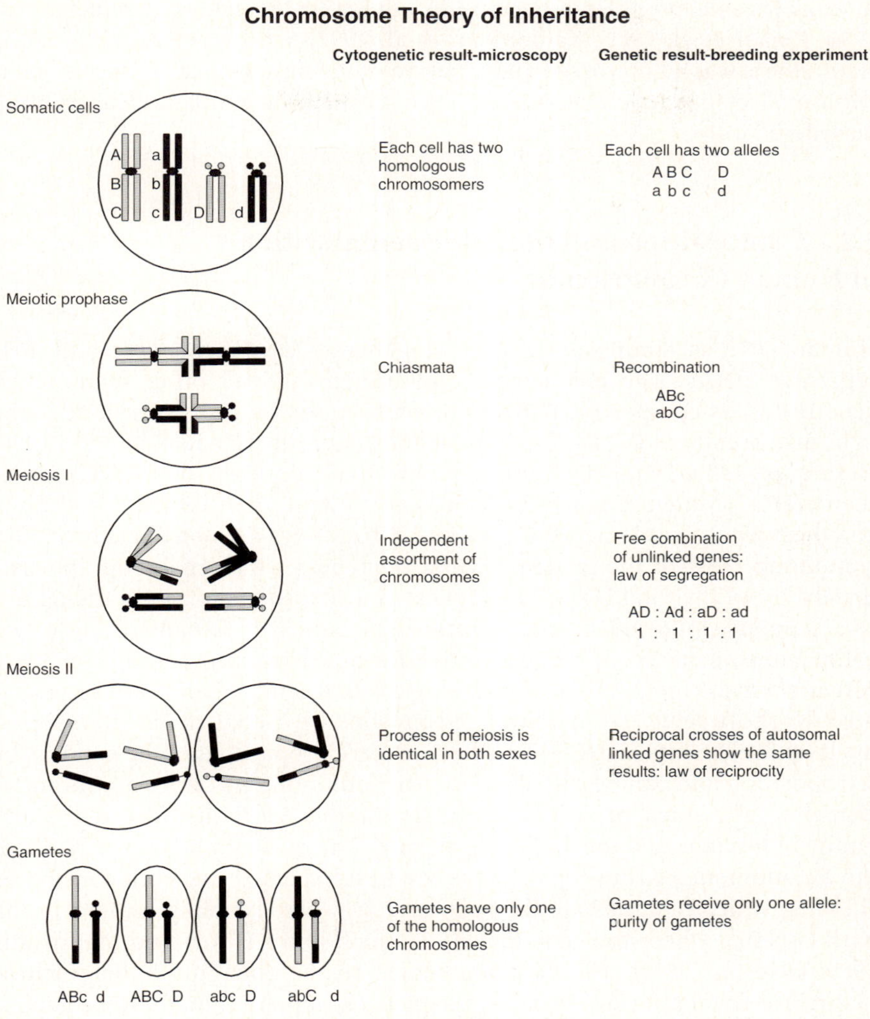

Fig. 1.1 Chromosome theory of inheritance (for details, see the text)

the 1960s and 1970s. The establishment of lymphocyte cultures as an easily accessible source for chromosome preparations (Nowell 1960) and the development of the differential banding methods in the 1970s led to considerable improvement in the fields of experimental and clinical cytogenetics (Caspersson et al. 1969; Drets and Shaw 1971; Dutrillaux and Lejeune 1971; Patil et al. 1971; Sumner et al. 1971). In the 1980s and 1990s, the field of conventional cytogenetics was again revolutionized by the introduction of molecular cytogenetic techniques such as fluorescence in situ hybridization (FISH), spectral karyotyping, and comparative genomic hybridization (CGH), leading to considerable progress especially in clinical cytogenetics (Cremer et al. 1986; Pinkel et al. 1986; Kallioniemi et al. 1992; Schrock et al. 1996). As a result, there was an increased appreciation of the importance of "subtle" constitutional cytogenetic aberrations, such as microdeletions and imprinting disorders.

1.2 Composition and Compartmentalization of Human Chromosomes

Human DNA is composed of 60% single-copy DNA sequences and 40% repetitive DNAs. The characteristic of human and all other mammalian genomes is its compartmentalization, which finds its expression under the light microscope as G-, R-, T-, and C-bands (Korenberg and Rykowski 1988; Holmquist 1989, 1992). The different chromatin domains differ not only in their AT/GC content but also in their gene content, their replication timing, and their repetitive elements. G-, R-, and T-bands are defined as euchromatin containing most of the protein coding sequences which make up approximately 2% of the total DNA. The highest gene density is found in the telomeric T-bands, followed by the Giemsa-light bands or R-bands. These two compartments harbor predominantly the housekeeping genes which are essential for the metabolism of each single cell. R- and T-bands are accompanied by short repetitive nonviral retroposons, so-called short interspersed nucleotide elements (SINEs). These repeats are propagated in the genome by retroposition and can be classified as either autonomous or nonautonomous elements. The most prominent nonautonomous SINE member is the Alu family in humans and the B1 family in mice. The most abundant members of the autonomous retroposons are the long interspersed nucleotide elements (LINEs), which are several kilobases long and preferentially located in the dark G-bands where the gene density is much lower (Chen and Manuelidis 1989; Ostertag 2001). The C-bands equate to the constitutive heterochromatin, are devoid of protein coding sequences, and are thought to be genetically inert. Constitutive heterochromatin is mainly composed of satellite DNA containing simple repetitive elements, organized as complex tandem arrays (Waye and Willard 1989; Willard 1989; Fig. 1.2).

The different compartments are characterized by differences in their replication timing during S phase. Incorporation of the synthetic thymidine analog bromodeoxyuridine during S phase leads to a characteristic banding pattern which shows that the T- and R-bands replicate in the first half of S phase, while the dark G-bands replicate in the second half. At the very end of S phase the constitutive heterochromatin is replicated as well as the inactivated X chromosome in mammalian females designated as facultative heterochromatin.

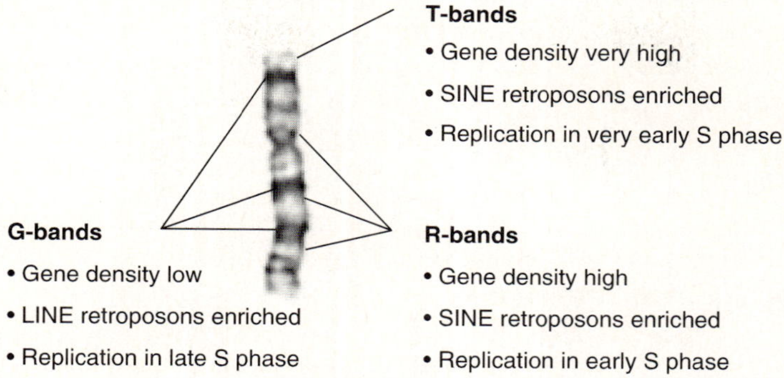

Fig. 1.2 Compartmentalization of human chromosomes. *LINE* long interspersed nucleotide element, *SINE* short interspersed nucleotide element

The differences in gene content of the different compartments can be impressively illustrated by the gene mapping data of the Human Genome Project (Fig. 1.4). The smallest human autosome, chromosome 21, which has a broad Giemsa-dark band comprises 46.9 Mbp of DNA and accommodates 352 genes. In contrast, chromosome 22 consists mainly of R-band material, it is 50 Mbp in physical length, only slightly bigger than chromosome 21, and harbors 742 genes (National Center for Biotechnology Information, Map Viewer, build 36).

1.3 The Human Karyotype and Clinical Cytogenetics

In humans, the normal diploid number of chromosomes is 46, consisting of 22 pairs of autosomal chromosomes numbered 1–22 and one pair of sex chromosomes (XX in females and XY in males) (Fig. 1.3). The genome is estimated to contain approximately 25,000 genes which are distributed along 23 chromosome pairs. As mentioned already, even the smallest autosome, chromosome 21, contains 352 genes. Thus, it is not surprising that duplications or deletions of chromosomes, or even small chromosome segments, have profound consequences on normal gene expression, leading to severe developmental and physiologic abnormalities.

Deviations in the number or structure of the 46 human chromosomes are astonishingly common, despite their severe deleterious consequences. Chromosomal disorders occur in an estimated 10–25% of all pregnancies. They are the leading cause of fetal loss and, among pregnancies continuing to term, are the leading known cause of birth defects and mental retardation. The most common chromosomal aberrations are trisomies for various chromosomes, indicating that chromosome segregation at meiosis is an extremely error-prone process in

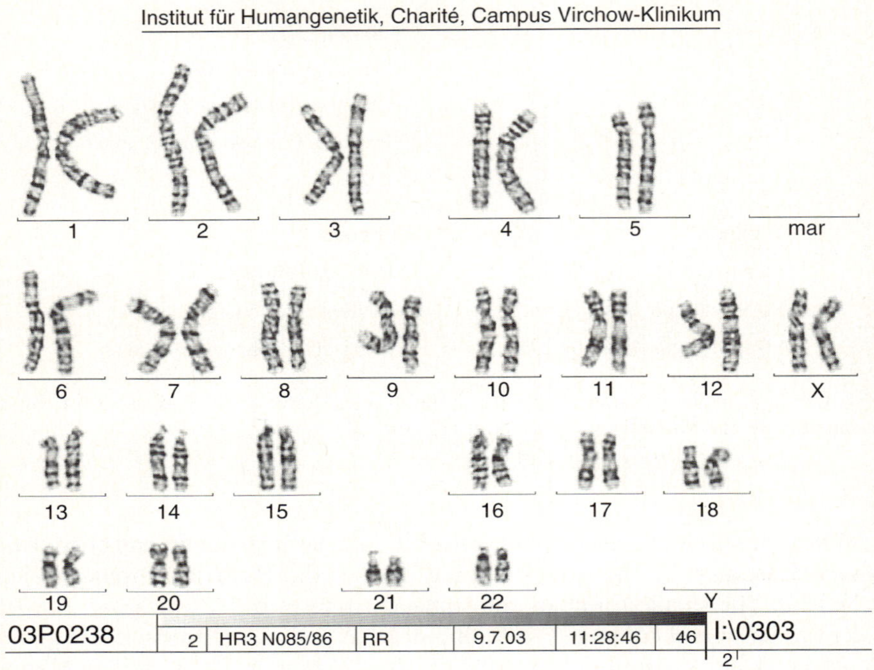

Fig. 1.3 Human female karyotype 46,XX after G-banding

humans. Most chromosomal aneuploidies originate from female meiosis and contribute significantly to pregnancy failures, particularly in women of advanced maternal age. The most common autosomal trisomy is trisomy 21, with a frequency of approximately 1 in 700 live births. In contrast, trisomy 22 results in spontaneous miscarriage due to the extent of imbalance since chromosome 22 carries twice as many genes (742 genes) as chromosome 21. In addition to trisomy 21, only two other autosomal trisomies with low gene density, trisomy 13 (551 genes) and trisomy 18 (432 genes), occur in live births with a prevalence of 1 in 10,000 and 1 in 20,000, respectively. However, most conceptions with trisomy 13 and trisomy 18 result in fetal loss during pregnancy or are associated with death in infancy, typically occurring during the first year of life. Autosomal trisomies of all other chromosomes are not compatible with survival to term, indicating that dose imbalance of the number of autosomal genes has a severe effect on developmental processes and survival (Fig. 1.4). This is further illustrated by the fact that all full autosomal monosomies do not survive to term. Karyotype–phenotype correlation studies on the basis of clinical findings demonstrate that monosomies for an autosomal segment cause more, and more severe alterations to the phenotype and restrict survival more than do trisomies for the same segment.

In contrast to autosomal imbalances, sex chromosome trisomies such as 47,XXX and 47,XXY have few phenotypic complications owing to the

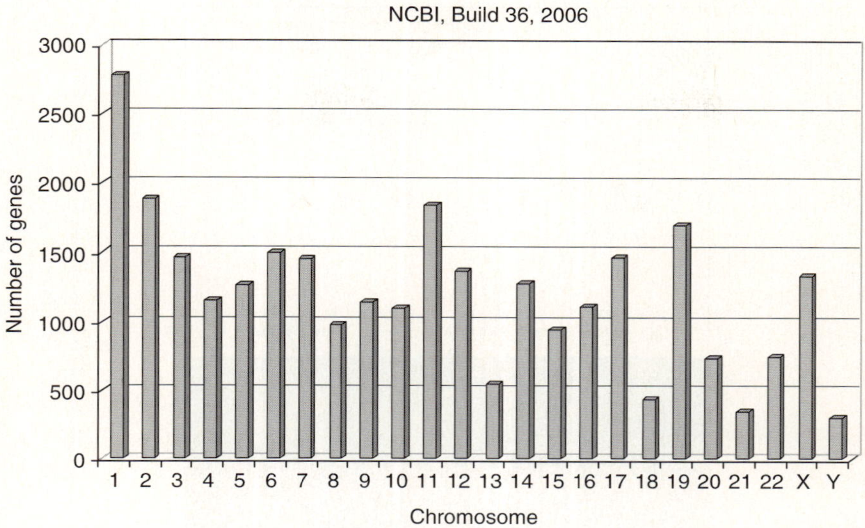

Fig. 1.4 Number of genes mapped to human chromosomes (National Center for Biotechnology Information, *NCBI*, Map Viewer, build 36)

mechanism of X chromosome inactivation, or as in 47,XYY owing the low number of Y-linked genes, most of which are involved in testicular development or spermatogenesis.

The introduction of molecular cytogenetic techniques such as FISH revolutionized the field of conventional cytogenetics, leading to considerable progress especially in clinical cytogenetics (Cremer et al. 1986; Pinkel et al. 1986; Kallioniemi et al. 1992; Schrock et al. 1996; Fig. 1.5). Subsequently, numerous deletion and duplication syndromes have been described that are too small to be detected under the microscope using conventional cytogenetic methods, such as G-banding. The molecular cytogenetic methods have expanded the possibilities for precise genetic diagnoses, which are extremely important for clinical management of patients and appropriate counseling of their families. Depending on the size of the deletion or duplication, specific FISH probes are employed to identify the aberration (Tonnies 2002).

Since most microdeletion/microduplication syndromes are defined by a common deleted/duplicated region, the abnormal dose of genes located within these regions can explain the phenotypic similarities among individuals with a specific syndrome. Consequently, detailed genotype–phenotype correlations provide a unique resource towards the genetic dissection of complex phenotypes such as congenital heart defects, mental and growth retardation, and specific cognitive and behavioral components of humans. For example, the Williams–Beuren syndrome (WBS), a neurodevelopmental disorder caused by a microdeletion at 7q11.23, provides one of the most convincing models of a relationship that links genes with human cognition and behavior. Detailed

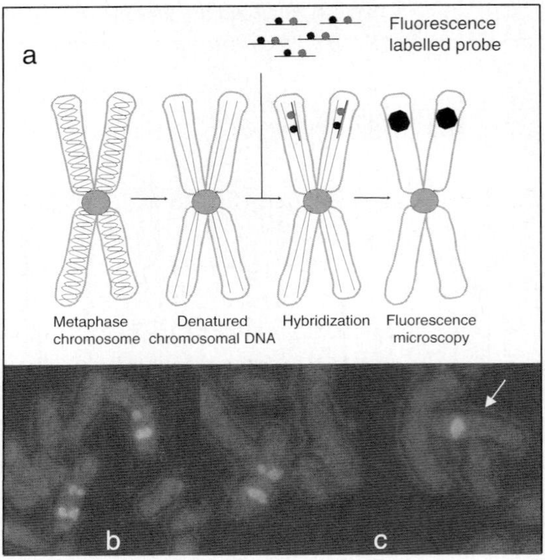

Fig. 1.5 **a** Principle of fluorescence in situ hybridization (FISH), **b** FISH with the Elastin probe for Williams–Beuren syndrome (WBS) and a control region in a normal individual, and **c** in a patient with WBS deletion showing the missing signal in one of the homologous chromosomes (*arrow*)

molecular characterization of the deletion alongside well-defined cognitive profiling in WBS provides a unique opportunity to investigate the neuromolecular basis of complex cognitive behavior and develop integrated approaches to study gene function and genotype–phenotype correlations (Osborne 1999; Mervis and Klein-Tasman 2000).

Furthermore, the molecular analysis of these chromosomal aberrations has led to a growing understanding of their mechanisms of origin, indicating that certain regions of the human genome are especially prone to structural rearrangements due to the presence of repetitive sequence elements. Interaction between chromosome-specific repetitive DNA leads to gain, loss, or inversion of the intervening sequence by nonallelic homologous recombination between misaligned repetitive elements. Where a particular region contains dose-sensitive or imprinted genes, this can lead to a specific genetic disease: loss of 7q11.23 results in Williams syndrome (Osborne 1999), loss of 22q11 results in Di George syndrome/velocardiofacial syndrome (Cuneo 2001), and loss of 15q11–q13 results in either Prader–Willi syndrome or Angelman syndrome (Cassidy et al. 2000; Horsthemke and Buiting 2006; Thomas et al. 2006). The available data demonstrate that the majority of rearrangements, approximately three or four, are interchromosomal. Therefore, they are likely to have arisen as the result of unequal meiotic crossing over between repetitive elements on different chromosome homologues. The remaining intrachromosomal rearrangements are also likely to be meiotic, although for these cases a postzygotic error during mitosis cannot be excluded (Thomas et al. 2006).

1.4 Cell Cycle and Chromosome Cycle

The cell cycle consists of four distinct phases: G1 phase, S phase, G2 phase, and M phase, or mitosis. G1, S, and G2 comprise together the interphase. G1 phase is a growth phase with high metabolic activity increasing the amount of cytoplasm and important organelles for preparing the cell for duplicating its DNA in S phase. In G2 phase, the cell continues with growth and metabolism in preparation for undergoing mitosis. During M phase the replicated chromatids segregate to the two daughter cells.

The chromosomes at G1, S, and G2 phases can be directly visualized after premature chromosome condensation (PCC) (Rao et al. 1982; Fig. 1.6). Fusion between mitotic and interphase cells results in a rapid chromosome condensation, with dissolution of the nuclear membrane due to the activity of the mitosis-promoting factor (MPF) in the interphase cells. The morphology of PCC chromosomes varies according to the stage of the interphase cell at the time of fusion. Thus, the PCC at G1 phase are very long with single chromatids and those at G2 phase are elongated with slender double chromatids. PCC chromosomes at S phase are characterized by their fragmented, pulverized appearance. The gaps of S-PCC represent the sites of DNA replication.

In all phases of the cell cycle, the surveillance of the chromosomal integrity is crucial for the genetic processes. Therefore, a complex network of checkpoints has evolved to monitor chromosome structure and coordinate cell cycle progression with DNA repair and spindle formation. Checkpoints are coordinated series of responses that delay progress through the cell cycle at a particular phase or transition in response to the lack of appropriate conditions

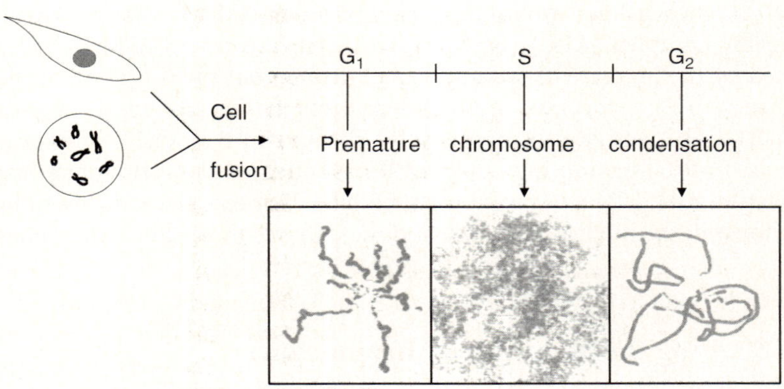

Fig. 1.6 Visualization of G1-, S-, and G2-phase chromosomes after premature chromosome condensation

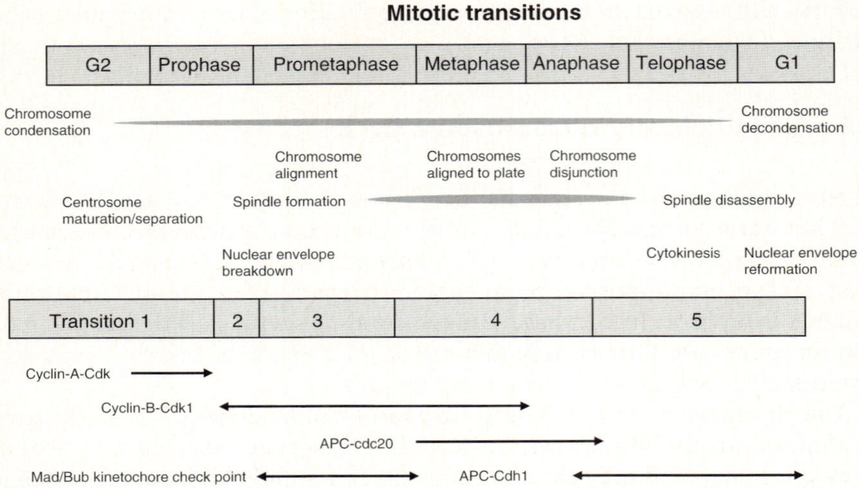

Fig. 1.7 Mitotic transitions according to the classification of Pines and Rieder (2001)

for progression. Delay allows enough time for crucial processes such as DNA repair and spindle attachment to be completed before continuing the cell-division cycle or initiating apoptosis (Hartwell and Weinert 1989; Hartwell and Kastan 1994; Nurse 1997; Clarke and Gimenez-Abian 2000; Morrison and Rieder 2004).

Especially, cell entry into mitosis is under the control of a tightly regulated network of protein kinases, cyclins, and protein phosphatases. According to Pines and Rieder (Pines and Rieder 2001) G2 phase and mitosis can be subdivided into five transitional phases which are characterized not only by the structural and behavioral changes of the chromosomes and the spindle, but also, at the molecular level, by the activation and inactivation of cell cycle regulators such as the cyclin-dependent kinases (Cdks) and the anaphase-promoting complex (APC) (Fig. 1.7). In vertebrates, the G2–M transition is initiated by the increase of cyclin-A-Cdk2 throughout the G2 phase of the cell cycle, resulting in chromosome condensation in the absence of significant cyclin-B1-Cdk1 activity. Subsequently, the cyclin-B1-Cdk1 complex, also known as mitosis-promoting factor (MPF), is activated as a result of its dephosphorylation by Cdc25 and rapidly accumulates in the nucleus, followed by the breakdown of the nuclear envelope and the entry of the cell into metaphase.

1.5 Shaping the Metaphase Chromosome

Chromosomes are complex and highly dynamic structures containing DNA, histones, and non-histone proteins. Understanding the structural and molecular basis of mitotic chromosome condensation remains a basic challenge in cell biology. During cell cycle progress towards mitosis, the chromosome undergoes progressive morphological conversion. From prophase to metaphase, apparently amorphous interphase chromatin is reorganized into individual chromosomes, with a pair of separate sister chromatids. This process, referred to as chromosome assembly, implying chromosome condensation and sister-chromatid resolution, is an essential prerequisite for the faithful segregation of duplicated genetic information into two daughter cells.

The human genome contains 6×10^9 bp of DNA per diploid cell, corresponding to approximately 1.7 m of DNA which is organized in chromatin fibers. The basic unit of chromatin is the nucleosome, which consists of 146 bp of DNA wound around an octamer of histone proteins, two each of histones H2A, H2B, H3, and H4 accounting for the first sixfold to sevenfold linear compaction of the DNA (Bednar et al. 1998).

The binding of linker histone H1 to linker DNA sequences, so-called scaffold-associated regions (SARs), localized between nucleosomes leads to further chromatin compaction – the 30-nm fiber – generating another sixfold to sevenfold compaction. It has been proposed that the 30-nm chromatin fiber is accomplished by a unique three-dimensional zigzag folding pattern rather than by supercoiling. In mitosis, the 30-nm fiber must compact another 200- to 500-fold to achieve the final 10,000- to 20,000-fold linear compaction of the mitotic chromosome (Swedlow and Hirano 2003; Fig. 1.8). Early studies suggested the presence of a chromosome scaffold composed of non-histone proteins that serve as the backbone of the mitotic chromosome. Two major scaffold proteins, topoisomerase II and SC2, a structural maintenance of chromosomes (SMC) family member, have been shown to be part of the scaffold and involved in condensation. *cis* sites for condensation are expected to lie along the chromosomal axis in vivo. Scaffold models propose that the SARs are the *cis*-acting DNA sequences that serve as binding sites for DNA topoisomerase II, a major component of the chromosome scaffold, resulting in radial DNA loops of approximately 50–100 kb in length (Earnshaw et al. 1985; Earnshaw and Heck 1985). Yet the radial loop model has remained highly controversial.

The highly organized chromatin can be modified further by various mechanisms, such as posttranslational modifications of histones, ATP-dependent chromatin remodeling, and the exchange of histone proteins.

1.6 Cohesion and Condensins

The discovery of SMC proteins led to rapid progress in our understanding of chromosome organization and behavior. The identification and characterization of cohesion and the condensins, SMC-containing complexes, demonstrated them to be key regulators that function in chromosome assembly and segregation during mitosis (Hirano 2005, 2006; Losada and Hirano 2005; Nasmyth and Haering 2005). SMC proteins are ubiquitous in organisms from

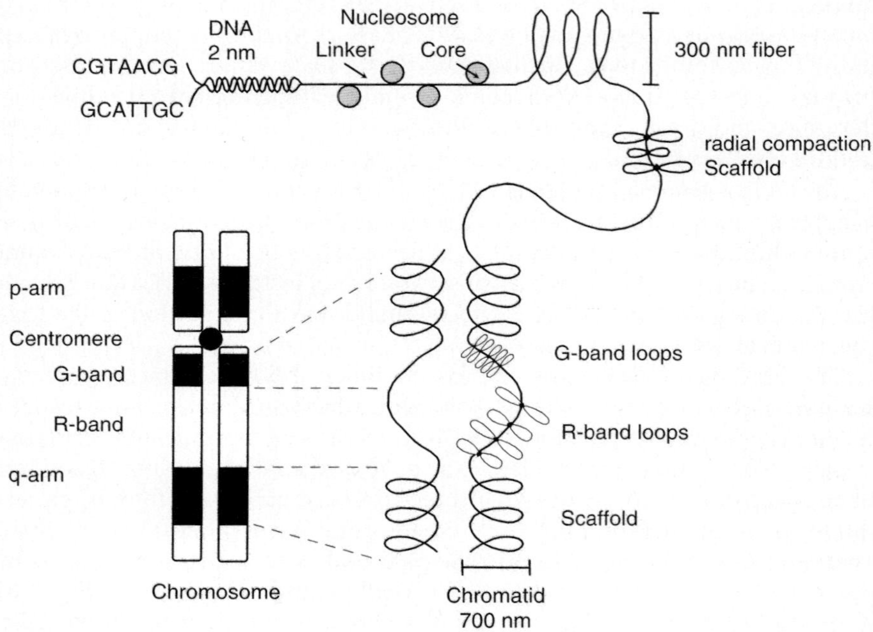

Fig. 1.8 Model of chromatin compaction from DNA double helix to metaphase chromosome

bacteria to humans. The SMC proteins SMC1 and SMC3 constitute the core of the cohesin complex and bind the non-SMC proteins Scc1 and Scc3 to form a ringlike structure that mediates sister-chromatid cohesion. A model of the interaction of cohesion predicts that the establishment of sister-chromatid cohesion is accomplished when a replication fork passes through the cohesin ring that is preloaded during the G1 phase of the cell cycle. The sister chromatids generated by DNA replication become aligned along the entire length of their arms and at the kinetochore. In vertebrates, most cohesin dissociates from chromatin at prophase, and only a small population, enriched in the pericentromeric region, remains on the chromosomes until metaphase. Cohesion is essential for the congression and alignment of chromosomes from prometaphase to metaphase. At the onset of anaphase, loss of cohesion triggers the separation of sister chromatids, allowing them to be pulled apart to opposite poles of the cell.

The core component of the two condensin complexes is the monomers SMC2 and SMC4 (Hirano 2005). The SMC cores are bound by different sets of regulatory subunits forming the functional complexes. Vertebrates have two different condensin complexes, condensin I and condensin II, each containing a unique set of regulatory subunits. CAP-G, CAP-D2, and CAP-H bind to the condensin core to form condensin I (Hirano et al. 1997), whereas condensin II is defined by its regulatory subunits CAP-G2, CAP-D3, and CAP-H2 (Ono et al. 2003). Condensin I (and possibly condensin II as well) has the ability to introduce positive helical tension into double-stranded DNA in vitro (Kimura and Hirano 1997; Kimura et al. 1999; Bazett-Jones et al. 2002; Hagstrom et al. 2002). While the two complexes cooperate to assemble metaphase chromosomes (Ono et al. 2003), their behaviors are regulated differently during the cell cycle (Hirota et al. 2004; Ono et al. 2004; Trimborn et al. 2006). Condensin II is nuclear throughout the cell cycle and participates in an early stage of prophase chromosome condensation within the nucleus, whereas condensin I gains access to chromosomes only after the breakdown of the nuclear envelope. Both complexes finally bind to the central chromatid axes in an alternate pattern. The molecular mechanism underlying the differential regulation of the two condensin complexes remains to be determined, but it was proposed that sequential activation of cyclin A/Cdk and cyclin B/Cdk could be responsible for the successive loading of condensin I and condensin II (Hirano 2005). The loading of condensins is a prerequisite for the proper assembly and segregation of metaphase chromosomes. The diverse functions of the SMC complexes, however, range far beyond chromosome segregation and may include gene regulation and DNA repair.

1.7 DNA Repair

The DNA damage checkpoints cause cell cycle delay before or during the decisive cell cycle transitions of replication and mitosis (G1/S, intra-S, G2/M checkpoints) (Sancar et al. 2004). This involves a number of highly conserved proteins that sense the damage and signal the cell cycle machinery. Central to this network are two protein kinases, ataxia telangiectasia mutated (ATM) and ATM-Rad3 related (ATR). These kinases sense the DNA damage and start signaling cascades which finally result in cell cycle arrest or induce apopoptic pathways. ATM kinase is primarily activated by DNA double-strand breaks (DSBs) induced by ionizing irradiation, whereas ATR kinase responds to UV-induced and replication-specific DNA damage. ATR kinase activation demands the association with the protein ATRIP and two additional complexes RAD17 and 9-1-1. The MRE11-NBS-RAD50 (MNR) complex plays a crucial role in the ATM kinase mediated answer on DNA DSBs. It was generally thought that ATM kinase and ATR kinase work independently. Recent reports, however, indicate that ATM kinase and nuclease activity of meiotic recombination 11 (MRE11) are required for the processing of DNA DSBs to generate the replication protein A (RPA) coated single-stranded DNA that is needed for ATR kinase recruitment (Watson and Crick 1953; Adams et al. 2006; Jazayeri et al. 2006). Critical for the signal transduction following the damage detection are a vast number of mediators, many of them (e.g., TOPBP1, 53BP1, and BRCA1) containing BRCT domains. By means of these transducers and the checkpoint kinases CHK1 and CHK2, the signal is transferred to the target/effector proteins like CDC25 phophatases, p53, or SMC1 (Lee 2002; Kastan and Bartek 2004; Sancar et al. 2004; Li and Zou 2005). DNA DSBs are repaired by two distinct but connected pathways: nonhomologous end-joining (NHEJ) and homologous repair (HR). NHEJ rejoins the two ends of a DSB by simple ligation in an error-prone process, while HR uses a homologous template to copy and restore the information disrupted by the break. This promotes error-free repair. Since the information is usually copied from an intact sister chromatid, HR is the preferred pathway in S and G2 phases, while NHEJ is the predominant pathway in the G1 phase (reviewed in Lisby and Rothstein 2004).

1.8 Mendelian Disorders Affecting Chromosome Integrity

1.8.1 Chromosome Instability Disorders

Our insights into DNA repair processes in normal cells have been considerably improved by the identification of the underlying genetic defects of chromosome instability syndromes. Chromosome instability syndromes are a group of inherited disorders associated with high genomic instability and a markedly increased cancer risk. In the following, we focus on two instability syndromes, one implicated in the repair of DSBs and one involved in the repair of DNA cross-links.

Patients affected by Nijmegen breakage syndrome (NBS) have biallelic mutations in the *NBS1* gene, mapped on chromosome 8q21. It encodes a 95-kDa protein called nibrin, a member of the hMre11hRad50 protein complex, involved in the ATM-dependent DNA damage signaling pathway of cellular response to DSBs. The affected patients present with microcephaly, a distinct facial appearance, growth retardation, immunodeficiency, cytogenetic abnormalities, radiosensitivity, and high susceptibility to lymphoid malignancy. In 40% of patients, a malignancy occurs before the age of 21 (Varon et al. 1998, 2001, Varon et al. 2003; Kitagawa and Kastan 2005; Kruger et al. 2007).

Fanconi anemia is a genetically heterogeneous, autosomal recessive or X-recessive chromosome instability disorder with increased hypersensitivity to cross-linking agents. At least 12 genetic complementation groups have been described (FA-A, FA-B, FA-C, FA-D1, FA-D2, FA-E, FA-F, FA-G, FA-I, FA-J, FA-L, FA-M) and all except FA-I have been linked to a distinct gene. All Fanconi anemia proteins act in a single pathway involved in DNA cross-link repair. Most Fanconi anemia proteins form a complex that activates the FANCD2 protein via monoubiquitination, which is prerequisite for the activation of *BRCA2*, a gene which was originally identified in families with increased breast and ovarian cancer susceptibility. Disruption of any of the Fanconi anemia proteins results in an increased chromosomal instability (Fig. 1.9a). Fanconi anemia patients have a high risk for bone marrow failure, aplastic anemia, myelodysplastic syndrome, acute myeloid leukemia, and, later in life, epithelial malignancies. The most life-threatening early event in most complementation groups is bone marrow failure, which occurs typically during the first decade of life (Wang and D'Andrea 2004; Kennedy and D'Andrea 2005; Bagby and Alter 2006; Lyakhovich and Surralles 2006).

These two examples might be sufficient to demonstrate that any mechanism impairing chromosome surveillance and chromosomal integrity has profound effects on cancer formation and progression.

1.8.2 PCC Syndrome

The first description of a disorder in humans affecting the fundamental process of chromosome condensation was reported in 2002 (Neitzel et al. 2002). The patients' chromosomes display PCC (PCC syndrome) in early G2 phase commencing as soon as 1 h after completion of S phase and also delayed decondensation after mitosis (Fig. 1.9b).

In 2004, it was demonstrated that PCC syndrome is caused by mutations in the *MCPH1* gene encoding microcephalin (Trimborn et al. 2004). Microcephalin encompasses 835 amino acids and contains one N-terminal und two C-terminal BRCT domains (BRCA1 C-terminus) linking its function to DNA checkpoint control and/or DNA repair. The clinical phenotype is characterized by microcephaly, growth retardation, and mental retardation. These findings implicated microcephalin as a novel regulator of chromosome condensation and linked the apparently disparate fields of neurogenesis and chromosome biology.

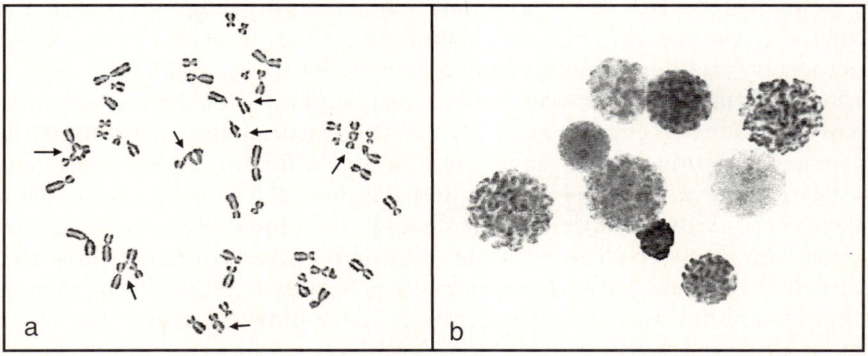

Fig. 1.9 Chromosomal instability syndromes: **a** increased chromosomal breakage in lymphocytes of a Fanconi anemia patient; **b** aberrant chromosome condensation in a patient with *MCPH1* autosomal recessive primary microcephaly

The misregulation of chromosome condensation in *MCPH1* deficiency is mediated by the SMC protein condensin II (Trimborn et al. 2006).

In patient cells with *MCPH1* deficiency, knockdown of condensin II subunits leads to a pronounced reduction of cells with the condensation defects in both G1 and G2 phases of the cell cycle. In contrast, knockdown of condensin I subunits does not reverse the cellular phenotype. Consistently, condensin I stays in the cytoplasm in the prophaselike cells of *MCPH1* patients. These results offer a molecular explanation for the aberrant chromosome condensation in *MCPH1* deficiency. In normal cells, microcephalin acts as a negative regulator of condensin II which prevents PCC until the onset of prophase and allows timely decondensation after mitosis.

1.8.3 Further Syndromes Affecting Structural Maintenance of the Chromosome

In the last few years a growing number of genes that regulate genome surveillance and cell cycle progression have been linked to developmental and progressive neurological diseases. Some of these are involved in chromosome dynamics, spindle formation, and the centrosome cycle. Mutations in the gene *NIPBL*, the human counterpart of *Scc2*, were shown to cause Brachmann/de Lange syndrome (Krantz et al. 2004; Tonkin et al. 2004), which is associated with growth retardation, microcephaly, and limb malformations. Scc2 is crucial for sister-chromatid cohesion and replication licensing (Furuya et al. 1998; Ciosk et al. 2000; Gillespie and Hirano 2004). Expression of *NIPBL* in developing limbs of human embryos was shown by in situ hybridization (Tonkin et al. 2004). A dual role for Nipped-B, the *Drosophila* homologue of *NIPBL*, in sister-chromatid cohesion and developmental regulation connected with limb formation was recently confirmed (Rollins et al. 2004). Mutations in yet another cohesin factor gene ESCO2 cause Roberts syndrome (Vega et al. 2005), characterized by premature centromere separation. Brachmann/de Lange syndrome and Roberts syndrome share clinical symptoms, such as growth retardation, microcephaly, and intriguingly limb malformations.

BUB1B is mutated in mosaic variegated aneuploidy (MVA), an autosomal recessive disorder characterized by mosaic aneuploidies, predominantly trisomies and monosomies, involving multiple different chromosomes and tissues (Hanks et al. 2004; Hanks and Rahman 2005). *BUB1B* encodes a key protein in the mitotic spindle checkpoint (Sudakin et al. 2001) and is involved in sister-chromatid cohesion (Kitajima et al. 2005). Affected individuals present with severe intrauterine growth retardation, various congenital abnormalities, microcephaly, developmental delay, and a high risk of malignancy.

These few examples illustrate the relevance of genes involved in the surveillance of the chromosomal integrity, cell cycle progression, and DNA repair. Deficiencies of these genes due to mutation do not only result in multiple congenital abnormalities of almost all tissues but also in a high risk for the development of various malignancies. Furthermore, these examples demonstrate impressively the importance of the field of human cytogenetics: in all cases the cytogenetic observation of distinctive chromosomal features, such as increased chromosomal instability, PCC, increased somatic nondisjunction, or premature centromere division, preceded the identification of the underlying gene defects.

Hsu's annotation in 1979 that chromosomes are "hypnotically beautiful objects" still holds true. Beyond it, the thorough observation of chromosomal changes by many researchers during the last four decades paved the way for our understanding of the underlying mechanisms of many congenital disorders, as well as of chromosome surveillance, DNA repair, and cancer susceptibility.

References

Adams KE, Medhurst AL, Dart DA, Lakin ND (2006) Recruitment of ATR to sites of ionising radiation-induced DNA damage requires ATM and components of the MRN protein complex. Oncogene 25:3894–3904

Bagby GC, Alter BP (2006) Fanconi anemia. Semin Hematol 43:147–156

Bazett-Jones DP, Kimura K, Hirano T (2002) Efficient supercoiling of DNA by a single condensin complex as revealed by electron spectroscopic imaging. Mol Cell 9:1183–1190

Bednar J, Horowitz RA, Grigoryev SA, Carruthers LM, Hansen JC, Koster AJ, Woodcock CL (1998) Nucleosomes, linker DNA, and linker histone form a unique structural motif that directs the higher-order folding and compaction of chromatin. Proc Natl Acad Sci USA 95:14173–14178

Caspersson T, Zech L, Modest EJ, Foley GE, Wagh U, Simonsson E (1969) Chemical differentiation with fluorescent alkylating agents in Vicia faba metaphase chromosomes. Exp Cell Res 58:128–140

Cassidy SB, Dykens E, Williams CA (2000) Prader-Willi and Angelman syndromes: sister imprinted disorders. Am J Med Genet 97:136–146

Chen TL, Manuelidis L (1989) SINEs and LINEs cluster in distinct DNA fragments of Giemsa band size. Chromosoma 98:309–316

Ciosk R, Shirayama M, Shevchenko A, Tanaka T, Toth A, Nasmyth K (2000) Cohesin's binding to chromosomes depends on a separate complex consisting of Scc2 and Scc4 proteins. Mol Cell 5:243–254

Clarke DJ, Gimenez-Abian JF (2000) Checkpoints controlling mitosis. BioEssays 22:351–363

Cremer T, Landegent J, Bruckner A, Scholl HP, Schardin M, Hager HD, Devilee P, Pearson P, van der Ploeg M (1986) Detection of chromosome aberrations in the human interphase nucleus by visualization of specific target DNAs with radioactive and non-radioactive in situ hybridization techniques: diagnosis of trisomy 18 with probe L1.84. Hum Genet 74:346–352

Cuneo BF (2001) 22q11.2 deletion syndrome: DiGeorge, velocardiofacial, and conotruncal anomaly face syndromes. Curr Opin Pediatr 13:465–472

Drets ME, Shaw MW (1971) Specific banding patterns of human chromosomes. Proc Natl Acad Sci USA 68:2073–2077

Dutrillaux B, Lejeune J (1971) (A new technique of analysis of the human karyotype). C R Acad Sci Hebd Seances Acad Sci D 272:2638–2640

Earnshaw WC, Heck MM (1985) Localization of topoisomerase II in mitotic chromosomes. J Cell Biol 100:1716–1725

Earnshaw WC, Halligan B, Cooke CA, Heck MM, Liu LF (1985) Topoisomerase II is a structural component of mitotic chromosome scaffolds. J Cell Biol 100:1706–1715
Flemming W (1882) Zellsubstanz, Kern und Zelltheilung. Vogel, Leipzig
Furuya K, Takahashi K, Yanagida M (1998) Faithful anaphase is ensured by Mis4, a sister chromatid cohesion molecule required in S phase and not destroyed in G1 phase. Genes Dev 12:3408–3418
Gillespie PJ, Hirano T (2004) Scc2 couples replication licensing to sister chromatid cohesion in Xenopus egg extracts. Curr Biol 14:1598–1603
Hagstrom KA, Holmes VF, Cozzarelli NR, Meyer BJ (2002) C. elegans condensin promotes mitotic chromosome architecture, centromere organization, and sister chromatid segregation during mitosis and meiosis. Genes Dev 16:729–742
Hanks S, Rahman N (2005) Aneuploidy-cancer predisposition syndromes: a new link between the mitotic spindle checkpoint and cancer. Cell Cycle 4:225–227
Hanks S, Coleman K, Reid S, Plaja A, Firth H, Fitzpatrick D, Kidd A, Mehes K, Nash R, Robin N, Shannon N, Tolmie J, Swansbury J, Irrthum A, Douglas J, Rahman N (2004) Constitutional aneuploidy and cancer predisposition caused by biallelic mutations in BUB1B. Nat Genet 36:1159–1161
Hartwell LH, Kastan MB (1994) Cell cycle control and cancer. Science 266:1821–1828
Hartwell LH, Weinert TA (1989) Checkpoints: controls that ensure the order of cell cycle events. Science 246:629–634
Hirano T (2005) Condensins: organizing and segregating the genome. Curr Biol 15:R265–275
Hirano T (2006) At the heart of the chromosome: SMC proteins in action. Nat Rev Mol Cell Biol 7:311–322
Hirano T, Kobayashi R, Hirano M (1997) Condensins, chromosome condensation protein complexes containing XCAP-C, XCAP-E and a Xenopus homolog of the Drosophila Barren protein. Cell 89:511–521
Hirota T, Gerlich D, Koch B, Ellenberg J, Peters JM (2004) Distinct functions of condensin I and II in mitotic chromosome assembly. J Cell Sci 117:6435–6445
Holmquist GP (1989) Evolution of chromosome bands: molecular ecology of noncoding DNA. J Mol Evol 28:469–486
Holmquist GP (1992) Chromosome bands, their chromatin flavors, and their functional features. Am J Hum Genet 51:17–37
Horsthemke B, Buiting K (2006) Imprinting defects on human chromosome 15. Cytogenet Genome Res 113:292–299
Hsu TC (1979) Human and mammalian cytogenetics. An historical perspective. Springer, New York
Jazayeri A, Falck J, Lukas C, Bartek J, Smith GC, Lukas J, Jackson SP (2006) ATM- and cell cycle-dependent regulation of ATR in response to DNA double-strand breaks. Nat Cell Biol 8:37–45
Kallioniemi A, Kallioniemi OP, Sudar D, Rutovitz D, Gray JW, Waldman F, Pinkel D (1992) Comparative genomic hybridization for molecular cytogenetic analysis of solid tumors. Science 258:818–821
Kastan MB, Bartek J (2004) Cell-cycle checkpoints and cancer. Nature 432:316–323
Kennedy RD, D'Andrea AD (2005) The Fanconi anemia/BRCA pathway: new faces in the crowd. Genes Dev 19:2925–2940
Kimura K, Hirano T (1997) ATP-dependent positive supercoiling of DNA by 13S condensin: a biochemical implication for chromosome condensation. Cell 90:625–634
Kimura K, Rybenkov VV, Crisona NJ, Hirano T, Cozzarelli NR (1999) 13S condensin actively reconfigures DNA by introducing global positive writhe: implications for chromosome condensation. Cell 98:239–248
Kitagawa R, Kastan MB (2005) The ATM-dependent DNA damage signaling pathway. Cold Spring Harbor Symp Quant Biol 70:99–109
Kitajima TS, Hauf S, Ohsugi M, Yamamoto T, Watanabe Y (2005) Human Bub1 defines the persistent cohesion site along the mitotic chromosome by affecting Shugoshin localization. Curr Biol 15:353–359

Korenberg JR, Rykowski MC (1988) Human genome organization: Alu, lines, and the molecular structure of metaphase chromosome bands. Cell 53:391–400

Krantz ID, McCallum J, DeScipio C, Kaur M, Gillis LA, Yaeger D, Jukofsky L, Wasserman N, Bottani A, Morris CA, Nowaczyk MJ, Toriello H, Bamshad MJ, Carey JC, Rappaport E, Kawauchi S, Lander AD, Calof AL, Li HH, Devoto M, Jackson LG (2004) Cornelia de Lange syndrome is caused by mutations in NIPBL, the human homolog of Drosophila melanogaster Nipped-B. Nat Genet 36:631–635

Kruger L, Demuth I, Neitzel H, Varon R, Sperling K, Chrzanowska KH, Seemanova E, Digweed M (2007) Cancer incidence in Nijmegen breakage syndrome is modulated by the amount of a variant NBS protein. Carcinogenesis 28:107–111

Lee EY (2002) BRCA1 and Chk1 in G2/M checkpoint: a new order of regulation. Cell Cycle 1:178–180

Lejeune J, Turpin R, Gautier M (1959) (Chromosomic diagnosis of mongolism). Arch Fr Pediatr 16:962–963

Li L, Zou L (2005) Sensing, signaling, and responding to DNA damage: organization of the checkpoint pathways in mammalian cells. J Cell Biochem 94:298–306

Lisby M, Rothstein R (2004) DNA damage checkpoint and repair centers. Curr Opin Cell Biol 16:328–334

Losada A, Hirano T (2005) Dynamic molecular linkers of the genome: the first decade of SMC proteins. Genes Dev 19:1269–1287

Lyakhovich A, Surralles J (2006) Disruption of the Fanconi anemia/BRCA pathway in sporadic cancer. Cancer Lett 232:99–106

Mervis CB, Klein-Tasman BP (2000) Williams syndrome: cognition, personality, and adaptive behavior. Ment Retard Dev Disabil Res Rev 6:148–158

Meselson M, Stahl FW (1958) The replication of DNA. Cold Spring Harbor Symp Quant Biol 23:9–12

Morrison C, Rieder CL (2004) Chromosome damage and progression into and through mitosis in vertebrates. DNA Repair (Amst) 3:1133–1139

Nasmyth K, Haering CH (2005) The structure and function of SMC and kleisin complexes. Annu Rev Biochem 74:595–648

Neitzel H, Neumann LM, Schindler D, Wirges A, Tonnies H, Trimborn M, Krebsova A, Richter R, Sperling K (2002) Premature chromosome condensation in humans associated with microcephaly and mental retardation: a novel autosomal recessive condition. Am J Hum Genet 70:1015–1022

Nowell PC (1960) Phytohemagglutinin: an initiator of mitosis in cultures of normal human leukocytes. Cancer Res 20:462–466

Nurse P (1997) Checkpoint pathways come of age. Cell 91:865–867

Ono T, Losada A, Hirano M, Myers MP, Neuwald AF, Hirano T (2003) Differential contributions of condensin I and condensin II to mitotic chromosome architecture in vertebrate cells. Cell 115:109–121

Ono T, Fang Y, Spector DL, Hirano T (2004) Spatial and temporal regulation of Condensins I and II in mitotic chromosome assembly in human cells. Mol Biol Cell 15:3296–3308

Osborne LR (1999) Williams-Beuren syndrome: unraveling the mysteries of a microdeletion disorder. Mol Genet Metab 67:1–10

Ostertag EM, Kazazian HH Jr (2001) Biology of mammalian L1 retrotransposons. Annu Rev Genet 35:501–538

Patil SR, Merrick S, Lubs HA (1971) Identification of each human chromosome with a modified Giemsa stain. Science 173:821–822

Pines J, Rieder CL (2001) Re-staging mitosis: a contemporary view of mitotic progression. Nat Cell Biol 3:E3–6

Pinkel D, Straume T, Gray JW (1986) Cytogenetic analysis using quantitative, high-sensitivity, fluorescence hybridization. Proc Natl Acad Sci USA 83:2934–2938

Rao PN, Johnson RT, Sperling K (1982) Premature chromosome condensation. Academic, New York

Rollins RA, Korom M, Aulner N, Martens A, Dorsett D (2004) Drosophila nipped-B protein supports sister chromatid cohesion and opposes the stromalin/Scc3 cohesion factor to facilitate long-range activation of the cut gene. Mol Cell Biol 24:3100–3111

Sancar A, Lindsey-Boltz LA, Unsal-Kacmaz K, Linn S (2004) Molecular mechanisms of mammalian DNA repair and the DNA damage checkpoints. Annu Rev Biochem 73:39–85

Schrock E, du Manoir S, Veldman T, Schoell B, Wienberg J, Ferguson-Smith MA, Ning Y, Ledbetter DH, Bar-Am I, Soenksen D, Garini Y, Ried T (1996) Multicolor spectral karyotyping of human chromosomes. Science 273:494–497

Sudakin V, Chan GK, Yen TJ (2001) Checkpoint inhibition of the APC/C in HeLa cells is mediated by a complex of BUBR1, BUB3, CDC20, and MAD2. J Cell Biol 154:925–936

Sumner AT, Evans HJ, Buckland RA (1971) New technique for distinguishing between human chromosomes. Nat New Biol 232:31–32

Swedlow JR, Hirano T (2003) The making of the mitotic chromosome: modern insights into classical questions. Mol Cell 11:557–569

Thomas NS, Durkie M, Potts G, Sandford R, Van Zyl B, Youings S, Dennis NR, Jacobs PA (2006) Parental and chromosomal origins of microdeletion and duplication syndromes involving 7q11.23, 15q11-q13 and 22q11. Eur J Hum Genet 14:831–837

Tjio JH, Levan A (1956) The chromosome number of man. Hereditas 42:1–6

Tonkin ET, Wang TJ, Lisgo S, Bamshad MJ, Strachan T (2004) NIPBL, encoding a homolog of fungal Scc2-type sister chromatid cohesion proteins and fly Nipped-B, is mutated in Cornelia de Lange syndrome. Nat Genet 36:636–641

Tonnies H (2002) Modern molecular cytogenetic techniques in genetic diagnostics. Trends Mol Med 8:246–250

Trimborn M, Bell SM, Felix C, Rashid Y, Jafri H, Griffiths PD, Neumann LM, Krebs A, Reis A, Sperling K, Neitzel H, Jackson AP (2004) Mutations in microcephalin cause aberrant regulation of chromosome condensation. Am J Hum Genet 75:261–266

Trimborn M, Schindler D, Neitzel H, Hirano T (2006) Misregulated chromosome condensation in MCPH1 primary microcephaly is mediated by condensin II. Cell Cycle 5:322–326

Varon R, Vissinga C, Platzer M, Cerosaletti KM, Chrzanowska KH, Saar K, Beckmann G, Seemanova E, Cooper PR, Nowak NJ, Stumm M, Weemaes CM, Gatti RA, Wilson RK, Digweed M, Rosenthal A, Sperling K, Concannon P, Reis A (1998) Nibrin, a novel DNA double-strand break repair protein, is mutated in Nijmegen breakage syndrome. Cell 93:467–476

Varon R, Reis A, Henze G, von Einsiedel HG, Sperling K, Seeger K (2001) Mutations in the Nijmegen Breakage Syndrome gene (NBS1) in childhood acute lymphoblastic leukemia (ALL). Cancer Res 61:3570–3572

Varon R, Schoch C, Reis A, Hiddemann WC, Sperling K, Schnittger S (2003) Mutation analysis of the Nijmegen breakage syndrome gene (NBS1) in nineteen patients with acute myeloid leukemia with complex karyotypes. Leuk Lymphoma 44:1931–1934

Vega H, Waisfisz Q, Gordillo M, Sakai N, Yanagihara I, Yamada M, van Gosliga D, Kayserili H, Xu C, Ozono K, Jabs EW, Inui K, Joenje H (2005) Roberts syndrome is caused by mutations in ESCO2, a human homolog of yeast ECO1 that is essential for the establishment of sister chromatid cohesion. Nat Genet 37:468–470

Wang X, D'Andrea AD (2004) The interplay of Fanconi anemia proteins in the DNA damage response. DNA Repair (Amst) 3:1063–1069

Watson GD, Crick FHC (1953) Molecular structure of nucleic acids: a structure for deoxyribose nucleic acid. Nature 171:737–738

Waye JS, Willard HF (1989) Human beta satellite DNA: genomic organization and sequence definition of a class of highly repetitive tandem DNA. Proc Natl Acad Sci USA 86:6250–6254

Willard HF (1989) The genomics of long tandem arrays of satellite DNA in the human genome. Genome 31:737–744

2 DNA Lesions Induced by Ionizing Radiation

CLEMENS VON SONNTAG

Abstract

When ionizing radiation is absorbed in cells, a fraction of the absorbed energy is deposited in their DNA ("direct effect") and in the water surrounding it ("indirect effect," mainly owing to the formation of $^{\bullet}OH$ radicals). These ionization events induce a series of free-radical reactions that eventually show up in biological end points such as mutations and reproductive cell death. Precursors of these biological end points are various kinds of damaged DNA sites, predominantly damaged bases, single-stand breaks, double-strand breaks and DNA–protein cross-links. Molecular oxygen enhances and thiols reduce the severity of the DNA damage. The mechanisms leading to these lesions are discussed in the light of our present knowledge of the underlying free-radical chemistry.

2.1 Introduction

For biological effects such as point mutations and chromosomal mutations induced by ionizing radiation, DNA is the obvious target, but there is also a report that mutations are induced when the cytoplasm is hit by an α-particle (Wu et al. 1999). For other biological end points such as cell death, DNA is also considered to be the main target, but there is a phenomenon, the bystander effect (a cell is hit by a single particle of ionizing radiation – a neighboring cell dies by apoptosis; Shao et al. 2004), that cannot be accounted for by radiation-induced DNA damage.

The cell's repair enzymes cope with most of the radiation-induced DNA damage very effectively, and what we see as radiation-induced DNA damage on the cellular level is the damage that remains after the much greater DNA damage originally caused has been dealt with by the interplay of "chemical repair" (see later) and enzymatic repair. To give the reader an idea of the number of lesions and of the efficiency of the repair machinery, the DNA damage caused in a mammalian cell by a dose of 1 Gy is shown in Table 2.1.

At this dose, the lethality is only near 50%. Yet, as many as several thousand damaged bases, 1,000 single-strand breaks (SSBs), 150 DNA–protein cross-links

Günter Obe and Vijayalaxmi (Eds.)
Chromosomal Alterations: Methods, Results and Importance in Human Health

Table 2.1 Some of the damage in a mammalian cell nucleus from 1 Gy of low linear energy transfer radiation (Ward 1988; Goodhead 1994)

Initial physical damage	
Ionizations in the cell nucleus	~100,000
Ionizations directly in DNA	~2,000
Excitations directly in DNA	~2,000
Selected biochemical damage	
Single-strand breaks	1,000
8-Oxo-A (typical single-base damage)	700
Double-strand breaks	40
DNA–protein cross-links	150
Selected cellular effects	
Lethal events	~0.2–0.8
Chromosome aberrations	~1
Hprt mutations	$\sim 10^{-5}$

and 40 double-strand breaks (DSBs) have been caused. Obviously, most of this is repaired effectively. For example, a DSB as such is not yet a sufficiently severe type of DNA damage to cause cell death, and it has been concluded that there must be a much more complex type of damage responsible for cell death. There is evidence that clustered damage beyond a DSB is responsible for or is at least a major contributor to cell death by ionizing radiation. Yet, inadequately repaired DSBs are thought to be the origin of chromosomal mutations and induction of cancer; for a detailed review on the production and the fidelity of repair of endogenous and ionizing-radiation-induced DSBs see Vilenchik and

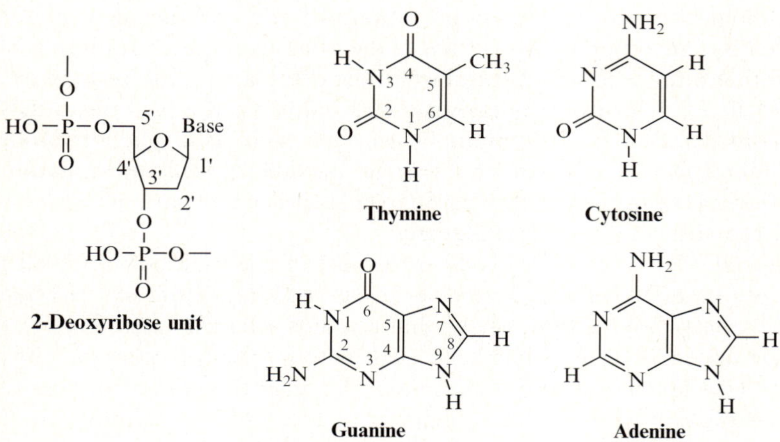

Fig. 2.1 Chemical structures of the DNA backbone (2-deoxyribose unit) and of the nucleobases thymine, cytosine, guanine and adenine

Knudson (2003) and for the details on endogenous and ionizing-radiation-induced DNA base damage see De Bont and van Larebeke (2004).

While UV radiation (at $\lambda > 200$ nm) practically only damages the pyrimidines thymine and cytosine, ionizing radiation also damages to a major extent the purines guanine and adenine as well as the 2-deoxyribose moiety (Fig. 2.1). Thus, SSBs and DSBs are also important lesions induced by ionizing radiation (Table 2.1).

The free-radical-induced DNA damage has been dealt with recently in a monograph (von Sonntag 2006) in much more detail then space allows here. Only some salient points can be addressed, and for further reading the reader is referred to that book and the original literature cited therein.

2.2 Energy Deposition by Ionizing Radiation

The energy of high-energy photons such as γ-rays and X-rays below 1 MeV is absorbed by the Compton effect and the photo effect. Above 1.01 MeV (twice the rest mass of the electron), pair formation, i.e., creation of an electron–positron pair, starts to play a role. Compton and photo effects generate high-energy electrons. The energy loss of charged particles is described by the Bethe equation which relates the energy loss (linear energy transfer) of the charged particle to its mass, charge and energy. Multiply charged and heavy particles (e.g., α-particles) lose their energy more effectively (densely ionizing radiation) than singly charged light particles such as high-energy electrons (β-particles). As their energy decreases, electrons also become more densely ionizing, i.e., they cause further ionizations in the very neighborhood of their formation. Considering these mechanisms of energy absorption, it is understood why high-energy electrons (1–10 MeV) and γ-rays (^{60}Co γ-rays 1.3 MeV, ^{137}Cs γ-rays 0.5 MeV) have practically the same linear energy transfer.

Upon energy loss of high-energy electrons, approximately 100 eV of energy is deposited per ionization event (on average), whereby a radical cation and an electron are generated. Reactions 1 and 2 show this for the two relevant targets that we are interested in here, the DNA and the water that surrounds it (the cell consists of about 70% water, and DNA is highly hydrated). The electron may still contain enough energy to cause further ionizations in the very near neighborhood. These areas, containing a number of ionization and occasionally also electronic-excitation events (reactions 3 and 4), are called spurs. In the case of sparsely ionizing radiation, these spurs do not overlap. In densely ionizing radiation, however, they form cylinders of spurs that are called tracks.

$$\mathrm{DNA} \xrightarrow[\text{radiation}]{\text{ionizing}} \mathrm{DNA}^{\bullet+} + e^- \quad (1)$$

$$\mathrm{H_2O} \xrightarrow[\text{radiation}]{\text{ionizing}} \mathrm{H_2O}^{\bullet+} + e^- \quad (2)$$

$$DNA \xrightarrow[\text{radiation}]{\text{ionizing}} DNA^* \quad (3)$$

$$H_2O \xrightarrow[\text{radiation}]{\text{ionizing}} H_2O^* \quad (4)$$

The water radical cation, produced in reaction 2, is a very strong acid and immediately loses a proton to neighboring water molecules thereby forming an $^{\bullet}OH$ radical (reaction 5). The electron becomes hydrated by water (reaction 6). Electronically excited water can decompose into $^{\bullet}OH$ and $H^{\bullet}$ (reaction 7). Thus, three kinds of free radicals are formed side by side in the spurs, $^{\bullet}OH$, e_{aq}^- and $H^{\bullet}$. To match the charge of the electrons, an equivalent amount of H^+ is also present.

$$H_2O^{\bullet+} \rightarrow {}^{\bullet}OH + H^+ \quad (5)$$

$$e^- + nH_2O \rightarrow e_{aq}^- \quad (6)$$

$$H_2O^* \rightarrow {}^{\bullet}OH + H^{\bullet} \quad (7)$$

A spur may contain more than one reactive intermediate, and they can interact with one another (e.g., $^{\bullet}OH + e_{aq}^- \rightarrow OH-$; $H^+ + e_{aq}^- \rightarrow H^{\bullet}$; $^{\bullet}OH + H^{\bullet} \rightarrow H_2O$; $2HO^{\bullet} \rightarrow H_2O_2$; $2e_{aq}^- + 2H^+ \rightarrow H_2$; $2H^{\bullet} \rightarrow H_2$) in competition with a reaction with a neighboring solute such as DNA.

The radiation-chemical yield is expressed by the *G* value. Originally, it was given in the units of molecules formed (or destroyed) per 100 eV of absorbed energy. The electronvolt is not an SI unit and hence is usually no longer in use. These two units may be converted into one another taking into account that 1 molecule $(100\ eV)^{-1} = 1.036 \times 10^{-7}\ mol\ J^{-1}$.

2.3 Direct and Indirect Effect

As depicted in reactions 1 and 3, the energy of the ionizing radiation may be absorbed by DNA itself, and this part of the DNA damage by ionizing radiation is called the *direct effect*. When the ionizing radiation is absorbed by the water that surrounds DNA (reactions 2 and 4), reactive intermediates, notably $^{\bullet}OH$ (reaction 5), are generated that can also damage DNA. This contribution is called the *indirect effect*. There have been a number of attempts to disentangle the two contributions by the addition of $^{\bullet}OH$ scavengers. The results are not really clear-cut (von Sonntag 2006), but it is obvious that there must be these two contributions. It is hoped that microdosimetric approaches using Monte Carlo type calculations may sharpen our view on this topic; for a review on the contribution of quantum-mechanical and Monte Carlo calculations in understanding DNA damage, see von Sonntag (2007).

2.4 Cerenkov Radiation

High-energy electrons that pass through a dense medium such as a cell lose a small fraction of their energy as Cerenkov radiation. This UV radiation is also absorbed by DNA and causes the well-known UV lesions, notably pyrimidine dimers and 6,4-photoproducts. The studies that are concerned with this effect were discussed in detail by von Sonntag (1987), and the relevant literature is also mentioned in von Sonntag (2006).

2.5 Hole and Electron Transfer Through DNA

In reaction 1, a DNA radical cation is formed. There is a rapid long-range hole transfer through DNA. The radical cation ends up at G, the nucleobase with the lowest ionization potential. The ionization potentials of clustered Gs, GG and GGG, are even lower than that of a single G, and the hole ends up preferentially at a GGG site (Saito et al. 1998). In competition with this rapid hole transfer, there are other reactions such as the reaction of the hole with water or with traps other than GGG. For example, the DNA radiation product 8-oxo-G may serve as a trap on account of its very low ionization potential (Shafirovich et al. 2001). This hole transfer is presently a very hot topic in DNA free-radical research, and a large number of excellent experimental as well as theoretical studies have been reported over the last few years. For details, the reader is referred to some reviews (Giese 2000; Schuster 2000; O'Neill and Barton 2004; Nakatani and Saito 2004; Berlin et al. 2004; von Sonntag 2006).

Electron transfer can also occur over some distance, whereby T, the nucleobase that has the highest electron affinity, is the trapping site (von Sonntag 2006). Our understanding of electron transfer through DNA is not yet as detailed as that of the corresponding hole transfer (Wagenknecht 2003, 2006).

2.6 Formation of Base Radicals

Radicals (often referred to as free radicals) are species with unpaired electrons. These unpaired electrons give the radicals a pronounced reactivity. Yet, depending on the type of radical, the reactivity may vary considerably (many orders of magnitude). The $^{\bullet}OH$ radicals that are generated in reaction 5 are among the most reactive free radicals. With DNA, they react preferentially with the nucleobases, and only a minor fraction abstracts a hydrogen atom from the sugar moiety (see later). Despite its high reduction potential (1.9 V at pH 7; Wardman 1989), the $^{\bullet}OH$ radical undergoes electron transfer reactions only very reluctantly. The radicals rather react by addition and by hydrogen abstraction. Although the $^{\bullet}OH$ radical reacts with the nucleobases at close to diffusion-controlled rates, there is a remarkable regioselectivity (Fujita and Steenken 1981). This reaction is controlled by kinetics rather than by thermodynamics; for a quantum-chemical approach to rationalize this, see von Sonntag (2006). For example, in thymine it adds preferentially to the C6 position (60%, reaction 8) rather than to the C5 position (30%, reaction 10), although the latter reaction is thermodynamically favored by 46 kJ mol^{-1} (S. Naumov and C. von Sonntag 2006, unpublished results). Only a small fraction abstracts a hydrogen atom from the methyl group (approximately 10%, reaction 9), despite the fact that this reaction seems to be thermodynamically the most favored one.

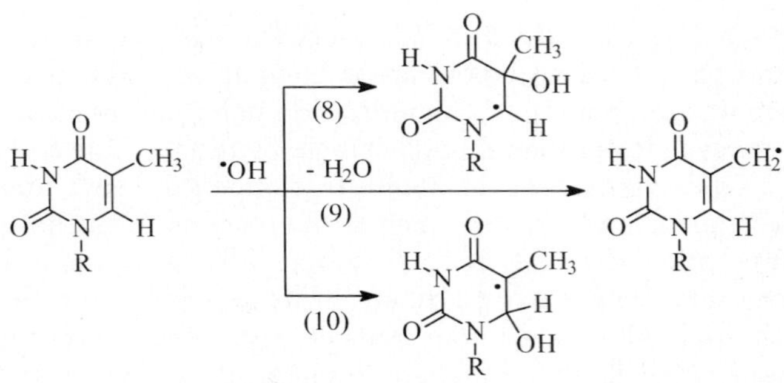

In the purines, C4 and C8 are the preferred sites for $^{\bullet}OH$ radical attack, and formation of the thermodynamically strongly favored guanyl and adenyl radicals ($G^{\bullet}$; $A^{\bullet}$) only occurs upon release of water from the $^{\bullet}OH$ adduct radicals (e.g., reaction 11).

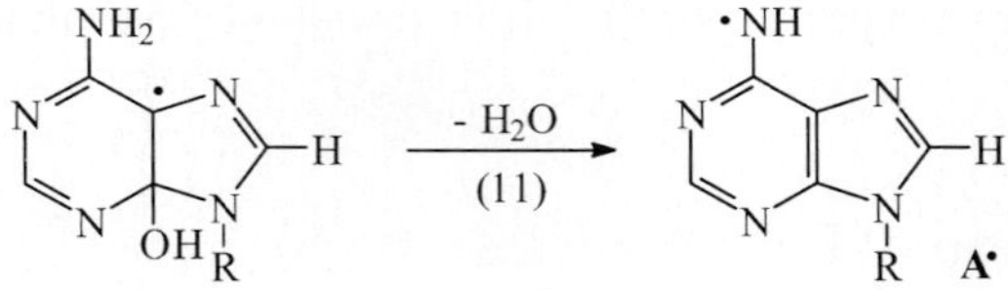

An addition to the C8 position gives rise to a radical that readily opens the ring (e.g., reaction 12), and the ring-open radicals are the precursors of the well-documented FAPY products (FAPY-A is 4,6-diamino-5-formamidopyrimidine from adenine; FAPY-G is 2,6-diamino-4-hydroxy-5-formamidopyrimidine from guanine; e.g., reaction 13). Mechanistic details of this rapid ring-opening reaction have not yet been fully elucidated, and reaction 12 shows only one of the various possible intermediates.

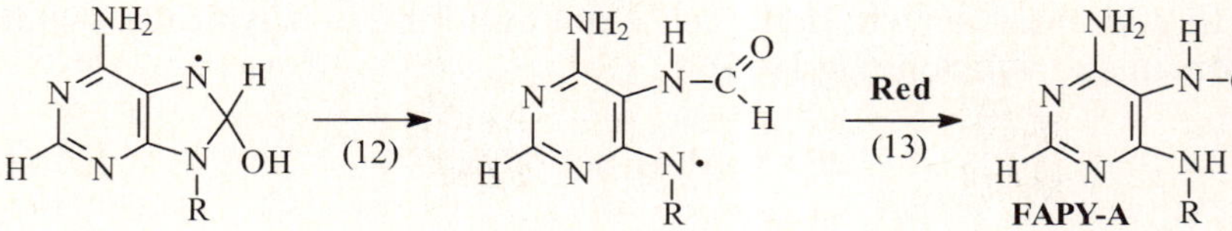

The nucleobase radical cations formed according to reaction 1 release a heteroatom-bound proton quite readily, and in the case of A and G this leads to the formation of the thermodynamically favored $G^{\bullet}$ and $A^{\bullet}$ radicals (e.g., reaction 14).

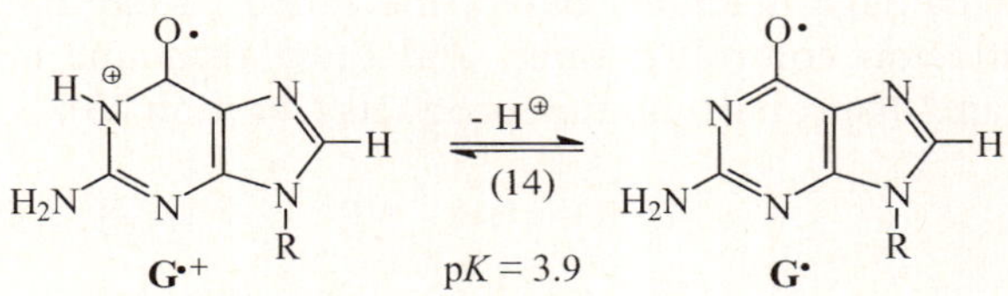

In thymidine, the N3 radical that is formed upon deprotonation of the radical cation (equilibrium 16) is not the thermodynamically favored species and eventually is converted into the more stable allylic radical (reaction 15).

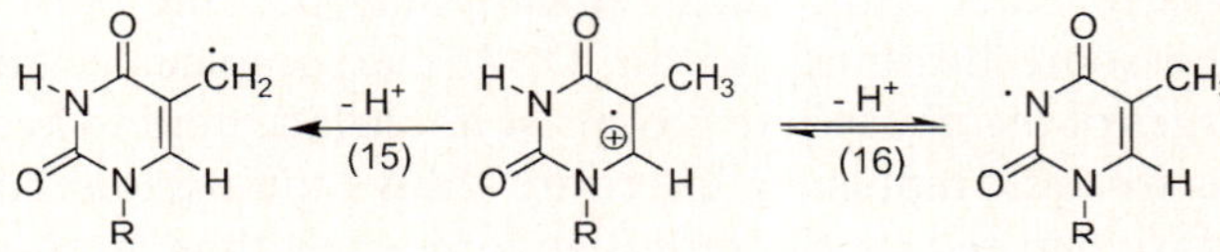

2.7 Base Damage

For the conversion of a radical into a final diamagnetic product, the radical must be either reduced or oxidized. In this respect, oxygen is not a good oxidant, since it usually converts the radical only into the corresponding peroxyl radical (reaction 17), i.e., the radical character is not lost.

$$R^{\bullet} + O_2 \rightarrow ROO^{\bullet} \tag{17}$$

In certain cases, transition metal ions such as Fe^{3+}/Fe^{2+} may act as oxidants/reductants and thiols such as glutathione (GSH) can undergo hydrogen-transfer reactions (reaction 18). GSH is the most abundant thiol in cells. Its concentration within the cell is not uniform but is typically 1–2 mM (Wardman and von Sonntag 1995).

$$R^{\bullet} + GSH \rightarrow RH + GS^{\bullet} \quad (18)$$

In DNA free-radical chemistry, reaction 18 has been termed "chemical repair," although it is obvious that a hydrogen donation to radicals such as those formed in reactions 8 and 10 do not cause these radicals to revert into thymine but rather gives rise to C5OH,C6H-dihydrothymine and C5H,C6OH-dihydrothymine, respectively. This is of importance when discussing reactions 17 and 18 in a cellular environment (see later).

In model systems on which our knowledge of the mechanism of free-radical-induced DNA damage is largely based, the radials decay bimolecularly at practically diffusion controlled rates, and these reactions lead to dimers (reaction 19) and disproportionation products (reaction 20).

$$2HR^{\bullet} \rightarrow HR - HR \quad (19)$$

$$2HR^{\bullet} \rightarrow RH_2 + R \quad (20)$$

It is a particular feature of ionizing radiation that with a high probability more than one radical is formed within a very small volume. If these radicals are sufficiently close to one another, they may decay bimolecularly. Much more problematic is the situation when only one radical is formed (or remains when the close-by ones have decayed) on the DNA without another nearby radical to react with. Such radicals may become very long lived. There may be some flexibility of the DNA macromolecule, although the negative charges of the phosphate groups at its surface tend to keep it largely rodlike by electrostatic repulsion. There are always counterions condensed at the DNA surface that reduce the repulsive forces and thus increase the flexibility. In man-made polymers that have been studied as model systems, bimolecular decay processes also take place in charged polymers, albeit with a slower rate than in uncharged coiled polymers. For DNA, we lack detailed information on the rates, but the DNA radical lifetimes are not very long also in the absence of any added external oxidant/reductant. The techniques that are available at present for the determination of lifetimes of DNA radicals, pulse radiolysis and electron paramagnetic resonance, require radical concentrations that generate many radicals per DNA macromolecule. Under such conditions, the decay kinetics do not follow simple second-order kinetics (Ulanski et al. 1994; von Sonntag 2006), and extrapolation to a few radicals per macromolecule is fraught with very large errors. An additional problem is the fact that in cells the DNA is wrapped up by the nucleohistones, and this situation will further lower the mobility of DNA radicals. In the absence of O_2, this increases the probability of reaction 18 with respect to a bimolecular decay of DNA radicals. In the presence of O_2, i.e., when DNA peroxyl radicals

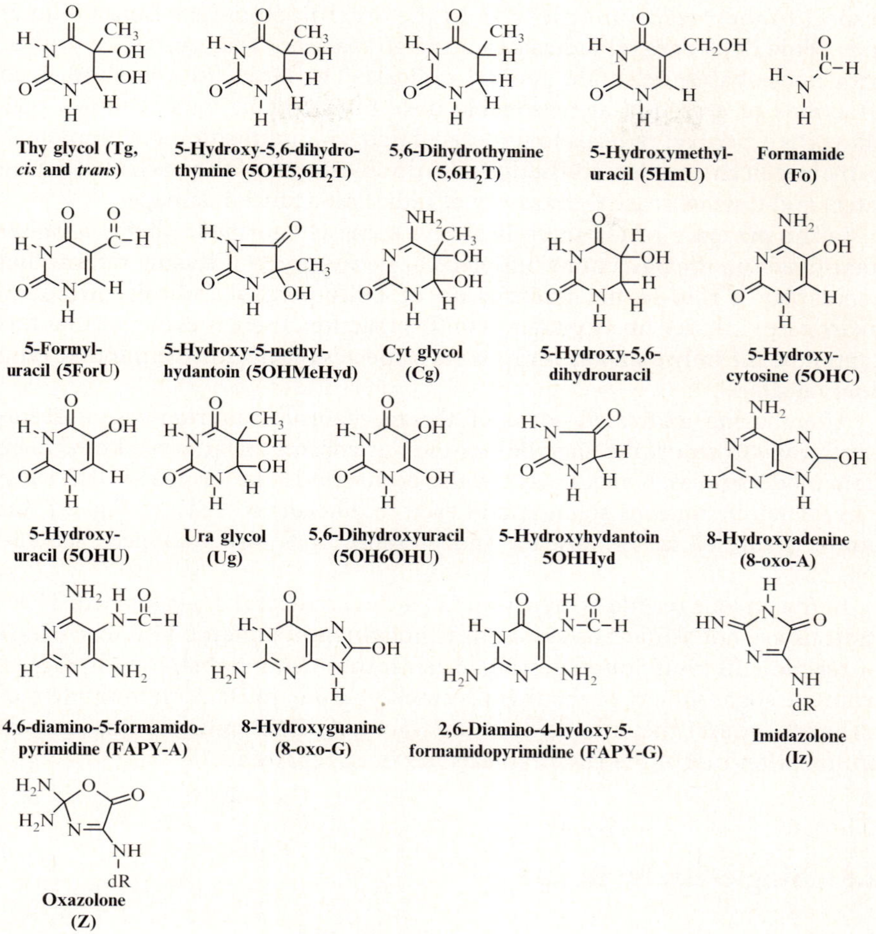

Fig. 2.2 DNA base lesions as known from model systems and from DNA γ-irradiated in aqueous solution

dominate, one is confronted with the same problem of the rate of the bimolecular decay of the DNA peroxyl radicals as only a few, at the most, can eliminate $HO_2^{\bullet}/O_2^{\bullet-}$. The rate for reaction of GSH with peroxyl radicals is slow. An originally reported high rate constant (Schulte-Frohlinde et al. 1986) has been revised to a much lower value (Hildenbrand and Schulte-Frohlinde 1997). The rate constant for reaction of $HO_2^{\bullet}$ with dithiothreitol has been measured at 120 M^{-1} s^{-1} (Lal et al. 1997), and as a first approximation one may assume that the rate constant for reaction of DNA peroxyl radicals with GSH is of the same order of magnitude. With a GSH concentration of 1 mM, this gives a rate of decay near 0.1 s^{-1} and a lifetime of the DNA peroxyl radicals with respect to their reaction with GSH on the order of 10 s. Such long lifetimes may allow DNA peroxyl radicals to undergo reactions that cannot be studied with nucleobase/nucleotide peroxyl radicals. The latter are too short-lived (fractions of a second at reasonable dose rates). Their long lifetimes may allow DNA peroxyl radicals to undergo damage-amplification reactions such as hydrogen abstraction and addition to double bonds of the nucleobases (see later) and thus increase the severity of radiation-induced damage.

In the presence of O_2, there is always some $O_2^{\bullet-}$ formed. This is a rather long-lived (on the basis of its bimolecular decay) freely diffusing radical that could play a role in the termination of DNA peroxyl radicals. Although hydroperoxides are likely products of this reaction, there is evidence that this cannot be the only pathway for how such species react with one another (von Sonntag 2006).

After having addressed some of the mechanistic uncertainties, it seems worthwhile to report the alteration of the nucleobases as far as we know them from model studies. A major part has also been found to be formed in DNA when γ-irradiated in aqueous solution and even in γ-irradiated cells. In Fig. 2.2, the products formed in the absence and presence of O_2 are compiled (see also Fig. 2.3).

In the "repair reaction" (reaction 18) a thiyl radical (GS$^{\bullet}$) is formed. These radicals are not as inert as commonly thought, and undergo also the reverse of reaction 18 (von Sonntag 2006). Even if they do not abstract a hydrogen from the sugar moiety or the thymine methyl group in DNA, in multiple radical sites (spurs) they are likely to interact with DNA radicals that are still around after one of the DNA radicals has been repaired.

2.8 Single-Strand Breaks

We have only little information as to the involvement of the direct effect on the formation of SSBs in cellular systems (it must be of considerable importance). However, there is a wealth of information on the reactions of the $^{\bullet}$OH radical. To induce a SSB, the $^{\bullet}$OH radical must abstract a hydrogen atom from the sugar moiety. In DNA in aqueous solution, SSBs occur with a marked preference at local structures where the minor groove is wide and the sugar moiety is more accessible to the $^{\bullet}$OH radical. The situation has been successfully modeled by Monte Carlo type calculations to a perfection that even the protective effects of bound proteins are well reflected (Begusova et al. 2001); for further references, see von Sonntag (2006).

Mechanistically, strand breakage as a consequence of hydrogen abstraction at C4′ is understood best (Dizdaroglu et al. 1975). A neighboring phosphate group is eliminated heterolytically (e.g., reaction 21). The resulting radical cation may react with water (e.g., reactions 22 and 23; for a more

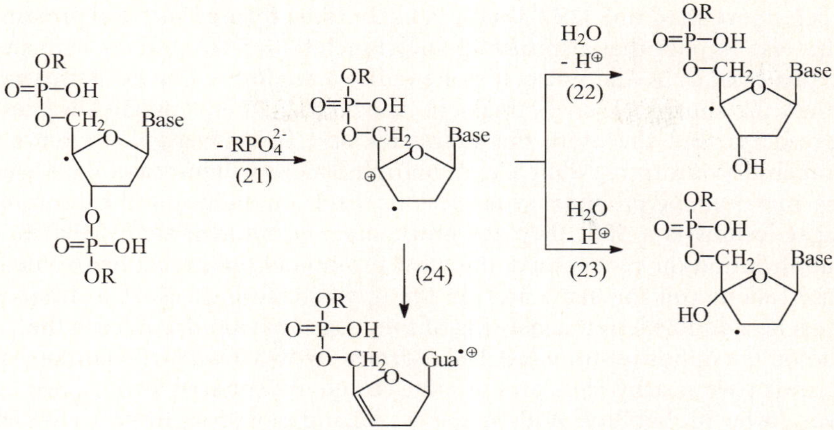

detailed description, see von Sonntag 2006) or oxidize a nearby guanine (reaction 24) with subsequent hole transfer through DNA (see earlier).

In the presence of O_2, the peroxyl radical at C4′ also gives rise to a strand break. This lesion is characterized by a 3′-phosphoglycolate end group (reactions 25–27; for details, see von Sonntag 2006).

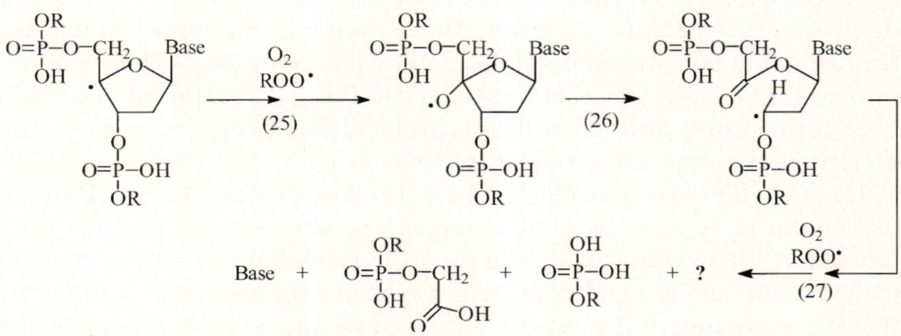

A similar sequence of events may be written for the C5′ peroxyl radical (von Sonntag 2006). In both cases, an interaction with a second peroxyl radical is required. The implication of this was discussed above.

2.9 Double Lesions

It has already been mentioned that the deposition of ionizing radiation occurs in packages of, on average, about 100 eV. Thus, several radicals may be generated in close proximity in these spurs. When the spurs overlap, such as in the tracks of densely ionizing radiation, the number of nearby radicals confined in a small volume increases. When the spurs/tracks hit the DNA in a cell, more than one DNA lesion is likely to be formed in close proximity. This also implies that a spur/track may touch both DNA and nucleohistone. In this case, DNA and nucleohistone radicals are formed in close proximity. When the spur/track only streaks at its edge the DNA, a single DNA lesion may occur, but the more the spur/track and DNA overlap, the more the probability of forming double and multiple lesions will increase. Base lesions are more abundant than sugar lesions, and since one of the prominent sugar lesions is a SSB, they are much more prominent than SSBs. As far double lesions are concerned, the most prominent ones must be two nearby base lesions (on the same and on the opposite strand). Next in frequency must be a double lesion consisting of a damaged base and a SSB (on the same and on the opposite strand). A DSB, formed by two nearby SSBs on opposite strands (two nearby SSBs on the same strand are counted as one SSB), is of even lower probability. With sparsely ionizing radiation, there are 40 DSBs versus 1,000 SSBs (Table 2.1). With densely ionizing radiation, this ratio is considerably higher, and clustered lesions beyond a double lesion are also more prominent (see later). These multiple lesions when processed by repair enzymes may lead to severer lesions than the processing of single lesions. For example, excision repair of a damaged base opposite a SSB will result in a DSB. There is increasing evidence that double lesions can also be formed as a consequence of the attack of a single radical, in contrast to the spur effect just discussed. Such reactions may have to be taken into account when discussing the formation of endogenous DSBs. Some aspects will be dealt with now.

It has been suggested that the sugar radical that remains after a SSB has been formed may abstract with low probability a hydrogen from the sugar moiety of the opposite strand thereby causing a DSB (Siddiqi and Bothe 1987). This flip-over concept has been used to explain the low DSB yields (3%) induced by photons whose energies are just sufficient to cause a single ionization (Prise et al. 2000). The flip-over mechanism has been challenged (Ward 1988), and as an alternative it has been suggested (von Sonntag 2006) that the formation of the second SSB could be due to an epithermal electron that is released in the photoionization process (reactions 28–30) (Li et al. 2003; Becker et al. 2003).

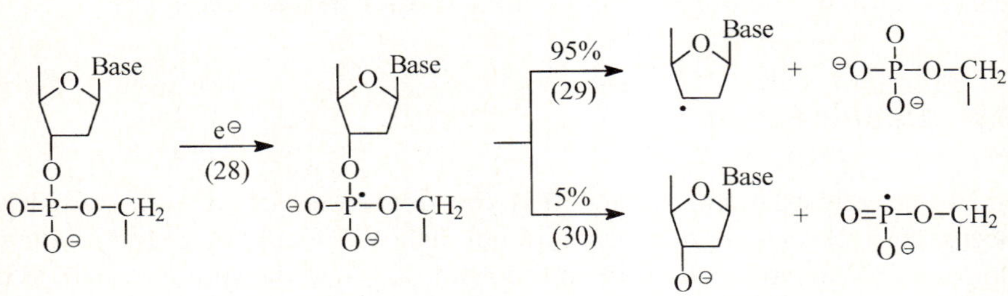

The flip-over mechanism is one example where two damaged sites may be generated from one single radical originally formed. There are other examples that are based on a much better footing. The well-documented 5′, 8-cyclonucleotide lesions cA and cG are another example. Although they may originate from a C5′ radical, there is evidence from model systems that base radicals are likely precursors, although the mechanistic details are still to be elucidated (von Sonntag 2006). There are a number of vicinal lesions that belong to this group (Fig. 2.3; for mechanistic aspects and further details, see von Sonntag 2006).

Further studies on such damage-amplification reactions will hopefully shed some light on the formation of endogenous DSBs and other higher-order lesions that may be responsible for chromosomal mutations.

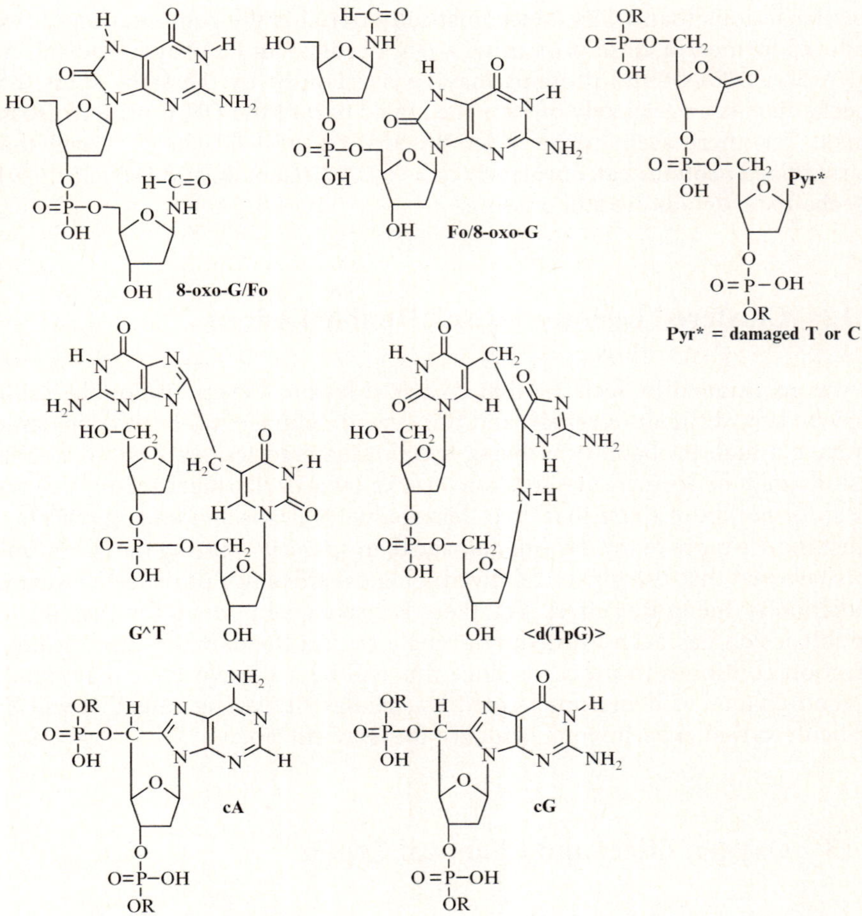

Fig. 2.3 Double lesions, partially vicinal lesions induced by a single •OH radical, as detected in DNA or relevant model systems

2.10 DNA–DNA Cross-Links

There must be a given likelihood that two radicals on opposite strands formed within a spur within a very close distance recombine. Such a process leads to an interstrand DNA–DNA cross-link. There are, however, also single-radical reactions that give rise to such a lesion (Hong and Greenberg 2005).

2.11 DNA–Protein Cross-Links

As seen from Table 2.1, DNA–protein cross-links are even more abundant cellular lesions than DSBs. Mechanistically trivial is the combination of two radicals formed in close proximity, a DNA radical and a protein radical. As DNA and DNA-bound proteins may be overlapped by the same spur, this mechanism is a very likely one for the major part of the DNA–protein cross-links. Yet, there seem to be other precursors, as it has been shown that peroxidized proteins can covalently link to DNA (Gebicki and Gebicki 1999). Mechanistic details are still missing.

2.12 Clustered Lesions Beyond Double Lesions

It was mentioned in Sect. 2.1 that clustered lesions beyond a double lesion may be very difficult to repair, and this type of damage has been considered to have a high probability of being lethal. Clustered lesions beyond double lesions may be formed in spurs and tracks by a trivial mechanism, i.e., the local formation of more than two damaged sites. Densely ionizing radiation must induce more of this type of damage than sparsely ionizing radiation, and it is believed that the higher relative biological efficiency of densely ionizing radiation is due to this effect. Yet, there is also a given possibility that after a double lesion has been formed by a single radical the damage-amplification reaction continues to proceed. Since this will be a slow process that stands a good chance of being intercepted by cellular thiols (see later), it will be difficult to find experimental evidence for its occurrence.

2.13 Oxygen Effect and Chemical Repair

In cells, radiation-induced DNA damage as expressed by SSB/DSB formation or by biological end points such as reproductive cell death are enhanced in the presence of oxygen. This phenomenon has been termed "oxygen effect" and been attributed to the competition of damage fixation by O_2 (reaction 17)) and "chemical repair" by the cell's thiol pool (reaction 18) (Alper 1983). At the time, it was not fully clear that the chemical repair rarely restores the former integrity of DNA. Thus, it is somewhat too simplistic to call reaction 18 a "repair" reaction. Yet, reduction by GSH may lower the severity of the damage. For example, let us take the situation of a multiply damaged DNA site that in the presence of O_2 would give rise to a DSB. The reduction of one of the two sugar radicals by GSH prevents the formation of one of the two close-by SSBs that would give rise to a DSB, i.e., the damage is lowered from the DSB to the SSB level. That DSBs largely originate from two close-by SSBs has been confirmed by measuring the oxygen enhancement ratio (Prise et al. 1993) and the repair rates for these two lesions (Prise et al. 1999). The prevention of a SSB by hydrogen donation is an atypical case, as here a proper repair is not unlikely. Thus, there must also be a benefit when a base radical is inadequately repaired. Earlier, it was mentioned that most damage-amplification reactions are induced by peroxyl radicals. As GSH "repair" prevents their formation, such damage-amplification reactions are intercepted. As a consequence, the level of damage is lowered and the subsequent enzymatic repair facilitated.

2.14 DNA Damage by UV-C and Ionizing Radiation – a Comparison

UVC radiation damages mainly the pyrimidines, and the major products by far are pyrimidine cyclobutane dimers and the 6,4-photoproducts (Cadet 1990). In the terminology used here, they are doubly damaged base lesions. There are only very low yields of singly damaged base sites such as FAPY-A and 8-oxo-G and of SSBs. The presence of radical scavengers such as O_2 or GSH has no or only a negligible effect. In contrast, ionizing radiation produces many singly damaged base sites with a large variety of chemical changes in pyrimidines and purines alike. SSBs are abundant lesions. Double base lesions are not confined to the same strand as those induced by UVC radiation, but also occur on opposite strands. DSBs, DNA–DNA interstrand and DNA–protein cross-links are other double lesions. Clustered lesions beyond a double lesion are another typical feature of ionizing radiation, as are the effects of O_2 and radiation quality.

In this comparison, it is worth mentioning that when in DNA T is substituted by 5BrU the photochemistry changes dramatically, and free-radical processes now dominate. Among others, SSBs and DNA–protein cross-links are observed (for references and mechanistic aspects, see von Sonntag 2006). The observation, that sister-chromatid exchanges are induced led to the conclusion that interstrand DNA cross-links may also be an important lesion (Wojcik et al. 2003) and this has now been substantiated experimentally (Wojcik et al. 2006).

The large variety of damaged sites induced by ionizing radiation is a different challenge for cellular repair systems from UV damage. It seems that highly specific enzymes such as the photoreactivating enzyme that repairs cyclobutane-type dimers effectively do not exist for ionizing-radiation damage. In many aspects, ionizing radiation appears to resemble the damage caused by "oxygen stress," which is considered to be free radical in nature as well. "Oxygen stress" and possibly also ionizing radiation have, like UV radiation, a long evolutionary carrier, and cells have adapted their repair system to this challenge. To a large extent, such repair enzymes have to be generalists and have to be able to cope with a large set of lesions. With this in mind, the large variation of products formed by ionizing radiation may not be that much of a problem, but the interesting questions remain at what rate such lesions are repaired and what are the reasons for and the products of a misrepair. These are the key questions in our attempt to understand the effects of ionizing radiation on biological end points such as chromosomal mutations and reproductive cell death.

References

Alper T (1983) Oxygen as radiosensitizer: methods of analysis. Int J Radiat Biol 44:313–314

Becker D, Bryant-Friedrich A, Trzasko C-A, Sevilla MD (2003) Electron spin resonance study of DNA irradiated with an argon-ion beam: evidence for formation of sugar phosphate backbone radicals. Radiat Res 160:174–185

Begusova M, Spotheim-Maurizot M, Sy D, Michalik V, Charlier M (2001) RADACK, a stochastic simulation of hydroxyl radical attack on DNA. J Biomol Struct Dyn 19:141–158

Berlin YA, Kurnikov IV, Beratan D, Ratner MA, Burin AL (2004) DNA electron transfer processes: some theoretical notions. Top Curr Chem 237:1–36

Cadet J (1990) The photochemistry of nucleic acids. In: Morrison H (ed) Photochemistry and the nucleic acids. Wiley-Interscience, New York, pp 1–272

De Bont R, van Larebeke N (2004) Endogeneous DNA damage in humans: a review of quantitative data. Mutagenesis 19:169–185

Dizdaroglu M, von Sonntag C, Schulte-Frohlinde D (1975) Strand breaks and sugar release by γ-irradiation of DNA in aqueous solution. J Am Chem Soc 97:2277–2278

Fujita S, Steenken S (1981) Pattern of OH radical addition to uracil and methyl- and carboxyl-substituted uracils. Electron transfer of OH adducts with *N,N,N′,N′*-tetramethyl-p-phenylenediamine and tetranitromethane. J Am Chem Soc 103:2540–2545

Gebicki S, Gebicki JM (1999) Crosslinking of DNA and proteins induced by protein hydroperoxides. Biochem J 338:629–636

Giese B (2000) Long-distance charge transport in DNA: the hopping mechanism. Acc Chem Res 33:631–636

Goodhead DT (1994) Initial events in the cellular effect of the ionizing radiations: clustered damage in DNA. Int J Radiat Biol 65:7–17

Hildenbrand K, Schulte-Frohlinde D (1997) Time-resolved EPR studies on the reaction rates of peroxyl radicals of poly(acrylic acid) and of calf thymus DNA with glutathione. Re-examination of a rate constant for DNA. Int J Radiat Biol 71:377–385

Hong IS, Greenberg MM (2005) Efficient DNA interstrand cross-link formation from a nucleotide radical. J Am Chem Soc 127:3692–3693

Lal M, Rao R, Fang X, Schuchmann H-P, von Sonntag C (1997) Radical-induced oxidation of dithiothreitol in acidic oxygenated aqueous solution: a chain reaction. J Am Chem Soc 119:5735–5739
Li X, Sevilla MD, Sanche L (2003) Density functional theory studies of electron interaction with DNA: can zero eV electrons induce strand breaks? J Am Chem Soc 125:13668–13699
Nakatani K, Saito I (2004) Charge transport in duplex DNA containing modified nucleotide bases. Top Curr Chem 236:163–186
O'Neill MA, Barton JK (2004) DNA-mediated charge transport chemistry and biology. Top Curr Chem 236:67–115
Prise KM, Davies S, Michael BD (1993) Evidence for induction of DNA double-strand breaks at paired radical sites. Radiat Res 134:102–106
Prise KM, Gillies NE, Michael BD (1999) Further evidence for double-strand breaks originating from a paired radical precursor from studies of oxygen fixation processes. Radiat Res 151:635–641
Prise KM, Folkard M, Michael BD, Hopkirk A, Munro IH (2000) Critical energies for SSB and DSB induction in plasmid DNA by low-energy photons: action spectra for strand-break induction in plasmid DNA irradiated in vacuum. Int J Radiat Biol 76:881–890
Saito I, Nakamura T, Nakatani K, Yoshioka Y, Yamaguchi K, Sugiyama H (1998) Mapping of the hot spots for DNA damage by one-electron oxidation: Efficacy of GG doublets and GGG triplets as a trap in long-range hole migration. J Am Chem Soc 120:12868–12687
Schulte-Frohlinde D, Behrens G, Önal A (1986) Lifetime of peroxyl radicals of poly(U), poly(A) and single- and double-stranded DNA and the rate of their reaction with thiols. Int J Radiat Biol 50:103–110
Schuster GB (2000) Long-range charge transfer in DNA: transient structural distortions control the distance dependence. Acc Chem Res 33:253–260
Shafirovich V, Cadet J, Gasparutto D, Dourandin A, Huang W, Geacintov NE (2001) Direct spectroscopic observation of 8-oxo-7,8-dihydro-2′-deoxyguanosine radicals in double-stranded DNA generated by one-electron oxidation at a distance by 2-aminopurine radicals. J Phys Chem B 105:586–592
Shao C, Folkard M, Michael BD, Prise KM (2004) Targeted cytoplasmic irradiation induces bystander responses. Proc Nat Acad Sci USA 101:13495–13500
Siddiqi MA, Bothe E (1987) Single- and double-strand break formation in DNA irradiated in aqueous solution: Dependence on dose and OH radical scavenger concentration. Radiat Res 112:449–463
Ulanski P, Bothe E, Rosiak JM, von Sonntag C (1994) OH-Radical-induced crosslinking and strand breakage of poly(vinyl alcohol) in aqueous solution in the absence and presence of oxygen. A pulse radiolysis and product study. Macromol Chem Phys 195:1443–1461
Vilenchik MM, Knudson AG (2003) Endogeneous DNA double-strand breaks: Production, fidelity of repair, and induction of cancer. Proc Nat Acad Sci USA 100:12871–12876
von Sonntag C (1987) The chemical basis of radiation biology. Taylor and Francis, London
von Sonntag C (2006) Free-radical-induced DNA damage and its repair. A chemical perspective. Springer, Berlin
von Sonntag C (2007) Free-radical-induced DNA damage as approached by quantum mechanics and Monte Carlo calculations: an overview from the standpoint of an experimentalist. Adv Quantum Chem 52:5–20
Wagenknecht H-A (2003) Reductive electron transfer and transport of excess electrons in DNA. Angew Chem Int Ed Engl 42:2454–2460
Wagenknecht H-A (2006) Electron transfer processes in DNA: mechanisms, biological relevance and applications in DNA analytics. Nat Prod Rep 23:973–1006
Ward JF (1988) DNA damage produced by ionizing radiation in mammalian cells: identities, mechanisms of formation, and repairability. Prog Nucleic Acid Res Mol Biol 35:95–125
Wardman P (1989) Reduction potentials of one-electron couples involving free radicals in aqueous solution. J Phys Chem Ref Data 18:1637–1755

Wardman P, von Sonntag C (1995) Kinetic factors that control the fate of thiyl radicals in cells. Methods Enzymol 251:31–45

Wojcik A, von Sonntag C, Obe G (2003) Application of the biotin-dUTP chromosome-labelling technique to study the role of 5-bromo-2′-deoxyuridine in the formation of UV-induced sister chromatid exchanges in CHO cells. J Photochem Photobiol B 69:139–144

Wojcik A, Bochenek A, Lankoff A, Lisowska H, Padjas A, Szumiel I, von Sonntag C, Obe G (2006) DNA interstrand crosslinks are induced in cells prelaballed with 5-bromo-2′-deoxyuridine and exposed to UVC radiation. J Photochem Photobiol B 84:15–20

Wu LJ, Randers-Pehrson G, Xu A, Waldren CA, Yu Z, Yu Z, Hei TK (1999) Targeted cytoplasmic irradiation with alpha particles induces mutations in mammalian cells. Proc Nat Acad Sci USA 96:4959–4964

3 Effects of Ultraviolet Radiation on DNA

JÜRGEN KIEFER

Abstract

UV radiation is selectively absorbed by DNA, mainly in the UV-B and the UV-C regions. Excitation of DNA leads to typical photoproducts, the most important being pyrimidine dimers and the so-called 6-4-photoproduct. The yield of strand breaks is very low directly after exposure, but they may be formed indirectly upon further incubation. Although DNA absorbs only weakly in the UV-A region, it may be damaged indirectly via endogenous photosensitisers. Photoproducts are subject to repair processes which are genetically determined. Photoreactivation splits selectively pyrimidine dimers but it is limited to lower eukaryotes and does not occur in mammalian cells. Nucleotide excision repair acts on various types of damage. UV-induced damage manifests itself also at the level of chromosomes. Chromatid-type aberrations are more frequent with UV irradiation than chromosomal aberrations. The action spectrum for their formation demonstrates that an indirect mechanism plays an important role in the UV-A region. Also micronuclei are found after broadband UV exposure which are presumably due to indirectly formed double-strand breaks. UV radiation is a strong inducer of sister-chromatid exchanges, in contrast to ionising radiation, where the yields are low. This suggests that strand breaks play only a minor role.

3.1 Introduction

Ultraviolet (UV) radiation is a part of optical radiation and covers the wavelength region between 100 and 480 nm, corresponding to quantum energies from about 2.5 to 12.5 eV. The UV region is commonly subdivided as given in Table 3.1.

The most energetic part is called the "Schumann UV" or "vacuum UV" because it is strongly absorbed by atmospheric gases. It plays little practical role on Earth but may have some significance in open space. The most important natural source of UV radiation is the sun, which emits UV radiation over the whole range. The ozone layer absorbs most of the UV-C radiation, so only UV-B and UV-A radiation seem to be of significance. The destruction of

Günter Obe and Vijayalaxmi (Eds.)
Chromosomal Alterations: Methods, Results and Importance in Human Health

Table 3.1 Wavelength regions of UV radiation and their common designations

Wavelength (nm)	Quantum energy(eV)	Name
100–200	12.5–6	Vacuum UV, Schumann UV
200–280	6–4.4	UV-C
280–315	4.4–4	UV-B
315–380	4–3.3	UV-A

stratospheric ozone reduces its shielding effect, so the short-wavelength edge of terrestrial solar radiation is shifted towards higher quantum energies, leading to increased hazards for humans, animals and plants.

It is clear from the second column of Table 3.1 that the quantum energies are not sufficient to cause ionisations. Interaction of UV radiation with matter proceeds via selective absorption and excitation. The absorption spectra of biological molecules are, therefore, of great importance for the understanding of action of UV radiation because they determine the probability of energy deposition. Strongly absorbing molecules are also called "chromophores". If radiation is absorbed in essential biomolecules (e.g. DNA) their excitation may lead to photochemical alterations causing biological effects like cell inactivation or mutations. But even if biomolecules are not directly excited because they do not absorb the impinging radiation, they may be indirectly affected via energy transfer from excited states of absorbing "photosensitisers". An important example of this process is the "photodynamic action".

The wavelength dependence of a biological effect can be used to identify the chromophore from where it initiated. This principle of "action spectroscopy" (see later) constitutes an important analytical tool and played some role in the history of molecular biology although this was not really appreciated at the time. As long ago as 1930, Gates (1930) showed that the spectral dependence of bactericidal action of UV radiation followed closely the absorption spectrum of nucleic acids and was distinctly different from that of proteins, which at that time were thought to be the building blocks of the genetic material. A few years later, similar results were obtained for mutagenesis (Knapp et al 1939; Emmons and Hollaender 1939), demonstrating the importance of nucleic acids before this was biochemically proven (and honoured by award of the Nobel Prize) by Avery et al. (1944).

Figure 3.1 displays the absorption spectrum of DNA. It has a maximum around 265 nm (in the UV-C region) but there is still also substantial absorption of UV-B radiation, while the molecule is virtually translucent for UV-A radiation.

Although frequently used, the term "dose", meaning "deposited energy per unit mass", is misleading in connection with UV radiation exposure. In most cases it stands for "energy fluence" (*E*), i.e. impinging energy per unit area, which is measured in joules per square metre. In many health-related publications also "weighted doses" are used. They are obtained by weighting all wavelengths according to their efficiency to cause a certain effect

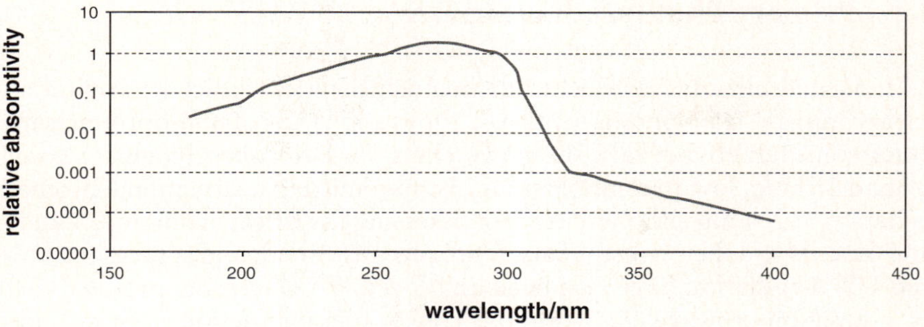

Fig. 3.1 Absorption spectrum of DNA. (Data from Sutherland and Griffin 1981)

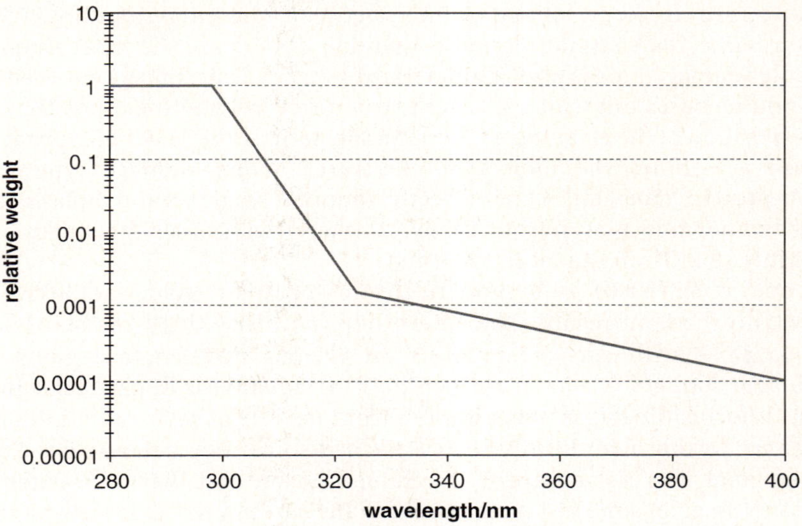

Fig. 3.2 The CIE-weighting function for erythema induction (McKinley and Diffey 1987)

(e.g. sunburn) and summing over the whole spectrum. These weighted doses are unfortunately given in the same units, mostly joules per square metre, which may lead to misunderstandings. The most important weighting function is that for erythema induction. Figure 3.2 shows it in a standardised form as defined by the Commission Internationale de l'Eclairage; International Commission on Illumination (CIE) (McKinley and Diffey 1987).

In photochemistry "photon fluence" (Φ), i.e. the number of photons per unit area, is more customary. The relationship between energy and photon fluence is simple:

$$E = h\nu\Phi,$$

where h is Planck's constant and ν the frequency.

3.2 Primary Photoproducts in DNA

As indicated already, one has to distinguish between direct UV action and energy transfer from photosensitisers. Types and yields of photoproducts may differ considerably between these two cases. As DNA absorbs most strongly around 265 nm, low-pressure mercury lamps emitting nearly monochromatic radiation at 254 nm are often used for assessing the direct action of UV radiation on cells and their constituents. More recently broadband sources for UV-B and UV-A radiation have also been employed. To determine product yields as a function of wavelength for the construction of action spectra, either broadband sources coupled to monochromators or tunable lasers are used.

The spectrum of DNA alterations induced by UV radiation is distinctly different from that seen with ionising radiation. The most important photoproducts are *cis-syn* cyclobutane dimers between adjacent pyrimidines (CPD) on the same DNA strand; thymine–thymine dimers are the most frequent. Another damage type is also characterised by a covalent linkage between two pyrimidine bases but with a different structure involving the 6-position and the 4-position of the ring. It is thus commonly termed the "6-4-photoproduct". Figure 3.3 shows the molecular structures. Other damage types, like DNA–protein cross-links and directly induced single-strand breaks, play only a minor role as seen from Table 3.2, where yields of DNA changes upon exposure to 254-nm radiation are listed.

Douki et al. (2000) using modern chromatographic techniques arrived at essentially the same results. The same group summarised UV-mediated damage to DNA with special emphasis on various wavelength regions and simulated sunlight (Cadet et al. 2005). UV-A radiation is of particular interest as it forms the largest contribution from terrestrial solar radiation but is only very weakly absorbed by DNA. This had sometimes prompted the conclusion that it is not hazardous to human health and that even extensive exposure (e.g. on sunbeds) does not carry any cancer risk. This view has now been revised. The yields of CPDs and the 6-4-photoproducts are much smaller in the UV-A region compared with those in the UV-B region but a new class of damage is seen with longer wavelengths, namely oxidative products and strand breaks. This situation is comparable to that with low

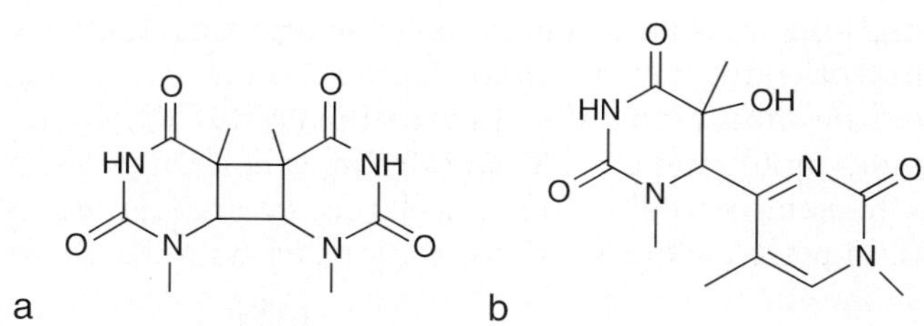

Fig. 3.3 UV photoproducts in DNA. **a** Cyclobutane-type pyrimidine dimer, **b** 6-4-photoproduct

Table 3.2 Yields of photoproducts in mammalian cells with exposure to 254-nm UV radiation

Product	Yield $(10^8\ u^a)^{-1}\ (J\ m^{-2})^{-1}$	Relative yield	Reference
CPD	2.0	1	Rosenstein and Mitchell (1987)
6-4-Product	0.4	0.2	Rosenstein and Mitchell (1987)
DNA–protein cross-links	1.3×10^{-4}	6.6×10^{-5}	Peak et al. (1985)
DNA single-strand breaks	4.1×10^{-4}	2×10^{-4}	Rosenstein and Ducore (1983)

CPD cis-syn cyclobutane dimers between adjacent pyrimidines
[a] u is the atomic mass unit

Table 3.3 DNA lesions in UV-A exposed and γ-irradiated human monocytes (Cadet et al. 2005)

Lesion	Yield per kJ m^{-2} UV-A radiation	Yield per Gy γ-rays
8-Oxo-7,8-dihydo-2′-deoxyguanosine	0.98	11
DNA strand breaks (single-strand breaks, double-strand breaks, alkali labile sites)	0.9	130
Endonuclease III sensitive sites[a]	0.3	53
FPG-sensitive sites[b]	1.9	48

FPG formamidopyrimidine DNA glycosylase
[a] Mostly oxidised pyrimidine bases
[b] Mostly modified purine bases

linear energy transfer ionising radiation and suggests a free-radical mechanism. Table 3.3 lists some of the products in cells exposed to UV-A radiation in comparison with those in cells after γ-irradiation. It is seen that there is a great similarity in the kind of products, but relative yields are different. While DNA breaks figure most prominently with γ-rays, modified purine bases are most important with UV-A radiation.

Nevertheless, the formation of strand breaks whose yield is negligible with short-wavelength UV is very remarkable as they are also held responsible for most of the hazardous effects of ionising radiation. This is strictly true, however, only for DNA double-strand breaks (DSBs). It is not clear at present whether they are also induced directly by UV-A radiation.

DSBs may, however, be formed indirectly when photoproducts remain unrepaired in cellular DNA. This was first suggested by experiments in a yeast mutant conditionally defective for DSB repair. (Kiefer 1987; Frankenberg-Schwager et al. 1987). It was previously assumed that the accumulation of excision repair related single-strand breaks may also lead to DSBs (Bradley 1981; Bradley and Taylor 1983), but it was later shown that UV-induced DSBs were also involved in excision-defective cells (Kiefer and Feige 1993). Experiments with transformed human fibroblasts defective in the ability to bypass pyrimidine dimers during DNA replication (XP variants) showed that DSBs originate at stalled replication forks (Limoli et al. 2002). DSBs can be specifically determined by an immunofluorescence assay for the phosphorylated histone γH2AX which is formed at DSB sites in DNA (Rogakou et al. 1998, 1999). A complex of recombination proteins (Mre11 complex) localises at stalled replication forks (Limoli et al. 2000). Simultaneous immunofluorescence staining both for γH2AX and for the Mre11 complex showed that both types of foci colocalise in UV-exposed cells. This is in line with an increased induction of sister-chromatid exchanges (SCEs) in these repair-deficient cells.

3.3 Action Spectra

The dependence of a given biological effect on the wavelength of the impinging radiation is often called an "action spectrum". The fundamental interest in action spectra lies in the possibility to identify the molecule responsible for a photobiological action by comparing the action spectrum with the absorption spectra of candidate cellular constituents. In order to reach this goal, a number of experimental precautions have to be taken but these cannot be discussed in detail here (see e.g. Kiefer 1990). A necessary condition, however, which must be mentioned in the context of this chapter is that the radiation must reach the intracellular target without being modulated by intervening material. This is easy to achieve with in vitro experiments with cell cultures (provided the cell suspension or the cell layer is not too dense), but is rather difficult in vivo. A prominent example is the so-called action spectrum for erythema induction by UV radiation. The cells responsible lie in the skin's stratum germinativum at a depth of some micrometres covered by the epidermis. This layer absorbs strongly in the UV-C region and partly also in the UV-B region (Parrish et al. 1978), which causes a strong distortion of the emission spectrum of an external light source. The resultant erythema action spectrum is thus a convolution of skin absorption and the action spectrum for cellular effects in the stratum germinativum.

Figures 3.4 and 3.5 show action spectra for the formation of some of the photoproducts mentioned already. The induction efficiency of CPDs and 6-4-photoproducts follows closely DNA absorption, which is not the case for single-strand breaks and DNA–protein cross-links, where a second peak is seen between 350 and 450 nm. This behaviour indicates that another chromophore is involved as discussed already.

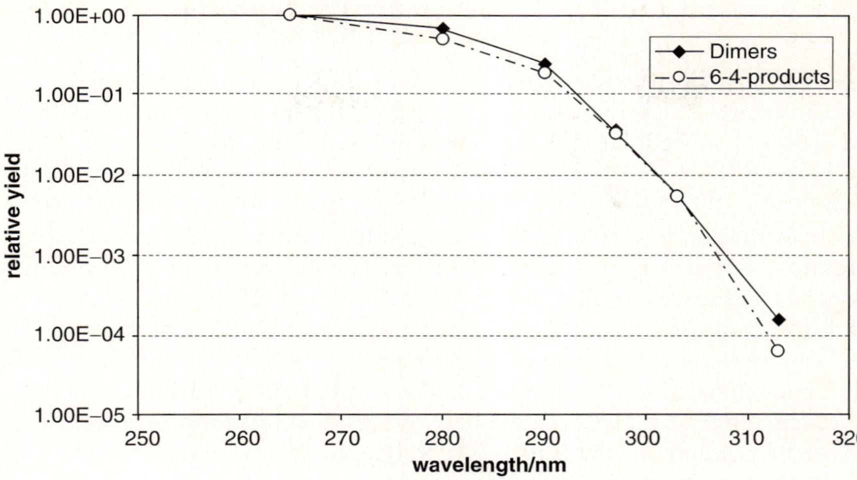

Fig. 3.4 Action spectra for the formation of thymine dimers and 6-4-photoproducts. (After Rosenstein and Mitchell 1987)

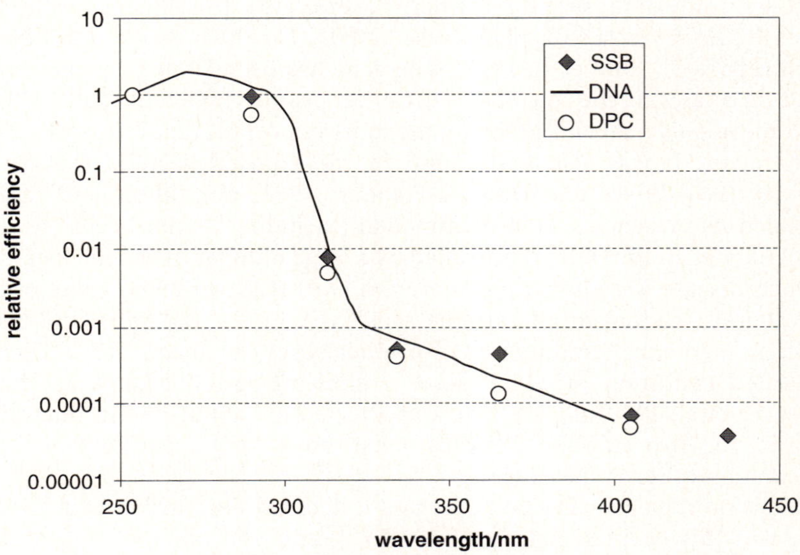

Fig. 3.5 Action spectra for the formation of single-strand breaks (*SSB*) and DNA–protein cross-links (*DPC*). The *line* shows DNA absorption for comparison. (After Peak and Peak 1991)

3.4 Reversal and Repair of Primary Photoproducts

The history of processes by which genetic damage is repaired or its effect circumvented is tightly linked to the study of the interaction of UV radiation on cells and subcellular systems. There is a wealth of literature on this topic, and even a short review is outside the scope of this chapter (see e.g. Walker et al. 1995). CPDs play here a leading role because of their specific properties. Their formation is a steady-state process: they are not only induced by UV radiation but also split at the same time. The efficiency for both pathways depends on the wavelength: the relative maximum for formation is at 280 nm, that for splitting is at 240 nm. This provides a "photochemical switch" between induction and reversal of these photoproducts. As a consequence of this behaviour, the CPD-induction dose–effect curve saturates at higher fluences, mirroring a steady state between product formation and reversal. It has to be pointed out that this is a merely photochemical process and is not related to biological repair phenomena.

There is, however, also a light-driven enzymatic reaction by which CPDs can be split. It is called "photoreactivation" and has an efficiency maximum at longer wavelengths, around 365 nm. The underlying molecular processes are now quite well understood. The enzyme now termed "photolyase" is not active in all cells, especially not in higher mammalian systems, but in marsupialia, birds and reptiles as well as in most microbes (reviewed in Weber 2005). Photoreactivation was assumed to be unique for CPDs, so one had a rather simple tool to judge their contribution to photobiological effects but, of course, only in those systems where the enzyme is present and active. More recently, however, also a photolyase activity for the 6-4-photoproduct has been reported (Todo et al. 1993). The gene has been cloned and its product used for analysing the role of specific photoproducts (Otoshi et al. 2000).

A more universal and probably the most important repair process is excision repair. It was also first found in UV-exposed bacteria (Setlow and Carrier 1964; Boyce and Howard-Flanders 1964) but subsequent research revealed its presence also in mammalian (including human) cells (reviewed in Walker et al. 1995). It is controlled by a large number of genetic loci. In the human disease *Xeroderma pigmentosum* (XP) (Cleaver 1968) some of them are altered, leading to an extreme sensitivity to UV radiation and also to proneness to skin tumours and also other types of cancer. This recessively inherited condition of which seven complementation groups are known (Cleaver et al. 1999) is very rare and affects only about one in one million people. In vitro cultured XP cells constitute a very valuable tool for the elucidation of repair processes.

Excision repair is very complex and a detailed description is outside the scope of this chapter. Briefly, the DNA strand with altered nucleotides is incised and the damaged part removed. The single-strand gap is then sealed by repair replication, with the remaining undamaged DNA strand serving as a template. Nucleotide excision repair (NER), as it is termed now, is thus essentially error-free (apart from the background error rate during replication) and its action is not restricted to UV-induced lesions but virtually to all types of altered nucleotides.

It was found later that the efficiency of NER is not the same over the genome and depends on transcriptional activity in the affected part. Transcription coupled repair (TCR), dealing exclusively with lesions on the actively transcribed strand, is considerably faster than global genome repair, which acts on damage in the whole cellular genome (Mellon et al. 1987; reviewed in van Zeeland et al. 2005). TCR is controlled by genes which are defective in the human disease Cockayne's syndrome (CS). Two complementation groups are known (Cleaver et al. 1999).

Although NER is the most important repair process in UV-exposed cells, other processes may also contribute significantly to lesion removal or circumvention, e.g. homologous recombination (HR) and lesion bypass. The latter plays a role when the density of lesions on both DNA strands is so high that excision gaps may overlap and thus repair replication can no longer work efficiently (Lehmann 2006).

As pointed out already, DNA DSBs are not directly induced at least by UV-B and UV-C radiation, but they may be formed indirectly. It is well known from studies with ionising radiation that they can be repaired by essentially two pathways, namely HR, which is essentially error-free, and nonhomologous end-joining (NHEJ), which may lead to chromosomal misarrangements. These pathways may also be important for DSBs formed indirectly upon UV exposure.

A recent review of all relevant repair pathways has been provided by Christmann et al. (2003) and should be consulted for further details.

3.5 Cytological Effects

As with other mutagenic or clastogenic agents, various cytological end points have been investigated with UV radiation, including structural chromosomal aberrations, micronuclei and SCEs.

3.5.1 Chromosomal Aberrations

Chromosomal aberrations in mammalian cells upon UV exposure were first studied in a systematic manner by Chu (1965a, b). He first used unsynchronised exponential cultures of Chinese hamster cells and found both chromatid as well as chromosome aberrations whose yields, however, depended on the time interval between exposure and scoring. Chromosome aberrations were seen only 42 h after exposure, and their number was always distinctly lower than those of chromatid-type aberrations. It can be assumed that chromosome-type aberrations are derived from chromatid-type aberrations during the subsequent cell cycle. The action spectrum (Fig. 3.6) follows in the short-wavelength region essentially DNA absorption, but there is a clear deviation above 320 nm, where breakage efficiency is much higher than expected on the basis of DNA absorption. This was not noted at the time but suggests a different action mechanism for the UV-A region.

In the work described in the second paper (Chu 1965b), synchronised populations were used. Although there was an indication that S phase is most sensitive in terms of the induction of aberrations by UV radiation, definite conclusions could not be drawn. When cells are exposed to short-wavelength UV radiation, mainly aberrations of the chromatid type are induced. Pyrimidine dimers and other photoproducts confined to a single DNA strand block replication in S phase and strand breaks are formed, leading to chromosome aberrations. UV radiation is thus considered as an S-phase-dependent agent (Bender et al. 1973; Natarajan et al. 1980).

There is general agreement that DNA DSBs are the principal lesions in the formation process of chromosomal aberrations (Obe et al. 2002). DSBs are, however, not directly formed by UV-B and UV-C radiation if the fluences are limited to physiological ranges. Nevertheless, all types of chromosomal aberrations are found in mammalian cells after UV exposure. This suggests that DSBs are created either by secondary mechanisms acting on primary photoproducts or as a result of stalled replication. DNA single-strand gaps

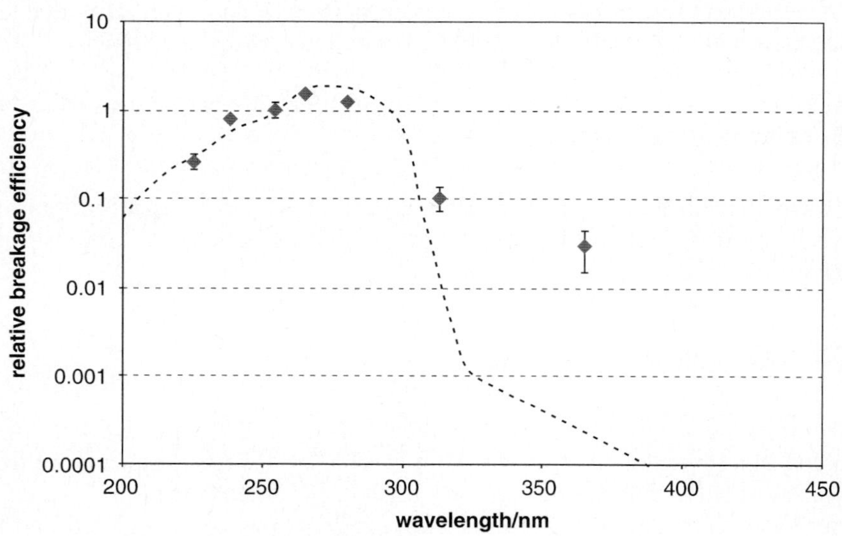

Fig. 3.6 Action spectrum for the formation of chromatid aberrations. The *dotted line* shows DNA absorption. (After Chu 1965a)

are part of NER. If their number is high, they may overlap and lead to DSBs. As the sequence of incision and subsequent repair replication is well coordinated, the probability for this to occur is low with not too high fluences. If gap filling is inhibited by specific chemicals, e.g. by ara-C or aphidicolin, single strands accumulate and DSBs are created (Bradley and Taylor 1983). Holmberg and Gumauskas (1990) demonstrated that in unstimulated human lymphocytes the yield of chromosomal aberrations after UV-C exposure is very low but is substantially increased when ara-C is added. The fluence–effect curve for dicentrics could be fitted to a purely quadratic function, indicating that they are formed via the interactions of overlapping single-strand gaps whose number is proportional to that of the primary photoproducts.

The process just described does not seem to play an important part in cells with normal and uninhibited repair capacities. In these cases DSBs are formed when stalled replication forks are disentangled. Their number depends thus on the number of photolesions still remaining in the DNA at the beginning of S phase and increases linearly with UV radiation fluence (Limoli et al. 2002).

The yield of chromosomal aberrations after UV exposure is inversely linked with the cell's capacity for the removal of photoproducts from the DNA before it enters S phase. Consequently deficiencies in NER lead to increased numbers of cells carrying chromosomal aberrations (Chipchase and Melton 2002). TCR seems to play an important part in this respect as shown in studies with various Chinese hamster cell lines homologous to human CS cells (Proietti de Santis et al. 2001). It is estimated that only 5–8% of the eukaryotic genome is actively transcribed. TCR is at first sight not expected to have a great impact on DNA lesions compared with the remainder of the genome. The experiments demonstrate, however, a drastic increase in UV sensitivity. This suggests that even relatively few remaining lesions in actively transcribed strands are critical for cell survival. This assumption is borne out by the fact that TCR-deficient cells undergo apoptosis in the second cell cycle after exposure. This raises the possibility that the accumulation of chromosomal aberrations may serve as a signal for apoptosis (Proietti de Santis et al. 2001).

3.5.2 Micronuclei

Micronuclei are assumed to be formed as a result of chromosome breakage. They can be assessed either by microscopical observation of interphase cells or by cytofluorometric measurements (Fenech 2000). The determination of micronuclei is simpler and less time consuming than the observation of structural chromosomal aberrations, but is also open to experimental pitfalls (Fenech 2000). So far it has been mainly employed to study the differential effects of various wavelength ranges, initially concentrating on the UV-B range. Keulers et al. (1998) showed in human peripheral blood lymphocytes that broadband UV-B radiation induces micronuclei in a fluence-dependent manner following a linear–quadratic relationship. More recently also UV-A radiation, which is a major component of sunlight, was investigated in more detail. As already pointed out, DNA absorption is very weak in this region and it has to be assumed that DNA is not damaged by direct photon absorption but via photosensitised processes. There are no studies on the whole range of structural chromosomal aberrations with UV-A radiation and most of the studies are on micronuclei induction. Rapp and Greulich (2004) showed that in human keratinocytes DSBs induced by UV-A radiation if not repaired lead to micronuclei. If these cells are exposed to 100 kJ m^{-2} of broadband UV-A radiation their survival is reduced to only about 70%, but the number of cells with micronuclei increases with further incubation until day 28, which is about the natural lifespan of these cells in culture (Phillipson et al. 2002). This suggests that UV-A radiation induces genomic instability similarly to ionising radiation (reviewed in Kronenberg 1994).

The micronuclei assay can also be used to assess the effect of UV-B radiation treatment of patients with skin diseases (Ferahbas et al. 2004). It could be shown that in lymphocytes isolated from their blood exposure to narrowband UV-B radiation in several treatment sessions leads to significant levels of damage.

3.5.3 Sister-Chromatid Exchanges

UV radiation is a potent inducer of SCEs. The classic way to evaluate SCEs is to label one DNA strand selectively with bromodeoxyuridine, which causes fainting of Giemsa staining so that the different strands can be differentiated in the microscopic picture. It was not realised initially that the bromine atom may be photochemically active and interfere with the SCE formation process. This problem was overcome by the introduction of another labelling method, namely the use of biotin-16-2′-deoxyuridine-5′-triphosphate, which is photochemically inert (Wojcik et al. 2004). Since this technique is comparatively new, older results have to be treated with some reservation. Wolff et al. (1974) observed that SCEs are only formed after UV exposure if the cells pass through S phase. They originate from long-lived lesions which may be induced in any of the cell cycle phases. The action spectrum below 320 nm coincides with the DNA absorption spectrum (Nussbaum 1995), but this does not give any information about the initiating lesion as the different photoproducts cannot be differentiated on the basis of short-wavelength action spectra (Figs. 3.4, 3.5). Wojcik et al. (2004) suggested that DNA interstrand cross-links may be the starting point of SCEs because chemicals like mitomycin C are also very potent SCE inducers (Kaina and Aurich 1985), but this, of course, is only indirect evidence and further clarification is required. Recent experiments by Wojcik et al. (2006) with a modified comet assay showed that cross-links may be formed as a result of radiation in 5-bromo-2′-deoxyuridine labelled cells, which supports the original assumption. It seems, however, that DNA DSBs do not play a major role as initiators because ionising radiation has comparatively low SCE yields (Wojcik et al. 2004).

3.6 Conclusions

Recent years have seen the development of powerful and sensitive analytical techniques to identify and quantify UV-induced photoproducts in DNA. The results demonstrate that not only the UV-B and UV-C regions, where DNA absorbs strongly, play a role for photobiological and photomedical effects but also the UV-A region, where DNA is virtually translucent. This points to the importance of indirect damage pathways presumably involving photosensitised reactions. Cytogenetic studies may be helpful for the understanding of the underlying basic mechanism of damage recognition and repair as well as for elucidation of the development of UV-related diseases like skin cancer. They may in future also aid the identification of genetically determined individual sensitivities.

References

Avery OT, MacLeod CM, McCarthy M (1944) Studies on the chemical nature of the substance inducing transformation of pneumococcal types. J Exp Med 79:137–158

Bender MA, Griggs HG, Walker PL (1973) Mechanisms of chromosomal aberration production I. Aberration induction by ultraviolet light. Mutat Res 20:387–402

Boyce R, Howard-Flanders P (1964) Release of ultraviolet light-induced thymine dimers from DNA in *E. coli* K-12. Proc Natl Acad Sci USA 51:293–300

Bradley MO (1981) Double-strand breaks in DNA caused by repair of damage due to ultraviolet light. J Supramol Struct Cell Biochem 16:337–343

Bradley MO, Taylor VI (1981) DNA double-strand breaks induced in normal human cells during the repair of ultraviolet light damage. Proc Natl Acad Sci USA 78:3619–3623

Bradley MO, Taylor VI (1983) Repair-induced DNA double strand breaks after ultraviolet-light and either aphidicolin or 1-β-D-abinofuranosylcytosine/hydroxyurea. Carcinogenesis 4:1533–1537

Cadet J, Sage E, Douki T (2005) Ultraviolet radiation-mediated damage to cellular DNA. Mutat Res 571:3–17

Chipchase MD, Melton DW (2002) The formation of UV-induced chromosome aberrations involves ERCC1 and XPF but not other nucleotide excision repair genes. DNA Repair (Amst) 1:335–340

Christmann M, Tomicic MT, Roos WP, Kaina B (2003) Mechanisms of human DNA repair: an update. Toxicology 193:3–34

Chu EHY (1965a) Effects of ultraviolet radiation on mammalian cells. I. Induction of chromosome aberrations. Mutat Res 2:75–94

Chu EHY (1965b) Effects of ultraviolet radiation on mammalian cells. II. Differential UV and x-ray sensitivity of chromosomes to breakage in 5-aminouracil synchronized cell populations. Genetics 52:1279-1294

Cleaver JE (1968) Defective repair replication of DNA in xeroderma pigmentosum. Nature 218:652–656

Cleaver JE, Thompson LH, Richardson AS, States JC (1999) A summary of mutations in the UV-sensitive disorders: xeroderma pigmentosum, Cockayne syndrome, and trichothiodystrophy. Hum Mutat 14:9–22

Douki T, Court M, Sauvaigo S, Odin F, Cadet J (2000) Formation of the main UV-induced thymine dimeric lesions within isolated and cellular DNA as measured by high performance liquid chromatography-tandem mass spectrometry. J Biol Chem 275: 11678–11685

Emmons W, Hollaender A (1939) The action of ultraviolet radiation on dermatophytes. 11. Mutations induced in cultures of dermatophytes by exposure of spores to monochromatic ultraviolet radiation. Am J Bot 26:467–475

Fenech M (2000) The in vitro micronucleus technique. Mutat Res 455:81–95

Ferahbas A, Donmez-Altuntas H, Hamurcu Z, Aktas E, Utas S (2004) Micronucleus evaluation in mitogen-stimulated lymphocytes of narrow-band (311 nm TL01) UVB-treated patients. Photodermatol Photoimmunol Photomed 20:81–85

Frankenberg-Schwager M, Frankenberg D, Harbich R (1987) Possible occurrence of DNA double-strand breaks during repair of u.v.-induced damage in yeast. Int J Radiat Biol 52:107–113

Gates FL (1930) A study of bactericidal action of ultraviolet light. II. The absorption of ultraviolet light by bacteria. J. Gen Physiol 14:31–42

Holmberg M, Gumauskas E (1990) Chromosome-type exchange aberrations are induced by inhibiting repair of UVC-induced DNA lesions in quiescent human lymphocytes. Mutat Res 232:261–266

Kaina B, Aurich O (1985) Dependency of the yield of sister-chromatid exchanges induced by alkylating agents on fixation time. Possible involvement of secondary lesions in sister-chromatid exchange induction. Mutat Res 149:451–461

Keulers RA, de Roon AR, de Roode S, Tates AD (1998) The induction and analysis of micronuclei and cell killing by ultraviolet-B radiation in human peripheral blood lymphocytes. Photochem Photobiol 67:426–432

Kiefer J (1987) UV response of the temperature-conditional rad 54 mutant of the yeast *Saccharomyces cerevisiae*. Mutat Res 191:9–12

Kiefer J (1990) Biological radiation action. Springer, Berlin

Kiefer J, Feige M (1993) The significance of DNA double-strand breaks in the UV inactivation of yeast cells. Mutat Res 299:219–224

Knapp E, Reuss,A, Risse A, Schreiber H (1939) Quantitative Analyse der mutationsauslosenden Wirkung monochromatischen u.v.-Lichtes. Naturwissenschaften 27:304

Kronenberg A (1994) Radiation-induced genomic instability. Int J Radiat Biol 66:603–609

Lehmann A (2006) Translesion synthesis in mammalian cells. Exp Cell Res 312:2673–2676

Limoli CL, Giedzinski E, Morgan WF, Cleaver JE (2000) Polymerase η deficiency in the xeroderma pigmentosum variant uncovers an overlap between the S phase checkpoint and double-strand break repair. Proc Natl Acad Sci USA 97:7939–7946

Limoli CL, Giedzinski E, Bonner WM, Cleaver JE (2002) UV-induced replication arrest in the Xeroderma variants leads to DNA double strand breaks, g-H2AX formation and Mre11 relocalisation. Proc Natl Acad Sci USA 99:233–238

McKinley AF, Diffey BL (1987) A reference spectrum for ultraviolet induced erythema in human skin. CIE J 6:17–22

Mellon I, Spivak G, Hanawalt PC (1987) Selective removal of transcription-blocking DNA damage from the transcribed strand of the mammalian DHFR gene. Cell 51:241–249

Natarajan AT, Kihlmann BA, Obe G (1980) Use of the 5-bromodeoxyuridine-labelling technique for exploring mechanisms involved in the formation of chromosomal aberrations. Mutat Res 73:307–317

Nussbaum A (1995) Aktionsspektroskopie über die Induktion von Schwesterchromatifaustäuschen (SCEs) an V79-Hamsterzellen und ein neues Modell der SCE-Entstehung. PhD thesis, University of Giessen

Obe G, Pfeiffer P, Savage JRK, Johannes C, Goedecke W, Jeppesen P, Natarajan AT, Martinez-Lopez W, Folle GA, Drets ME (2002) Chromosomal aberrations: formation, identification and distribution. Mutat Res 504:17–36

Otoshi E, Yagi T, Mori T, Matsunaga T, Nikaido O, Kim ST, Hitomi K, Ikenaga M, Todo T (2000) Respective roles of cyclobutane pyrimidine dimers, (6-4)photoproducts, and minor photoproducts in ultraviolet mutagenesis of repair-deficient xeroderma pigmentosum A cells. Cancer Res 60:1729–1735

Parrish JA, Anderson RR, Urbach J, Pitts D (1978) UV-A. Biological effects of ultraviolet radiation with emphasis on human responses to longwave ultraviolet. Plenum, New York

Peak JG, Peak, MJ (1991) Comparison of initial yields of DNA-to-protein crosslinks and single strand breaks induced in cultured human cells by far- and near-ultraviolet light, blue light and X-rays. Mutat Res 246:187–191

Peak JG, Peak MJ, Sikorski RS, Jones CA (1985) Induction of DNA-protein crosslinks in human cells by ultraviolet and visible radiations: action spectrum. Photochem. Photobiol 41:295–302

Phillipson RP, Tobi SE, Morris JA, McMillan TJ (2002) UV-A induces persistent genomic instability in human keratinocytes through an oxidative stress mechanism. Free Radic Biol Med 32:474–480

Proietti de Santis L, Garcia CL, Balajee AS, Latini P, Pichierri P, Nikaido O, Stefanini M, Palitti F (2002) Transcription coupled repair efficiency determines the cell cycle progression and apoptosis after UV exposure in hamster cells. DNA Repair (Amst) 28:209–223

Rapp A, Greulich KO (2004) After double-strand break induction by UV-A, homologous recombination and nonhomologous endjoining cooperate at the same DSB if both systems are available. J Cell Sci 117:4935–4945

Rogakou EP, Pilch DR, Ort AH, Ivanova VS, Bonner WM (1998) DNA double-stranded breaks induce histone H2AX phosphorylation on serine 139. J Biol Chem 273:5858–5868

Rogakou EP, Boon C, Redon C, Bonner WM (1999) Megabase chromatin domains involved in DNA double-strand breaks in vivo. J Cell Biol 146:905–916

Rosenstein BS, Ducore JM (1983) Induction of DNA strand breaks in normal human fibroblasts exposed to monochromatic ultraviolet and visible wavelengths in the 240-546 nm range. Photochem Photobiol 38:51–55

Rosenstein BS, Mitchell DL (1987) Action spectra for the induction of pyrimidine(6-4)pyrimidone photoproducts and cyclobutane pyrimidine dimers in normal human skin fibroblasts. Photochem Photobiol 45:775–780

Setlow RB, Carrier W (1964) The disappearance of thymine dimmers from DNA: an error correcting mechanisms. Proc Natl Acad Sci USA 51:226–231

Sutherland JC, Griffin KP (1981) Absorption spectrum of DNA for wavelengths longer than 320 nm. Radiat Res 86:399–409

Todo T, Takemori H, Ryo H, Ihara M, Matsunaga T, Nikaido O, Sato K, Nomura, T (1993) A new photoreactivating enzyme that specifically repairs ultraviolet light-induced (6-4)photoproducts. Nature 361:371–374

van Zeeland AA, Vreeswijk MPG, de Gruijl F, van Kranen HJ, Vrieling H, Mullenders LFH (2005) Transcription-coupled repair: Impact on UV-induced mutagenesis in cultures rodent cells and mouse skin tumours. Mutat Res 577:170–178

Walker GC, Siede W, Friedberg EC (1995) DNA repair and mutagenesis. American Society for Microbiology, Washington

Weber S (2005) Light-driven enzymatic catalysis of DNA repair: a review of recent biophysical studies on photolyase. Biochim Biophys Acta 1707:1–23

Wojcik A, Bruckmann E, Obe G (2004) Insights into the mechanisms of sister chromatid exchange formation. Cytogenet Genome Res 104:304–309

Wojcik A, Bochenek A, Lankoff H, Lisowska A, Padjas I, Szumiel C, von Sonntag, Obe G (2006) DNA interstrand crosslinks are induced in cells prelabelled with 5-bromo-2′-deoxyuridine and exposed to UVC radiation. J Photochem Photobiol B 84:15–20

Wolff S, Bodycote J, Painter RB (1974) Sister chromatid exchanges induced in Chinese hamster cells by UV irradiation of different stages of the cell cycle: the necessity for cells to pass through S. Mutat Res 25:73–81

4 Double Strand Break Repair Mechanisms in Mammalian Cells

Wolfgang Goedecke

Abstract

Double-strand breaks (DSBs) represent major threats in chromosomal DNA. They arise either as intermediate structures during recombination, replication and repair events or as potentially lethal lesions introduced by ionising radiation or drugs. In any case, DSBs have to be eliminated immediately, because of the recombinogenic capacity of the DNA ends generated by the DSB that increases the risk for undesired chromosomal aberrations (CAs). In order to cope with DSBs, cells exhibit two different sets of repair activities, namely non-homlogous end joining (NHEJ) and homologous recombinational repair (HRR). Both pathways are regulated during the cell cycle with peak activities either in G1 phase (NHEJ) or in late S and G2 phases (HRR). While NHEJ has the capacity to join arbitrary DNA ends together, HRR depends on the presence of a second undamaged sequence in the genome providing a template for the reconstitution of the sequence at the DSB site. While most of the proteins identified so far participate in one of the two repair pathways, a few proteins are known to participate in both pathways. These last group of proteins is supposed to be involved in the selection process of the DNA repair pathway. Mutations in one DSB repair system result in the accumulation of DSBs and increased levels of CAs. As a result, genome instability is observed in cells with impaired DSB repair functions.

4.1 Introduction

Double-strand breaks (DSBs) in chromosomal DNA are considered to represent intermediate structures which exist transiently during replication, recombination and repair of the DNA. They occur either spontaneously based on the cellular DNA metabolism or by exogeneous damage created by ionising radiation or DNA-damaging drugs. Therefore it is obvious that threats generated by DSBs are hardly to be avoided and DSBs from exogenous sources increase previously existing levels. Since DSBs are the ultimate lesions for the generation of chromosomal aberrations (CAs), accumulation of DSBs will raise the probability of the creation of CAs. Generally about 50 DSBs are induced by

Günter Obe and Vijayalaxmi (Eds.)
Chromosomal Alterations: Methods, Results and Importance in Human Health

2 Gy of sparsely ionising irradiation and nearly the same number of DSBs occur spontaneously in each mammalian cell during a single cell division (Vilenchik and Knudson 2003).

The severe effects caused by DSBs result mainly from their abilities to generate DNA structures with free ends. These free DNA ends are potent inducers of various kinds of recombinational activities, such as rearrangements of the immunoglobulin genes, for example. Balanced translocations which are frequently seen in tumour cells are considered to be the result of uncontrolled or accidental recombination events; therefore, it is a great achievement for each cell to control the number of DSBs and to ensure genomic stability by preventing CAs. This is achieved by DSB repair mechanisms which are able to eliminate most of the DSBs immediately after they have been generated.

During evolution, two strategies of DSB repair were established, namely the non-homologous end-joining (NHEJ) and the homologous recombinational repair (HRR) systems. Both appeared early during evolution and are present in probacteria or bacteria, underlining their significance for survival of living cells (Bowater and Doherty 2006; Cavalier-Smith 2002). NHEJ is characterised by the ability to join free DNA ends, irrespective of their sequence and structure. The activity is a universal pathway for the elimination of DSBs. Since nucleolytic degradation and non-templated addition of nucleotides occur frequently at free DNA ends, this repair pathway comprises a high risk for the generation of de novo mutations at sites of former DSBs. This susceptibility to mutagenesis is a direct consequence of the NHEJ system to join any pair of free DNA ends, because sequence alterations that are attached to ends are incorporated into the junction of the joined molecule.

In contrast to NHEJ, HRR is able to restore the original sequence of the broken DNA molecule. This decreases the risk for the generation of small deletions and insertions at the site of the former DSB. HRR activity is nonmutagenic at sites of DSBs but only a small fraction of breaks is accessible for repair by this mechanism. In fact, HRR is limited to those DSBs which are located within sequences having at least one further and undamaged copy somewhere else in the genome (Obe et al. 2002).

4.2 NHEJ Reaction

4.2.1 NHEJ Mechanism and Junction Formation

Most of our knowledge about the mechanisms underlying the NHEJ reaction results from studies with cell-free extracts. Several organisms and cell types have been used for the preparation of such extracts (Pfeiffer et al. 2005). Since NHEJ is the dominant DSB repair activity in vertebrate cells, most of our knowledge is derived from mammalian cell cultures.

The NHEJ activity is assayed by either circle formation or concatenation of linear DNA substrate molecules. Making use of restriction endonucleases, we can easily generate them from plasmid molecules and they permit the generation of substrates with defined and non-compatible combinations of DNA ends. It should be emphasised that DNA ends created with restriction enzymes provide structures with either blunt or 3′ or 5′ protruding single strands (PSS). These end structures differ from those generated by ionising radiation. Nevertheless, it was shown that DNA ends generated by ionising radiation are also suitable substrates for NHEJ reactions (Odersky et al. 2002).

Depending on the nature of the DNA ends, two modes of junctions may be observed, as outlined in Fig. 4.1. The first is designated as the overlap type and utilises matches of a few base pairs to align DNA ends for end-joining reactions. Substrates preferentially processed by the overlap type are ends with PSS of the same polarity (3′-PSS vs. 3′-PSS and 5′-PSS vs. 5′-PSS). The second pathway is the fill-in type. This mode is used to join blunt ends, PSS of opposing polarity (3′-PSS vs. 5′-PSS) or combinations of blunt and PSS ends. Only a minor fraction of junctions are associated with a more pronounced loss of nucleotides during junction formation. Interestingly, the 3′-OH of one of the DNA ends can be used to prime DNA synthesis on a template of the other end (Thode et al. 1990). DNA ends with PSS longer than ten nucleotides are not used as substrates for the NHEJ mechanism; these structures may be converted into joinable DNA ends by successive degradation of the PSS tail (Beyert et al. 1994; Daley and Wilson 2005).

The mechanisms described above have been confirmed by in vivo data from different species and cell types. If linear substrate molecules are introduced into cells by either transfection or injection and the sequences of their

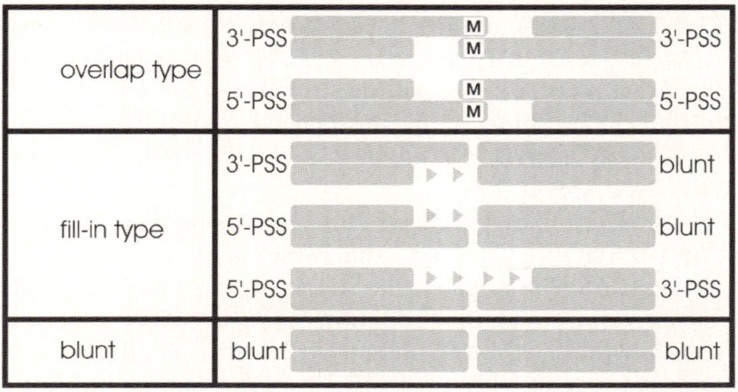

Fig. 4.1 The two modes of non-homologous end-joining (NHEJ) reactions depending on DNA end configurations. Ends of the same polarity are joined by overlap formation and ends of opposite polarity or with at least one blunt end involved are joined by the fill-in mode. White squares labelled with *M* indicate bases that match arbitrarily. *PSS* protruding single strand

junctions are reinvestigated after DSB repair, they look similar to those generated in vitro (Roth and Wilson 1986). Another type of evidence comes from investigating the junctions of DSBs generated by the rare cutting endonuclease *Sce*I. This enzyme from *Saccharomyces cerevisiae*, where it functions in intron homing, has a long recognition sequence of 18 base pairs, which is absent in mammalian genomes. In genetically engineered cell lines, where this site is introduced into the mammalian genome, introduction of *Sce*I allows site-specific cleavage at a single site and the junctions observed after NHEJ resemble those found by the joining in cell-free extracts (Dahm-Daphi et al. 2005).

4.2.2 Proteins Involved in NHEJ

Both types of the NHEJ pathway, the one using arbitrary matches between complementary base pairs (overlap type) and the one conserving PSS (fill-in type), tend to avoid formation of de novo mutations. They are therefore called the "high-fidelity" NHEJ pathways. The proteins involved in the high-fidelity end joining are called core proteins. They are DNA-PK, Artemis, XRCC4 and Ligase IV (O'Driscoll and Jeggo 2006). In addition to these core proteins, the polymerases Polμ and Polλ are both recruited for the gap filling by templated addition of nucleotides. It is assumed, that both polymerases play a redundant role during gap filling, but in contrast to Polλ, only Polμ is able to perform gap filling across the DSB (Nick McElhinny et al. 2005). Terminal desoxytransferase (TdT) is a further DNA polymerase acting at DSBs. The protein is involved in the non-templated addition of nucleotides, which is occasionally observed in cells of the immune system (Paull 2005).

All these proteins assemble at sites of DSBs. DNA-PK is a DNA-dependent phosphokinase, composed of three subunits, the DNA-PK catalytic subunit (DNA-PKcs) and a DNA-binding heterodimer formed by the two subunits Ku70 and Ku80. The DNA-PK complex is a DNA-dependent kinase, which is recruited under physiological conditions to DNA ends owing to its binding to the Ku70/Ku80 heterodimer (Dip and Naegeli 2005). The function of the Artemis nuclease is to convert DNA termini into ligatable structures (Drouet et al. 2006; Niewolik et al. 2006).

Other proteins are known that assemble at sites of DSBs. These are ataxia telangiectasia mutated (ATM) and the components of the MRN complex, MRE11, RAD50 and NBS1. ATM, like DNA-PKcs, is another member of the protein kinase family and is a component of the MRN complex, namely MRE11 was shown to possess NHEJ activity in vitro (Paull and Gellert 2000).

Some evidence suggests that the MRN complex is involved in the second pathway of NHEJ reactions. Since this pathway is characterised by extended deletions that occur as a result of loss of terminal nucleotides at the site of the DSBs, this pathway is called "low-fidelity" NHEJ. It is tempting to assume that junctions formed by low-fidelity end joining are processed by a

MRN-dependent activity, representing a second NHEJ pathway associated with a different set of proteins.

This model is supported by the fact that mutants defective in one of the core components are still able to perform NHEJ reactions, but junctions resulting from these activities are usually characterised by elevated levels of terminal nucleotide losses and an increased usage of patch homologies to align the juxtaposed strands. Data from MRE11-depleted extracts of the clawed frog *Xenopus laevis* and MRE11 knockout chicken DT40 cells show that in these systems the Ku-dependent high-fidelity NHEJ pathway remains unaffected. This indicates that in these cells the high-fidelity pathway is dominant over the low-fidelity pathway (Evans and Hearing 2005).

4.3 DSB Repair by HRR

HRR mechanisms have been extensively studied in procaryotes and lower eucaryotes such as the budding yeast *Saccharomyces cerevisiae*. A reason that might account for the overrepresentation of data from these organisms is the predominance of HRR compared with that in vertebrates. In order to explain the frequently observed association of gene conversions with crossover events especially during meiotic recombination, the DSB repair model was established as a standard model for HRR (Szostak et al. 1983). Nevertheless it should be noted that there are alternative models and that no single model is able to explain all aspects of HRR assays. Other models are the synthesis dependent strand annealing (SDSA) model and the break induced replication (BIR) model. For a comprehensive description of the different HRR mechanisms, see Paques and Haber (1999).

A common feature of DSB repair by HRR is the existence of an undamaged template sequence somewhere else in the genome that serves as a donor for the sequence information of the broken molecule. In somatic cells this donor sequence is usually provided by the sister chromatids. Compared with templates from other locations (ectopic sites) in mammalian cells, the sister chromatid is used several orders of magnitude more frequently. This is a clear difference between mammalian cells and *Saccharomyces cerevisae*, where the difference in recombination between sister chromatids and ectopic sites is much less pronounced (Johnson and Jasin 2001). The strong bias towards sister-chromatid preference suggests that HRR takes place predominantly during late S phase and probably early G2 phase. In contrast to this, NHEJ mainly occurs in prereplicative G1 cells (Fig. 4.2).

The homologous chromosome is efficiently used as a donor sequence during prophase I of meiosis, when both homologous chromosomes are aligned by the synaptonemal complex. In contrast to meiotic recombination, which is characterised by gene conversions associated with crossovers, HRR in somatic cells is generally not associated with crossovers (Johnson and Jasin 2001).

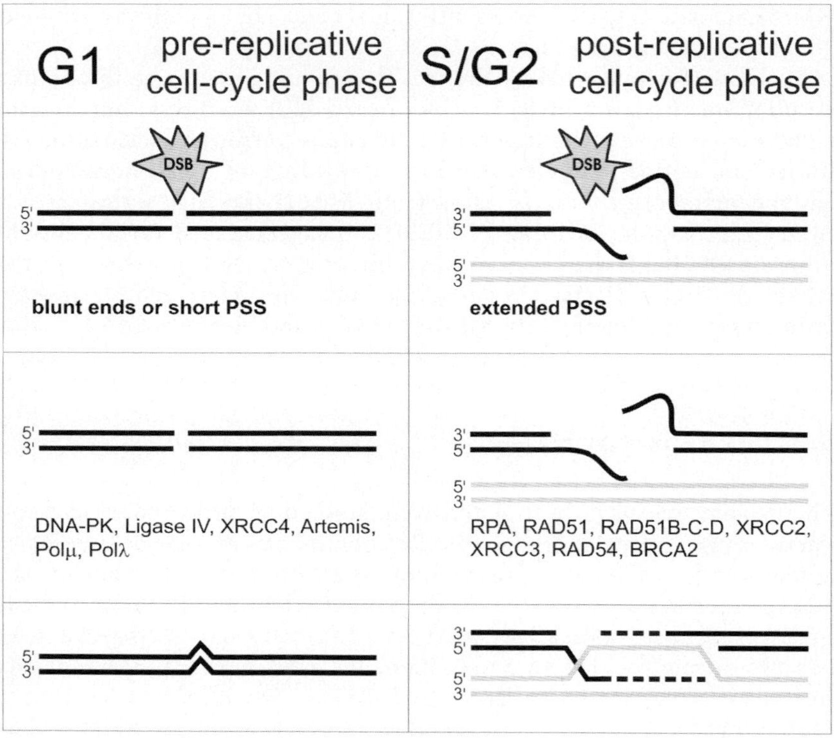

Fig. 4.2 NHEJ is the dominant pathway in prereplicative cells. The single-strand protrusions (*PSS*) should not be longer than ten nucleotides. In contrast, homologous recombinational repair is able to process longer PSS tails and is dominant in postreplicative cells. Enzymes involved in one of the two mechanisms are indicated and explained in the text. The *kink* symbolises the high risk of mutation formation during NHEJ, while the indicated double Holliday structure indicates the precise restoration of the break. *DSB* double-strand break

The underlying mechanisms for HRR in meiosis and somatic cells may differ. The DNA ends used in meiotic recombination differ from those breaks that are generated by irradiation or replication errors, since they are tagged by the spo11 protein, an enzyme which is generates the DSBs during prophase I of meiosis (Neale et al. 2005). It is therefore problematic to discuss meiotic DSBs as a model for DSB repair, because the DSBs used to initiate HRR during meiosis do not resemble the free ends of DNA, which are the major threats of DSBs in somatic cells. Meiotic recombination represents a highly specialised system to guarantee homologous exchange of DNA molecules, and it should not be taken as a model system for the repair of accidental DSBs.

4.3.1 HRR Mechanisms

The common scheme of HRR involves copying of sequence information from the unbroken template molecule to the site of the DSB. HRR events are carried out by a series of successive steps, namely (1) presynapsis, in which extended protruding 3′-tails are generated which are much longer than the PSS appearing during NHEJ, (2) synapsis, where strand invasion of these tails into the undamaged template sequence generates a structure called a displacement loop (D-loop) and (3) postsynapsis, which includes branch migration and resolution of the double Holliday junctions (dHJ).

The main function of DSB repair by HRR is the restoration of stalled replication forks. This is supported by a specific activation of HRR during S-phase, which is not the case for the NHEJ pathway (Rothkamm et al. 2003; Segurado et al. 2002). The tight correlation between HRR and replication is further indicated by genetic evidence. Recombinogenic ends are generated by either stalled or collapsed replication forks (Michel et al. 2001). Mutants like ATM and ataxia-telangiectasia related (ATR), which are defective in S-phase control, are also impaired in HRR, indicating the close relationship between cell cycle and DSB repair. The same holds for mutants deficient for sister-chromatid cohesion. Loss of sister-chromatid cohesion abolishes postreplicative DSB repair. This indicates that sister chromatids generated during replication are required as template molecules for HRR repair in somatic cells.

4.3.2 Proteins Involved in HRR

The components of the MRN complex and exonucleases like Exo1 are key components for the generation of long single-stranded 3′-tails needed for strand invasion. Immediately after generation of the 3′-tails, they become stabilised by the heterotrimeric replication factor A (RPA) complex. Strand transfer and D-loop formation require several further RAD52 proteins, namely RAD52 and Rad51B, Rad51C, Rad51D, Xrcc2, Xrcc3 and BRCA2. While the proteins involved in these early recombination steps are quite well known, the subsequent steps are less clear. Among other things, they require the function of the RAD54 protein and, in order to resolve the dHJ, the action of BLM, topo3α, sgs1 and others (Heyer et al. 2006).

4.4 Impaired DSB Repair and Chromosomal Aberrations

As outlined already, impaired DSB repair leads to elevated levels of DSBs and to the persistence of DSBs over longer periods of time (Iliakis et al. 2004). Since DSBs themselves have a recombinogenic potential, one would expect increased frequencies of CAs in cells with impaired DSB repair systems.

Inherited diseases are known which show the direct link between DSB repair and CAs. Around 125 genes are involved in overall DNA-repair activity, and those participating in DSB repair show elevated levels of CAs, if they are mutated (Wood et al. 2001; Ronen and Glickman 2001).

The effect of NHEJ proteins on CAs was shown from patient data as well as from mouse models. Screening of peripheral lymphocytes of patients suffering from untreated cancer showed elevated levels of CAs. The number of CAs correlated with the activity of DNA-PK in these cells. This result indicate that, in addition to impaired DSB repair on the DNA level, there is also inefficiency of DSB repair on the chromosomal level (Someya et al. 2006). Translocations, which are a hallmark of aberrant DSB repair, are frequently observed in transgenic mice, knocked out for NHEJ genes. Mouse embryonic fibroblasts from Ku70 and Ligase IV deficient mice show a dramatic spontaneous increase of chromosomal fragmentations and non-reciprocal translocations, even in the absence of genotoxic drugs or irradiation. This result is consistent with elevated levels of unprocessed DSBs, acting as a primary source for CAs (Ferguson et al. 2000).

Elevated frequencies of CAs are also observed in cells with disturbed functions in HRR. Cell lines derived from Chinese hamster ovary cells which are deficient in members of the RAD51 family (XRCC2 and XRCC3) accumulate CAs such as breaks, deletions and unbalanced translocations (Cui et al. 1999). Together with the previously mentioned results from mutants in the NHEJ system, these data clearly show that problems in one of the two DSB repair systems lead to CAs and genome instability. Although there is striking evidence that DSBs are the ultimate lesions that are converted into CAs by an incorrect action of the aforementioned repair pathways ("misrepair"), the exact mechanisms of how this happens are not quite clear.

4.5 Fragile Sites Are Associated with DNA DSBs

Common fragile sites (CFSs) are specific chromosomal loci that appear as constrictions or gaps on metaphase chromosomes. They are induced in cultured cells treated with drugs that partially block DNA replication or stimulate premature chromosome condensation (El Achkar et al. 2005). Examples of these two groups are aphidicolin and calyculin A, respectively. In contrast to the CFSs, which are present in all individuals, rare fragile sites (RFSs) only appear in in cells with extended amplifications of repetitive elements at a certain chromosomal locus and are therefore not present in all individuals. This rather depends on the allele of the repetitive element, which may change from one individual to the other owing to dynamic mutations that change the length of the allele (Pearson et al. 2005). CFSs and RFSs are both characterised by stretches of repetitive DNA of either AT-rich or GC-rich sequences underlining their common basis on the level of the DNA sequence (Zlotorynski et al. 2003).

The fact that replication problems as well as premature chromosome condensation lead to the induction of fragile sites indicates that the basis of this phenomenon is the disturbance of the chronology of chromosome replication and condensation during the cell cycle. Therefore proteins regulating the cell cycle, like the ATR protein, also affect fragile site stability (Tibbetts et al. 2000).

The observed correlation of chromosomal breakpoints of balanced translocations involved in cancer and the high frequency of sister-chromatid exchanges associated with CFSs lead to the suggestion that DSBs are involved in the establishment of fragile sites (Arlt et al. 2003). This proposal has recently been supported by the detection of DSB repair proteins at CFSs (Schwartz et al. 2005). Further evidence comes from downregulation of DSB repair proteins involved in HRR (RAD51) and NHEJ (DNA-PK and Ligase IV). In the absence of these proteins, a high number of constrictions are induced, in agreement with the model that proteins from both DSB repair pathways are involved in stabilising fragile sites (Schwartz et al. 2005). Obviously, DSBs generated in repetitive DNA are converted into fragile sites, if they fail to be repaired before the onset of the next cycle of chromosome condensation.

4.6 Conclusions

Free DNA ends generated by DSBs can be joined to other free ends present in the genome or they are able to invade intact DNA molecules, initiating recombination events. Since these activities bear the risk for the generation of CAs, DSBs, if possible, are eliminated immediately after they are created. Two major pathways are available, the universal NHEJ pathway and the HRR pathway. The latter one is restricted to DSBs with extended sequence similarities between the broken DNA molecule and other sequences in the genome. Usually the unbroken DNA molecule is provided by the sister chromatid. This is the reason why HRR takes place predominantly in postreplicative cells. Consistently NHEJ is detected in prereplcative cell cycle phases.

In cases of "misrepair", the repair mechanisms fail to eliminate DSBs by joining of the originally linked DNA ends. In these cases, DSB repair pathways create new junctions that are detected as translocations on the chromosomal level. Therefore, a consequence of unmeant DSB repair is the generation of chromosomal aberrations.

References

Arlt MF, Casper AM, Glover TW (2003) Common fragile sites. Cytogenet Genome Res 100:92–100

Beyert N, Reichenberger S, Peters M, Hartung M, Gottlich B, Goedecke W, Vielmetter W, Pfeiffer P (1994) Nonhomologous DNA end joining of synthetic hairpin substrates in Xenopus laevis egg extracts. Nucleic Acids Res 22:1643–1650

Bowater R, Doherty AJ (2006) Making ends meet: repairing breaks in bacterial DNA by nonhomologous end-joining. PLoS Genet 2:e8
Cavalier-Smith T (2002) Origins of the machinery of recombination and sex. Heredity 88:125–141
Cui X, Brenneman M, Meyne J, Oshimura M, Goodwin EH, Chen DJ (1999) The XRCC2 and XRCC3 repair genes are required for chromosome stability in mammalian cells. Mutat Res 434:75–88
Dahm-Daphi J, Hubbe P, Horvath F, El Awady RA, Bouffard KE, Powell SN, Willers H (2005) Nonhomologous end-joining of site-specific but not of radiation-induced DNA double-strand breaks is reduced in the presence of wild-type p53. Oncogene 24:1663–1672
Daley JM, Wilson TE (2005) Rejoining of DNA double-strand breaks as a function of overhang length. Mol Cell Biol 25:896–906
Dip R, Naegeli H (2005) More than just strand breaks: the recognition of structural DNA discontinuities by DNA-dependent protein kinase catalytic subunit. FASEB J 19:704–715
Drouet J, Frit P, Delteil C, de Villartay JP, Salles B, Calsou P (2006) Interplay between Ku, Artemis, and the DNA-dependent protein kinase catalytic subunit at DNA ends. J Biol Chem 281:27784–27793
El Achkar E, Gerbault-Seureau M, Muleris M, Dutrillaux B, Debatisse M (2005) Premature condensation induces breaks at the interface of early and late replicating chromosome bands bearing common fragile sites. Proc Natl Acad Sci USA 102:18069–18074
Evans JD, Hearing P (2005) Relocalization of the Mre11-Rad50-Nbs1 complex by the adenovirus E4 ORF3 protein is required for viral replication. J Virol 79:6207–6215
Ferguson DO, Sekiguchi JM, Chang S, Frank KM, Gao Y, DePinho RA, Alt FW (2000) The nonhomologous end-joining pathway of DNA repair is required for genomic stability and the suppression of translocations. Proc Natl Acad Sci USA 97:6630–6633
Heyer WD, Li X, Rolfsmeier M, Zhang XP (2006) Rad54: the Swiss Army knife of homologous recombination? Nucleic Acids Res 34:4115–4125
Iliakis G, Wang H, Perrault AR, Boecker W, Rosidi B, Windhofer F, Wu W, Guan J, Terzoudi G, Pantelias G (2004) Mechanisms of DNA double strand break repair and chromosome aberration formation. Cytogenet Genome Res 104:14–20
Johnson RD, Jasin M (2001) Double-strand-break-induced homologous recombination in mammalian cells. Biochem Soc Trans 29:196–201
Michel B, Flores MJ, Viguera E, Grompone G, Seigneur M, Bidnenko V (2001) Rescue of arrested replication forks by homologous recombination. Proc Natl Acad Sci USA 98:8181–8188
Neale MJ, Pan J, Keeney S (2005) Endonucleolytic processing of covalent protein-linked DNA double-strand breaks. Nature 436:1053–1057
Nick McElhinny SA, Havener JM, Garcia-Diaz M, Juarez R, Bebenek K, Kee BL, Blanco L, Kunkel TA, Ramsden DA (2005) A gradient of template dependence defines distinct biological roles for family X polymerases in nonhomologous end joining. Mol Cell 19:357–366
Niewolik D, Pannicke U, Lu H, Ma Y, Wang LC, Kulesza P, Zandi E, Lieber MR, Schwarz K (2006) DNA-PKcs dependence of artemis endonucleolytic activity: differences between hairpins and 5′ or 3′ overhangs. J Biol Chem 10:33900–33909
O'Driscoll M, Jeggo PA (2006) The role of double-strand break repair – insights from human genetics. Nat Rev Genet 7:45–54
Obe G, Pfeiffer P, Savage JR, Johannes C, Goedecke W, Jeppesen P, Natarajan AT, Martinez-Lopez W, Folle GA, Drets ME (2002) Chromosomal aberrations: formation, identification and distribution. Mutat Res 504:17–36
Odersky A, Panyutin IV, Panyutin IG, Schunck C, Feldmann E, Goedecke W, Neumann RD, Obe G, Pfeiffer P (2002) Repair of sequence-specific 125I-induced double-strand breaks by nonhomologous DNA end joining in mammalian cell-free extracts. J Biol Chem 277:11756–11764
Paques F and Haber JE (1999) Multiple pathways of recombination induced by double-strand breaks in Saccharomyces cerevisiae. Microbiol Mol Biol Rev 63:349–404
Paull TT (2005) Saving the ends for last: the role of pol mu in DNA end joining. Mol Cell 19:294–296

Paull TT, Gellert M (2000) A mechanistic basis for Mre11-directed DNA joining at microhomologies. Proc Natl Acad Sci USA 97:6409–6414

Pearson CE, Edamura KN, Cleary JD (2005) Repeat instability: mechanisms of dynamic mutations. Nat Rev Genet 6:729–742

Pfeiffer P, Feldmann E, Odersky A, Kuhfittig-Kulle S, Goedecke W (2005) Analysis of DNA double-strand break repair by nonhomologous end joining in cell-free extracts from mammalian cells. Methods Mol Biol 291:351–371

Ronen A, Glickman BW (2001) Human DNA repair genes. Environ Mol Mutagen 37:241–283

Roth DB, Wilson JH (1986) Nonhomologous recombination in mammalian cells: role for short sequence homologies in the joining reaction. Mol Cell Biol 6:4295–4304

Rothkamm K, Kruger I, Thompson LH, Lobrich M (2003) Pathways of DNA double-strand break repair during the mammalian cell cycle. Mol Cell Biol 23:5706–5715

Schwartz M, Zlotorynski E, Goldberg M, Ozeri E, Rahat A, le Sage C, Chen BP, Chen DJ, Agami R, Kerem B (2005) Homologous recombination and nonhomologous end-joining repair pathways regulate fragile site stability. Genes Dev 19:2715–2726

Segurado M, Gomez M, Antequera F (2002) Increased recombination intermediates and homologous integration hot spots at DNA replication origins. Mol Cell 10:907–916

Someya M, Sakata K, Matsumoto Y, Yamamoto H, Monobe M, Ikeda H, Ando K, Hosoi Y, Suzuki N, Hareyama M (2006) The association of DNA-dependent protein kinase activity with chromosomal instability and risk of cancer. Carcinogenesis 27:117–122

Szostak JW, Orr-Weaver TL, Rothstein RJ, Stahl FW (1983) The double-strand-break repair model for recombination. Cell 33:25–35

Thode S, Schafer A, Pfeiffer P, Vielmetter W (1990) A novel pathway of DNA end-to-end joining. Cell 60:921–928

Tibbetts RS, Cortez D, Brumbaugh KM, Scully R, Livingston D, Elledge SJ, Abraham RT (2000) Functional interactions between BRCA1 and the checkpoint kinase ATR during genotoxic stress. Genes Dev 14:2989–3002

Vilenchik MM, Knudson AG (2003) Endogenous DNA double-strand breaks: production, fidelity of repair, and induction of cancer. Proc Natl Acad Sci USA 100:12871–12876

Wood RD, Mitchell M, Sgouros J, Lindahl T (2001) Human DNA repair genes. Science 291:1284–1289

Zlotorynski E, Rahat A, Skaug J, Ben Porat N, Ozeri E, Hershberg R, Levi A, Scherer SW, Margalit H, Kerem B (2003) Molecular basis for expression of common and rare fragile sites. Mol Cell Biol 23:7143–7151

5 Backup Pathways of Nonhomologous End Joining May Have a Dominant Role in the Formation of Chromosome Aberrations

George Iliakis, Wenqi Wu, Minli Wang, Georgia I. Terzoudi,
and Gabriel E. Pantelias

Abstract

Chromosome aberrations are large-scale chromatin rearrangements produced in cells after exposure to ionizing radiation and other DNA-damaging agents. They have been directly implicated in cell killing, mutation induction and oncogenic transformation and have spearheaded a great number of investigations focused on their qualitative and quantitative analysis. The central aim of modern cytogenetics is the description at the molecular level of the observations at the chromosome level. There is consensus that the majority of chromosome aberrations are the ultimate consequence of errors in the repair of DNA double-strand breaks (DSBs). Here, we outline the molecular aspects of DSB repair pathways and discuss them vis-à-vis the two prominent theories of chromosome aberration formation. Although both homology-dependent and homology-independent pathways contribute to DSB repair, homology-independent pathways are predominantly responsible for the initial removal of the majority of DSBs from the genome. From the homology-independent pathways, both DNA-PK-dependent nonhomologous end joining (D-NHEJ) and backup nonhomologous end joining (B-NHEJ) have the potential of forming chromosome aberrations, but it is likely that most of the aberrations derive from a combination of low-frequency misjoining events during D-NHEJ, and from much higher frequency misjoining events during B-NHEJ. Further elucidation of the mechanism of DSB repair at the level of chromatin will allow a more complete molecular description of the process.

5.1 Introduction

Chromosome aberrations are large-scale chromatin rearrangements produced in cells after exposure to ionizing radiation (IR) and other DNA-damaging agents (Cornforth 2006; Cornforth and Bedford 1993; Evans 1962; Hlatky et al. 2002; Natarajan 2002; Obe et al. 2002; Revell 1974). Such rearrangements, which are directly observable at metaphase, have been directly implicated in cell killing, mutation induction and oncogenic transformation and have spearheaded a great number of investigations focused on their

Günter Obe and Vijayalaxmi (Eds.)
Chromosomal Alterations: Methods, Results and Importance in Human Health

qualitative and quantitative analysis (Savage 1975). The complex nature of chromosomal organization at the molecular level and the lack of information on the underlying repair mechanisms, together with the diversity in the mode of action between agents producing similar chromosomal changes, have led to the adoption in the field of a descriptive terminology that is still in use today. The central aim of modern cytogenetics is the description at the molecular level of the observations at the chromosome level.

Efforts to interpret cytogenetic results from early studies culminated in the formulation of two seemingly incompatible theories of aberration formation (Revell 1974; Savage 1998; Savage and Harvey 1991): The classical "breakage-and-reunion" theory and the "exchange" theory. Both theories are based on generalizations and abstractions put forward at a time when our understandings of DNA repair and chromatin structure were not nearly as well developed as they are today. Probably as a result of this limitation, and owing to their individual success in describing distinct sets of data, it has not been possible to conclusively refute one of these theories in favor of the other.

The breakage-and-reunion theory postulates that radiation induces a distinct physical break in the chromosome (chromonema) that leads to the generation of open ends. While the majority (over 90%) of these ends restitute (rejoin), some remain unrejoined and directly lead to terminal deletions. Others rejoin illegitimately with ends of a neighboring break(s), causing chromosome exchanges. The break in this model is at the chromosome level (not the DNA level), and the ends are assumed to be free to move around (flapping ends) and to participate in promiscuous rejoining events that lead to chromosome exchanges.

In the exchange theory, the primary lesion is not a chromosome break (i.e., not a breakage in the chromonema), but an unstable (repairable) recombinogenic lesion of undefined nature that has the potential of interacting pairwise with a similar lesion to cause chromosome exchanges. The mediation of such, presumably nonrandom, interactions in the absence of chromosome breaks is an essential feature of the model according to which chromatid (or chromosome) breaks arise only indirectly from incomplete exchanges. Exchanges in this model are the result of error-prone processing and explaining complex exchanges will require additional assumptions.

Although the breakage-and-reunion theory is clearly favored among researchers, the arguments for the exchange theory have not been conclusively refuted, and when challenged they are "rescued" by invoking scoring limitations (Revell 1974; Savage 1998; Savage and Harvey 1991); as a result, the debate continuous. In many ways the ongoing debate of these theories reflects the fact that despite large strides in our understanding of DNA metabolism, our knowledge of the chromosomal organization of DNA remains incomplete and the same holds for some aspects of DNA metabolism, including repair, in the context of chromatin. As a result, the formulation of a molecular theory of chromosome aberration formation that would settle the debate is lacking.

On the positive side, there is consensus that the majority of chromosome aberrations are the ultimate consequence of errors in the repair of DNA double-strand breaks (DSBs). By focusing on a concrete and measurable DNA lesion, a solid foundation is generated for reexamining the fundamental assumptions of the aforementioned theories and for comparing them with our present knowledge of DSB repair and chromatin structure. Since both theories require two lesions for the formation of an exchange aberration, the issues that need to be explained include:

1. Are DSBs breaking the chromosome and if not under what conditions can a DNA DSB lead to a chromosome break?
2. If DSBs do not break the chromosome directly, can they be considered as the undefined recombinogenic lesion of the exchange theory? Or, can a DSB that does not break the chromosome still lead to the formation of a chromosome aberration?
3. Are DSBs "flapping" around "searching" for exchange partners? Under what conditions could this happen and which repair pathways are then involved in the formation of chromosome aberrations?
4. Is processing required before a DSB can lead to chromosome aberrations and if so which pathway of DSB repair is involved?
5. Are certain DSB repair pathways more likely than others to cause formation of chromosome aberrations?

To begin addressing these questions, we briefly outline salient features of the repair pathways implicated in the removal of DSBs, and discuss their potential involvement in the formation of chromosome aberrations.

5.2 Pathways of DSB Repair

Here, we outline molecular aspects of the DSB repair pathways that are relevant to chromosome aberration formation but refrain from an exhaustive molecular description of the participating proteins and their properties. For a more detailed description of these pathways, the reader is referred to Chap. 4 by Goedecke.

It is well established that cells of higher eukaryotes possess efficient pathways for repairing DSBs (Jackson 2002; Lieber et al. 2003; Valerie and Povirk 2003). These include two homology-dependent pathways, homologous recombination repair (HRR) and single-strand annealing (SSA), and two homology-independent pathways, DNA-PK-dependent nonhomologous end joining (D-NHEJ) and backup nonhomologous end joining (B-NHEJ) (Fig. 5.1).

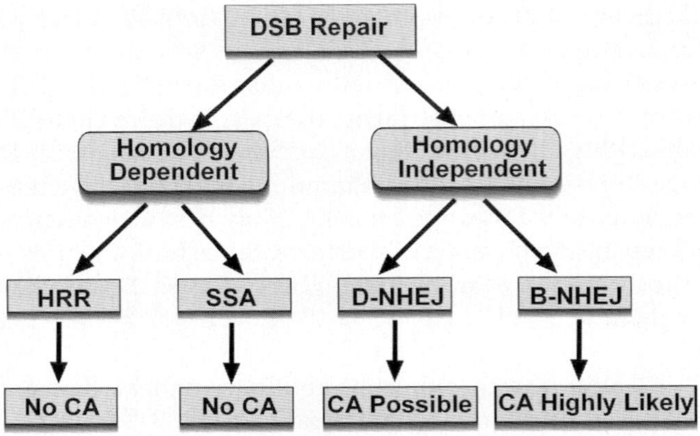

Fig. 5.1 In cells of higher eukaryotes double-strand breaks (*DSBs*) can in principle be removed by homology-dependent and homology-independent pathways. Homologous recombination repair (*HRR*) and single-strand annealing (*SSA*) are the main homology-dependent pathways and are unlikely to cause chromosome aberrations (*CA*). Homology-independent pathways are classified as DNA-PK-dependent nonhomologous end joining (*D-NHEJ*) and backup nonhomologous end joining (*B-NHEJ*) and have a varying potential regarding the formation of chromosome aberrations (see text)

5.2.1 Homology-Dependent Pathways of DSB Repair

In genetics homologous recombination describes the process through which reciprocal exchanges occur between homologous chromosomes during meiosis. This process is typically initiated by the programmed generation of DSBs and occurs without loss or gain of genetic material. In mitotic cells, homologous recombination can be employed to repair randomly induced DSBs in a process that utilizes the sister chromatid and is therefore thought to be active mainly in the late S phase and the G2 phase of the cell cycle. This repair pathway is well characterized at the molecular level and utilizes the genes of the Rad52 epistasis group of genes (Griffin and Thacker 2004; West 2003). By design HRR, if correctly initiated, is error-free and will not lead to chromosome aberrations. However, aberrant initiation of HRR is possible and has been documented in certain genetic diseases in humans. In a similar manner it is therefore possible that aberrant initiation of homologous recombination between repeat sequences in ectopic locations causes genomic rearrangements that manifest as chromosome aberrations (Griffin and Thacker 2004). Yet, chromosome aberration junctions fail to show the sequence homology that would have been expected had they been initiated by aberrant homologous recombination events. It is therefore unlikely that this repair pathway has a significant contribution to the formation of chromosome aberrations.

SSA is a variant of HRR (Griffin and Thacker 2004) and can in principle operate when a DSB is induced between two flanking homologous repeat sequences (Fig. 5.1). During SSA one of the two repeats and the intervening sequence are lost, thus making the process error-prone. While the molecular requirements of SSA are in general similar to those of HRR, distinct differences have been noted. As with HRR, canonical use of SSA will not cause genomic rearrangements that will manifest as chromosome aberrations. Aberrant use may of course cause genomic rearrangements, but here again the expected homology in the sequences around the break is not found.

Thus, homology-dependent pathways of DNA DSB repair are unlikely to be among the main players in the formation of chromosome aberrations and will not be considered further in the discussion that follows (Fig. 5.1).

5.2.2 Homology-Independent Pathways of DSB Repair

In their simplest manifestation, homology-independent pathways of DSB repair remove DSBs from the DNA by simply ligating the two ends (Lieber et al. 2003; Valerie and Povirk 2003). As the term implies, there are no homology requirements for the rejoining event, and because nucleotides are often removed from or added to the DNA ends (also to make them ligatable), sequence changes in the vicinity of the break are likely. Thus, homology-independent pathways satisfy only one aspect of DNA DSB repair, i.e., the restoration in the integrity of the DNA molecule, but do not ensure preservation of sequence; they should therefore be classified as error-prone at this level.

Although DSB repair by a homology-independent pathway was first characterized in higher eukaryotes, where it appears to dominate DSB removal, it is now clear that the pathway is evolutionarily conserved and active from bacteria to humans (Lieber et al. 2003; Valerie and Povirk 2003).

One characteristic of homology-independent pathways of DSB repair is that there is no "proofreading" mechanism incorporated to ensure the exclusive rejoining of the original ends – synonymous to the restoration of the DNA chain in each chromosome. DNA ends from different DSBs can be joined and although this can cause large-scale genomic rearrangements, there is no known cellular surveillance system registering and responding to this type of error. It appears that elimination of DNA ends, even through misrejoining, is the primary goal of these repair pathways, raising the question as to how the adverse consequences of misrejoining are minimized within the cell. As we discuss later, the short half life of the DSBs, the privilege of spatial proximity of the original ends within one DSB, and the possible restriction in motion imposed upon the DNA ends by the coordinated action of DNA repair proteins are parameters shifting the balance towards the joining of the original ends (correct rejoining) – at least for certain repair pathways. As a result of these intrinsic properties and limitations, it comes as no surprise that these repair pathways are considered as the major sources of chromosome aberrations (Iliakis et al. 2004).

In recent years the term nonhomologous end joining (NHEJ) has been predominantly used to describe homology-independent pathways of DSB repair and it will also be used here. While the term NHEJ was initially introduced to describe a DSB repair pathway utilizing, among other proteins, DNA-dependent protein kinase (DNA-PK; see later), evidence is being accumulated for the operation of a DNA-PK independent pathway of NHEJ that operates as a backup to the DNA-PK dependent pathway (see later). To distinguish between the two pathways of NHEJ, the term D-NHEJ was introduced to describe the DNA-PK dependent pathway and the term B-NHEJ to describe the backup DNA-PK independent pathway (Wang et al. 2001a, 2003, 2006); these terms will also be used here. In the following sections we describe the molecular characteristics of D-NHEJ and B-NHEJ and discuss their ramifications for chromosome aberration formation.

5.2.2.1 DNA-PK-Dependent NHEJ

This DSB repair pathway was the first homology-independent pathway to be characterized and a great deal of information regarding its molecular details is now known (Lieber et al. 2003; Valerie and Povirk 2003). Following exposure to IR and the induction of random DSBs in the genome, the first protein thought to interact with the generated DNA ends is Ku. Ku is a heterodimer of Ku70 and Ku80, with strong DNA binding properties and a toroidal structure that allows the DNA strand to be tightly threaded and stabilized (Fig. 5.2). Ku is very abundant, with nearly half a million molecules per cell, which is in vast excess to the 20–40 DSBs induced per gray of X-rays. The high equilibrium constant of Ku for DNA ends and its high abundance ensure that Ku, when available, will quickly capture the DNA ends. To avoid blocking and to enable the further end processing, Ku diffuses inwards along the DNA strand by an ATP-independent mechanism.

Binding of Ku to DNA ends is quickly followed by the recruitment of another protein, DNA-PK catalytic subunit – DNA-PKcs, with DNA binding properties and DNA-dependent serine/threonine kinase activity that are strongly stimulated by Ku. Although DNA-PKcs is thought to be approximately fivefold less abundant than Ku, ample amounts are present to support processing of a staggering number of DSBs. Binding of this large protein (molecular mass 469 kDa), together with Ku, to each of the DNA ends at a DSB is thought to generate a landing pad for the recruitment of components required for further processing. In addition, the reported association of the DNA-PKcs with the nuclear matrix (Mauldin et al. 2002) raises the possibility that the rejoining apparatus benefits from further stabilization provided by this nuclear structure. Although there is evidence that Ku and DNA-PKcs can mediate DNA end-to-end associations, the exact mechanism by which the synapsis of DNA ends is achieved in preparation for the rejoining step remains largely unknown. This is of course a critical question with direct ramifications for our mechanistic understanding of chromosome aberration formation.

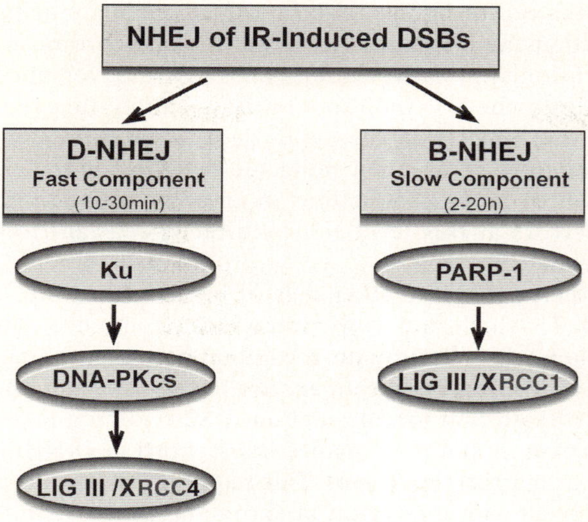

Fig. 5.2 One can distinguish between two pathways of NHEJ. D-NHEJ is fast and utilizes DNA-PK and DNA ligase IV, whereas B-NHEJ is slow and may utilize PARP-1 and DNA ligase III. See Fig. 5.1 for an outline of the contribution of each pathway to the formation of chromosome aberrations. *IR* ionizing radiation

Before final rejoining, and because IR also damages the sugar–phosphate backbone (Lieber et al. 2003; Valerie and Povirk 2003), processing is required to make ends ligatable. The nucleases and DNA polymerases suspected to operate at this step are discussed elsewhere in this volume. It is at this step that alterations in the original sequence around the DSB occur with the frequent loss or addition of nucleotides. This is the first level of error-prone operation for D-NHEJ. However, since sequence errors introduced at this stage are not causing chromosome aberrations, they will not be further discussed.

Ligation of the processed ends and restoration of DNA integrity is provided by DNA ligase IV, which functions as a complex with the stabilizing factor XRCC4 (Fig. 5.2). The XRCC4–DNA ligase IV complex does not bind stably to DNA, but is probably recruited to the DSB site by Ku–DNA interactions.

Studies in intact cells have shown that D-NHEJ is the dominant pathway utilized for the repair of DSBs in cells of higher eukaryotes (Wang et al. 2001a). One of its most salient features is the speed of operation. Nearly 90% of DSBs induced by sparsely IR are removed with half lives of 10–15 min, suggesting a highly efficient repair apparatus that captures, processes and rejoins DNA ends through precisely regulated functional coordination of its constituents (DiBiase et al. 2000). The biochemical properties of the participating proteins further suggest that naked DNA ends will not be present in the cell for more than a few seconds after their induction and that protein captured ends will be removed from the genome within minutes.

How does the cell recognize that the correct ends are being rejoined by D-NHEJ and that the original DNA fibers in each chromosome are restituted? As mentioned already, there is no known surveillance mechanism incorporated into the D-NHEJ apparatus to satisfy this requirement. In contrast, D-NHEJ may rejoin DNA ends detected in sufficient spatial proximity independently of the DNA molecule they belong to. This is the second level at which D-NHEJ functions in an error-prone manner, and it is evident that errors at this level will be directly relevant to chromosome aberration formation.

How is it then possible that in cells of higher eukaryotes a DSB repair system evolved, which not only introduces sequence alterations but which also allows for illegitimate recombination events that can lead to chromosome aberrations? Since an answer to this question remains elusive, it provides fertile ground for speculation. DSB rejoining takes place in the context of chromatin and may involve interactions of D-NHEJ constituents with the nuclear matrix (see before). This requirement confines topologically the rejoining event within a region of chromatin in the vicinity of the DSB. Since DSBs are rare events in the normal life cycle of a cell (when induced, e.g., by reactive oxygen species generated as by-products of metabolic processes), the probability that more than one DSB will be induced in the same region of chromatin is very small and therefore illegitimate recombination is an unlikely event. This fact together with the extremely high speed of D-NHEJ operation generate conditions for predominantly correct rejoining.

Can we speculate that D-NHEJ is a predominantly error-free pathway of DSB repair with insignificant contribution to chromosome aberration formation? Such a postulate may be plausible for the approximately 90% of DSBs that are repaired with the fast kinetics (if other DSBs are not present in the immediate vicinity; see later). However, it is well established that a small percentage (10–20%) of IR-induced DSBs are rejoined with slow kinetics, similar to those of interphase chromosome breaks (measured by the technique of premature chromosome condensation). Since this subset of DSBs appears to break chromosomes upon premature condensation (Cornforth and Bedford 1993) and is rejoined slowly, it may consist of DSBs whose rejoining by D-NHEJ is somehow compromised. To the extent that the same D-NHEJ proteins operate on these DSBs, the reduction in rejoining efficiency could either reflect a requirement for more extensive processing, or a failure in the synapsis of DNA ends (see before). Although extensive end processing will somewhat delay ultimate rejoining, it is unlikely to be responsible for the 5–10 times slower rejoining kinetics frequently observed, unless it interferes with the function of Ku. Therefore, the possibility of synapsis failure becomes particularly attractive as it also allows for illegitimate recombination events causing the observed chromosome aberrations. Circumstances leading to synapsis failure are discussed later.

Thus, D-NHEJ is an efficient pathway of DSB repair with a high potential of capturing, processing and ultimately rejoining DNA ends, restoring the original DNA molecules, albeit with possible sequence alterations in the vicinity of the DSB. Possible failure of this system at the synapsis step is accompanied by a drastic reduction in the rejoining kinetics and may be associated with illegitimate recombination events that lead to chromosome aberrations.

5.2.2.2 Backup NHEJ

Is D-NHEJ the only means a cell has to rejoin DSBs by homology-independent processes? Frequently in the literature, postulates to this effect are made, implicitly or explicitly, based on the limited DSB rejoining observed under certain (not always optimal) conditions in cells with defects in components of D-NHEJ. Our laboratory studied this question extensively as part of an effort to quantify the contribution of homology-directed processes in DSB repair and their coordination with homology-independent processes (Wang et al. 2001a, 2003, 2006).

Human tumor cells deficient in DNA-PKcs, studied under conditions optimal for repair (DiBiase et al. 2000), rejoin, like wild-type cells, IR-induced DSBs with kinetics that displays a fast and a slow component. While wild-type cells rejoin the majority of DSBs with fast kinetics, leaving only an estimated 10% for rejoining with slow kinetics, DNA-PKcs deficient cells rejoin a large proportion of DSBs with slow kinetics. Notably, the half lives of the fast and the slow component of rejoining as well as the total number of DSBs rejoined are similar in DNA-PKcs mutant and wild-type cells. Nearly complete rejoining of DSBs was also observed in cells deficient in Ku (Wang et al. 2001a) or DNA ligase IV (Wang et al. 2001b), as well as after treatment with DNA-PK inhibitors (Wang et al. 2001b). In addition, cells of different origins showed similar responses, suggesting that these observations are not species-specific.

These results strongly suggest that cells must be endowed with a DSB repair system operating with slow kinetics and handling a large proportion of IR-induced DSBs whenever D-NHEJ is genetically or chemically compromised. Although at first the slower kinetics of operation raised the possibility that homology-directed repair underlies this component of DSB repair, a study utilizing mutants with defects in diverse components of homologous recombination failed to generate evidence for a significant contribution by these pathways (Wang et al. 2001a). We postulated therefore that slow rejoining in cells with compromised D-NHEJ reflects the functions of an alternative pathway operating independently of homology and serving as a backup to D-NHEJ, thus termed backup NHEJ – B-NHEJ (Wang et al. 2001a). The observation that slow DSB rejoining is error-prone (Loebrich et al. 1995) is in line with the properties of homology-independent pathways of DSB repair but is incompatible with the mostly error-free operating HRR. Operation of such a NHEJ pathway in cells is also supported by the presence of end-joining activities in cellular extracts that do not rely on D-NHEJ components (Cheong et al. 1999; Perrault et al. 2004; Wang et al. 2003).

Genetic and biochemical studies have provided evidence that ligation during B-NHEJ is carried out by DNA ligase III (Wang et al. 2005) and that its operation is aided by PARP-1 (Wang et al. 2006). PARP-1 has DNA end binding properties similar to those of Ku and can potentially play a role in DSB repair (Fig. 5.2). However, in irradiated wild-type cells the higher affinity of Ku for DSBs and an excessive number of single-strand breaks and base damages that recruit PARP-1 limit its contribution to DSB repair. When essential components of D-NHEJ are absent, PARP-1 can be recruited for DSB repair, particularly in the absence of Ku. Thus, the repair module PARP-1/DNA ligase III/XRCC1 (PLX), hitherto regarded as central in repair of single-strand breaks, is implicated in the repair of DSBs (Wang et al. 2006). This backup function does not require novel activities, because a DSB can be seen as two single-strand breaks, and repair of single-strand breaks is the area of specialization of the PLX complex.

What makes B-NHEJ slow? The most plausible explanation is that this pathway of DSB repair lacks the efficient synapsis characteristics provided in D-NHEJ by the coordinated function of its constituents on the nuclear matrix. As a result, rejoining relies on synapsis by random encounters of the freely diffusing DNA ends, and relatively weak interactions of the participating proteins with the DSB (DNA ligase III, unlike DNA ligase IV, binds DNA) that make rejoining inefficient. There is evidence that junctions generated when D-NHEJ is compromised show more extensive deletions and increased incidence of microhomologies (Roth and Wilson 1986; Verkaik et al. 2002), suggesting end stabilization by hydrogen bonding. Inefficient rejoining increases the half life of DNA ends inside the nucleus and the risk of nuclease-mediated degradation and/or illegitimate recombination. Indeed both of these aspects have been experimentally documented (Lieber et al. 2003; Valerie and Povirk 2003).

The above postulate suggests similarities between slow rejoining observed for a few DSBs in wild-type cells and slow rejoining of a much larger proportion of DSBs in D-NHEJ deficient cells. Slow rejoining of DSBs in wild-type cells, to the extent that it remains DNA-PK dependent, must therefore be enzymatically distinct from B-NHEJ. The similarity in the kinetics suggests that synapsis problems can have similar consequences for the enzymatic apparatus of the D-NHEJ or the B-NHEJ. However, we speculate that in the case of B-NHEJ difficulty in DNA end synapsis is the consequence of suboptimal enzymatic processing, whereas in wild-type cells it derives from the characteristics of the lesion that abrogate the normal function of the pathway. From the perspective of chromosome aberration formation, both methods of synapsis abrogation may have similar consequences.

Cytogenetic data provide direct support for the operation of backup pathways of NHEJ and indicate their propensity for illegitimate recombination. Thus, cells deficient in DNA-PKcs show large increases in the incidence of exchanges, scored as dicentrics and complex exchanges (Ferguson et al. 2000; Martin et al. 2005; Virsik-Köpp et al. 2003, 2005). The presence of chromosome and chromatid exchanges in these cells is direct evidence for DNA-PKcs independent rejoining activity and their substantial increase in these cells is an indication that this repair pathway is prone to illegitimate recombination events (Fig. 5.1).

5.3 Homology-Independent End Joining and Chromosome Aberration Formation

Assuming that chromosome aberrations are the visual manifestation of errors during DSB repair, the following speculative ideas can be offered regarding the contribution of the already discussed repair pathways in their formation.

The fast component of D-NHEJ has the highest probability of error-free operation in the sense that illegitimate recombination events are rare. This is supported by studies evaluating DSB-rejoining fidelity using a combination of pulsed-field gel electrophoresis and Southern blotting (Loebrich et al. 1995). Furthermore, there is evidence that exchange-type aberrations are forming with half lives on the order of hours, different from what would be expected from a contribution of this repair component (Cornforth 2006; Cornforth and Bedford 1993; Evans 1962; Hlatky et al. 2002; Natarajan 2002; Obe et al. 2002; Revell 1974). DSBs removed by the slow component of D-NHEJ (assuming that D-NHEJ processes slowly repairing DSBs), which may also be responsible for chromosome breakage upon premature condensation (Cornforth and Bedford 1993), are the most likely contributors of chromosome aberrations in wild-type cells. The presumed deficiency in the synapsis of DNA ends owing to the nature of the underlying lesion delays rejoining and increases the probability of interaction with other DSBs, which may result in events of illegitimate recombination causing exchange-type aberrations. Some of these DSBs may also remain unrejoined, giving rise to acentric fragments.

As a potential contributor to chromosome aberrations in wild-type cells, and the major contributor of chromosome aberrations in D-NHEJ deficient cells, one should consider B-NHEJ. This backup repair pathway is normally suppressed by D-NHEJ (Perrault et al. 2004; Wang et al. 2003) but comes to the fore whenever D-NHEJ is compromised. Suboptimally regulated synapsis results in a low speed of rejoining and allows with high probability illegitimate recombination events. Despite the high probability for misjoining, B-NHEJ must be advantageous to the cell. This is indicated by the fact that the survival curves of D-NHEJ deficient cells have slopes suggesting accumulation of many (10–20) DSBs before registering a lethal event (DiBiase et al. 2000). In contrast, yeast deficient in homologous recombination dies upon registering approximately one DSB.

The operation of backup pathways of NHEJ is not limited to radiation-induced DSBs but it extends to other phenotypic manifestations where processing of DSBs is expected to occur. Thus, backup pathways of DSB repair bring together the *c-myc* and *Igh* locus and cause B-cell lymphomas in mice with defects in *Ku*, *LIG4* or *XRCC4* (Difilippantonio et al. 2000; Ferguson et al. 2000; Frank et al. 2000; Gao et al. 2000; Karanjawala et al. 1999; Ramiro et al. 2006; Zhu et al. 2002). Functionally equivalent pathways may generate the aberrant junctions manifesting chromosome instability in the same mutants. Similar pathways of end joining generate the few V(D)J junctions observed in cells with defects in D-NHEJ (Bogue et al. 1997; Lee et al. 2004; Li et al. 1995) and are implicated in antibody class switching occurring under the same conditions (Ramiro et al. 2006). Thus, B-NHEJ, when utilized, helps the cell to restore its genome and thus presumably to avert cell death. However, its error-prone nature causes genomic instability and cancer in the affected organism.

While prominent when the classic pathway is compromised, B-NHEJ may be utilized to a limited degree in cells with functional D-NHEJ as well. This can occur when a DSB is induced in a nuclear region where Ku or other key components of D-NHEJ are not immediately available; e.g., because they are already recruited to a neighboring DSB. Although this type of event is expected to be rare overall, its contribution to the total chromosome aberration load in a wild-type cell may be significant. B-NHEJ may also come to the fore when D-NHEJ becomes somehow compromised during processing of a DSB.

5.4 DSBs, NHEJ and Theories of Chromosome Aberration Formation

Here, we reconsider the fundamental postulates of the breakage-and-reunion and the exchange theories and discuss them under the assumption that DSBs, if processed incorrectly by homology-independent pathways of DSB repair, will initiate chromosome aberrations.

The fundamental assumption of the breakage-and-reunion theory is that the initiator lesion breaks the chromonema (Revell 1974; Savage 1998; Savage and Harvey 1991). Are DSBs capable of directly breaking the chromosome? The extremely complex folding patterns and levels of organization of DNA in the metaphase chromosome make it seem intuitively quite unlikely that a DSB will break the chromosome directly. And there is direct evidence to support this assumption. Irradiation of cells at metaphase, or of interphase cells after premature chromosome condensation, almost never breaks chromosomes (Cornforth and Bedford 1993). Thus, the fundamental postulate of the breakage-and-reunion theory that the underlying lesion directly breaks the chromonema is not tenable when the entirety of IR-induced DSBs are considered.

On the other hand, premature chromosome condensation clearly demonstrates that chromosome breakage can be a consequence of a DSB if chromatin condensation is initiated before repair can been completed (Cornforth and Bedford 1993). It can be postulated therefore that although DSBs by themselves must not break the chromosome, they nevertheless have the potential of being "translated" to a chromosome break through the action of chromatin-reorganizing processes. However, under these conditions, it is not the DSB that ultimately breaks the chromosome but the fortuitously concurrent, repair-independent process of chromatin condensation (decondensation may also have a similar outcome). This is a major departure from the direct chromosome breakage requirement of the breakage-and-reunion theory.

Can all IR-induced DSBs be "transformed" to chromosome breaks by the induction of chromatin conformation changes, such as premature chromosome condensation? Data suggest that only a relatively small fraction of IR-induced DSBs can be translated to interphase chromosome breaks (about 10–20%) even when precautions are taken to minimize repair during chromosome condensation (Okayasu and Iliakis 1993; Okayasu et al. 1993). Can we consider this fraction of DSBs as the only relevant fraction for chromosome aberration formation and apply to it the breakage-and-reunion theory? To do this we will have to assume that the remaining DSBs are incapable of participating in exchanges, but as we have discussed already this cannot be considered as generally true.

The above analysis imposes the status of chromatin and its conformational change as collateral for the consequences of a DSB on the integrity of the chromosome. The same parameters are also likely to have consequences for the repair outcome. For a DSB repair pathway operating with the limitations of D-NHEJ (inaccurate DNA sequence restoration at the DSB and more importantly the lack of a mechanism for correct rejoining-partner selection) the best means for minimizing interference is speed. Yet, the level of chromatin condensation at the site of the DSB as a result of ongoing metabolic processes (transcription or replication) and changes in that structure shortly thereafter are parameters that should be considered in the formulation of generalized theories of chromosome aberration formation.

The recognition that DSBs do not by themselves break the chromosome makes them good candidates for the undefined recombinogenic (potential of illegitimate recombination) lesions of the exchange theory. The exchange theory postulates that the initial damage is not a break, but an unstable lesion, termed the primary event of damage, which tends to decay towards a normal or undetectable state (Revell 1974; Savage 1998; Savage and Harvey 1991). If such events are brought together in pairs, they may take part in an exchange process, and so give rise to the aberrations seen at metaphase. Incomplete exchange will lead to a discontinuity that will be seen as a break. Breaks are therefore secondary.

Both pathways of NHEJ can in principle be recombinogenic if more than one DSB is induced in close proximity. But the interaction does not have to be limited to pairs and will not require active movement. Spatially proximate DSBs can be induced by random ionization events initiated by one or more electron traversals and will be located in the DNA of the same chromosome (highest probability event), or in the DNA of adjacently organized chromosomes.

The ends of each DSB are a priori spatially privileged in the sense that they have the highest possible initial proximity, and this will favor restitution. However, within an interaction radius, which will depend on the characteristics of the repair pathways involved, interaction with other DSBs is possible. On the basis of the high speed and the well-coordinated operation of the enzymatic apparatus, we anticipate small interaction radii (i.e., only very closely located DSBs will have a chance to interact) for DSBs handled by the fast component of D-NHEJ (Fig. 5.2). A significantly greater radius of interaction will be available for DSBs handled by the slow component of D-NHEJ. The maximum radius of interaction will be available for DSBs handled slowly by B-NHEJ. Increase of linear energy transfer will increase the probability of multiple DSBs within the sphere of interaction. Furthermore, changes in chromatin reorganization after DSB induction, taking place independently of repair, may change the proximity of the DNA ends and may bring close together "remotely" induced DSBs, thus modulating repair and the probability for illegitimate recombination.

In this model, all DSBs are potentially recombinogenic despite the fact that they are not capable of breaking the chromosome. The geometric size of the interaction radius is determined by the repair pathway active at the site. The domain organization of the chromosomes inside the nucleus makes it plausible that when two or more DSBs are induced within a given interaction radius, these will belong predominantly to the same chromosome and may even be closely located within the DNA molecule. This means that many illegitimate events may not be visible cytogenetically. This has direct consequences in the analysis of acentric fragments and is discussed below.

The recombinogenic potential of DSBs is a direct consequence of the repair pathways involved and does not require that DSBs break the chromosome directly. Since the ultimate goal of these repair pathways is the removal of DNA ends, the outcome when successful (albeit in an error-prone manner) will be complete exchanges. Incomplete exchanges indicate that interference after the determining initial misjoining event may prevent rejoining of remaining ends. It is noteworthy in this regard that activation of checkpoints in G1 and G2 phases of the cell cycle strives, among other things, the delay of cell cycle progression associated, scheduled alterations in chromatin conformation (Iliakis 1997; Iliakis et al. 2003).

If the initial chromosome break is only an unlikely outcome of the induction of a DSB, the idea of "flapping" ends "searching" for exchange partners (Savage 1998) should be considered unrealistic at the chromosome level. Can this idea be transferred to DSBs? On the basis of the above discussion of DSB repair pathways, freedom of movement for DNA ends can be considered as significant only for the small percentage of DSBs rejoined with slow kinetics by D-NHEJ, and probably for the majority of DSBs rejoined by B-NHEJ. (Note that when D-NHEJ is compromised, a small percentage of DSBs still rejoin with fast kinetics, suggesting efficient synapsis by alternative mechanisms, probably as a result of their location in chromatin.) This freedom of movement defines the spheres of interaction introduced above to discuss the formation of exchanges. Thus, "flapping" DNA, but not chromosome, ends may be generated under certain conditions and may be contributing to the formation of chromosome aberrations.

The key discrepancy between the breakage-and-reunion theory and the exchange theory concerns the origin of acentric fragments. While in the exchange theory acentric fragments are only produced by incomplete exchanges, in the breakage-and-reunion theory they are the direct product of the lack of chromosome end processing (Revell 1974; Savage 1998; Savage and Harvey 1991). A direct consequence of the underlying assumptions of acentric fragment formation by the two theories is a linear dose–effect relationship for the breakage-and-reunion theory and a linear–quadratic dose–effect relationship for the exchange theory. Interestingly, both functional dependencies have been observed in different organisms, making it difficult to settle the issue on this basis. This issue is further complicated by the often discussed difficulties in scoring acentric fragments (Revell 1974; Savage 1998; Savage and Harvey 1991) and limitations in the detection of recombination events associated with relatively small fragments of DNA (Savage 2004).

The breakage-and-reunion theory of course does not deny the existence of breaks arising from incomplete exchanges, but assumes that the contribution from this source is negligible when compared with those arising from failed restitution of single breaks. This raises the question as to the processes that could cause failed restitution of DSBs and the most likely source here again is changes in chromatin conformation shortly after induction of the DSB that may pull the DNA ends apart, thus compromising subsequent rejoining by NHEJ. Mechanistically, the dispute is translated to the question as to when the cell gives up looking for a partner for a DNA end given that even after chromatin reorganization the two ends will remain reasonably close together. A high degree of persistence will favor the exchange theory and will make the first exchange the "point of no return" for exchange aberration formation, whereas a low degree of persistence will support the breakage-and-reunion theory and will make the "pulling apart" of ends the point of no return.

It would be an omission to leave the subject without mentioning achromatic lesions (Savage 2004). This form of chromosomal change has generated much controversy and has introduced many errors in the analysis of acentric fragments. Yet, the question as to whether achromatic lesions contribute to the formation of chromosome aberrations remains open. The reduced staining in these regions of the chromosomes suggests changes in chromatin structure, and one wonders whether they reflect chromatin processing associated with DSB repair. Their frequent occurrence in cells irradiated in the G2 phase, but not in the G1 phase, and their complete disappearance upon repair possibly makes them intriguing manifestations of ongoing repair processes. Molecular characterization of achromatic lesions and connection with known molecular processes such as generation of γ-H2AX or homologous recombination will be mostly instructive.

The question whether IR induces qualitative different types of DSBs remains a central question not only in the field of cytogenetics but in radiation biology in general. Different "quality" may be dictated indirectly by the context, e.g., the state of condensation and chromatin organization in the vicinity of the break, or directly by the type of DSB.

The most common qualitative distinction between DSBs is that of a complex DSB, understood as a combination between a simple DSB with base damages and/or single-strand breaks a few nucleotides apart (Ward 1985, 1988). It is thought that complex DSBs impose additional requirements on the DSB repair pathways that reduce their efficiency or inhibit their function (Riballo et al. 2004). As mentioned above though, such proximal lesions must not always present great difficulty to the DNA end processing systems, and additional damaged nucleotides may be removed by mechanisms similar to those developed to process DNA ends.

A different level of complication may be generated when two or more DSBs are induced in close proximity in the DNA, with proximity here ranging from a few base pairs to several million base pairs (Johnston et al. 1998). The local destabilization of DNA and chromatin produced under these conditions may generate a major obstacle for the repair system and may complicate synapsis of the DNA ends. This form of "complex" breaks may have a higher probability of forming chromosome aberrations, both acentric fragments and exchange aberrations.

In summary, although both homology-dependent and homology-independent pathways contribute to the ultimate repair of DSBs, homology-independent pathways are responsible for the initial removal of the vast majority of DSBs from the genome. From the homology-independent pathways of DSB repair both D-NHEJ and B-NHEJ have the potential of forming chromosome aberrations, but it is likely that most of the aberrations derive from misjoining events associated with the slow repair by D-NHEJ of a subset of DSBs, and with the majority of DSBs rejoined by B-NHEJ. DSBs do not break the chromosomes directly, and as a consequence of the characteristics of NHEJ are highly recombinogenic. This satisfies the essential requirements of the exchange theory that becomes breakage-and-reunion theory at the DNA level. Further elucidation of the mechanism of DSB repair at the level of chromatin will allow the description of processes leading to chromosome aberrations at the molecular level and will settle a dispute that has challenged the field intellectually for many years. Down to the basic concepts, both theories appear to have made correct assumptions and like so often in life, the truth may lie somewhere in the middle.

Acknowledgements: This work was supported by grants from the German Research Foundation (DFG), the German Ministry of Education and Research (BMBF) and the EU.

References

Bogue MA, Wang C, Zhu C, Roth DB (1997) V(D)J Recombination in Ku86-deficient mice: distinct effects on coding, signal, and hybrid joint formation. Immunity 7:37–47

Cheong N, Perrault AR, Wang H, Wachsberger P, Mammen P, Jackson I, Iliakis G (1999) DNA-PK-independent rejoining of DNA double-strand breaks in human cell extracts *in vitro*. Int J Radiat Biol 75:67–81

Cornforth MN (2006) Perspectives on the formation of radiation-induced exchange aberrations. DNA Repair (Amst) 5:1182–1191

Cornforth MN, Bedford JS (1993) Ionizing radiation damage and its early development in chromosomes. Adv Radiat Biol 17:423–496

DiBiase SJ, Zeng Z-C, Chen R, Hyslop T, Curran WJ, Jr., Iliakis G (2000) DNA-dependent protein kinase stimulates an independently active, nonhomologous, end-joining apparatus. Cancer Res 60:1245–1253

Difilippantonio MJ, Zhu J, Chen HT, Meffre E, Nussenzweig NC, Max EE, Ried T, Nussenzweig A (2000) DNA repair protein Ku80 suppresses chromosomal aberrations and malignant transformation. Nature 404:510–514

Evans HJ (1962) Chromosome aberrations induced by ionizing radiations. Int Rev Cytol 13:221–321

Ferguson DO, Sekiguchi JM, Chang S, Frank KM, Gao Y, DePinho RA, Alt FW (2000) The nonhomologous end-joining pathway of DNA repair is required for genomic stability and the suppression of translocations. Proc Natl Acad Sci USA 97:6630–6633

Frank KM, Sharpless NE, Gao Y, Sekiguchi JM, Fergson DO, Zhu C, Manis JP, Horner J, DePinho RA, Alt FW (2000) DNA ligase IV deficiency in mice leads to defective neurogenesis and embryonic lethality via the p53 pathway. Mol Cell 5:993–1002

Gao Y, Ferguson DO, Xie W, Manis JP, Sekiguchi JA, Frank KM, Chaudhuri J, Horner J, DePinho RA, Alt FW (2000) Interplay of p53 and DNA-repair protein XRCC4 in tumorigenesis, genomic stability and development. Nature 404:897–900

Griffin CS, Thacker J (2004) The role of homologous recombination repair in the formation of chromosome aberrations. Cytogenet Genome Res 104:21–27

Hlatky L, Sachs RK, Vazquez M, Cornforth MN (2002) Radiation-induced chromosome aberrations: insights gained from biophysical modeling. BioEssays 24:714–723

Iliakis G (1997) Cell cycle regulation in irradiated and non-irradiated cells. Semin Oncol 24:602–615

Iliakis G, Wang Y, Guan J, Wang H (2003) DNA damage checkpoint control in cells exposed to ionizing radiation. Oncogene 22:5834–5847

Iliakis G, Wang H, Perrault AR, Boecker W, Rosidi B, Windhofer F, Wu W, Guan J, Terzoudi G, Pantelias G (2004) Mechanisms of DNA double strand break repair and chromosome aberration formation. Cytogenet Genome Res 104:14–20

Jackson SP (2002) Sensing and repairing DNA double-strand breaks. Carcinogenesis 23:687–696

Johnston PJ, MacPhail SH, Stamato TD, Kirchgessner CU, Olive PL (1998) Higher-order chromatin structure-dependent repair of DNA double-strand breaks: involvement of the V(D)J recombination double-strand break repair pathway. Radiat Res 149:455–462

Karanjawala ZE, Grawunder U, Hsieh C-L, Lieber MR (1999) The nonhomologous DNA end joining pathway is important for chromosome stability in primary fibroblasts. Curr Biol 9:1501–1504

Lee GS, Neiditch MB, Salus SS, Roth DB (2004) RAG proteins shepherd double-strand breaks to a specific pathway, suppressing error-prone repair, but RAG nicking initiates homologous recombination. Cell 117:171–184

Li Z, Otevrel T, Gao Y, Cheng W-L, Seed B, Stamato TD, Taccioli GE, Alt FW (1995) The XRCC4 gene encodes a novel protein involved in DNA double-strand break repair and V(D)J recombination. Cell 83:1079–1089

Lieber MR, Ma Y, Pannicke U, Schwarz K (2003) Mechanism and regulation of human nonhomologous DNA end-joining. Nat Rev Mol Cell Biol 4:712–720

Loebrich M, Rydberg B, Cooper PK (1995) Repair of x-ray-induced DNA double-strand breaks in specific *Not* I restriction fragments in human fibroblasts: Joining of correct and incorrect ends. Proc Natl Acad Sci USA 92:12050–12054

Martin M, Genesca A, Latre L, Jaco I, Taccioli GE, Egozcue J, Blasco MA, Iliakis G, Tusell L (2005) Postreplicative joining of DNA double-strand breaks causes genomic instability in DNA-PKcs-deficient mouse embryonic fibroblasts. Cancer Res 65:10223–10232

Mauldin SK, Getts RC, Liu We, Stamato TD (2002) DNA-PK-dependent binding of DNA ends to plasmids containing nuclear matrix attachment region DNA sequences: evidence for assembly of a repair complex. Nucleic Acids Res 30:4075–4087

Natarajan AT (2002) Chromosome aberrations: past, present and future. Mutat Res 504:3–16

Obe G, Pfeiffer P, Savage JRK, Johannes C, Goedecke W, Jeppesen P, Natarajan AT, Martinez-Lopez W, Folle GA, Drets ME (2002) Chromosomal aberrations: formation, identification and distribution. Mutat Res 504:17–36

Okayasu R, Iliakis G (1993) Ionizing radiation induces two forms of interphase chromosome breaks in CHO cells that rejoin with different kinetics and show different sensitivity to treatment in hypertonic medium or b-araA. Radiat Res 136:262–270

Okayasu R, Varlotto J, Iliakis G (1993) Hypertonic treatment does not affect the radiation yield of interphase chromosome breaks in DNA double-strand break repair deficient *xrs-5* cells. Radiat Res 135: 171–177

Perrault R, Wang H, Wang M, Rosidi B, Iliakis G (2004) Backup pathways of NHEJ are suppressed by DNA-PK. J Cell Biochem 92:781–794

Ramiro AR, Jankovic M, Callen E, Difilippantonio S, Chen H-T, McBride KM, Eisenreich TR, Chen J, Dickins RA, Lowe SW, Nussenzweig A, Nussenzweig MC (2006) Role of genomic instability and p53 in AID-induced c-myc-Igh translocations. Nature 440:105–109

Revell SH (1974) The breakage-and-reunion theory and the exchange theory for chromosomal aberrations induced by ionizing radiation: a short history. Adv Radiat Biol 4:367–416

Riballo E, Kühne M, Rief N, Doherty A, Smith GCM, Recio M-J, Reis C, Dahm K, Fricke A, Krempler A, Parker AR, Jackson SP, Gennery A, Jeggo PA, Loebrich M (2004) A pathway of double-strand break rejoining dependent upon ATM, Artemis, and proteins locating to g-H2AX foci. Mol Cell 16:715–724

Roth DB, Wilson JH (1986) Nonhomologous recombination in mammalian cells: Role for short sequence homologies in the joining reaction. Mol Cell Biol 6:4295–4304

Savage JRK (1975) Classification and relationships of induced chromosomal structural changes. J Med Genet 12:103–122

Savage JRK (1998) A brief survey of aberration origin theories. Mutat Res 404:139–147

Savage JRK (2004) On the nature of visible chromosomal gaps and breaks Cytogenet Genome Res 104:46–55

Savage JRK, Harvey AN (1991) Revell revisited. Mutat Res 250:307-317

Valerie K, Povirk LF (2003) Regulation and mechanisms of mammalian double-strand break repair. Oncogene 22:5792–5812

Verkaik NS, Esveldt-van Lange REE, van Heemst D, Brüggenwirth HT, Hoeijmakers JHJ, Zdzienicka MZ, van Gent DC (2002) Different types of V(D)J recombination and end-joining defects in DNA double-strand break repair mutant mammalian cells. Eur J Immunol 32:701–709

Virsik-Köpp P, Rave-Fränk M, Hofman-Hüther H, Schmidberger H (2003) Role of DNA-PK in the process of aberration formation as studied in irradiated human glioblastoma cell lines M059K and M059J. Int J Radiat Biol 79:61–68

Virsik-Köpp P, Hofman-Hüther H, Rave-Fränk M, Schmidberger H (2005) The Effect of Wortmannin on Radiation-Induced Chromosome Aberration Formation in the Radioresistant Tumor Cell Line WiDr. Radiat Res 164:148–156

Wang H, Zeng Z-C, Bui T-A, Sonoda E, Takata M, Takeda S, Iliakis G (2001a) Efficient rejoining of radiation-induced DNA double-strand breaks in vertebrate cells deficient in genes of the RAD52 epistasis group. Oncogene 20:2212–2224

Wang H, Zhao-Chong Z, Perrault AR, Cheng X, Qin W, Iliakis G (2001b) Genetic evidence for the involvement of DNA ligase IV in the DNA-PK-dependent pathway of non-homologous end joining in mammalian cells. Nucleic Acids Res 29:1653–1660

Wang H, Perrault AR, Takeda Y, Qin W, Wang H, Iliakis G (2003) Biochemical evidence for Ku-independent backup pathways of NHEJ. Nucleic Acids Res 31:5377–5388

Wang H, Rosidi B, Perrault R, Wang M, Zhang L, Windhofer F, Iliakis G (2005) DNA Ligase III as a candidate component of backup pathways of nonhomologous end joining. Cancer Res 65:4020–4030

Wang M, Wu W, Wu W, Rosidi B, Zhang L, Wang H, Iliakis G (2006) PARP-1 and Ku compete for repair of DNA double strand breaks by distinct NHEJ pathways. Nucleic Acids Res 34:6170–6182

Ward JF (1985) Biochemistry of DNA lesions. Radiat Res 104:S103–S111

Ward JF (1988) DNA damage produced by ionizing radiation in mammalian cells: Identities, mechanisms of formation, and reparability. Prog Nucl Acid Res 35:95–125

West SC (2003) Molecular views of recombination proteins and their control. Nat Rev Mol Cell Biol 4:1–11

Zhu C, Mills KD, Ferguson DO, Lee C, Manis J, Fleming J, Gao Y, Morton CC, Alt FW (2002) Unrepaired DNA breaks in p53-deficient cells lead to oncogenic gene amplification subsequent to translocations. Cell 109:811–821

6 Targeted and Non-Targeted Induction of Chromosomal Rearrangements After Exposure to Ionizing Radiation

William F. Morgan and Marianne B. Sowa

Abstract

Exposure to ionizing radiation can result in chromosomal rearrangements in a cell cycle dependent manner. This is presumably due to the deposition of energy in the nucleus of the irradiated cell causing DNA double-strand cleavage and the failure of the DNA repair machinery to faithfully restore the integrity of the genetic material. There is also increasing evidence that an irradiated cell can signal to a non-irradiated "bystander" cell and elicit cellular responses, including chromosomal rearrangements, in these non-targeted cells. These signals can be communicated between targeted (irradiated) cells and non-targeted (non-irradiated) cells by both cell-to-cell gap junction communication mechanisms as well as soluble factors secreted into the culture medium. In this chapter we review the evidence for targeted and non-targeted induction of chromosomal rearrangements after exposure to ionizing radiation and speculate about potential mechanisms and their significance in radiation risk assessment.

6.1 Introduction

Analysis of chromosomal rearrangements has been used for many years to determine whether an individual has been exposed to ionizing radiation, and to provide a reliable estimate of the exposure dose (reviewed in Cornforth 1998). A rich and well-documented literature indicates that cytogenetic analysis can be used to determine exposures to radiation doses as low as 15–20 cGy, and thus provides the most sensitive and reliable assay for radiation exposure. Implicit in interpreting these types of analysis is that the induced cytogenetic alterations result from the deposition of energy by ionizing radiation and subsequent cellular responses. Specifically, radiation exposure leads to DNA damage, namely, single and double stranded DNA breaks, DNA–DNA and DNA–protein cross-links, and DNA base damages (Ward 1988). The cell then responds to this damage by initiating changes in gene expression, inducing cell cycle checkpoint control strategies and signal transduction pathways, and activating DNA repair processes.

Günter Obe and Vijayalaxmi (Eds.)
Chromosomal Alterations: Methods, Results and Importance in Human Health

The general consensus in the field has long been that chromosomal rearrangements result from the non-rejoining or misrejoining of radiation-induced DNA double-strand breaks (King et al. 1994). That is, deletion-type aberrations can result from the non-rejoining of an induced break or acentric rings induced by two double-strand breaks. Inversions result in misrejoining of breaks within a chromosome and exchange-type alterations result from the misrejoining of breaks on two or more different chromosomes, leading to insertions, rings, polycentric chromosomes, and/or reciprocal translocations. Given that in most radiation-exposure scenarios cells from irradiated individuals would be in G0/G1 phase, the dominant molecular process giving rise to misrejoining events would likely be non-homologous end joining.

In the present context, chromosomal rearrangements can be equated with the "breakage first hypothesis" (Sax 1938) that predicts breaks must be "open" at the same time and within close enough proximity to interact. While certain fundamental aspects of the formation of chromosomal rearrangements remain contentious, breakage followed by either non-rejoining of the break or misrejoining of induced breaks can satisfactorily describe most of the observed cytogenetic alterations in irradiated cells. In the past decade however, a host of different laboratories have described a number of non-targeted effects associated with exposure to ionizing radiation. In contrast to the traditional paradigm that a cell must be irradiated, or "targeted," to exhibit those detrimental effects associated with radiation exposure, these non-targeted effects occur in cells that were not subject to energy-deposition events because they were not irradiated. These non-targeted effects include radiation-induced chromosomal instability, bystander effects, the death-inducing effect, clastogenic factors, as well as abscopal effects described in the clinical literature as effects occurring outside the directly irradiated tissue volume (reviewed in Morgan 2003a,b). They occur in cells that were not irradiated but that either were neighbors of irradiated cells or received soluble signals from irradiated cells. A schematic summarizing the concept of targeted and non-targeted effects associated with observed chromosomal rearrangements is presented in Fig. 6.1. In this chapter we will review the evidence for non-targeted radiation-induced chromosomal instability, and discuss how factors produced by irradiated cells may communicate the legacy of radiation exposure to non-targeted cells resulting in DNA double strand break formation and ultimately chromosomal rearrangements.

6.2 Radiation-Induced Chromosomal Instability

First described by Kadhim et al. (1992), radiation-induced chromosomal instability refers to the induction of chromosomal rearrangements in the progeny of an irradiated cell. Instability was observed in cells that had undergone multiple cell division cycles after exposure to ionizing radiation. This

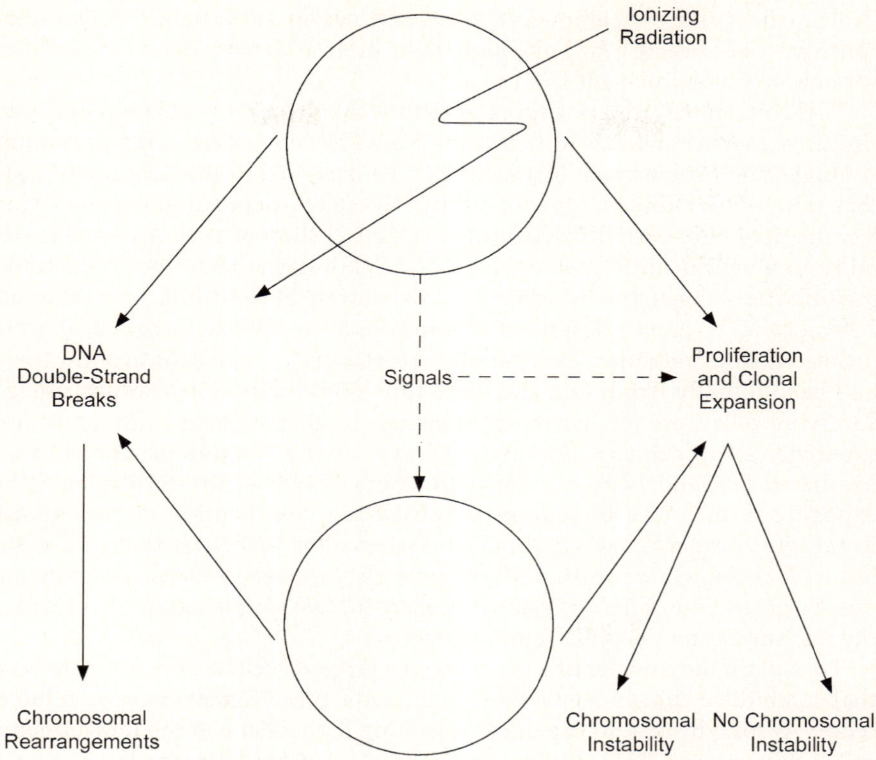

Fig. 6.1 A schematic summarizing the concept of targeted and non-targeted chromosomal aberration induction after ionizing radiation. A cell is irradiated (targeted) breaks are induced that can lead to cytogenetic damage. An irradiated cell can also send a signal to a non-irradiated (non-targeted) cell via cell-to-cell gap junction communication and/or soluble shed/secreted factors. These factors can elicit DNA cleavage leading to the induction of chromosomal rearrangements. Clonal expansion of both targeted and non-targeted cells can also result in chromosomal instability in the progeny multiple cell generations after the initial challenge and/or signal

demonstration of delayed cytogenetic damage has since been confirmed by many investigators (reviewed in Morgan 2003a), and suggests that ionizing radiation can elicit a memory in an irradiated cell that can manifest as cytogenetic changes weeks, even months, after cellular irradiation. At the time as these chromosomal alterations are observed, the radiation-induced DNA lesions are presumably long gone, either as a consequence of DNA repair processes or by dilution during successive cell division cycles. Thus, the memory of ionizing radiation must include the delayed induction of genetic lesions in the progeny of an irradiated cell. Radiation-induced chromosomal instability has been described in vivo in both animal studies (Plumb et al. 1998; Watson et al. 2000, 2001) and in exposed human populations

(Nakanishi et al. 1999; Salomaa et al. 1998); however, these data are somewhat controversial and currently the subject of intense debate (Tawn et al. 2000; Whitehouse and Tawn 2001).

This conceptual understanding of the mechanism(s) of radiation-induced chromosomal instability is further complicated by convincing evidence demonstrating that chromosomal instability can be observed in the progeny of cells that were not irradiated, i.e., not subject to energy-deposition events. Using low fluences of α-particles, Kadhim et al. (1992) showed chromosomal instability in a significantly greater number of clonogenic survivors than could possibly have been hit by radiation. Extending these initial observations, Lorimore et al. (1998) described chromosomal instability in cells that were protected from radiation exposure by a metal grid, while cells around them had been lethally irradiated and were thus non-viable. These non-targeted surviving cells were likely subject to signals from irradiated cells communicated via cell-to-cell gap junctions, secreted or shed factors produced by an irradiated cell and being available to interact with a non-targeted cell by exposure in the culture medium, or possibly a combination of both (Sowa Resat and Morgan 2004). It should be stressed at this stage that radiation-induced chromosomal instability in vivo also appears to have a significant non-targeted component (Watson et al. 2000, 2001) that might also involve the transmission of soluble signaling factors.

To explain the transfer of a signal from a targeted cell to a non-targeted cell that can induce chromosomal rearrangements, these signals must be soluble, relatively long lived, and capable of reaching the nucleus of the non-targeted cell and inducing a DNA double-strand break. A precedence for this has been described in vivo in humans occupationally (Gemignani et al. 1999), therapeutically (Hollowell and Littlefield 1968; Littlefield et al. 1969), or accidentally exposed to ionizing radiation (Goh and Sumner 1968; Pant and Kamada 1977). Blood plasma isolated from these irradiated individuals can induce chromosomal breakage when mixed and co-cultured with peripheral blood lymphocytes from non-irradiated individuals. These so-called clastogenic factors (Parsons et al. 1954) indicate that soluble factors can be induced in irradiated humans and these can cause cytogenetic damage in normal healthy lymphocytes.

6.3 The Death-Inducing Effect

We were interested in whether secreted soluble factors in the culture medium could provide a mechanism for perpetuating chromosomal instability in our GM10115 human hamster hybrid cell system. We knew that one characteristic of our unstable cell clones was that they demonstrated persistently elevated levels of reactive oxygen species (ROS; Limoli et al. 1998, 2003). Furthermore, addition of antioxidants could reduce the levels of induced chromosomal instability (Limoli et al. 2001b). While this could be explained in part by dysfunctional mitochondria (Kim et al. 2006), it was difficult to reconcile how a single radiation exposure could induce mitochondrial abnormalities that expressed themselves at delayed times after irradiation and persisted over time. We proposed that the initial radiation exposure and subsequent induction of ROS initiated cycles of cytokine secretion that in turn stimulated ROS production producing cytokine secretion – ROS production, etc. (Morgan et al. 2002). This would provide a source for the mitochondrial dysfunction (Kim et al. 2006) and persistently elevated levels of ROS (Limoli et al. 2003), as well as an extracellular mechanism driving the generation of the novel chromosome rearrangements over time that characterize our chromosomally unstable clones.

If this indeed were the case, then medium from our chromosomally unstable cell clones should be capable of generating chromosomal instability in cells that had never been exposed to ionizing radiation. This would be similar to the transmission of factors from irradiated to non-irradiated cells leading to chromosome destabilization as described by Kadhim et al. (1992) and Lorimore et al. (1998). To test this hypothesis Nagar et al (2003a) selected three chromosomally unstable cell clones derived from irradiated GM10115 cells, one irradiated but chromosomally stable derivative of GM10115 cells, and parental GM10115 cells for investigation. All clones were grown to confluence, the medium was replaced with fresh medium for 48 h, after which time it was decanted from the cells, filtered, and transferred to flasks containing 100 GM10115 cells plated 4 h earlier. The idea was to clonally expand surviving cells and investigate chromosomal instability in these clones, the logic being that if factors secreted by unstable cells do lead to cytogenetic rearrangements, then instability should be observed in colonies cultured in medium from unstable clones, but not in colonies expanded in medium from either parental GM10115 cells or the irradiated but chromosomally stable clone. However, rather than exhibiting chromosomal instability, Nagar et al. (2003a) found that medium from two (clones LS-12 and Fe-10-3) of the three unstable cells was almost completely cytotoxic to GM10115 cells. This cytotoxic effect after transfer of medium from these unstable clones was called the death-inducing effect (DIE). The DIE was not observed after medium transfer from parental cells or the irradiated but chromosomally stable clone.

Nagar and colleagues then investigated how the DIE killed non-irradiated GM10115 cells. They observed that after transfer of medium from unstable clones there was a rapid induction of γH2AX foci formation in recipient cells (Nagar et al. 2003b). Formation of double-strand breaks induces phosphorylation of histone H2AX and this phosphorylated form, γH2AX, forms foci at the sites of DNA cleavage (Rogakou et al. 1998). Presumably then, DIE medium contained soluble factors secreted by unstable cells that could induce DNA double-strand cleavage. This was a significant observation because it indicates a source for the DNA double-strand breakage that would drive the chromosomal instability observed in our unstable clones. Given the preceding discussion, this lends credence to the hypothesis that the DIE factor(s) are responsible for perpetuating chromosomal instability in the unstable clone by providing a source for the genetic damage that leads to chromosomal alterations in the progeny of the irradiated cell. In those naive cells cultured in DIE medium, the induced DNA cleavage led to micronuclei formation and ultimately cell death by apoptosis.

So why do chromosomally unstable clones not die when cultured in this medium? It was rapidly apparent when initially expanding these unstable clones that many showed a general failure to thrive. In fact, one well-described characteristic of radiation-induced genomic instability is delayed lethal mutations (Seymour and Mothersill 1997) or delayed reproductive cell death (Chang and Little 1991). This is observed as a reduced plating efficiency in some cell clones surviving radiation exposure. Many of our unstable GM10115 cells also show reduced clonogenic survival (Limoli et al. 1997; Marder and Morgan 1993), and this generally manifests itself as persistently high levels of apoptosis (Limoli et al. 1998; Nagar et al. 2005). We had originally thought that this high level of apoptosis might contribute to perpetuating chromosomal instability because lytic products from apoptosing cells might also cause DNA strand breakage. Analysis of the apoptotic profiles of two of our unstable clones Fe-10-3 and LS-12 that showed the DIE led us to conclude that because of the significantly different levels of apoptosis in these two clones, lytic products associated with apoptosis are unlikely to provide the driving force for the DIE (Nagar et al. 2005). This conclusion was supported by a series of mixing experiments: We took a flask of GM10115 cells grown to confluence and lethally irradiated it with 20 Gy of X-rays. The irradiated cells plus medium were mixed with 100 non-irradiated GM10115 cells and survival of the non-irradiated cells was determined by colony formation. Survival in the non-irradiated cells was normal and was not affected by filtering out the dead/dying cells and just transferring the medium, or irradiation of the medium alone. Consequently it is our hypothesis that the DIE results from soluble factor(s) released by unstable cells that is/are cytotoxic to non-irradiated cells.

Despite the observed delayed reproductive cell death in many of our unstable clones, the majority of cells are reproductively viable and unstable clones have been cultured off and on for more than 4 years following irradiation. So why is DIE medium not cytotoxic to these cells? It is our hypothesis that this is due to a combination of reasons. Firstly, it takes unstable cells some time in culture to secrete sufficient amounts of DIE factor(s) into the culture medium to produce an almost complete cytotoxic response (Nagar et al. 2003a). Therefore our standard protocol of subculturing 5×10^5 cells twice a week provides ample time for them to attach, replicate, and divide before they really begin to secrete/shed the DIE factor(s). By the time the DIE is fully expressed, there are a large number of target cells that will soon be subcultured again. Secondly, we hypothesize that the unstable clones have "adapted" to this medium. Nevertheless they are susceptible to breakage induced by DIE factors that leads to persistent chromosomal instability and elevated apoptosis. To this end we examined whether this "adaptation" to a potentially cytotoxic environment conferred a more radiation resistant phenotype to our unstable clones (Limoli et al. 2001a). Chromosomal instability and radiation sensitivity were compared in 30 independently isolated clones. We observed a broader distribution of radiosensitivity and a higher mean surviving fraction after 2 Gy of X-rays in chromosomally unstable clones. Cytogenetic and DNA double strand break rejoining assays suggested that sensitivity was a function of DNA repair efficiency. In the unstable population, the more radioresistant clones also had significantly lower plating efficiencies. We proposed that the process of chromosome breakage and recombination that accompanies chromosome instability might provide some selective pressure for more radioresistant variants (Limoli et al. 2001a).

6.4 Chromosomal Instability Induced by DIE Medium from Unstable Clones

Despite the surprisingly effective cytotoxic effect of DIE medium, occasionally single cell clones can survive and form colonies. Nagar and Morgan (2005) expanded 20 clonal isolates of GM10115 cells surviving in DIE medium from either LS-12 or Fe-10-3 unstable cell clones. Surviving colonies were expanded in a 1:1 mixture of DIE medium to fresh medium, a ratio that we have previously demonstrated to be cytotoxic to non-irradiated GM10115 cells (Nagar et al. 2003a). Cytogenetic analysis of these clones revealed that one of 20 clones surviving in medium from LS-12 cells and three of 20 clones cultured in medium from Fe-10-3 cells showed chromosomal instability. This indicates that DIE medium can elicit the unstable phenotype, but why the significant majority of surviving clones do not manifest chromosomal instability is not immediately apparent. However, with time in culture in DIE medium, those clones demonstrating instability demonstrated increasing levels and complexity of delayed chromosomal rearrangements (Nagar and Morgan 2005).

6.5 A Role for Non-Targeted Bystander Effects in Radiation-Induced Chromosomal Instability

Non-targeted bystander effects refer to those responses occurring in cells that were not subject to energy-deposition events following ionizing radiation. Instead, these bystander cells may have been neighbors of irradiated cells, or physically separated, but susceptible to secreted signals from irradiated cells. This signal produced by an irradiated cell can result in a number of bystander responses in non-targeted cells. These responses include increased sister-chromatid exchanges, chromosomal rearrangements, micronuclei, gene mutations, apoptosis, genomic instability, transformation, altered cellular proliferation, and a variety of damage-inducible stress responses (reviewed in Morgan 2003a, b). A role for both gap junction communication (Azzam et al. 2001) and secreted soluble factors (Mothersill and Seymour 1998) in communicating the bystander response has been described (Sowa Resat and Morgan 2004). Consequently the transmission of radiation-induced chromosomal instability might well be considered a bystander-like response (Lorimore et al. 2003; Morgan et al. 2002). Candidate molecules responsible for the bystander signal include ROS and reactive nitrogen species as well as proteins associated with cellular stress responses (Azzam et al. 2003; Lehnert and Goodwin 1997; Mothersill et al. 2000; Narayanan et al. 1997, 1999; Shao et al. 2002, 2003).

Recent studies indicate that the signal from an irradiated cell to the non-targeted bystander cell can induce DNA cleavage in the recipient cell as indicated by phosphorylation of histone H2AX (Sokolov et al. 2005; Yang et al. 2005). This suggests that DNA double-strand breakage may be responsible for the observed bystander effects, which in turn would explain why the well-documented bystander effects in non-targeted cells are the same as the deleterious effects described in directly irradiated cells. A role for DNA breaks in mediating the bystander effect is also supported by studies indicating that the bystander response is significantly enhanced in repair deficient xrs5 cell lines following exposure to low fluences of α-particles (Nagasawa and Little 2002; Nagasawa et al. 2003). Nagasawa and Little (2002) have proposed that the primary DNA damage in bystander Chinese hamster ovary cells is oxidative base damage. This leads to a relatively small bystander effect for gross chromosomal aberrations as compared with mutations or sister-chromatid exchanges in wild-type cells, but to a significant bystander effect in xrs-5 cells because of non-repaired DNA strand breaks caused by opposed oxidative lesions.

6.6 Implications of Non-Targeted Effects for Cytogenetic Biodosimetry

It is premature at present to speculate on the significance of non-targeted effects of radiation on the use of chromosome analysis following human exposure to ionizing radiation. Despite of a large body of in vitro information, there continues to be considerable uncertainty regarding the causal relationship between non-targeted effects and the observed effects occurring in exposed individuals that might be attributable to radiation. The available information provides some support for potential associations, but not causation, and further research focusing on mechanistic information regarding effects occurring at delayed times after radiation exposure should help clarify this issue.

6.7 Conclusions

It is well established that exposure to ionizing radiation results in chromosomal rearrangements in a cell cycle independent manner. It is generally agreed that potential cytogenetic effects in irradiated (targeted) cells should be analyzed at the first metaphase following exposure to avoid complications associated with the formation of derived aberrations. Over the last decade, reports of chromosomal instability occurring in cells multiple generations after irradiation have indicated that a legacy of radiation insult can be transmitted from an irradiated cell to a non-irradiated (non-targeted) bystander cell and/or the progeny of that irradiated or non-targeted cell. Whatever the mechanism, these data indicate that chromosomal rearrangements can occur in non-irradiated cells. These non-targeted cells were either bystanders at the time of irradiation, or the progeny of irradiated cells. It appears that a signal from an irradiated cell can elicit molecular changes and subsequent responses in non-irradiated cells that results in chromosomal rearrangements. Furthermore, once chromosomal instability is initiated it can be perpetuated over time by bridge–breakage–fusion cycles (Kaplan et al. 1997; Marder and Morgan 1993) and DIE-type factors secreted by unstable cells (Nagar and Morgan 2005). Clearly identification and characterization of the signal(s) responsible for the DIE and other non-targeted bystander effects is critical for understanding the molecular mechanism(s) leading to chromosomal rearrangements in cells not directly exposed to DNA-damaging agents.

Acknowledgement: This research was supported by the Office of Science (BER), US Department of Energy, grant no. DE-FG02-05ER64082.

References

Azzam EI, de Toledo SM, Little JB (2001) Direct evidence for the participation of gap junction-mediated intercellular communication in the transmission of damage signals from alpha - particle irradiated to nonirradiated cells. Proc Natl Acad Sci USA 98:473–478

Azzam EI, de Toledo SM, Little JB (2003) Oxidative metabolism, gap junctions and the ionizing radiation-induced bystander effect. Oncogene 22:7050–7057

Chang WP, Little JB (1991) Delayed reproductive cell death in X-irradiated Chinese hamster ovary cells. Int J Radiat Biol 60:483–496

Cornforth MN (1998) Radiation-induced damage and the formation of chromosomal aberrations. In: Hoekstra MF, Nickoloff JA (eds) DNA damage and repair: DNA repair in higher eukaryotes, vol 2. Humana, Totawa, pp 559–585

Gemignani F, Ballardin M, Maggiani F, Rossi AM, Antonelli A, Barale R (1999) Chromosome aberrations in lymphocytes and clastogenic factors in plasma detected in Belarus children 10 years after Chernobyl accident. Mutat Res 446:245–253

Goh K, Sumner H (1968) Breaks in normal human chromosomes: are they induced by a transferable substance in the plasma of persons exposed to total-body irradiation? Radiat Res 35:171–181

Hollowell JG Jr, Littlefield LG (1968) Chromosome damage induced by plasma of x-rayed patients: an indirect effect of x-ray. Proc Soc Exp Biol Med 129:240–244

Kadhim MA, Macdonald DA, Goodhead DT, Lorimore SA, Marsden SJ, Wright EG (1992) Transmission of chromosomal instability after plutonium alpha-particle irradiation. Nature 355:738–740

Kaplan MI, Limoli CL, Morgan WF (1997) Perpetuating radiation-induced chromosomal instability. Radiat Oncol Invest 5:124–128

Kim GK, Chandrasekaran K, Morgan WF (2006) Mitochondrial dysfunction, persistently elevated levels of reactive oxygen species, and radiation induced genomic instability: a review. Mutagenesis 21:361–367

King JS, Phillips JW, Morgan WF (1994) The role of DNA double strand break rejoining in chromosome damage and repair. In: Obe G, Natarajan AT (eds) Chromosomal alterations: origin and significance. Springer, Berlin, pp 64–75

Lehnert BE, Goodwin EH (1997) Extracellular factor(s) following exposure to alpha particles can cause sister chromatid exchanges in normal human cells. Cancer Res 57:2164–2171

Limoli CL, Kaplan MI, Corcoran J, Meyers M, Boothman DA, Morgan WF (1997) Chromosomal instability and its relationship to other end points of genomic instability. Cancer Res 57:5557–5563

Limoli CL, Hartmann A, Shephard L, Yang CR, Boothman DA, Bartholomew J, Morgan WF (1998) Apoptosis, reproductive failure, and oxidative stress in Chinese hamster ovary cells with compromised genomic integrity. Cancer Res 58:3712–3718

Limoli CL, Corcoran JJ, Jordan R, Morgan WF, Schwartz JL (2001a) A role for chromosomal instability in the development of and selection for radioresistant cell variants. Br J Cancer 84:489–492

Limoli CL, Kaplan MI, Giedzinski E, Morgan WF (2001b) Attenuation of radiation-induced genomic instability by free radical scavengers and cellular proliferation. Free Radic Biol Med 31:10–19

Limoli CL, Giedzinski E, Morgan WF, Swarts SG, Jones GD, Hyun W (2003) Persistent oxidative stress in chromosomally unstable cells. Cancer Res 63:3107–3111

Littlefield LG, Hollowell JG Jr, Pool WH Jr (1969) Chromosomal aberrations induced by plasma from irradiated patients: an indirect effect of X radiation. Further observations and studies of a control population. Radiology 93:879–886

Lorimore SA, Kadhim MA, Pocock DA, Papworth D, Stevens DL, Goodhead DT, Wright EG (1998) Chromosomal instability in the descendants of unirradiated surviving cells after alpha-particle irradiation. Proc Natl Acad Sci USA 95:5730–5733

Lorimore SA, Coates PJ, Wright EG (2003) Radiation-induced genomic instability and bystander effects: inter-related nontargeted effects of exposure to ionizing radiation. Oncogene 22:7058–7069

Marder BA, Morgan WF (1993) Delayed chromosomal instability induced by DNA damage. Mol Cell Biol 13: 6667–6677

Morgan WF (2003a) Non-targeted and delayed effects of exposure to ionizing radiation: I. Radiation-induced genomic instability and bystander effects in vitro. Radiat Res 159:567–580

Morgan WF (2003b) Non-targeted and delayed effects of exposure to ionizing radiation: II. Radiation-induced genomic instability and bystander effects in vivo, clastogenic factors and transgenerational effects. Radiat Res 159:581–596

Morgan WF, Hartmann A, Limoli CL, Nagar S, Ponnaiya B (2002) Bystander effects in radiation-induced genomic instability. Mutat Res 504:91–100

Mothersill C, Seymour CB (1998) Cell-cell contact during gamma irradiation is not required to induce a bystander effect in normal human keratinocytes: evidence for release during irradiation of a signal controlling survival into the medium. Radiat Res 149:256–262

Mothersill C, Stamato TD, Perez ML, Cummins R, Mooney R, Seymour CB (2000) Involvement of energy metabolism in the production of 'bystander effects' by radiation. Br J Cancer 82:1740–1745

Nagar S, Morgan WF (2005) The death-inducing effect and genomic instability. Radiat Res 163:316–323
Nagar S, Smith LE, Morgan WF (2003a) Characterization of a novel epigenetic effect of ionizing radiation: the death inducing effect. Cancer Res 63:324–328
Nagar S, Smith LE, Morgan WF (2003b) Mechanisms of cell death associated with death-inducing factors from genomically unstable cell lines. Mutagenesis 18:549–560
Nagar S, Smith LE, Morgan WF (2005) Variation in apoptosis profiles in radiation-induced genomically unstable cell lines. Radiat Res 163:324–331
Nagasawa H, Little J (2002) Bystander effect for chromosomal aberrations induced in wild-type and repair deficient CHO cells by low fluences of alpha particles. Mutat Res 508:121
Nagasawa H, Huo L, Little JB (2003) Increased bystander mutagenic effect in DNA double-strand break repair-deficient mammalian cells. Int J Radiat Biol 79:35–41
Nakanishi M, Tanaka K, Shintani T, Takahashi T, Kamada N (1999) Chromosomal instability in acute myelocytic leukemia and myelodysplastic syndrome patients among atomic bomb survivors. J Radiat Res 40:159–167
Narayanan PK, Goodwin EH, Lehnert BE (1997) Alpha particles initiate biological production of superoxide anions and hydrogen peroxide in human cells. Cancer Res 57:3963–3971
Narayanan PK, LaRue KE, Goodwin EH, Lehnert BE (1999) Alpha particles induce the production of interleukin-8 by human cells. Radiat Res 152:57–63
Pant GS, Kamada N (1977) Chromosome aberrations in normal leukocytes induced by the plasma of exposed individuals. Hiroshima J Med Sci 26: 149–154
Parsons WB, Watkins CH, Pease GL, Childs DS (1954) Changes in sternal bone marrow following roentgen-ray therapy to the spleen in chronic granulocytic leukaemia. Cancer 7:179–189
Plumb M, Cleary H, Wright E (1998) Genetic instability in radiation-induced leukaemias: mouse models. Int J Radiat Biol 74:711–720
Rogakou EP, Pilch DR, Orr AH, Ivanova VS, Bonner WM (1998) DNA double-stranded breaks induce histone H2AX phosphorylation on serine 139. J Biol Chem 273:5858–5868
Salomaa S, Holmberg K, Lindholm C, Mustonen R, Tekkel M, Veidebaum T, Lambert B (1998) Chromosomal instability in in vivo radiation exposed subjects. Int J Radiat Biol 74:771–779
Sax K (1938) Chromosome aberrations induced by X-rays. Genetics 23:494–516
Seymour CB, Mothersill C (1997) Delayed expression of lethal mutations and genomic instability in the progeny of human epithelial cells that survived in a bystander-killing environment. Radiat Oncol Invest 5:106–110
Shao C, Furusawa Y, Aoki M, Matsumoto H, Ando K (2002) Nitric oxide-mediated bystander effect induced by heavy-ions in human salivary gland tumour cells. Int J Radiat Biol 78:837–844
Shao C, Stewart V, Folkard M, Michael BD, Prise KM (2003) Nitric oxide-mediated signaling in the bystander response of individually targeted glioma cells. Cancer Res 63:8437–8442
Sokolov MV, Smilenov LB, Hall EJ, Panyutin IG, Bonner WM, Sedelnikova OA (2005) Ionizing radiation induces DNA double-strand breaks in bystander primary human fibroblasts. Oncogene 24:7257–7265
Sowa Resat MB, Morgan WF (2004) Radiation-induced genomic instability: a role for secreted soluble factors in communicating the radiation response to non-irradiated cells. J Cell Biochem 92:1013–1019
Tawn EJ, Whitehouse CA, Martin FA (2000) Sequential chromosome aberration analysis following radiotherapy – no evidence for enhanced genomic instability. Mutat Res 465:45–51
Ward JF (1988) DNA damage produced by ionizing radiation in mammalian cells: identities, mechanisms of formation, and reparability. Prog Nucleic Acid Res Mol Biol 35:95–125
Watson GE, Lorimore SA, Macdonald DA, Wright EG (2000) Chromosomal instability in unirradiated cells induced in vivo by a bystander effect of ionizing radiation. Cancer Res 60: 5608–5611
Watson GE, Pocock DA, Papworth D, Lorimore SA, Wright EG (2001) In vivo chromosomal instability and transmissible aberrations in the progeny of haemopoietic stem cells induced by high- and low-LET radiations. Int J Radiat Biol 77:409–417

Whitehouse CA, Tawn EJ (2001) No evidence for chromosomal instability in radiation workers with in vivo exposure to plutonium. Radiat Res 156:467–475
Yang H, Asaad N, Held KD (2005) Medium-mediated intercellular communication is involved in bystander responses of X-ray-irradiated normal human fibroblasts. Oncogene 24:2096–2103

7 DNA Methylation Damage: Formation, Repair and Biological Consequences

Markus Christmann, Wynand P. Roos, and Bernd Kaina

Abstract

DNA methylating agents are powerful mutagens and carcinogens. They are endogenously formed, are present in food, beverages and tobacco smoke and are being used in cancer therapy. Methylating agents attack DNA at 13 sites. The main mutagenic and carcinogenic DNA lesion is O^6-methylguanine, whereas both N-methylation lesions and O^6-methylguanine contribute to sister-chromatid exchanges, chromosomal aberrations and cell death. The contribution depends on the relative level of adducts induced (S_N1 versus S_N2 agents), the alkyltransferase (MGMT) and mismatch repair activity as well as base excision repair capacity. The repair of the lesions and the genotoxic pathways triggered by them are discussed.

7.1 Methylating Agents as Environmental Carcinogens

Many compounds found in the environment methylate DNA. This methylation occurs on different sites of the DNA. Depending on how these lesions are processed by the cell, the lesions can give rise either to mutation or to cell death. The carcinogenic properties of methylating agents are due to the fact that they induce mutations. The most abundant methylating agents found in the environment are *N*-nitroso compounds.

The sources of *N*-nitroso compounds are food and tobacco smoke. For food, humans are mainly exposed to *N*-nitroso compounds through the consumption of cured meats and different fish products (Fiddler 1975). These compounds can also be formed in the body when nitrite or nitrogen oxides are ingested along with a nitrosable substrate like a secondary amine (Wogan and Tannenbaum 1975). Food contains non-volatile and volatile *N*-nitroso compounds, but it is thought that only the volatile compounds in food are mutagenic and therefore carcinogenic. A large group of volatile *N*-nitroso compounds found in food is the carcinogenic *N*-nitrosamines. The most abundant *N*-nitrosamines, in order of decreasing carcinogenicity, are *N*-nitrosodiethylamine (NDEA), *N*-nitrosodimethylamine (NDMA), *N*-nitrosopyrrolidine (NPYR) and *N*-nitrosopiperidine (NPIP) (Jagerstad and Skog 2005). Although

Günter Obe and Vijayalaxmi (Eds.)
Chromosomal Alterations: Methods, Results and Importance in Human Health

the non-volatile food *N*-nitroso compounds are not thought to be carcinogenic, they can be converted to carcinogenic volatile *N*-nitrosamines by heat; an example is *N*-nitrosoproline that is formed while frying bacon owing to the nitrosation of the amino acid proline that then becomes decarboxylated to form NPYR (Walker 1990).

N-Nitrosamines found in tobacco smoke are normally divided into two groups, namely, non-specific volatile nitrosamines and non-volatile tobacco-specific nitrosamines (TSNAs). The non-specific *N*-nitrosamines include NDMA, NDEA, *N*-nitrosoethylmethylamine, *N*-nitrosodiethanolamine, NPYR, and *N*-nitroso-n-butylamine (Mitacek et al. 1999). The four most widely reported TSNAs are *N*-nitrosoanabasine, *N*-nitrosoanabatine, 4-(methylnitrosamino)-1-(3-pyridyl)-1-butanone (NNK), and *N*-nitrosonornicotine (NNN).

N-Nitrosamines require metabolic activation to become mutagenic and carcinogenic. The hydroxylation is catalysed mainly by CYP2E1 (Guengerich and Shimada 1991; Lai and Arcos 1980; Lin et al. 1999), and to a lesser extent by CYP2A6 (Guengerich and Shimada 1991; Kamataki et al. 1999). Because of the involvement of the cytochrome P450 monooxygenase system in the activation of these compounds, it mainly occurs in the liver, but also to a lesser extent in other tissue types.

The DNA lesion mainly responsible for the mutagenicity, and therefore the carcinogenicity, of methylating agents is O^6-methylguanine (O6MeG). O6MeG forms GC-AT point mutations in proliferating cells if it is not repaired by O6MeG–DNA methyltransferase (MGMT). For a more mechanistic discussion on the mutagenicity of methylating agents see Sect. 7.5. The carcinogenic properties of dietary *N*-nitrosamines are strongly supported by experiments in animal models (Bartsch and Montesano 1984). Also, the carcinogenic properties of the TSNAs NNN and NNK have been shown in animals (Hecht et al. 1980).

7.2 Anticancer Drugs

As stated already, depending on how the DNA lesions are processed by the cell, methylating agents can give rise to either mutation or cell death. The death-provoking properties (cytotoxicity) of methylating agents have found an application in cancer therapy. Thus, methylating agents are widely used as anticancer chemotherapeutics as they exhibit antitumour properties. The best known DNA methylating anticancer chemotherapeutics today are procarbazine (PCB), dacarbazine (DIC), streptozotocin (STZ) and temozolomide (TMZ).

PCB(PCZ, N-methyl hydrazine, Natulan®, Matulane®) is used for the treatment of Hodgkin's disease, non-Hodgkin's lymphoma and less frequently for the treatment of brain tumours. PCB requires metabolic activation in order to accomplish its antitumour function. It is metabolized by the cytochrome

P450 isoforms CYP1A and CYP2B to the DNA reactive methyl radicals (Goria-Gatti et al. 1992).

DIC (imidazole carboxamide, dimethyltriazenoimidazole carboxamide, DTIC®-Dome) is also used for the treatment of Hodgkin's disease. Apart from Hodgkin's disease, it is used for the treatment of malignant melanoma, neuroblastoma and soft-tissue sarcomas. DIC is inactive until it is metabolized in the liver by cytochromes P450 to form the reactive N-demethylated species 5-(3-hydroxymethyl-3-methyltriazen-1-yl)-imidazole-4-carboxamide and 5-(3-methyltriazen-1-yl)-imidazole-4-carboxamide (MTIC) (Hill 1975; Skibba et al. 1970). This reaction is predominantly catalysed by CYP1A2, by CYP2E1 at higher substrate concentrations, and CYP1A1 contributes only a little (Reid et al. 1999).

STZ (NSC 85998, Zanosar®) is primarily used for the treatment of pancreatic (islet cell) cancer. The STZ molecule is structurally similar enough to glucose that it is transported into the cell by the glucose transport protein GLUT2. As the islet cells are particularly rich in GLUT2, it is clear why STZ is used for the treatment of this cancer (Brentjens and Saltz 2001; Wang and Gleichmann 1998). It is less frequently used in the treatment of carcinoid tumours, colon cancer, hepatic cancer, Hodgkin's disease, pancreatic (ductal) cancer and prostate cancer. STZ does not require metabolic activation and is known to directly methylate DNA (Bolzan and Bianchi 2002).

TMZ (SCHS2.365, NSC 362856, Temodal®, Temodar®) is the newest of the methylating chemotherapeutics. As it can pass the blood–brain barrier, it is used in the treatment of astrocytoma, glioblastoma and brain metastasis from solid tumours (Newlands et al. 1997). TMZ is also used for the treatment of melanoma. After oral administration, TMZ is rapidly absorbed with almost 100% bioavailability and undergoes spontaneous hydrolysis at physiological pH into its active metabolite, MTIC (Ostermann et al. 2004), and does therefore not need activation.

The cytotoxic action of the methylating agent chemotherapeutics is due to the DNA lesion O6MeG. As MGMT repairs this lesion, tumours with high MGMT will be resistant to these agents. A second requirement for the cytotoxicity of these agents is mismatch repair (MMR). A more detailed explanation on the mechanism of cytotoxicity of methylating agents will be given later in this chapter.

7.3 DNA Lesions, Sites of Attack

N-Nitrosamines and *N*-nitrosamides are the methylating agents with the highest biological relevance. *N*-Nitrosamides are unstable, undergo spontaneous hydrolysis and form reactive electrophilic species (e.g. carbenium cations). In contrast, *N*-nitrosamines are stable and need metabolic activation in order to form reactive electrophilic species. These simple methylating agents reacts with 13 different sites on the DNA (Fig. 7.1) via

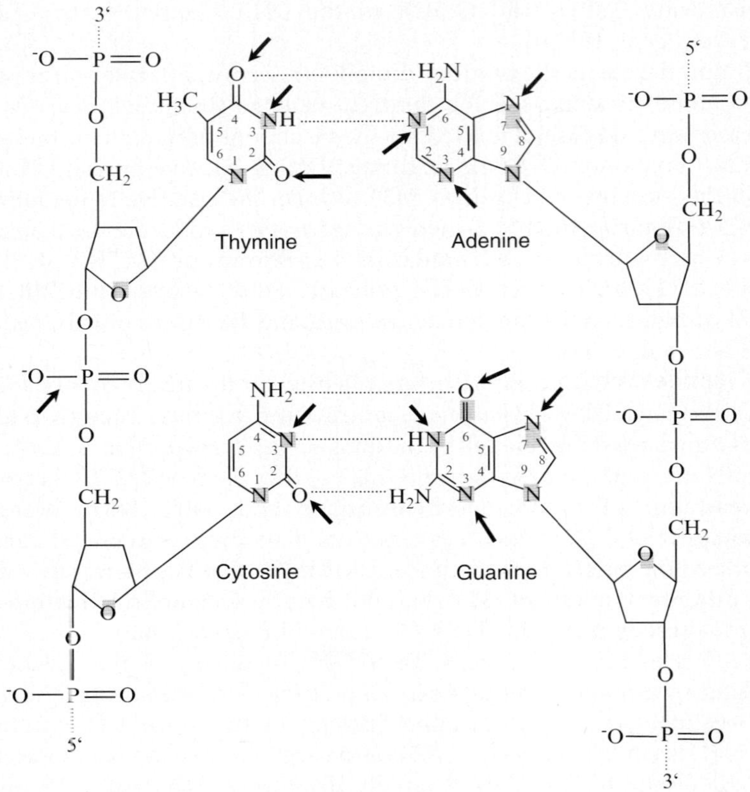

Fig. 7.1 Sites of attack of methylating agents

nucleophilic substitution reactions, leading to the formation of N-methylated and O-methylated purines and pyrimidines as well as phosphotriesters (reviewed in Beranek 1990).

Depending on the electrophilicity of the methylating agents, they react via either S_N1 or S_N2 mechanisms. S_N1 reactions follow first-order kinetics that is dependent on the formation of an electrophilic carbocation. This intermediate is covalently bound by the nucleophilic centre on the DNA. S_N2 reactions are bimolecular and dependent on steric accessibility, involving a transition state where the electrophile attacks the nucleophilic centre. This leads to formation of a transition complex, which thereafter forms the methylated product and a free leaving group. Therefore, the reactivity of different electrophilic methylating agents depends on the stability of the carbocation and/or the strength of the leaving group. Thus, nitrosoureas such as *N*-methyl-*N'*-nitro-*N*-nitrosoguanidine (MNNG) preferentially bind to oxygen via S_N1 reactions, whereas methyl methanesulfonate (MMS) binds to the most nucleophilic site (N7 position of guanine) and reacts via S_N2 reaction. This different reactivity explains the strong variation between different methylating agents concerning methylation of the O6 and N7 positions of guanine. Because of this, upon MNNG treatment 7–9% of all methylation products are O6MeG, whereas for MMS treatment only 0.3% are O6MeG (Beranek 1990).

Methyl DNA adducts have different stabilities. Therefore *N3*-methylpurines and *N7*-methylpurines are easily depurinated at neutral pH. Only some adducts, such as *N7*-methylguanine (N7MeG), *N3*-methyladenine (N3MeA) and O6MeG are stable for longer times ($T_{1/2}$ in vitro 40–80, 7–16 and 20 h, respectively). Besides representing the most stable lesions, N7MeG (80–85%) and N3MeA (8–20%) are also the most frequent methylation lesions. However, the biologically most critical lesion is O6MeG. Another premutagenic methylation lesion is O^4-methylthymine (O4MeT), which is induced in very small amounts (less than 0.4%). O6MeG and O4MeT are repaired by MGMT, N7MeG and N3MeA are repaired by the base excision repair (BER) system and *N1*-methyladenine (N1MeA) and *N3*-methylcytosine (N3MeC) are repaired by the three AlkB homologous (ABH1, ABH2, ABH3) proteins (Fig. 7.2).

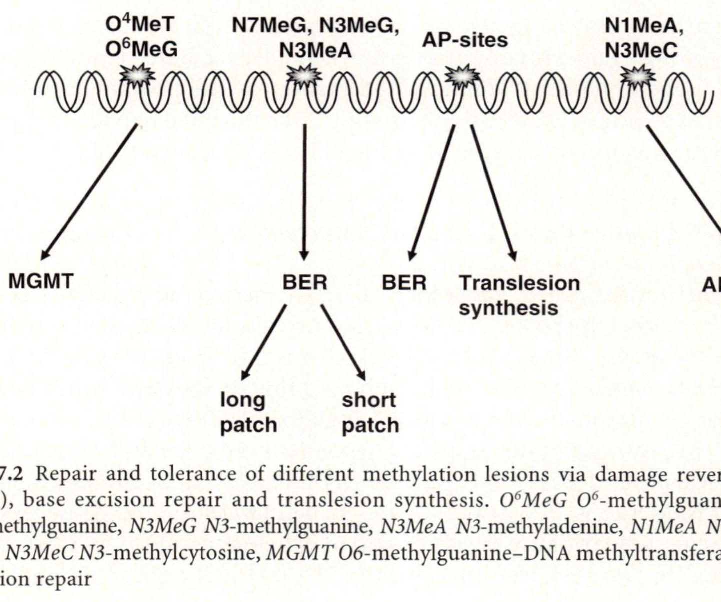

Fig. 7.2 Repair and tolerance of different methylation lesions via damage reversal (MGMT, ABH), base excision repair and translesion synthesis. *O⁶MeG* O^6-methylguanine, *N7MeG N7*-methylguanine, *N3MeG N3*-methylguanine, *N3MeA N3*-methyladenine, *N1MeA N1*-methyladenine, *N3MeC N3*-methylcytosine, *MGMT O6*-methylguanine–DNA methyltransferase, *BER* base excision repair

7.4 Repair of DNA Methylation Damage

7.4.1 Reversal Repair

7.4.1.1 Single-Step Repair by MGMT

In *Escherichia coli*, resistance to methylating agents is mediated by the induction of the four genes *Ada, alkA, alkB* and *aidB* (reviewed in Sedgwick and Lindahl 2002). Ada repairs O6MeG, O^4-methylthymine and methyl-phosphotriester. In human cells, O^6-methylation lesions are repaired in a single-step reaction by MGMT (also known as ATase, AGT, AGAT), which is the mammalian homologue of the Ada gene product of *E. coli*. The human MGMT protein transfers the methyl group from the methylated guanine in a one-step reaction onto an internal cysteine residue in its active centre (reviewed in Pegg et al. 1995), leading to irreversible inactivation of the MGMT protein. This one-step reaction restores the guanine in the DNA to its unmethylated state. MGMT-deficient cells are unable to repair O6MeG and are therefore sensitive to methylating agents (Day et al. 1980a; Yarosh et al. 1983), while MGMT transfection provokes expression-dependent resistance to O^6-methylating agents (Kaina et al. 1991). MGMT knockout mice are viable, show a higher frequency of spontaneous tumours and are sensitive to treatment with methylating agents (Tsuzuki et al. 1996). Conversely, MGMT overexpressing mice show a reduced frequency of tumours in liver, thymus or skin (Becker et al. 1996; Dumenco et al. 1993; Nakatsuru et al. 1993; Zhou et al. 2001). MGMT overexpression specifically protects against tumour initiation without affecting tumour promotion in two-stage carcinogenesis experiments (Becker et al. 1996, 1997). It also protects against tumour progression (Becker et al. 2003) upon exposure to O^6-methylating agents. MGMT expression is highly regulated by methylation of both the gene and the promoter. Methylation of the promoter provokes inhibition of expression (Qian et al. 1995), which was shown to happen frequently during tumour formation (Hegi et al. 2004, 2005; Toyooka et al. 2001).

7.4.1.2 DNA Damage Reversal by AlkB Homologues

As already mentioned, in *E. coli* repair of DNA methylation damage is also performed by AlkB. Expression of AlkB in human cells increases their resistance to methylating agents (Chen et al. 1994). AlkB specifically repairs methylation damage in both single-stranded and double-stranded DNA and binds preferentially to single-stranded DNA in vitro (Dinglay et al. 2000). AlkB belongs to a superfamily of 2-oxoglutarate- and iron-dependent dioxygenases (Falnes et al. 2002; Trewick et al. 2002). AlkB repairs DNA methylation damage such as N1MeA and N3MeC (Fig. 7.1) in an oxygen-, ketoglutarate- and Fe(II)-dependent reaction, by coupling oxidative decarboxylation of ketoglutarate to hydroxylation of methylated bases (reviewed in Falnes et al. 2002). In humans, eight AlkB homologues were identified, and were designated as ABH1–ABH8 (Duncan et al. 2002; Kurowski et al. 2003). Whereas for ABH1 and ABH4–ABH8 the biological function is unknown, ABH2 and ABH3 belong, similarly to AlkB, to the superfamily of ketoglutarate- and Fe(II)-dependent dioxygenases and repair DNA methylation damage N1MeA and N3MeC (Aas et al. 2003; Duncan et al. 2002). During this repair, ABH2 and ABH3 differ in their template specificity; ABH3 repairs RNA and single-stranded DNA, whereas ABH2 preferentially repairs single-stranded and double-stranded DNA. It is thought that ABH2 and ABH3 remove the methylation in an α-ketoglutarate-dependent mechanism similar to the AlkB protein, involving oxidative demethylation leading to direct damage reversal and restoration of the undamaged base. By generation of ABH2- and ABH3-knockout mice, it was shown recently that mouse ABH2 acts as the primary oxidative demethylase for repairing N1MeA and N3MeC lesions in DNA (Ringvoll et al. 2006).

7.4.2 Mismatch Repair

The MMR system is responsible for removal of DNA base mismatches caused by spontaneous and induced base deamination, oxidation, methylation and replication errors (Modrich and Lahue 1996; Umar and Kunkel 1996). The main targets of MMR are base mismatches such as G/T, arising from deamination of 5-methylcytosine, G/G, A/C and C/C (Fang and Modrich 1993).

The recognition of mismatches is performed by the so-called MutSα complex, which binds to the lesions. MutSα is composed of the bacterial MutS homologous proteins MSH2 (Fishel et al. 1993; Leach et al. 1993) and MSH6 (Palombo et al. 1995). For the efficient binding to mismatches, phosphorylation of the MutSα complex is required (Christmann et al. 2002). MMR removes bases specifically from the newly synthesized DNA strand; however, the exact mechanism of the strand discrimination is unclear, but it involves the formation of an ADP/ATP-dependent sliding clamp (Gradia et al. 1999). In this translocation model, MutSα uses the energy gained by ATP hydrolysis to actively translocate along the DNA from the site of mismatch recognition to a site responsible for signalling the strand specificity (most likely a DNA single-strand break), where the assembly of the MutLα complex occurs (Blackwell et al. 2001). MutLα consists of the MutL homologous MMR proteins MLH1 and PMS2 (Li and Modrich 1995; Nicolaides et al. 1994; Papadopoulos et al. 1994). The excision of the DNA strand containing the mispaired base is performed by exonuclease I (Genschel et al. 2002) and new synthesis of DNA is performed by Polδ (Longley et al. 1997).

MMR does not play a direct role in the repair of methylated DNA; it rather counteracts the protective effect of MGMT by converting O6MeG into a cytotoxic secondary lesion. It was shown that cells which have a defect in MMR are resistant to methylating agents (Branch et al. 1993; Dosch et al. 1998). In MMR-deficient cells, the so-called tolerant phenotype, O6MeG induces mutations owing to the loss of its cytotoxic properties. As mentioned above, MMR is able to convert O6MeG into a cytotoxic lesion. This is initiated by binding of MutSα to O6MeG paired with cytosine or thymine (Duckett et al. 1996). During the subsequent repair of these faulty base pairs, MMR inserts thymine opposite O6MeG. During the following replication cycles, this leads to futile cycles of DNA repair (Karran and Bignami 1994) leading to the generation of double-strand breaks (DSBs), which represent the ultimate cytotoxic DNA lesions (Kaina et al. 1997).

In addition, the binding of MMR to O6MeG paired with thymine signals different mechanisms such as cell cycle regulation and apoptosis. It was shown that treatment with MNNG arrested only MMR-proficient cells in the G2-phase after treatment, which was based on activation of ATR and CHK1 (Stojic et al. 2004). ATR directly interacts with MSH2 (Wang and Qin 2003) and is recruited to O6MeG/T mismatches (Yoshioka et al. 2006). In addition, processing of O6MeG/T by MMR is also necessary for p38K activation, which can signal cell cycle arrest in glioma and colon cancer cell lines treated with TMZ (Hirose et al. 2003).

7.4.3 Base Excision Repair

While O6MeG is subject to direct reversal, lesions at the N-position (N7MeG, N3MeA and *N*3-methylguanine, N3MeG) are mainly repaired via the BER pathway. The BER mechanism, which is shown in Fig. 7.3, depends on the following steps: recognition, base removal and incision. The first step in BER is carried out by specific DNA glycosylases, which recognize and remove damaged bases by hydrolysing the N-glycosidic bond. In human cells, 12 different DNA glycosylases have been found which are characterized by different substrate specificities and their ability to cut the DNA backbone. Type I glycosylases remove modified bases leaving an AP site in DNA, whereas type II enzymes remove the base and subsequently cleave the AP site by an endogenous 3′ endonuclease activity giving rise to a single-strand break. For type I glycosylases, incision into the phosphodiester bond of the AP site occurs by AP endonuclease (APE1 alias APEX, Ref-1), resulting in 5′-deoxyribose-5-phosphate (5′-dRP) and 3′OH (Wilson and Barsky 2001). The glycosylase responsible for the removal of N-methylated bases is called methylpurine glycosylase [MPG, alias 3-alkyladenine DNA glycosylase (AAG) or alkylpurine DNA N-glycosylase (APNG)]. MPG is the homologue of *E. coli* AlkA and is a type I glycosylase (O'Connor 1993). It releases N3MeA, N7MeG, and N3MeG from DNA (Chakravarti et al. 1991) and imbalanced expression renders cells more sensitive to methylating agents (Coquerelle et al. 1995). MPG-knockout mice are viable and show an increased level of mutations following MMS treatment (Elder et al. 1998).

The insertion of the first nucleotide is not dependent on the chemical structure of the AP site and is performed by DNA polymerase β (Polβ), which

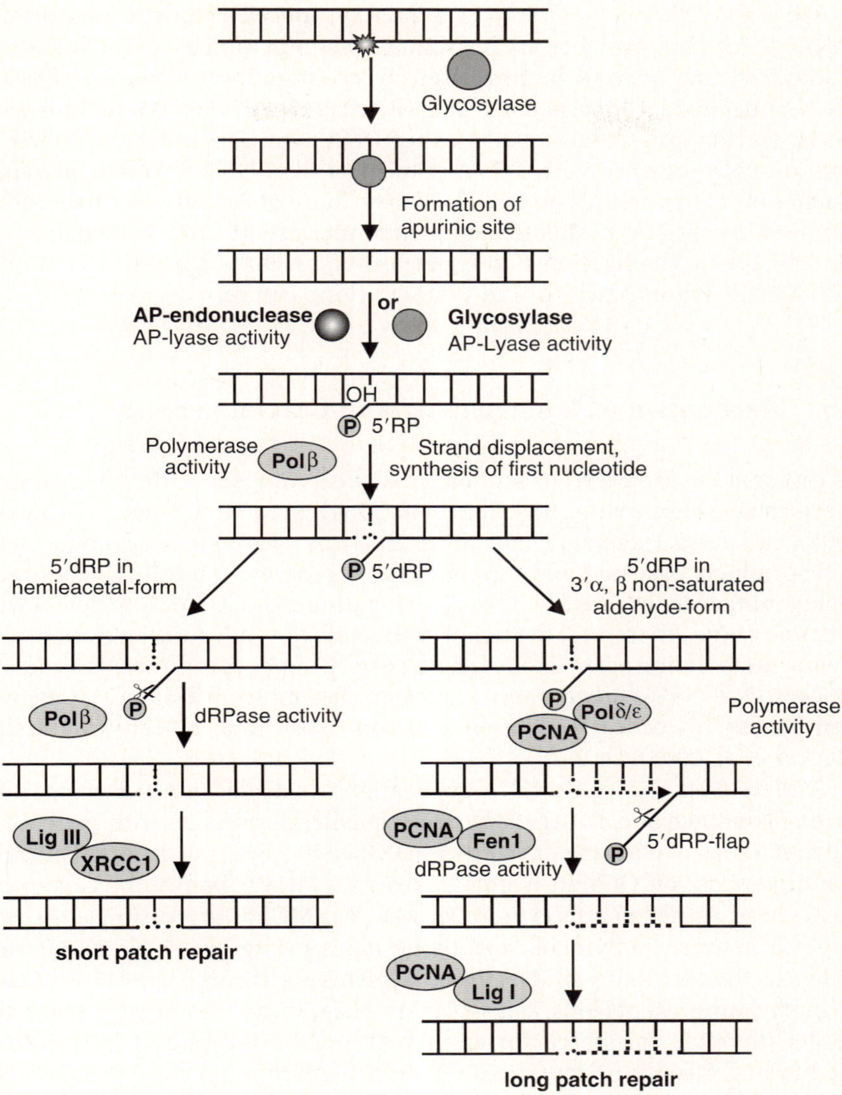

Fig. 7.3 Base excision repair: short-patch and long-patch pathway

inserts a single nucleotide (Dianov et al. 1999; Sobol and Wilson 2001). The critical step in the decision between short-patch and long-patch BER is the removal of 5′-dRP upon the insertion of the first nucleotide. Polβ has lyase activity and is able to catalyse the release of the hemiacetal form of 5′-dRP residues from incised AP sites by β-elimination (Matsumoto and Kim 1995;

Prasad et al. 1998; Sobol et al. 2000). Oxidized or reduced AP sites, 3′-unsaturated aldehydes or 3′-phosphates are resistant to β-elimination by Polβ (Nakamura et al. 2000) and have to be processed by proliferating cell nuclear antigen (PCNA) dependent long-patch repair (Frosina et al. 1996; Matsumoto et al. 1999). During long-patch repair further DNA synthesis is accomplished by Polε or Polδ together with PCNA (Stucki et al. 1998), resulting in repair patches of up to ten nucleotides. The deoxyribose phosphate flap structure is removed by PCNA-stimulated flap endonuclease (FEN1; Klungland and Lindahl 1997). The ligation step is performed by DNA ligase III in complex with XRCC1 (short patch) or DNA ligase I (long patch).

7.5 Mechanism of Mutagenicity and Carcinogenicity

As discussed in the previous sections, many carcinogens in the environment (Bartsch and Montesano 1984; Peto et al. 1984), in tobacco smoke (DeMarini 2004) and food (Jagerstad and Skog 2005) as well as endogenous metabolic products (Marnett and Burcham 1993) generate O6MeG. This lesion is highly mutagenic because of its mispairing properties. O6MeG mispairs with thymine and is, therefore, the major source of GC→AT point mutations following methylating agent exposure. It is a very important tumorigenic lesion (Becker et al. 1996) since point mutations arising from O6MeG contribute significantly to cancer induction (Gerson 2004) and cancer progression (Becker et al. 2003).

O6MeG does not cause DNA replication blockade, but is a mispairing lesion (Toorchen and Topal 1983). Consequently, it mispairs with thymine in the first S phase following exposure of DNA to methylating agents. This mispair only occurs if O6MeG is not repaired by MGMT before the cell enters the S phase. As not all cells express MGMT, [the MGMT expression levels vary 100-fold between individuals and tissue types (Margison et al. 2003)], and owing to the fact that MGMT can be depleted by chronic exposure of cells to methylating agents this can occur at a significant frequency. Once the O6MeG/thymine mismatch forms, it may not be recognized by MMR as the binding affinity of MutSα to O6MeG/thymine is much lower than the binding affinity of MutSα to guanine/thymine mismatches (Dosch et al. 1998). Therefore, it may remain in the DNA through the subsequent cell cycle phases and in the next S phase thymine will pair with adenine, while O6MeG will again mispair with thymine. Subsequent mispairs of O6MeG with thymine depend on the stability (half life) of O6MeG and the replication rate of the cells. MGMT is still able to remove the methyl group from guanine when it is mispaired with thymine. From the preceding discussion of the mechanism it is clear that O6MeG-mediated GC→AT point mutation formation can only occur in the absence of MGMT, and that replication is absolutely required for the formation of these mutations.

Mutations cause cancer owing to the inactivation of proteins that suppress cancer, or owing to hyperactivation of proteins that promote cancer. An example of the former is the cancer-suppressor p53 (Carson and Lois 1995). p53 plays important roles in cell cycle progression, DNA repair and apoptosis. By inactivation of this protein by mutation, the chance of a cell becoming cancerous increases significantly. An example where a mutation may lead to hyperactivation of a protein is when epidermal growth factor receptor (EGFR) is mutated and therefore signals continuously. This has been observed in many tumour types (Garcia de Palazzo et al. 1993; Ro et al. 1988; Xu et al. 1984).

Apart from the fact that humans are exposed to environmental and endogenous carcinogenic methylating compounds every day, methylating agents are also used in the clinic for the treatment of cancer. The cytotoxic properties of methylating agents will be discussed in the next section. The mutagenic properties of these chemotherapeutics may lead to therapy-based carcinogenesis by exactly the same mechanism described above. Not only may these methylating agents induce tumour formation in other sites of the body; they also induce mutations in the tumour itself (Lees et al. 2004).

7.6 Mechanism of Cytotoxicity of Methylating Agents by Apoptosis

PCB, DIC, STZ and TMZ are used in cancer therapy owing to the cytotoxicity of these methylating agents. The critical cytotoxic lesion is O6MeG (Day et al. 1980b; Glassner et al. 1999; Kaina et al. 1991; Qin et al. 1999; Sklar and Strauss 1981). Today, O6MeG provides perhaps the best understood example for a well-defined DNA lesion that triggers apoptosis. Evidence that O6MeG is not only a genotoxic lesion in cells and mice (Glassner et al. 1999; Kaina et al. 1991; Qin et al. 1999) but that it also triggers apoptosis was provided by comparing cells deficient and proficient for the DNA repair protein MGMT (Kaina et al. 1997; Meikrantz et al. 1998; Tominaga et al. 1997). Cells expressing MGMT show very low levels of apoptosis after treatment with O^6-methylating agents. This clearly shows that O6MeG induces apoptosis. However, the lesion does not directly induce apoptosis; it needs to be converted into a secondary apoptosis signalling lesion. Therefore, O6MeG is considered as a proapoptotic DNA lesion.

A requirement for apoptosis triggered by O6MeG is MMR (Hickman and Samson 1999; Pepponi et al. 2003). This is because O6MeG mispairs, during replication, with thymine. This gives rise to GC→AT point mutations in the absence of MMR. In the presence of MMR, the O6MeG/thymine mismatch is recognized by the heterodimeric protein complex MutSα, and MutSα binding has been shown to directly correlate with cell survival (Dosch et al. 1998). Thereafter, MutLα assembles to the complex and along with exonuclease I and Polδ removes the thymine opposite O6MeG. Because of the mispairing properties of O6MeG, thymine will be reinserted. At this stage it is reasonable to hypothesize that this results in a futile repair cycle (Karran and Bignami 1994) that ends up with the formation of severe secondary lesions (presumably gaped DNA) that significantly interfere with the subsequent DNA replication. Blocking of replication at this site may lead to collapse of replication forks, ending up with the formation of DSBs. This has indeed been shown to occur in MGMT-deficient MMR-proficient cells (Ochs and Kaina 2000). MMR-deficient cells are highly resistant to O6MeG DNA lesions because they do not undergo apoptosis. They survive (and thus tolerate O6MeG lesions) at the expense of a high level of GC→AT point mutations (Karran and Bignami 1994). Therefore, MMR-driven apoptosis is believed to eliminate premutated cells from the population, reducing the mutation pressure. It should be noted, however, that only a small amount of O6MeG (about 30 lesions) gives rise to detectable genotoxic effects (sister-chromatid exchanges, SCEs) without an apparent threshold dose, whereas apoptosis is induced at a higher dose level, displaying a threshold. This explains why a single lesion, O6MeG, is so powerful mutagenic and carcinogenic and also highly efficient in evoking the apoptotic pathway mediated by MMR.

The model of O6MeG-triggered apoptosis implies that DNA replication is essential. This has been shown experimentally. Using proliferating and non-proliferating human peripheral blood lymphocytes (PBLCs), Roos et al. (2004) showed that for clinically relevant doses of the methylating anticancer drug TMZ only the proliferating PBLCs underwent apoptosis. Thus, proliferation and MutSα-dependent MMR are essential components of O6MeG-triggered apoptosis. A wave of DSBs preceded apoptosis (Roos et al. 2004).

MutSα not only binds to O6MeG/thymine mispairs but also to UV-C (Hoffman et al. 2005), cisplatin (Claij and te Riele 2004), *N*-acetyl-2-aminofluorene and benzo[*a*]pyrene adducts in DNA (Li et al. 1996). On the basis of this finding it has been speculated that MutSα may directly signal downstream to the apoptotic machinery, thereby acting as a sensor for non-repaired DNA damage. Contradicting this "sensor" model are the findings that (1) there is no general cross-resistance of tolerant cells to UV-C, cisplatin, benzo[a]pyrene and other agents, (2) apoptosis is a late response occurring in the first to fourth post-treatment cell cycle and (3) DNA replication is required for O6MeG-triggered apoptosis. For isogenic embryonic stem cells expressing and not expressing MSH2, it has been shown that MMR does not impact on cisplatin resistance (Claij and te Riele 2004). Therefore, the model of futile MMR on O6MeG/thymine lesions causing DSBs at blocked replication forks that ultimately trigger apoptosis is favoured (Ochs and Kaina 2000). DSBs were indeed observed following O6MeG induction, and they preceded apoptosis in a replication-dependent manner (Ochs and Kaina 2000; Roos et al. 2004).

The apoptotic pathways employed by cells following the exposure to O^6-methylating agents differ depending on their p53 status. In cells mutated in the p53 gene, cells make use of the mitochondrial apoptotic pathway (Ochs and Kaina 2000). Consequently, following O6MeG induction, Chinese hamster ovary

(CHO) cells (mutated for p53) show a decline in Bcl-2, which increases the Bax to Bcl-2 ratio and allows for the release of cytochrome *c* from the mitochondria. This release of cytochrome *c* leads to the activation of the apoptosome consisting of Apaf-1, ATP, cytochrome *c* and procaspase-9. In turn, caspase-3 becomes activated and cleaves the inhibitor of caspase-activated DNase (ICAD), thereby freeing caspase-activated DNase (CAD). CAD finally degrades DNA, resulting in the characteristic apoptotic DNA fragmentation (Ochs and Kaina 2000). In human PBLCs and lymphoblastoid cell lines wild-type for p53, O6MeG was shown to activate the death receptor apoptotic pathway (Fas/CD95/Apo-1) (Dunkern et al. 2003; Roos et al. 2004). In line with this is the finding that execution of apoptosis can be attenuated by blocking Fas (Roos et al. 2004). Thus the data show that the same DNA lesion, O6MeG, is able to trigger both the mitochondrial and the death receptor (Fas, CD95, Apo-1) pathway, stressing the importance of the cellular background.

Since O^6-methylating agents (TMZ, DIC) are being used for the treatment of glioblastomas and malignant melanomas, the question arises as to what is the apoptotic mechanism evoked in these tumours. In gliomas, O6MeG was shown to trigger apoptosis by activating both the mitochondrial (in p53-mutated tumour cells) and the p53-controlled Fas-dependent pathway (Roos et al. 2007). MGMT and p53 are therefore main components of TMZ resistance of gliomas.

7.7 Mechanism of Formation of SCEs and Aberrations

7.7.1 Critical Lesions

The identification of methylation lesions responsible for recombinogenic and clastogenic effects of the agents was based on molecular dosimetry (van Zeeland et al. 1983; Vogel and Natarajan 1979) and later on the comparison of genetically engineered isogenic cell lines that are MGMT-deficient, MGMT- proficient and MGMT-tolerant (MGMT-deficient, MMR-impaired) (reviewed in Kaina et al. 1993). The data revealed that (1) agents like MMS and dimethyl sulfate (DMS) that predominantly induce N-methylations (S_N2 agents) exhibit high clastogenic but low point-mutagenic potency, indicating that N-methylation lesions are responsible for clastogenic effects; (2) agents like 1-methyl-1-nitrosourea (MNU) and MNNG that induce a higher level of O-methylations (S_N1 agents) are clastogenic and point-mutagenic, confirming that O-methylations are responsible for the induction of point-mutations; (3) cells deficient in MGMT are highly sensitive to the clastogenic and SCE-inducing effects of S_N1 agents (e.g. MNU and MNNG) compared with the same cells expressing MGMT. The same holds true for tolerant cells (Kaina 2004). From this it has been concluded that O6MeG (the minor lesion O4MeT is usually neglected because of its rarity) is a critical lesion leading to SCEs and chromatid-type aberrations. Additionally, it was shown that MGMT-deficient Chinese hamster cells treated with MNU or MNNG induced SCEs only in the second posttreatment cell cycle detectable in the second mitosis following treatment. In contrast, MMS is able to induce SCEs both in the treatment and in the posttreatment cell cycle (Kaina and Aurich 1985). Similar data were obtained for chromosomal aberrations for which the frequency after treatment with MNNG was low in the first mitosis and much higher in the second mitosis following treatment (Kaina 2004). Interestingly, if the same cells were transfected with a MGMT expression vector, aberrations in M2 cells were reduced to nearly background levels (Kaina 2004). These data indicated that O6MeG needs two replication cycles in order to be converted into SCEs and chromatid-type aberrations. The data also revealed that the methylation lesion O6MeG is converted during the first round of replication into a secondary DNA lesion that acts as the ultimate trigger for either SCEs or aberrations (Kaina 1985b; Kaina et al. 1993). In summary, the data showed that in cells deficient for MGMT, O6MeG is the predominant recombinogenic and clastogenic lesion, with the effect only being detectable in the second mitosis following treatment. In cells that are MGMT-proficient, N-methylation products (N7MeG, N3MeA, N3MeG) appear to be the predominant SCE and clastogenic lesions, and provoke the genotoxic effect already in the treatment cell cycle, visible as SCEs or aberrations in the first mitosis following treatment. It should be noted that the dose of the mutagenic agent is also important. If the dose of the methylating agent is high, N-methylations will also contribute to genotoxicity even in MGMT-expressing cells.

On the basis of these findings it becomes clear that in human fibroblasts and primary Chinese hamster cells, both expressing MGMT, the highest clastogenic response is observed in the first mitosis following treatment, which is due to non-repaired N-methylation lesions. In contrast, MGMT-deficient V79 and CHO cells (and many other established cell lines) display the highest clastogenic effect in the second mitosis following treatment, which is due to non-repaired O6MeG lesions (Kaina 2004). For S_N2 agents (MMS, DMS), the aberration maximum is expected to occur in the first mitosis following treatment because of the relatively high number of N-alkylations compared with the amount O6MeG the agents induce in DNA.

7.7.2 Involvement of MMR in Clastogenicity

MGMT-deficient cells defective in MMR display lower MNNG-induced aberration frequencies than MGMT-deficient cells not defective in MMR (Armstrong and Galloway 1996; Galloway et al. 1995; Kaina 2004; Kaina et al. 1995); therefore, it has been suggested that MMR is essential for the conversion of O6MeG lesions into SCEs and aberrations (Kaina et al. 1997). The process is similar to the one described above in explaining the killing effect of the lesion. The heterodimeric complex MutSα, composed of MSH2 and MSH6, binds to O6MeG/thymine mispairs, which are formed during replication of O6MeG lesions. MLH1 and PMS2 assemble to the lesion (Sibghat-Ullah et al. 2001) in order to repair the mismatch. This, however, is a futile process generating secondary lesions that interfere with the subsequent round of DNA replication, leading to DSBs (Kaina et al. 1997; Ochs and Kaina 2000). This model provides an explanation of why SCEs and aberrations are formed in response to O6MeG in a replication-dependent manner almost exclusively in the first posttreatment cell cycle (appearing in the second mitosis following treatment). Although this model still awaits further verification, it provides a firm basis for explaining most of the data obtained in this field.

7.7.3 Conversion Rates

The probability that O6MeG will be converted into SCEs has been determined in two independent studies. For MGMT-deficient CHO-9 cells, a correlation was found between 30 O6MeG adducts producing one SCE (30:1) (Kaina et al. 1993). For human glioblastoma cells, it was reported that 42 O6MeG adducts produce one SCE (42:1) (Rasouli-Nia et al. 1994). This is quite a high conversion frequency compared, for example, with that of cyclobutane dimers, for which a relation of 600:1 was estimated (De Weerd-Kastelein et al. 1977). It should be noted, however, that this relation might need to be reestimated in view of the fact that UV-C induced 5-bromo-2′-deoxyuridine (BrdU) photolysis may contribute to SCE formation (Wojcik et al. 2004). The conversion probabilities for chromosomal aberrations are much lower, and are estimated to be in the range of 22,000:1 and 147,000:1 for aberrations induced in the second and the first mitosis following treatment, respectively (Kaina et al. 1993). For lethal events, the values between 360:1 (Kaina et al. 1993) and 6,650:1 (Rasouli-Nia et al. 1994) have been calculated. The close correlation between O6MeG lesions and SCEs is striking. Taking into account that O6MeG mispairs in only a third of base pairings (provided equal amounts of deoxynucleotides are present) (Abbott and Saffhill 1979), the probability of the mispaired lesion O6MeG/thymine converting into SCEs would even be higher, i.e. about 10:1.

O6MeG/thymine mispaired lesions are obviously extremely potent in inducing SCEs. The reason for this may be found in the processing of O6MeG/thymine lesions by MMR. Thus, it is tempting to speculate that MMR-generated lesions in the DNA cause a severe block in DNA replication, which is circumvented by a recombination process that finally leads to a SCE (Fig. 7.4). There are many lesions that block DNA replication although they seem to induce SCEs at a much lower level, such as cyclobutane pyrimidine dimers. It may therefore be speculated that most of the presumed replication-blocking lesions, if they encounter a replication fork, are tolerated by translesion synthesis (*replication bypass*) accomplished by low-fidelity SOS polymerases (Lehman 2002), reading through non-instructive lesions without initiating the recombination process that leads to SCEs. The MMR-mediated lesion induced by O6MeG/thymine can presumably not be handled

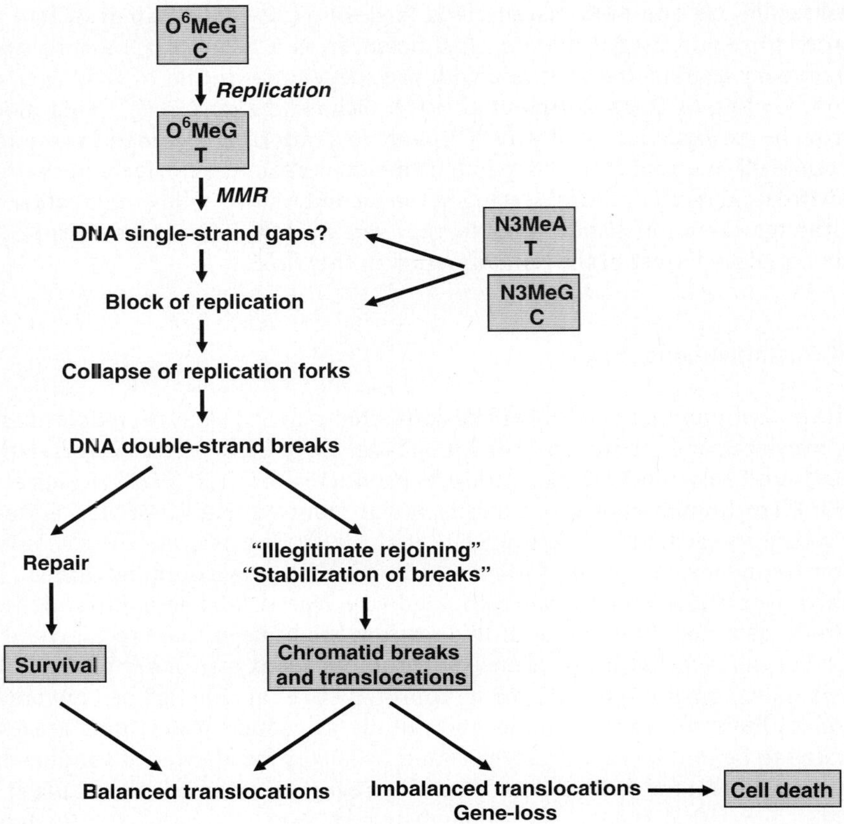

Fig. 7.4 Model of conversion of O^6-methylguanine lesions into DNA double-strand breaks and, further, into chromosomal aberrations and cell-killing effects. *MMR* mismatch repair

by SOS polymerases and, therefore, recombination bypass will be the preferred mode of lesion tolerance. It should be noted that the doses of MNNG and MNU inducing SCEs in the second mitosis following treatment do not cause any significant inhibition of DNA replication (as measured by thymidine incorporation and the delay in entry of BrdU-labelled cells into mitosis) (Kaina 1985a; Kaina and Aurich 1985), suggesting that recombination bypass leading to SCEs is a very efficient and fast process of lesion circumvention.

7.7.4 Critical Role of Inhibition of Replication Fork Movement

While SCEs are induced at very low dose levels, and are therefore a very sensitive indicator of DNA damage, higher doses of MNNG or MNU are required for the induction of chromosomal aberrations (Kaina 1985a). If we compare

SCEs and aberrations with DNA synthesis inhibition, it becomes clear that DNA synthesis is inhibited in the dose range that induces aberrations in the first mitosis following treatment (Kaina 1985a, 1998). On the basis of the overlap in dose–response of replication inhibition and aberrations in the first mitosis following treatment, it has been proposed that inhibition of replication fork movement causes nuclease attack at stalled replication forks that results in DSBs, which cause chromatid-type aberrations (Kaina 1998). Various N-methylation lesions such as N3MeA block DNA replication. Also, apurinic sites resulting from spontaneous hydrolysis of N-methylpurines such as N7MeG and N3MeA are critical candidates for blocking DNA replication, thus leading to DSBs. If they are not repaired, or are illegitimately recombined, these DSBs will be translated into chromatid-type aberrations in the first mitosis following treatment. A similar process may give rise to chromatid-type aberrations in the second mitosis following treatment if O6MeG/thymine lesions processed by MMR block DNA replication in the second cell cycle after treatment (Kaina et al. 1997). It will be an interesting topic of future research to figure out the molecular players and pathways that are involved in this highly vulnerable process.

Acknowledgement: This work was supported by the Deutsche Forschungsgemeinschaft, grants KA724/13-1 and 13-2.

References

Aas P.A, Otterlei M, Falnes PO, Vagbo CB, Skorpen F, Akbari M, Sundheim O, Bjoras M, Slupphaug G, Seeberg E et al (2003) Human and bacterial oxidative demethylases repair alkylation damage in both RNA and DNA. Nature 421:859–863

Abbott PJ, Saffhill R (1979) DNA synthesis with methylated poly(dC-dG) templates. Evidence for a competitive nature to miscoding by O6-methylguanine. Biochim Biophys Acta 562:51–61

Armstrong MJ, Galloway SM (1996) Mismatch repair provokes chromosome aberrations in hamster cells treated with methylating agents or 6-thioguanine, but not with ethylating agents. Mutat Res 373:167–178

Bartsch H, Montesano R (1984) Relevance of nitrosamines to human cancer. Carcinogenesis 5:1381–1393

Becker K, Dosch J, Gregel CM, Martin BA, Kaina B (1996) Targeted expression of human O(6)-methylguanine-DNA methyltransferase (MGMT) in transgenic mice protects against tumor initiation in two-stage skin carcinogenesis. Cancer Res 56:3244–3249

Becker K, Gregel CM, Kaina B (1997). The DNA repair protein O6-methylguanine-DNA methyltransferase protects against skin tumor formation induced by antineoplastic chloroethylnitrosourea. Cancer Res 57:3335–3338

Becker K, Gregel C, Fricke C, Komitowski D, Dosch J, Kaina B (2003) DNA repair protein MGMT protects against N-methyl-N-nitrosourea-induced conversion of benign into malignant tumors. Carcinogenesis 24:541–546

Beranek DT (1990) Distribution of methyl and ethyl adducts following alkylation with monofunctional alkylating agents. Mutat Res 231:11–30

Beranek DT, Weis CC, Swenson DH (1980) A comprehensive quantitative analysis of methylated and ethylated DNA using high pressure liquid chromatography. Carcinogenesis 1:595–606

Blackwell LJ, Bjornson KP, Allen DJ, Modrich P (2001) Distinct MutS DNA-binding modes that are differentially modulated by ATP binding and hydrolysis. J Biol Chem 276:34339–34347
Bolzan AD, Bianchi MS (2002) Genotoxicity of streptozotocin. Mutat Res 512:121–134
Branch P, Aquilina G, Bignami M, Karran P (1993) Defective mismatch binding and a mutator phenotype in cells tolerant to DNA damage. Nature 362:652–654
Brentjens R, Saltz L (2001) Islet cell tumors of the pancreas: the medical oncologist's perspective. Surg Clin North Am 81:527–542
Carson DA, Lois A (1995) Cancer progression and p53. Lancet 346:1009–1011
Chakravarti D, Ibeanu GC, Tano K, Mitra S (1991) Cloning and expression in Escherichia coli of a human cDNA encoding the DNA repair protein N-methylpurine-DNA glycosylase. J Biol Chem 266:15710–15715
Chen BJ, Carroll P, Samson L (1994) The Escherichia coli AlkB protein protects human cells against alkylation-induced toxicity. J Bacteriol 176:6255–6261
Christmann M, Tomicic MT, Kaina B (2002). Phosphorylation of mismatch repair proteins MSH2 and MSH6 affecting MutSalpha mismatch-binding activity. Nucleic Acids Res 30:1959–1966
Claij N, te Riele H (2004). Msh2 deficiency does not contribute to cisplatin resistance in mouse embryonic stem cells. Oncogene 23:260–266
Coquerelle T, Dosch J, Kaina B (1995) Overexpression of N-methylpurine-DNA glycosylase in Chinese hamster ovary cells renders them more sensitive to the production of chromosomal aberrations by methylating agents – a case of imbalanced DNA repair. Mutat Res 336:9–17
Day RS III, Ziolkowski CHJ, Scudiero DA, Meyer SA, Lubiniecki AS, Girardi AJ, Galloway SM, Bynum GD (1980a) Defective repair of alkylated DNA by human tumour and SV40-transformed human cell strains. Nature 288:724–727
Day RS III, Ziolkowski CHJ, Scudiero DA, Meyer SA, Mattern MR (1980b) Human tumor cell strains defective in the repair of alkylation damage. Carcinogenesis 1:21–32
DeMarini DM (2004) Genotoxicity of tobacco smoke and tobacco smoke condensate: a review. Mutat Res 567:447–474
De Weerd-Kastelein EA, Keijzer W, Rainaldi G, Bootsma D (1977) Induction of sister chromatid exchange in Xeroderma pigmentosum cells after exposure to ultraviolet light. Mutat Res 45:253–261
Dianov GL, Prasad R, Wilson SH, Bohr VA (1999) Role of DNA polymerase beta in the excision step of long patch mammalian base excision repair. J Biol Chem 274:13741–13743
Dinglay S, Trewick SC, Lindahl T, Sedgwick B (2000) Defective processing of methylated single-stranded DNA by E. coli AlkB mutants. Genes Dev 14:2097–2105
Dosch J, Christmann M, Kaina B (1998) Mismatch G-T binding activity and MSH2 expression is quantitatively related to sensitivity of cells to methylating agents. Carcinogenesis 19:567–573
Duckett DR, Drummond JT, Murchie AI, Reardon JT, Sancar A, Lilley DM, Modrich P (1996) Human MutSalpha recognizes damaged DNA base pairs containing O6-methylguanine, O4-methylthymine, or the cisplatin-d(GpG) adduct. Proc Natl Acad Sci USA 93:6443–6447
Dumenco LL, Allay E, Norton K, Gerson SL (1993) The prevention of thymic lymphomas in transgenic mice by human O6-alkylguanine-DNA alkyltransferase. Science 259:219–222
Duncan T, Trewick SC, Koivisto P, Bates PA, Lindahl T, Sedgwick B (2002) Reversal of DNA alkylation damage by two human dioxygenases. Proc Natl Acad Sci USA 99:16660–16665
Dunkern T, Roos W, Kaina B (2003) Apoptosis induced by MNNG in human TK6 lymphoblastoid cells is p53 and Fas/CD95/Apo-1 related. Mutat Res 544:167–172
Elder RH, Jansen JG, Weeks RJ, Willington MA, Deans B, Watson AJ, Mynett KJ, Bailey JA, Cooper DP, Rafferty JA et al (1998) Alkylpurine-DNA-N-glycosylase knockout mice show increased susceptibility to induction of mutations by methyl methanesulfonate. Mol Cell Biol 18:5828–5837
Falnes PO, Johansen RF, Seeberg E (2002) AlkB-mediated oxidative demethylation reverses DNA damage in Escherichia coli. Nature 419:178–182
Fang WH, Modrich P (1993) Human strand-specific mismatch repair occurs by a bidirectional mechanism similar to that of the bacterial reaction. J Biol Chem 268:11838–11844

Fiddler (1975) The occurrence and determination of N-nitroso compounds. Toxicol Appl Pharmacol 31:352–360
Fishel R, Lescoe MK, Rao MR, Copeland NG, Jenkins NA, Garber J, Kane M, Kolodner R (1993) The human mutator gene homolog MSH2 and its association with hereditary nonpolyposis colon cancer. Cell 75:1027–1038
Frosina G, Fortini P, Rossi O, Carrozzino F, Raspaglio G, Cox LS, Lane DP, Abbondandolo A, Dogliotti E (1996) Two pathways for base excision repair in mammalian cells. J Biol Chem 271:9573–9578
Galloway SM, Greenwood SK, Hill RB, Bradt CI, Bean CL (1995) A role for mismatch repair in production of chromosome aberrations by methylating agents in human cells. Mutat Res 346:231–245
Garcia de Palazzo IE, Adams GP, Sundareshan P, Wong AJ, Testa JR, Bigner DD, Weiner LM (1993) Expression of mutated epidermal growth factor receptor by non-small cell lung carcinomas. Cancer Res 53:3217–3220
Genschel J, Bazemore LR, Modrich P (2002) Human exonuclease I is required for 5′ and 3′ mismatch repair. J Biol Chem 277:13302–13311
Gerson SL (2004) MGMT: its role in cancer aetiology and cancer therapeutics. Nat Rev Cancer 4:296–307
Glassner BJ, Weeda G, Allan JM, Broekhof JL, Carls NH, Donker I, Engelward BP, Hampson RJ, Hersmus R, Hickman MJ (1999) DNA repair methyltransferase (Mgmt) knockout mice are sensitive to the lethal effects of chemotherapeutic alkylating agents. Mutagenesis 14:339–347
Goria-Gatti L, Iannone A, Tomasi A, Poli G, Albano E (1992) In vitro and in vivo evidence for the formation of methyl radical from procarbazine: a spin-trapping study. Carcinogenesis 13:799–805
Gradia S, Subramanian D, Wilson T, Acharya S, Makhov A, Griffith J, Fishel R (1999) hMSH2-hMSH6 forms a hydrolysis-independent sliding clamp on mismatched DNA. Mol Cell 3:255–261
Guengerich FP, Shimada T (1991) Oxidation of toxic and carcinogenic chemicals by human cytochrome P-450 enzymes. Chem Res Toxicol 4:391–407
Hecht SS, Chen CB, Ohmori T, Hoffmann D (1980) Comparative carcinogenicity in F344 rats of the tobacco-specific nitrosamines, N′-nitrosonornicotine and 4-(N-methyl-N-nitrosamino)-1-(3-pyridyl)-1-butanone. Cancer Res 40:298–302
Hegi ME, Diserens AC, Godard S, Dietrich PY, Regli L, Ostermann S, Otten P, Van Melle G, de Tribolet N, Stupp R (2004) Clinical trial substantiates the predictive value of O-6-methylguanine-DNA methyltransferase promoter methylation in glioblastoma patients treated with temozolomide. Clin Cancer Res 10:1871–1874
Hegi ME, Diserens AC, Gorlia T, Hamou MF, de Tribolet N, Weller M, Kros JM, Hainfellner JA, Mason W, Mariani L et al (2005) MGMT gene silencing and benefit from temozolomide in glioblastoma. N Engl J Med 352:997–1003
Hickman MJ, Samson LD (1999) Role of DNA mismatch repair and p53 in signaling induction of apoptosis by alkylating agents. Proc Natl Acad Sci USA 96:10764–10769
Hill DL (1975) Microsomal metabolism of triazenylimidazoles. Cancer Res 35:3106–3110
Hirose Y, Katayama M, Stokoe D, Haas-Kogan DA, Berger MS, Pieper RO (2003) The p38 mitogen-activated protein kinase pathway links the DNA mismatch repair system to the G2 checkpoint and to resistance to chemotherapeutic DNA-methylating agents. Mol Cell Biol 23:8306–8315
Hoffman PD, Wang H, Lawrence CW, Iwai S, Hanaoka F, Hays JB (2005) Binding of MutS mismatch repair protein to DNA containing UV photoproducts, "mismatched" opposite Watson–Crick and novel nucleotides, in different DNA sequence contexts. DNA Repair (Amst) 4:983–993
Jagerstad M, Skog K (2005) Genotoxicity of heat-processed foods. Mutat Res 574:156–172
Kaina B (1985a) The interrelationship between SCE induction, cell survival, mutagenesis, aberration formation and DNA synthesis inhibition in V79 cells treated with N-methyl-N-nitrosourea or N-methyl-N′-nitro-N-nitrosoguanidine. Mutat Res 142:49–54
Kaina B (1985b) Secondary lesions and the induction of SCEs and chromosomal aberrations. In: Proceedings of the 13th international congress on biochemistry, Amsterdam, p 31

Kaina B (1998) Critical steps in alkylation-induced aberration formation. Mutat Res 404:119–124
Kaina B (2004) Mechanisms and consequences of methylating agent-induced SCEs and chromosomal aberrations: a long road traveled and still a far way to go. Cytogenet Genome Res 104:77–86
Kaina B, Aurich O (1985) Dependency of the yield of sister-chromatid exchanges induced by alkylating agents on fixation time. Possible involvement of secondary lesions in sister-chromatid exchange induction. Mutat Res 149:451–461
Kaina B, Fritz G, Mitra S, Coquerelle T (1991) Transfection and expression of human O6-methylguanine-DNA methyltransferase (MGMT) cDNA in Chinese hamster cells: the role of MGMT in protection against the genotoxic effects of alkylating agents. Carcinogenesis 12:1857–1867
Kaina B, Fritz G, Coquerelle T (1993) Contribution of O6-alkylguanine and N-alkylpurines to the formation of sister chromatid exchanges, chromosomal aberrations, and gene mutations: new insights gained from studies of genetically engineered mammalian cell lines. Environ Mol Mutagen 22:283–292
Kaina B, Fritz G, Eschabch E, Preuss I (1995) Transfer of genes affecting resistance of mammalian cells to alkylating drugs – implication for cancer chemotherapy. In: Zeller WJ, Eisenbrand G, Hellmann K (eds) Reduction of anticancer drug toxicity. Pharmacological, biologic and gene therapeutic approaches. Karger, Basel, pp 125–141
Kaina B, Ziouta A, Ochs K, Coquerelle T (1997) Chromosomal instability, reproductive cell death and apoptosis induced by O6-methylguanine in Mex–, Mex+ and methylation-tolerant mismatch repair compromised cells: facts and models. Mutat Res 381:227–241
Kamataki T, Nunoya K, Sakai Y, Kushida H, Fujita K (1999) Genetic polymorphism of CYP2A6 in relation to cancer. Mutat Res 428:125–130
Karran P, Bignami M (1994) DNA damage tolerance, mismatch repair and genome instability. BioEssays 16:833–839
Klungland A, Lindahl T (1997) Second pathway for completion of human DNA base excision-repair: reconstitution with purified proteins and requirement for DNaseIV (FEN1). EMBO J 16:3341–3348
Kurowski MA, Bhagwat AS, Papaj G, Bujnicki JM (2003) Phylogenomic identification of five new human homologs of the DNA repair enzyme AlkB. BMC Genomics 4:48
Lai DY, Arcos JC (1980) Minireview: dialkylnitrosamine bioactivation and carcinogenesis. Life Sci 27:2149–2165
Leach FS, Nicolaides NC, Papadopoulos N, Liu B, Jen J, Parsons R, Peltomaki P, Sistonen P, Aaltonen LA, Nystrom-Lahti M et al (1993) Mutations of a mutS homolog in hereditary nonpolyposis colorectal cancer. Cell 75:1215–1225
Lees NP, Harrison KL, Hall CN, Margison GP, Povey AC (2004) Reduced MGMT activity in human colorectal adenomas is associated with K-ras GC→AT transition mutations in a population exposed to methylating agents. Carcinogenesis 25:1243–1247
Lehman AR (2002) Replication of damaged DNA in mammalian cells: new solutions to an old problem. Mutat Res 509:23–34
Li GM, Modrich P (1995) Restoration of mismatch repair to nuclear extracts of H6 colorectal tumor cells by a heterodimer of human MutL homologs. Proc Natl Acad Sci USA 92:1950–1954
Li GM, Wang H, Romano LJ (1996) Human MutSalpha specifically binds to DNA containing aminofluorene and acetylaminofluorene adducts. J Biol Chem 271:24084–24088
Lin HL, Parsels LA, Maybaum J, Hollenberg PF (1999) N-Nitrosodimethylamine-mediated cytotoxicity in a cell line expressing P450 2E1: evidence for apoptotic cell death. Toxicol Appl Pharmacol 157:117–124
Longley MJ, Pierce AJ, Modrich P (1997) DNA polymerase delta is required for human mismatch repair in vitro. J Biol Chem 272:10917–10921
Margison GP, Povey AC, Kaina B, Santibanez Koref MF (2003) Variability and regulation of O6-alkylguanine-DNA alkyltransferase. Carcinogenesis 24:625–635
Marnett LJ, Burcham PC (1993) Endogenous DNA adducts: potential and paradox. Chem Res Toxicol 6:771–785
Matsumoto Y, Kim K (1995) Excision of deoxyribose phosphate residues by DNA polymerase beta during DNA repair. Science 269:699–702

Matsumoto Y, Kim K, Hurwitz J, Gary R, Levin DS, Tomkinson AE, Park MS (1999) Reconstitution of proliferating cell nuclear antigen-dependent repair of apurinic/apyrimidinic sites with purified human proteins. J Biol Chem 274:33703–33708

Meikrantz W, Bergom MA, Memisoglu A, Samson L (1998) O6-Alkylguanine DNA lesions trigger apoptosis. Carcinogenesis 19:369–372

Mitacek EJ, Brunnemann KD, Hoffmann D, Limsila T, Suttajit M, Martin N, Caplan LS (1999) Volatile nitrosamines and tobacco-specific nitrosamines in the smoke of Thai cigarettes: a risk factor for lung cancer and a suspected risk factor for liver cancer in Thailand. Carcinogenesis 20:133–137

Modrich P, Lahue R (1996) Mismatch repair in replication fidelity, genetic recombination, and cancer biology. Annu Rev Biochem 65:101–133

Nakamura J, La DK, Swenberg JA (2000) 5′-nicked apurinic/apyrimidinic sites are resistant to beta-elimination by beta-polymerase and are persistent in human cultured cells after oxidative stress. J Biol Chem 275:5323–5328

Nakatsuru Y, Matsukuma S, Nemoto N, Sugano H, Sekiguchi M, Ishikawa T (1993) O^6-Methylguanine-DNA methyltransferase protects against nitrosamine-induced hepatocarcinogenesis. Proc Natl Acd Sci USA 90:6468–6472

Newlands ES, Stevens MF, Wedge SR, Wheelhouse RT, Brock C (1997) Temozolomide: a review of its discovery, chemical properties, pre-clinical development and clinical trials. Cancer Treat Rev 23:35–61

Nicolaides NC, Papadopoulos N, Liu B, Wei YF, Carter KC, Ruben SM, Rosen CA, Haseltine WA, Fleischmann RD, Fraser CM et al (1994) Mutations of two PMS homologues in hereditary nonpolyposis colon cancer. Nature 371:75–80

O'Connor TR (1993) Purification and characterization of human 3-methyladenine-DNA glycosylase. Nucleic Acids Res 21:5561–5569

Ochs K, Kaina, B. (2000) Apoptosis induced by DNA damage O6-methylguanine is Bcl-2 and caspase-9/3 regulated and Fas/caspase-8 independent. Cancer Res 60:5815–5824

Ostermann S, Csajka C, Buclin T, Leyvraz S, Lejeune F, Decosterd LA, Stupp R (2004) Plasma and cerebrospinal fluid population pharmacokinetics of temozolomide in malignant glioma patients. Clin Cancer Res 10:3728–3736

Palombo F, Gallinari P, Iaccarino I, Lettieri T, Hughes M, D'Arrigo A, Truong O, Hsuan JJ, Jiricny J (1995) GTBP, a 160-kilodalton protein essential for mismatch-binding activity in human cells. Science 268:1912–1914

Papadopoulos N, Nicolaides NC, Wei YF, Ruben SM, Carter KC, Rosen CA, Haseltine WA, Fleischmann RD, Fraser CM, Adams MD et al (1994) Mutation of a mutL homolog in hereditary colon cancer. Science 263:1625–1629

Pegg AE, Dolan ME, Moschel RC (1995) Structure, function, and inhibition of O6-alkylguanine-DNA alkyltransferase. Prog Nucleic Acid Res Mol Biol 51:167–223

Pepponi R, Marra G, Fuggetta MP, Falcinelli S, Pagani E, Bonmassar E, Jiricny J, D'Atri S (2003) The effect of O6-alkylguanine-DNA alkyltransferase and mismatch repair activities on the sensitivity of human melanoma cells to temozolomide, 1,3-bis(2-chloroethyl)1-nitrosourea, and cisplatin. J Pharmacol Exp Ther 304:661–668

Peto R, Gray R, Brantom P, Grasso P (1984) Nitrosamine carcinogenesis in 5120 rodents: chronic administration of sixteen different concentrations of NDEA, NDMA, NPYR and NPIP in the water of 4440 inbred rats, with parallel studies on NDEA alone of the effect of age of starting (3, 6 or 20 weeks) and of species (rats, mice or hamsters). IARC Sci Publ 627–665

Prasad R, Beard WA, Strauss PR, Wilson SH (1998) Human DNA polymerase beta deoxyribose phosphate lyase. Substrate specificity and catalytic mechanism. J Biol Chem 273:15263–15270

Qian X, von Wronski MA, Brent TP (1995) Localization of methylation sites in the human O^6-methylguanine-DNA methyltransferase promoter: correlation with gene suppression. Cacinogenesis 16:1385–1390

Qin X, Zhou H, Liu L, Gerson SL (1999) Transgenic expression of human MGMT blocks the hypersensitivity of PMS2-deficient mice to low dose MNU thymic lymphomagenesis. Carcinogenesis 20:1667–1673

Rasouli-Nia A, Sibghat-Ullah, Mirzayans R, Paterson MC, Day RS III (1994) On the quantitative relationship between O6-methylguanine residues in genomic DNA and production of sister-chromatid exchanges, mutations and lethal events in a Mer– human tumor cell line. Mutat Res 314:99–113

Reid JM, Kuffel MJ, Miller JK, Rios R, Ames MM (1999) Metabolic activation of dacarbazine by human cytochromes P450: the role of CYP1A1, CYP1A2, and CYP2E1. Clin Cancer Res 5:2192–2197

Ringvoll J, Nordstrand LM, Vagbo CB, Talstad V, Reite K, Aas PA, Lauritzen KH, Liabakk NB, Bjork A, Doughty RW et al (2006) Repair deficient mice reveal mABH2 as the primary oxidative demethylase for repairing 1meA and 3meC lesions in DNA. EMBO J 25:2189–2198

Ro J, North SM, Gallick GE, Hortobagyi GN, Gutterman JU, Blick M (1988) Amplified and overexpressed epidermal growth factor receptor gene in uncultured primary human breast carcinoma. Cancer Res 48:161–164

Roos W, Baumgartner M, Kaina B (2004) Apoptosis triggered by DNA damage O6-methylguanine in human lymphocytes requires DNA replication and is mediated by p53 and Fas/CD95/-Apo-1. Oncogene 23:359–367

Roos WP, Batista LF, Naumann SC, Wick W, Weller M, Menck CF, Kaina B (2007) Apoptosis in malignant glioma cells triggered by the temozolomide-induced DNA lesion O(6)-methylguanine. Oncogene 26:186–197

Sedgwick B, Lindahl T (2002) Recent progress on the Ada response for inducible repair of DNA alkylation damage. Oncogene 21:8886–8894

Sibghat-Ullah L, Day RS III, Dobler KD, Paterson MC (2001) Initiation of strand incision at G:T and O6-methylguanine:T base mismatches in DNA by human cell extracts. Nucleic Acids Res 29:2409–2417

Skibba JL, Beal DD, Ramirez G, Bryan GT (1970) N-demethylation the antineoplastic agent4(5)-(3,3-dimethyl-1-triazeno)imidazole-5(4)-carboxamide by rats and man. Cancer Res 30:147–150

Sklar RM, Strauss B (1981) Removal of O^6-methylguanine from DNA of normal and xeroderma pigmentosum-derived lymphoblastoid lines. Nature 289:417–420

Sobol RW, Wilson SH (2001) Mammalian DNA beta-polymerase in base excision repair of alkylation damage. Prog Nucleic Acid Res Mol Biol 68:57–74

Sobol RW, Prasad R, Evenski A, Baker A, Yang XP, Horton JK, Wilson SH (2000) The lyase activity of the DNA repair protein beta-polymerase protects from DNA-damage-induced cytotoxicity. Nature 405:807–810

Stojic L, Mojas N, Cejka P, Di Pietro M, Ferrari S, Marra G, Jiricny J (2004) Mismatch repair-dependent G2 checkpoint induced by low doses of SN1 type methylating agents requires the ATR kinase. Genes Dev 18:1331–1344

Stucki M, Pascucci B, Parlanti E, Fortini P, Wilson SH, Hubscher U, Dogliotti E (1998) Mammalian base excision repair by DNA polymerases delta and epsilon. Oncogene 17:835–843

Tominaga Y, Tsuzuki T, Shiraishi A, Kawate H, Sekiguchi M (1997) Alkylation-induced apoptosis of embryonic stem cells in which the gene for DNA-repair, methyltransferase, had been disrupted by gene targeting. Carcinogenesis 18:889–896

Toorchen D, Topal MD (1983) Mechanisms of chemical mutagenesis and carcinogenesis: effects on DNA replication of methylation at the O6-guanine position of dGTP. Carcinogenesis 4:1591–1597

Toyooka S, Toyooka KO, Maruyama R, Virmani AK, Girard L, Miyajima K, Harada K, Ariyoshi Y, Takahashi T. Sugio K et al (2001) DNA methylation profiles of lung tumors. Mol Cancer Ther 1:61–67

Trewick SC, Henshaw TF, Hausinger RP, Lindahl T, Sedgwick B (2002) Oxidative demethylation by Escherichia coli AlkB directly reverts DNA base damage. Nature 419:174–178

Tsuzuki T, Sakumi K, Shiraishi A, Kawate H, Igarashi H, Iwakuma T, Tominaga Y, Zhang S, Shimizu S, Ishikawa T et al. (1996) Targeted disruption of the DNA repair methyltransferase gene renders mice hypersensitive to alkylating agent. Carcinogenesis 17:1215–1220

Umar A, Kunkel TA (1996) DNA-replication fidelity, mismatch repair and genome instability in cancer cells. Eur J Biochem 238:297–307 van Zeeland AA, Mohn GR, Aaron CS, Glickman BW, Brendel M, de Serres FJ, Hung CY, Brockman HE (1983) Molecular dosimetry of the chemical mutagen ethyl methanesulfonate. Quantitative comparison of the mutagenic potency in Neurospora crassa and Saccharomyces cerevisiae. Mutat Res 119:45–54

Vogel E, Natarajan AT (1979) The relation between reaction kinetics and mutagenic action of mono-functional alkylating agents in higher eukaryotic systems. I. Recessive lethal mutations and translocations in Drosophila. Mutat Res 62:51–100

Walker R (1990) Nitrates, nitrites and N-nitrosocompounds: a review of the occurrence in food and diet and the toxicological implications. Food Addit Contam 7:717-768

Wang Y, Qin J (2003) MSH2 and ATR form a signaling module and regulate two branches of the damage response to DNA methylation. Proc Natl Acad Sci USA 100:15387–15392

Wang Z, Gleichmann H (1998) GLUT2 in pancreatic islets: crucial target molecule in diabetes induced with multiple low doses of streptozotocin in mice. Diabetes 47:50–56

Wilson DM, Barsky D (2001) The major human abasic endonuclease: formation, consequences and repair of abasic lesions in DNA. Mutat Res 485:283–307

Wogan GN, Tannenbaum SR (1975) Environmental N-nitroso compounds: implications for public health. Toxicol Appl Pharmacol 31:375–383

Wojcik A, Bruckmann E, Obe G (2004) Insights into the mechanisms of sister chromatid exchange formation. Cytogent Genome Res 104:301–309

Xu YH, Richert N, Ito S, Merlino GT, Pastan I (1984) Characterization of epidermal growth factor receptor gene expression in malignant and normal human cell lines. Proc Natl Acad Sci USA 81:7308–7312

Yarosh DB, Foote RS, Mitra S, Day RS 3rd (1983) Repair of O6-methylguanine in DNA by demethylation is lacking in Mer– human tumor cell strains. Carcinogenesis 4:199–205

Yoshioka K, Yoshioka Y, Hsieh P (2006) ATR kinase activation mediated by MutSalpha and MutLalpha in response to cytotoxic O6-methylguanine adducts. Mol Cell 22:501–510

Zhou ZQ, Manguino D, Kewitt K, Intano GW, McMahan CA, Herbert DC, Hanes M, Reddick R, Ikeno Y, Walter CA (2001) Spontaneous hepatocellular carcinoma is reduced in transgenic mice overexpressing human O6-methylguanine-DNA methyltransferase. Proc Natl Acad Sci USA 98:12566–12571

8 Adducts, Sister-Chromatid Exchanges and Mutations Following Benzo[*a*]pyrene Exposure: a Review of Quantitative Data Followed by Some Considerations Regarding Risk

RINNE DE BONT AND NIK VAN LAREBEKE

Abstract

We conducted a literature study (papers published until September 2004) concerning quantitative aspects of the formation of DNA adducts and the induction of sister-chromatid exchanges (SCE) and mutations following benzo[*a*]pyrene (BaP) exposure. Mean adduct levels per dose unit after exposure in vitro were higher in both human (11 studies) and animal (five studies) cells for doses below 5 μM (2.47 and 1.445 adducts per 10^6 nucleotides per μM, respectively) than for doses of 5 μM and above (0.104 and 0.106 adducts per 10^6 nucleotides per μM). In vivo in humans BaP adducts (0.054–0.918 adducts per 10^6 nucleotides per μM) occur in much lower concentrations than endogenous adducts. The total quantity of nine endogenous adducts in human cells in vivo amounts to between 4.97 and 69.05 adducts per 10^6 nucleotides. Mean SCE levels per dose unit after exposure in vitro were higher in both human and animal cells for doses below 5 μM BaP (1.48 and 13.32 SCEs per cell per μM, respectively) than for doses of 5 μM and above (0.18 and 0.97 SCEs per cell per μM). In human cells in vitro (three studies, all below 5 μM) a non-significant downward trend in mutation frequency per dose unit with increasing dose was reported, and in Chinese hamster ovary cells doses below 5 μM BaP resulted in a higher mean mutation frequency per micromole per litre (in the *HGPRT* or the *XPRT* gene) than doses of 5 μM and above (4.30 vs. 0.88 per 10^5 cells per micromole per litre). Calculations on cancer risk associated with a certain level of BaP adducts and on the relation between environmental exposures and a threshold level for induction of DNA repair are presented.

8.1 Introduction

The potent carcinogen benzo[*a*]pyrene (BaP), the most extensively studied of the polycyclic aromatic hydrocarbons (PAHs), is present in car exhaust fumes, coke oven emissions, cigarette smoke, environmental air and a variety of foods. Metabolic activation of BaP proceeds by two major mechanisms, monooxygenation to yield bay-region diolepoxides, the most important of which is BaP-7,8-dihydrodiol-9,10-epoxide (BPDE), giving rise to stable DNA

Günter Obe and Vijayalaxmi (Eds.)
Chromosomal Alterations: Methods, Results and Importance in Human Health

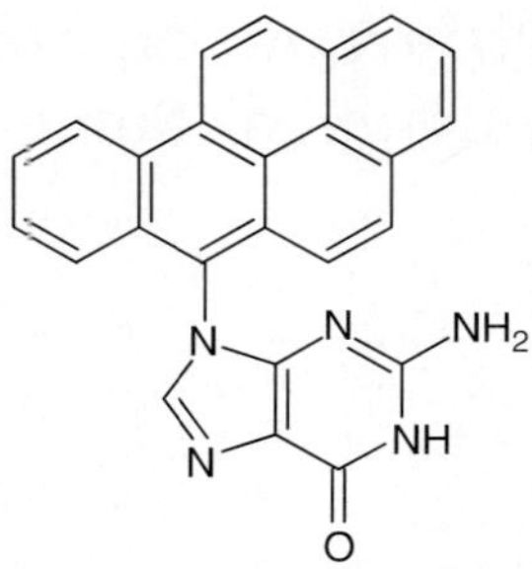

BaP-6-N7Gua

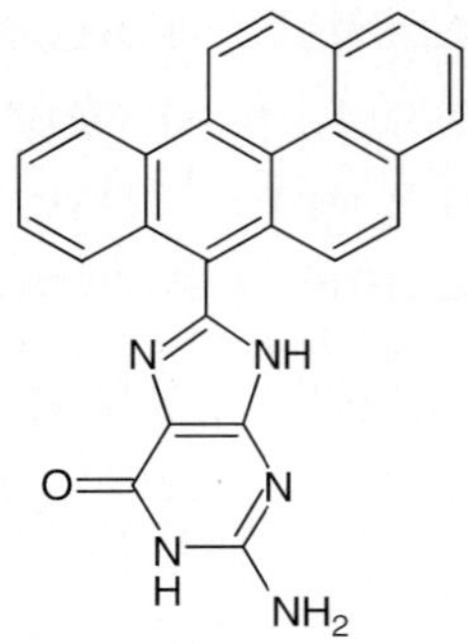

BaP-6-C8Gua

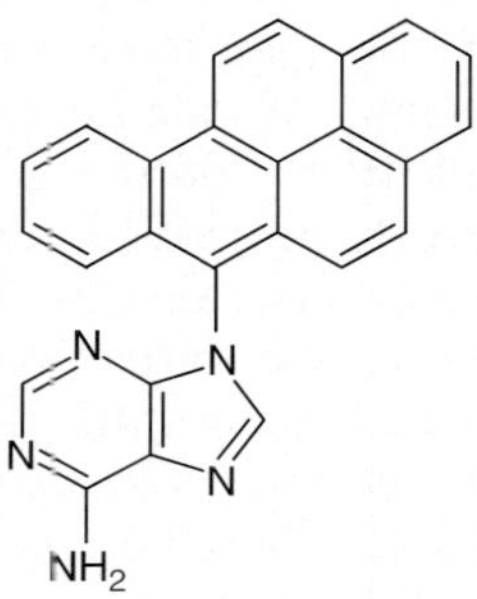

BaP-6-N7Ade

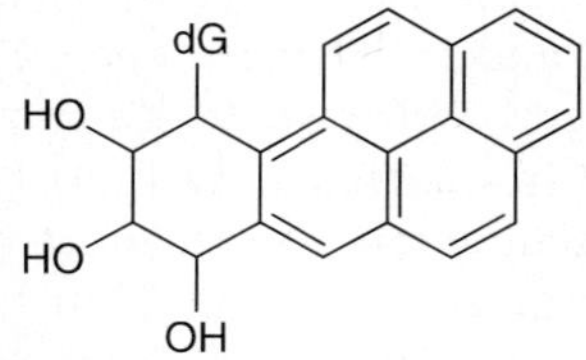

BPDE-10-N^2dG

Fig. 8.1 Structures of some benzo[*a*]pyrene (*BaP*) adducts

adducts, and one-electron oxidation with formation of radical cations leading to mostly unstable depurinating DNA adducts. The most important stable DNA adduct is the BPDE adduct to deoxyguanosine, BPDE-10-N2dG, but other adducts, including BPDE-N6dA and, to a lesser extent, BPDE-N4dC are also formed. The major depurinating adducts following the one-electron pathway are BaP-6-N7dG, BaP-6-C8dG and BaP-6-N7dA (Banasiewicz et al. 2004). Figure 8.1 shows the structure of some BaP DNA adducts.

To investigate the dose–effect relationship following exposure to BaP, we conducted a literature study concerning quantitative aspects of the formation of DNA adducts and the induction of sister-chromatid exchanges (SCEs) and mutations.

8.2 Methodological Approach

We sought all potentially informative peer-reviewed publications reporting studies on DNA adducts, SCEs and mutations following BaP exposure. We included in vitro studies and in vivo human observations. Using Medline, we searched for articles published until September 2004. Studies that used dose units not comparable and not transformable to the dose units we used in our analysis were excluded.

For the sake of comparison, DNA adduct levels are expressed in a standardized way, including the number of adducts per 10^6 nucleotides.

Possible confounding factors that complicate the comparison of different studies are the method of adduct measurement, the cell or tissue types investigated, the use of external metabolic activation and the duration of exposure. To correct for these factors we performed an ANOVA analysis using the statistical programme Statview.

Published data suggest that BPDE adducts constitute the vast majority of BaP adducts. From Shugart (1985) and Li et al. (1996) we calculated that BPDE-dG adducts constituted about 71.1% of all the BPDE adducts and from several other studies (Ashurst and Cohen 1980, 1981; DiGiovanni et al. 1985; Bodell et al. 1989; Ross et al. 1990; Van Schooten et al. 1991; Rogan et al. 1993; Devanesan et al. 1992; Roggeband et al. 1994a, b; Chen et al. 1996) that BPDE-dG adducts constitute about 77.2 % of all BaP adducts, so it is probable that BPDE adducts comprise almost all BaP adducts For further calculations, we assumed that BPDE adducts represent 100% of all BaP adducts.

We list quantitative in vivo data concerning BaP adducts in humans and compare these data with levels of endogenous DNA adducts (De Bont and van Larebeke 2004).

8.3 Results

8.3.1 Adducts

8.3.1.1 BaP Exposure in Vitro: Dose–Effect Relationship for Adduct Formation

The entire set of adduct levels measured in human cells (Table 8.1; 11 studies totalling 23 measured adduct levels) were plotted against the exposure dose (Fig. 8.2). The data were divided into adducts following low-dose exposure (below 5 μM BaP) and adducts following exposure to BaP doses of 5 μM and above. BaP. Linear regression shows an R^2 value of 0.579 at doses below 5 μM and an R^2 value of 0.719 at doses of 5 μM and above. In linear regression the slope of the curve for doses below 5 μM (1.676 with significance of the t value of 0.002) is more than 10 times higher than that for doses of 5 μM and above (0.105 with a significance of the t value of 0.004). We also calculated the mean adduct level per dose unit to be higher for doses below 5 μM (2.47 adducts per 10^6 nucleotides per μM) than for doses of 5 μM and above (0.104 adducts per 10^6 nucleotides per μM). Possible confounding factors that could complicate the comparison of different studies are the method of adduct measurement

Table 8.1 Benzo[*a*]pyrene (BaP) adducts following exposure of human cells to BaP in vitro

Method	Cell or tissue type	Dose (μM)	Adducts/10^6 nucleotides	Reference
HPLC	Monocytes	0.30	0.680	Rüdiger et al. (1985)
32P	Peripheral lymphocytes	8.00	0.140	Canova et al. (1996)
		20.00	0.200	
Technique based on radioactivity	Skin fibroblasts	0.25	2.260	Rüdiger et al. (1979)
32P	Lung fibroblasts	0.01	0.018	Binková et al. (2000)
		1.00	2.370	
HPLC	Mammary epithelial cells	2.00	5.440	Moore et al. (1987)
32P	White blood cells	10.00	2.200	Roggeband et al. (1994a)
		25.00	4.200	
		50.00	9.300	
		100.00	11.800	
HPLC	Epidermal cells	0.40	1.500	Theall et al. (1981)
		4.00	6.000	
HPLC	Bladder bronchus	1.00	6.400	Stoner et al. (1982)
		1.00	3.100	
32P	White blood cells	100.00	9.300	Roggeband et al. (1993)
32P	Lymphocytes	0.01	0.004	Wiencke et al. (1990)
		0.05	0.013	
		0.10	0.027	
		0.50	0.199	
		1.00	0.317	
		5.00	0.556	
32P	Fibroblasts	50.00	0.675	Hanelt et al. (1997)

HPLC high performance liquid chromatography, *32P* ^{32}P-postlabelling

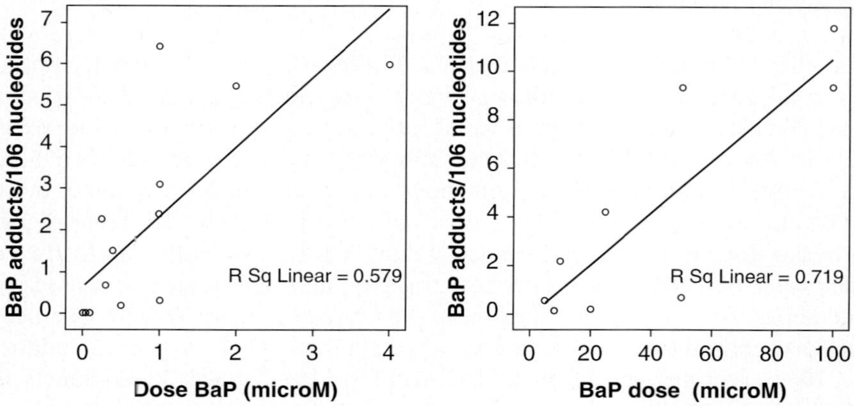

Fig. 8.2 BaP adducts in human tissues following BaP exposure in vitro: doses below 5 μM (*left*); doses of 5 μM and above (*right*)

(high-performance liquid chromatography or ^{32}P-postlabelling), the cell or tissue types investigated (blood, lung, skin, bladder, mammary epithelial tissue), the use of external metabolic activation (S9) and the duration of exposure (2–48 h). To correct for these factors we performed an ANOVA analysis after logarithmic transformation of the adducts-per-dose values using the statistical program Statview. The difference in the number of adducts per dose between the high-exposure group and the low-exposure group was statistically significant ($p = 0.0054$).

In vitro tests on animal cells (six studies totalling 11 measured adduct levels) showed that doses from 1 to 55 μM gave rise to adduct levels ranging from 0.1 to 11 adducts per 10^6 nucleotides (Table 8.2). The mean adduct level per dose unit was higher for doses below 5 μM (1.445 adducts per 10^6 nucleotides per μM) than for higher doses (0.10^6 adducts per 10^6 nucleotides) but there were too few data to test the significance of the difference in an ANOVA test taking into account the confounding factors.

8.3.1.2 *Adducts in Vivo*

Data on adduct levels after in vivo treatment of animals with BaP (seven mouse studies with 16 measured adduct levels and one rat study) are shown in Table 8.3. Owing to the limited number of data, it was not possible to compare the number of adducts per dose between high doses and low doses in an ANOVA analysis integrating all covariates.

Table 8.2 BaP adducts following exposure of animal cells to BaP in vitro

Method	Cell or tissue type	Dose (μM)	Adducts/10^6 nucleotides	Reference
32P	Mouse foetal liver	8.0	1.44	Canova et al. (1996)
		20.0	2.40	
		40.0	4.22	
HPLC	Rat dermal fibroblasts	1.2	3.15	Alexandrov et al. (1988)
	Rabbit dermal fibroblasts	1.2	1.68	
HPLC	Rat mammary epithelial cells	2.0	3.31	Moore et al. (1987)
HPLC	Rat hepatocytes	55.0	1.02	Ashurst et al. (1980)
		55.0	1.71	
Sephadex LH 20 chromatography	Rat lung	1.0	0.10	Deckers-Schmelzle et al. (1978)
32P	Rat trachea	19.8	3.70	Roggeband et al. (1994b)
	Hamster trachea	19.8	11.0	

Table 8.3 Animals: BaP adducts following BaP exposure in vivo

Method	Cell or tissue type	Dose (mg/kg)	Adducts/10^6 nucleotides	Reference
32P	Rat lung	10	0.031	Godschalk et al. (1997)
	Rat liver	10	0.029	
	Rat hart	10	0.063	
32P	Mouse skin	0.63[a]	0.125	Talaska et al. (1996)
ELISA	Mouse skin	1.6[a]	0.230	Albert et al. (1996)
		3.2[a]	0.393	
		6.4[a]	0.525	
[1,3-^{3}H]BPDE radiolabelling	Mouse liver	25[a]	0.864	Divi et al. (2002)
		25[a]	1.376	
32P		50[a]	0.114	
		50[a]	0.221	
DELFIA		25[a]	0.043	
		50[a]	0.079	
CIA		25[a]	0.110	
		50[a]	0.170	
HPLC	Mouse skin	10[a]	1.75	Bartley et al. (1982)
32P	Mouse skin	2.52[a]	1.07	Bodell et al. (1989)
32P	Mouse skin	2.52[a]	3.56	Rogan et al. (1993)
HPLC	Mouse skin	2.52[a]	3.15	DiGiovanni et al. (1985)

BDPE BaP-7,8-dihydrodiol-9,10-epoxide, *ELISA* enzyme-linked immunosorbent assay, *DELFIA* dissociation-enhanced lanthanide fluoroimmunoassay, *CIA* Chemiluminescence immunoassay
[a]Dose units transformed into milligrams per kilograms using a mean weight of mice of 20g

8.3.1.3 BPDE Adducts Following BPDE Exposure

We also considered five studies that quantify BPDE-dG adducts or total BPDE adducts following BPDE exposure (data not shown). On the basis of all available studies on BPDE adduct formation after in vitro exposure to BPDE or BaP, we calculated, by dividing the number of adducts by the dose, that 1 μM of BPDE forms the same number of BPDE adducts as 18.58 μM BaP. The number of adducts formed per microgram of BPDE at BPDE exposures equivalent to BaP exposures below 5 μM was lower than at BPDE exposures equivalent to BaP exposures of 5 μM and above. but taking the method of adduct measurement into account this difference was not significant

8.3.1.4 Human BaP Adducts in Vivo: Comparison with Endogenous Adducts

Table 8.4 provides quantitative in vivo data concerning BaP adducts in humans. BaP and BPDE adducts (0.054–0.918 and 0.013–0.787 adducts per 10^6

Table 8.4 BaP adducts in humans in vivo

Method	Cell or tissue type	Exposure type	Adducts/10^6 nucleotides	Reference
BaP adducts				
HPLC	Monocytes	Cancer patients	0.918	Nowak et al. (1992)
		Controls	0.698	
32P	Leukocytes	Controls: non-smokers	0.0054	Arnould et al. (1999)
		Smokers	0.0177	
		Factory workers	0.0057–0.389	
BPDE adducts				
IAC	Placenta		0.100	Manchester et al. (1988)
USERIA	Lymphocytes	Coke oven workers	0.558	Haugen et al. (1986)
USERIA	White blood cells	Factory workers	0.013–0.787	Shamsuddin et al. (1985)
ELISA	White blood cells	Factory workers	0.051	Van Schooten et al. (1990)
		Controls	0.027	
ELISA	Placenta	Smokers	0.066–0.393	Arnould et al. (1997)
		Non-smokers	0.066–0.087	
	Umbilical cord blood	Smokers	0.066–0.145	
		Non-smokers	0.066	
HPLC (1991)	Lung tissue		0.010–0.400	Weston and Bowman
HPLC	Lymphocytes	Coke oven workers	0.505	Weston et al. (1989)
32P	White blood cells	Controls	0.098–0.300	Rojas et al. (1994)
HPLC	Lymphocyte–monocyte fraction	Coke oven workers	0.044	Pavanello et al. (2005)
HPLC	Lung tissue	Cancer patients	0.006–0.099	Alexandrov et al. (1992)
HPLC	Lymphocyte–monocyte fraction	Controls	0.039	Pavanello et al. (1999)
		Anode aluminium workers	0.115	
		Chimney sweeps	0.055	
		Coke-oven workers	0.195	
		Psoriatic patients	0.023	

IAC immunoaffinity chromatography, *USERIA* ultrasensitive enzymatic radioimmunoassay

nucleotides, respectively) occur in much lower concentrations than several mainly endogenous adducts. DNA damage produced by reactive oxygen species (ROS) is the most frequently occurring damage. ROS are produced in part through exposure to certain agents, but are produced mainly through endogenous processes, including cellular respiration, biosynthetic and biodegrading processes of normal intermediary metabolism and activation of phagocytotic cells. Among the oxidatively modified bases, 8-oxo-dG is apparently the most abundant. Steady-state levels of this adduct range from 0.24 to 59.70 adducts per10^6 nucleotides in white blood cells (De Bont and van Larebeke 2004). Other abundant oxidatively modified bases comprise the formamido pyrimidine adducts of adenine and guanine and 5-hydroxyuracil and uracil glycol. Lipid peroxidation leads to propano adducts (up to 0.40 adducts per 10^6 nucleotides in humans), etheno adducts, including N^2-3-ethenodeoxyguanosine, whose levels were similar to those of propano adducts, and malondialdehyde adducts, the most abundant of which is the adduct to dG (M1G; pyrimido[1,2-α]purin-10(3*H*)-one) (up to 0.9 adducts per10^6 nucleotides in humans) (De Bont and van Larebeke 2004). Base propenals, arising from oxidative degradation of DNA, might even lead to more M1G adduct formation than malondialdehyde derived from lipid peroxidation (Dedon et al. 1998). Several endogenous oestrogen metabolites, particularly the catechol oestrogens, can be oxidized to quinones that lead to several types of DNA adducts (Liehr 2000). Endogenous alkylating agents can lead to alkyl adducts, the most frequent being 7-methylguanine (up to 1.35 adducts per 10^6 nucleotides in lymphocytes of non-smokers) (De Bont and van Larebeke 2004). Lipid peroxidation and glycation can lead to reactive carbonyl species such as glyoxal and methylglyoxal (Roberts et al. 2003). To get an idea of the total number of endogenous DNA adducts in white blood cells we considered the quantitative data regarding endogenous DNA adducts in white blood cells as summarized by De Bont and van Larebeke (2004). Only the following endogenous adducts were included: 5-hydroxy-2′-deoxycytidine, 5-hydroxy-2′-deoxyuridine,5,6-dihydroxy-5, 6-dihydro-2′-deoxyuridine, 8-oxo-7,8-dihydro-2′-deoxyguanosine, 7-methylguanine, O^6-methylguanine, 7-(2-hydroxyethyl)guanine, 7-alkylguanine and 7-ethyldeoxyguanosine. The total quantity of these endogenous adducts in human cells in vivo amounts to between 4.97 and 69.05 adducts per 10^6 nucleotides. The minimum level (4.97) was reached by adding up the lowest level observed for each of the endogenous adduct types included, while the maximum level (69.05) was based on the highest observed level.

8.3.2 Sister-Chromatid Exchanges

We found eight studies investigating SCEs following BaP exposure in vitro: five human studies and three animal studies (24 measurements) (Table 8.5). Seven of the eight studies determined more than two SCE levels and revealed a supralinear dose–response curve. Our analysis of the eight studies showed that, for humans, BaP doses below 5 μM induce a higher mean number of SCEs per micromole per litre of BaP than doses of 5 μM and above (1.48 vs. 0.18 SCEs per cell). The same can be concluded for animal studies (13.32 vs. 0.97 SCEs per cell). For the entire dataset this difference was statistically significant when corrected for species and the use of metabolic stimulation (after logarithmic transformation of the number of adducts per dose) ($p = 0.046$).

Table 8.5 Sister-chromatid exchanges (SCEs) following BaP exposure in human and animal cells in vitro

Cell or tissue type	Dose (μM)	SCEs/cell[a]	Reference
Humans			
Lymphocytes	50	3.40	Hartmann and Speit (1995)
	100	4.90	
	500	6.30	
Lymphocytes	1.00	0.08	Inoue et al. (1983)
	10.00	2.32	
	100.00	5.92	
Lymphocytes	39.60	3.44	Salama et al. (2001)
	198.00	3.94	
Lymphocytes	0.80	1.60	Jacono et al. (1992)
	20.00	8.80	
	40.00	12.20	
Lymphocytes	1.00	3.23	Dosaka et al. (1987)
	10.00	5.97	
	100.00	9.50	
Animals			
Rat liver epithelial cells	50	8.00	Cunningham and Ringrose (1983)
	200	23.50	
	400	27.70	
	2,000	35.60	
Chinese hamster ovary cells	0.25	7.37	Ray-Chaudhuri et al. (1980)
	2.50	13.67	
	25.00	14.07	
Rat liver epithelial cells	1.00	5.00	Tong et al. (1981)
	10.00	48.60	
	100.00	103.30	

[a]After subtraction of blank control value

8.3.3 Mutations

8.3.3.1 Mutations in Vivo in Animals Following BaP Exposure

We found only three in vivo mutation studies following BaP exposure and comparison between high and low doses was not possible (Table 8.6). Several other in vivo studies used dose units not comparable to the dose unit we used in our analysis (micromoles per litre) (Dean et al. 1998; Brooks et al. 1999). Brooks et al. (1999) showed a clear linear relationship between the dose and the mutation frequency. Hakura et al. (1998) measured the mutation frequency in 11 different tissues and showed a high variation in mutation frequencies (from 1.7×10^5 in brain to 267.6×10^5 in colon).

Table 8.6 Mutation frequency following BaP exposure in vivo

Cell or tissue type	Gene or locus	Dose (mg/kg)	Mutation frequency[a] ($\times 10^{-5}$)/cell	Reference
Mouse colon	*LacZ*	625	267.60	Hakura et al. (1998)
Mouse ilium		625	176.40	
Mouse forestomach		625	137.00	
Mouse bone marrow		625	87.00	
Mouse spleen		625	80.70	
Mouse glandular stomach		625	49.90	
Mouse liver		625	17.90	
Mouse lung		625	16.60	
Mouse kidney		625	11.70	
Mouse heart		625	8.90	
Mouse brain		625	1.70	
Mouse liver	*LacI*	120	6.40	Shane et al. (2000)
Mouse splenic T cells	*Hprt*	50	0.09	Monroe et al. (1998)
		150	0.91	
	LacI	50	4.40	
		150	13.80	
	CII	150	1.90	
	CI	50	1.70	
		150	0.50	

[a]After subtraction of blank control value

8.3.3.2 *Mutations in Vitro Following BaP Exposure*

We found three human studies (seven measuring points) and three animal studies (20 measuring points) measuring the mutation frequency in cells in vitro (Table 8.7). In human cells all experimental exposures were below 5 μM, demonstrating that mutations are already formed at low doses. The mutation frequency per dose unit showed a non-significant downward trend with increasing dose and the same trend was observed when only studies without metabolic activation were included. In Chinese hamster ovary cells doses below 5 μM BaP resulted in a higher mean mutation frequency per micromole per litre (in the *HGPRT* or the *XPRT* gene) than doses of 5 μM and above (4.30 vs. 0.88 per 10^5 cells per micromole per litre).

8.4 Discussion

PAHs present in ambient air, in tobacco smoke and in food contribute to the induction of cancer in the lung and in several other organs and to the induction of cardiovascular and airway disease (Ramos and Moorthy 2005;

Table 8.7 Mutation frequency following BaP exposure in vitro

Cell or tissue type	Gene or locus	Dose (μM)	Mutation frequency[a] ($\times 10^{-5}$)/cell	Reference
Humans				
Lymphoblastoid cells	*HGPRT*[b]	0.020	0.08	Danheiser et al. (1989)
		0.100	0.21	
		0.500	0.28	
		1.000	0.38	
Fibroblast without BER for BaP adducts	6-Thioguanine resistance[c]	0.100	6.50	Aust et al. (1980)
		0.200	9.90	
Epidermal keratinocytes	*HGPRT*[b]	3.960	3.08	Allen-Hoffmann and Rheinwald (1984)
Animals				
Chinese hamster ovary cells	*HGPRT*[b]	0.396	1.45	Machanoff et al. (1981)
		0.990	7.83	
		1.980	12.99	
		3.960	12.15	
		7.920	13.44	
		15.840	7.08	
Chinese hamster ovary cells	*XPRT*[d]	0.396	2.80	Oberly et al. (1993)
		1.980	2.80	
		3.960	2.10	
		7.920	6.20	
		15.840	13.50	
		31.680	18.90	
Chinese hamster V79 cells	Azaguanine resistance	0.100	5.00	Huberman et al. (1976)
		0.400	45.00	
		1.200	70.00	
		4.000	312.00	
	Ouabian resistance	0.200	0.30	
		0.400	6.80	
		1.200	18.80	
		4.00	106.80	

[a]After subtraction of blank control value
[b]Hypoxanthine-guanine phosphoribosyl transferase gene
[c]Mutations inactivating the *HGPRT* gene lead to resistance to 6-thioguanine
[d]Xanthine-guanine phosphoribosyl transferase gene

Miller et al. 2004). There are many different PAHs and each parental PAH can, for example through atmospheric reactions, give rise to several different derivatives. DNA adducts of BaP, the guide substance most often used to assess exposure to PAHs, have been found in human peripheral blood cells and in human tissues on most occasions when appropriate techniques have been applied. All humans are probably exposed to PAHs.

8.4.1 Cancer Risk Associated with a Certain Level of BaP Adducts

The data on the occurrence of BaP adducts and on the cancer incidence among workers in coke ovens and in other industries with exposure to PAHs can be used to estimate the cancer risk associated with a certain level of BaP adducts.

At least two considerations limit the meaning of such estimations. First, internal exposure can also originate from food sources and to a lesser extent from exposure through the skin. This might explain the fact that non-smoking controls, not exposed to coke oven emissions, show relatively high values of BaP adducts compared with coke oven workers or smokers (values for coke oven workers are a maximum 10 times higher). Exposures through food or skin have, however, a much lower risk for lung cancer than exposure through the airways (Culp et al. 2000) and although they might have the same risk for cancers not related to the airways, the total cancer risk will be substantially lower. So for people exposed through food or skin, the level of BaP adducts will lead to an overestimation of cancer risk. Second, both experimental and epidemiological studies have indicated that, for the induction of a certain lung tumour incidence, inhaled cigarette smoke contains about 100 times less BaP and inhaled diesel exhaust fumes contains about 1,000 times less BaP than the exhaust from coke ovens or heated tar pitch associated with the same risk of lung cancer (Pot and Heinrich 1990). So for smokers or people exposed to diesel exhaust fumes the level of BaP adducts will lead to an underestimation of cancer risk. Notwithstanding these and also other considerations, risk estimations based on data from workers in coke ovens and in other industries with exposure to PAHs can have some relevance as many people are exposed to similar complex mixtures and as clearly superior approaches are not available.

We calculated the mean BaP adduct level per dose unit (micrograms per cubic metre) in vivo in coke oven and other factory workers to estimate the cancer risk associated with the presence of BaP adducts based on the cancer risk experienced by these workers. Table 8.8 shows studies that measured BPDE or BaP adducts in white blood cells of coke oven or factory workers. We calculated the mean adduct level per dose unit to be 0.418 adducts per /10^6 nucleotides per microgram per cubic metre.

The unit risk for lung cancer from BaP (after 70 years of exposure) is $1.1–3.3 \times 10^{-3}$/μg BaP/m^3 (OEHHA 1994) so 0.418 adducts per 10^6 nucleotides could result in a unit lung cancer risk of $1.1–3.3 \times 10^{-3}$ after exposure to pure BaP. The number of endogenous DNA adducts in human DNA, which is on the order of 4.97– 69.05 adducts per 10^6 nucleotides, is far greater than the number of BaP adducts. This load of endogenous DNA adducts might underlie the spontaneous incidence of lung cancer, which is probably on the order of 6.6×10^{-3}, the cumulative incidence of lung cancer up to age 75 in Chennai, India, in the period 1978–1982 (Muir et al. 1987). This suggests that the pathogenic potency of exogenous high molecular weight adducts such as those

Table 8.8 BPDE or BaP adducts in white blood cells of coke oven or factory workers

Reference	Dose (μg/m³)	Adduct level (BPDE adducts/10^6 nucleotides)	Adducts per dose unit
Haugen et al. (1986)	2.5	0.210[b]	0.084
	8	0.255	0.032
Arnould et al. (1999)	0.64	0.034	0.053
Pavanello et al. (1999)	0.0289[a]	0.039	1.349
	0.0867[a]	0.115	1.326
	0.202[a]	0.055	0.272
	0.347[a]	0.195	0.562
Pavanello et al. (2005)	2.69[a]	0.044	0.016
Van Schooten et al. (1990)	0.8	0.051	0.064
			Mean 0.418

[a]Exposure doses were originally expressed in micromoles of 1-pyrenol per mole of creatine; 1 μmol 1-pyrenol/mol creatine = 0.289 μg BaP/m³

[b]Excluding the women in the study who had a very large number of adducts

formed by BaP is higher than that of most endogenous adducts. That this might be the case is supported by the finding that in male F344 rats, 0.85 aflatoxine adducts per 10^6 nucleotides are associated with a liver tumour incidence of 50% (Buss et al. 1990) at 2 years. The reported spontaneous liver tumour incidence in male F344 rats is only 1–2%. The level of endogenous DNA adducts as measured by Nath and Chung (1994), Chung et al. (1999, 2000), Wu et al. (1999), Walker et al. (1993), Föst et al. (1989), Nair et al. (1995), Loureiro et al. (2002), Schuler et al. (1997), Fraga et al. (1990) and Biancini et al. (1993), only including acrolein-dG, crotonaldehyde-dG, N^7-(2-hydroxyethyl)guanine, HNE-dG, etheno-dA, etheno-dC, 1,N^2-etheno-2′-dG, 8-oxo-dG and 7-methyldeoxyguanosine, amounts to 2.18–137.3 adducts per 10^6 nucleotides. The minimum level (2.18) was reached by adding up the lowest level observed for each of the endogenous adduct types considered, while the maximum level (137.3) was based on the highest level observed.

Occupational exposures involve complex mixtures of PAHs other than BaP. So BaP adducts can be considered as a measure of total PAH exposure, the impact of which on cancer incidence is much greater than that of the component BaP alone. We based the following calculations of the lung cancer risk associated with PAH exposure in industries measured as BaP on the meta-analysis of Armstrong et al. (2004). Using data from 39 studies, they calculated the unit relative risk (URR) of industrial PAH exposures with BaP as a guide substance. The URR gives the relative risk at 100 μg BaP/m³ years. From the data from all 39 studies the average estimated URR was 1.20. A calculation limited to the 22 studies on coke oven, gas works and aluminium production workers led to an average estimated URR of 1.17. As it is conceivable that extrapolation from high-exposure studies to much lower exposures leads to an underestimation of risk (see below), it is of interest to make a risk assessment restricted to studies in which exposures were not above 20 μg BaP/m^3 years. Based on seven studies on coke oven, gas works and aluminium production workers in which exposures were not above 20 μg BaP/m^3 years, Armstrong et al. (2004) found an average estimated URR of 1.87. Using the equation adapted from Armstrong et al. (2004), we converted the ULR factor into the unit lifetime risk:

$$\mathrm{URR} = 1 + \left[\mathrm{ULR}\left(\frac{1}{70}\right)\left(\frac{10}{23}\right)\left(\frac{230}{365}\right)\left(\frac{1}{0.09}\right)\right],$$

where ULR is the unit lifetime risk per nanogram per cubic metre continuous (23 m^3/day vs. 10 m^3/work day) exposure (365 days vs. 230 work days) over 70 years, assuming, as did Armstrong et al. (2004), a 9% baseline lifetime risk. Using a linear model, we calculated that 0.418 BaP adducts per 10^6 nucleotides as a marker of exposure to a mixture of PAHs such as coke oven emissions could be associated with a lung cancer risk of 39×10^{-3}–200×10^{-3} (risks derived from URRs equal to 1.17 and 1.87, respectively). It should be noted that Armstrong et al. (2004) found higher but less precisely estimated URRs when they included all 39 studies on cancer risk associated with PAH exposures in industry and did not restrict their calculations to coke oven, gas works and aluminium production workers. They found an estimated URR of 1.20 based on all 39 studies and an estimated URR of 4.54 based on 21 studies concerning exposures not exceeding 20 μg/m^3 years.

Besides lung cancer, exposure to coke oven emissions was also associated with kidney cancer, prostate cancer and genitourinary cancer in male workers. Mandell et al. (1995) found a relative risk of 1.7 for kidney cancer in male workers in the blast furnace or coke oven industry (95% confidence interval, CI, 1.1–2.7). Costantino et al. (1995) reported significantly elevated relative risks for lung cancer and for prostate cancer and genitourinary cancer in general (1.95, CI 1.59–2.33; 1.57, CI 1.09–2.30; 1.49, CI 1.11–2.01, respectively). From the number of cancer deaths and the relative risks reported by Costantino et al. (1995) it can be calculated that in a population of 5,321 coke oven workers the induction of 124 excess cases of lung cancer by exposure to coke oven emissions was accompanied by the induction of 28 cases of genitourinary cancers. So 0.418 BaP adducts per 10^6 nucleotides as a marker of exposure to a mixture of PAHs such as coke oven emissions could be associated with a genitourinary cancer risk of 9×10^{-3}–45×10^{-3} and a cancer risk of 48×10^{-3}–245×10^{-3} for lung and genitourinary cancer combined.

Based on a linear model, DNA adduct levels in peripheral blood cells of the order of 0.05 BaP adducts per 10^6 nucleotides (such as observed in non-smokers, Table 8.4) could, if they resulted from exposure to air polluted by PAHs, be associated with cancer risks on the order of 5×10^{-3}–25×10^{-3}. To the extent that BaP-DNA adduct levels are due to exposures through ingestion, lung cancer risk from PAHs will be lower. This will mainly be relevant for non-smokers.

8.4.2 Supralinear Dose–Effect Relationship

Observations by Danheiser et al. (1989) and most (but not all) data collected for this review suggest a supralinear dose–effect relationship (low-dose hypersensitivity) in vitro between BaP exposure, on the one hand, and damage to DNA and mutations, on the other. Such a supralinear dose–effect relationship is also found in epidemiologic observations on humans (Ostro 2004) and the review by Armstrong et al. (2004) shows that the calculation of the risk associated with exposure to PAHs not exceeding 20 μg BaP/m^3 years (corresponding to an exposure to 1 μg BaP/m^3 over 20 years) leads to a higher value than calculations also including more intensive exposures. It is interesting to consider carcinogenic effects of environmental or other low-intensity exposures to PAHs in terms of existing knowledge of the induction of DNA repair by genotoxic agents such as ionizing radiation. There is substantial evidence that exposure to a first dose (adaptive dose) of a few milligrays up to a few hundred milligrays (up to 500 mGy in some circumstances; Broome et al. 2002) leads to a reduction in the mutagenic effects of a second, higher dose (Wolf 1998; UNSCEAR 2000). This phenomenon called adaptation, which has also been described for genotoxic substances, is probably due to the induction of DNA repair mechanisms (Wolf 1998). The critical dose for induction of the adaptive response is probably on the order of 1 mGy, with no induction of DNA double strand break repair or adaptive response occurring below this dose (Broome et al. 2002; Rothkamm and Löbrich 2003). The adaptive response takes one or more hours to be established and disappears after about 40 h (Wolf 1998). Exposures to ionizing radiation in doses below the threshold level for an adaptive response could be, per unit dose, more efficient in the induction of mutations than higher doses (low dose hypersensitivity) (Joiner et al. 1996).

Lifelong exposure of a human population to ionizing radiation at a level corresponding to the threshold dose per time period during which the adaptive response persists, i.e. 1 mGy per 40 h or 15.3 Gy per 70 years, can be estimated to lead to a risk of death from cancer of 100% before age 70. Indeed, according to UNSCEAR (2000) the risk of death from all solid cancers combined amounts to 11% following an acute dose of 1 Sv, with total solid cancer incident risks roughly twice those for mortality, while chronic exposure might lead to only 50% of the aforementioned risk. That exposures to very low doses of ionizing radiation do lead to the induction of cancer in humans is demonstrated by observations concerning radon in houses (Doll 1998), by observations made after the Chernobyl disaster (Okeanov et al. 2004) and by the fact that low-dose γ-radiation induced cytogenetic effects in residents of buildings (Chang et al. 1997). Risks associated with environmental and most occupational exposures to BaP and complex PAH mixtures are far lower than the risk associated with exposure to ionizing radiation at the threshold dose per time period during which the adaptive response persists.

It is likely that these exposures to PAHs do not reach a threshold level for induction of DNA repair and that consequently a higher efficiency in the induction of biological and health effects (compared with high experimental or occupational doses) might be expected. This might explain supralinear effects suggested by some epidemiological data.

PAHs such as BaP play an important role in the induction of lung cancer by tobacco smoke and by particulate matter in ambient air (Denissenko et al. 1996; Petruzzelli et al. 1998). A recent study from Norway (Bjartveit and Tverdal 2005) found that smoking one to four cigarettes per day is associated with an almost threefold increase in relative risk of dying from cardiovascular disease or from lung cancer, whereas Vineis et al. (2000) found that the increase of lung cancer risk in heavy smokers levelled of above a relative risk of 20. According to the WHO, the response per dose unit in the induction of lung cancer and of cardiopulmonary disease by particulate matter in ambient air is higher for low exposures than for more intensive exposures (Ostro 2004).

Acknowledgement: This work was realized in the framework of the activities of the Focal Point Environmental Health (Steunpunt Milieu en Gezondheid) of the Flemish Community. This work does not, though, reflect the position of the Focal Point. It only reflects the opinion of the authors.

References

Albert RE, Miller ML, Cody TE, Talaska G, Underwood P, Andringa A (1996) Epidermal cytokinetics, DNA adducts, and dermal inflammation in the mouse skin in response to repeated benzo(a)pyrene exposures. Toxicol Appl Pharmacol 136:67–74

Alexandrov K, Rojas M, Geneste O, Castegnaro M, Camus AM, Pertuzzelli S, Giuntini C, Bartsch H (1992) An improved fluorometric assay for dosimetry of benzo(a)pyrene diol-epoxide-DNA adducts in smokers' lung: comparisons with total bulky adducts and aryl hydrocarbon hydroxylase activity. Cancer Res 52:6248–6253

Alexandrov K, Sala M, Rojas M (1988) Differences in the DNA adducts formed in cultured rabbit and rat dermal fibroblasts by benzo(a)pyrene and (–)-benzo(a)pyrene-7,8-diol. Cancer Res 48:7132–7139

Allen-Hoffmann BL, Rheinwald JG (1984) Polycyclic aromatic hydrocarbon mutagenesis of human epidermal keratinocytes in culture. Proc Natl Acad Sci USA 81:7802–7806

Armstrong B, Hutchinson E, Unwin J, Fletcher T (2004) Lung cancer risk after exposure to polycyclic aromatic hydrocarbons: a review and meta-analysis. Environ Health Perspect 112:970–978

Arnould JP, Verhoest P, Bach V, Libert JP, Belegaud J (1997) Detection of benzo(a)pyrene-DNA adducts in human placenta and umbilical cord blood. Hum Exp Toxicol 16:716–721

Arnould JP, Pfohl-Leszkowicz A, Bach V, Libert JP, Belegau, J (1999) Biological monitoring exposure of workers from plant producing carbon electrodes: quantification of benzo(a)pyrene DNA-adducts in leukocytes, by a 32P-postlabelling method and an immunoassay. Hum Exp Toxicol 18:314–321

Ashurst SW, Cohen GM (1980) A benzo(a)pyrene-7,8-dihydrodiol-9,10-epoxide is the major metabolite involved in the binding of benzo(a)pyrene to DNA in isolated viable rat hepatocytes. Chem Biol Interact: 29:117–127

Ashurst SW, Cohen GM (1981) In vivo formation of benzo(a)pyrene diol epoxide-deoxyadenosine adducts in the skin of mice susceptible to benzo(a)pyrene-induced carcinogenesis. Int J Cancer 27:357–364

Aust AE, Falahee KJ, Maher VM, McCormick J (1980) Human cell-mediated benzo(a)pyrene cytotoxicity and mutagenicity in human diploid fibroblasts. Cancer Res, 40:4070–4075

Banasiewicz M, Nelson G, Swank A, Grubor N, Ross J, Nesnow S, Köfeler H, Small J, Jankowiak R (2004) Identification and quantitation of benzo(a)pyrene-derived DNA adducts formed at low adduction level in mice lung tissue. Anal Biochem 334:390–400

Bartley J, Baratholomew JC, Stampfer MR (1982) Metabolism of benzo(a)pyrene by human epithelial and fibroblastic cells: metabolite patterns and DNA adduct formation. J Cell Biochem 18:135–148

Bianchini F, Montesano R, Shuker DEG, Cuzick J, Wild CP (1993) Quantification of 7-methyldeoxyguanosine using immunoaffinity purification and HPLC with electrochemical detection. Carcinogenesis 14:1677–1682

Binková B, Giguère Y, Rössner P Jr, Dostál M, Šrám RJ (2000) The effect of dibenzo(a,l)pyrene and benzo(a)pyrene on human diploid lung fibroblasts: the induction of DNA adducts, expression of p53 and $p21^{WAF1}$ proteins and cell cycle distribution. Mutat Res 471:57–70

Bjartveit K, Tverdal A (2005) Health consequences of smoking 1-4 cigarettes per day. Tob Control 14:315–320

Bodell WJ, Devanesan PD, Rogan EG, Cavalieri EL (1989) 32P-postlabeling analysis of benzo(a)pyrene-DNA adducts formed in vitro and in vivo. Chem Res Toxicol 2:312–315

Brooks RA, Gooderham NJ, Edwards RJ, Boobis A,. Winton DJ (1999) The mutagenicity of benzo(a)pyrene in mouse small intestine. Carcinogenesis 20:109–114

Broome EJ, Brown DL, Mitchel RE (2002) Dose responses for adaption to low doses of (60)Co gamma rays and (3)H beta particles in normal human fibroblasts. Radiat Res 158:181–186

Buss P, Caviezel M, Lutz WK (1990) Linear dose-response relationship for DNA adducts in rat liver from chronic exposure to aflatoxin B1. Carcinogenesis 11:2133–2135

Canova S, Viezzer C, Venier P, Navazio L, Revoltella R, Celotti L (1996) Metabolic activation of benzo(a)pyrene in two fetal mouse hepatocyte lines: induction of DNA adducts and micronuclei. Mutat Res 367:135–141

Chang WP, Hwang BF, Wang D, Wang JD (1997) Cytogenetic effect of chronic low-dose, low-dose-rate γ-radiation in residents of irradiated buildings. Lancet 350:330–333

Chen L, Devanesan PD, Higginbotham S, Ariese F, Jankowiak R, Small G., Rogan EG, Cavalieri EL (1996) Expanded analysis of benzo(a)pyrene-DNA adducts formed in vitro and in mouse skin: their significance in tumor initiation. Chem Res Toxicol 9:897–903

Chung FL, Zhang L, Ocando JE, Nath RG (1999) Role of 1,N^2-propanodeoxyguanonsine adducts as endogenous DNA lesions in rodents and humans. In: Singer B, Bartsch H (eds) IARC scientific publication no 150: Exocyclic DNA adducts in mutagenesis and carcinogenesis. International Agency on Cancer, Lyons

Chung FL, Nath RG, Ocando J, Nishikawa A, Zhang L (2000) Deoxyguanosine adducts of t-4-hydroxy-2-nonenal are endogenous DNA lesions in rodents and humans: detection and potential sources. Cancer Res 60:1507–1511

Costantino JP, Redmond CK, Bearden A (1995) Occupationally related cancer risk among coke oven workers: 30 years of follow-up. J Occup Environ Med 37:597–604

Culp SJ, Warbritton AR, Smith B., Li EE, Beland FA (2000) DNA adduct measurements, cell proliferation, and tumor mutation induction in relation to tumor formation in B6C3F1 mice fed coal tar or benzo(a)pyrene. Carcinogenesis 21:1433–1440

Cunningham ML, Ringrose PS (1983) Benzo(a)pyrene and aniline increase sister chromatid exchanges in cultured rat liver fibroblasts without addition of activating enzymes. Toxicol Lett 16:235–239

Danheiser SL, Liber SL, Thilly WG (1989) Long-term, low-dose benzo(a)pyrene-induced mutation in human lymphoblasts competent in xenobiotic metabolism. Mutat Res 201:143–147

Dean SW, Coates A, Brooks TM, Burlinson B (1998) Benzo(a)pyrene site of contact mutagenicity in skin of Muta™ mouse. Mutagenesis 13:515–518

De Bont R, van Larebeke N (2004) Endogenous DNA damage in humans: a review of quantitative data. Mutagenesis 19:169–185

Deckers-Schmelzle B, Klaus E, Kahl R, Kahl GF (1978) Binding of benzo(a)pyrene metabolites to cellular DNA in perfused rat lungs. Naunyn-Schmiedebert's Arch Pharmacol 303:303–307
Dedon PC, Plastaras JP, Rouzer CA, Marnett LJ (1998) Indirect mutagenesis by oxidative DNA damage: formation of the pyrimidopurinone adduct of deoxyguanosine by base propenal. Proc Natl Acad Sci USA 95:11113–11116
Denissenko MF, Pao A, Tang M, Pfeifer GP (1996) Preferential formation of benzo(a)pyrene adducts at lung cancer mutational hotspots in p53. Science 274:430–432
Devanesan PD, RamaKrishna NVS, Todorovic R, Rogan EG, Cavalieri EL, Jeong H, Jankowiak R, Small GJ (1992) Identification and quantitation of benzo(a)pyrene-DNA adducts formed by rat liver microsomes in vitro. Chem Res Toxicol 5:302–309
DiGiovanni J, Decina PC, Prichett WP, Fisher EP, Aalfs KK (1985) Formation and disappearance of benzo(a)pyrene DNA-adducts in mouse epidermis. Carcinogenesis 6:741–747
Divi RL, Beland FA, Fu PP, Von Tungeln LS, Schoket B, Camara JE, Ghei M, Rothman N, Sinha R, Poirier MC (2002) Highly sensitive chemiluminescence immunoassay for benzo(a)pyrene-DNA adducts: validation by comparison with other methods, and use in human biomonitoring. Carcinogenesis 23:2043–2049
Doll R (1998) Effects of small doses of ionising radiation. J Radiol Protect 18:163–174
Dosaka H, Abe S, Sasaka M, Miyamoto H, Kawakami Y (1987) Sister chromatid exchange induction by benzo(a)pyrene in cultured peripheral blood lymphocytes of lung cancer patients and healthy individuals with or without familial history of neoplasms. Int J Cancer 39:329–332
Föst U, Marczynski B, Kasemann R, Peter H (1989) Determination of 7-(2-hydroxyethyl)guanine with gas chromatography/mass spectrometry as a parameter for genotoxicity of ethylene oxide. Arch Toxicol Suppl 13:250–253
Fraga CG, Shigenaga MK, Park JW, Degan P, Ames BN (1990) Oxidative damage to DNA during aging: 8-hydroxy-2′-deoxyguanosine in rat organ DNA and urine. Proc Natl Acad Sci USA 87:4533–4537
Godschalk RWL, Vermeer TM, Kriek E, Floot B, Schilderman PAEL, Moonen EJC, Kleinjans JCS, van Schooten FJ (1997) Comparison of 32P-postlabeling and HPLC-FD analysis of DNA adducts in rats acutely exposed to benzo(a)pyrene. Chem Biol Interact 104:41–54
Hakura A, Tsutsui Y, Sonoda J, Kai J, Imade T, Shimada M, Sugihara Y, Mikami T (1998) Comparison between in vivo mutagenicity and carcinogenicity in multiple organs by benzo(a)pyrene in the lacZ transgenic mouse (Muta™ mouse). Mutat Res 398:123–130
Hanelt S, Helbig R, Hartmann A, Lang M, Seidel A, Speit G (1997) A comparative investigation of DNA adducts, DNA strand breaks and gene mutations induced by benzo(a)pyrene and (±)-anti-benzo(a)pyrene-7,8-diol 9,10-oxide in cultured human cells. Mutat Res 390:179–188
Hartmann A, Speit G (1995) Genotoxic effects of chemicals in the single cell gel (SCG) test with human blood cells in relation to the induction of sister-chromatid exchanges (SCE). Mutat Res 346:49–56
Haugen A, Becher G, Benestad C, Vahakangas K, Trivers GE, Newman MJ, Harris CC (1986) Determination of polycyclic aromatic hydrocarbons in the urine, benzo(a)pyrene diol epoxide–DNA adducts in lymphocyte DNA, and antibodies to the adducts in sera from coke oven workers exposed to measured amounts of polycyclic aromatic hydrocarbons in the work atmosphere. Cancer Res 46:4178–4183
Huberman E, Sachs L, Yang SK, Gelboin HV (1976) Identification of mutagenic metabolites of benzo(a)pyrene in mammalian cells. Proc Natl Acad Sci USA 73:607–611
Inoue K, Shibata T, Abe T (1983) Induction of sister-chromatid exchanges in human lymphocytes by indirect carcinogens with and without metabolic activation. Mutat Res 117:301–309
Jacono FL, Stecca C, Duverger M (1992) Mutagenic activation of benzo(a)pyrene by human red blood cells. Mutat Res 268:21–26
Joiner MC, Lambin P, Malaise EP, Robson T, Arrand JE, Skov KA, Marples B (1996) Hypersensitivity to very-low single radiation doses: its relationship to the adaptive response and induced radioresistance. Mutat Res 358:171–183
Li D, Wang M., Cheng L, Spitz MR, Hittelman WN, Wei Q (1996) In vitro induction of benzo(a)pyrene diol epoxide-DNA adducts in peripheral lymphocytes as a susceptibility marker for human lung cancer. Cancer Res 56:3638–3641

Liehr JG (2000) Is estradiol a genotoxic mutagenic carcinogen? Endocrine Rev 21:40–54
Loureiro APM, Marques SA, Garcia CCM, Di Mascio P, Medeiros HG (2002) Development of an on-line liquid chromatography-electrospray tandem mass spectrometry assay to quantitatively determine 1,N^2-etheno-2′-deoxyguanosine in DNA. Chem Res Toxicol 15:1302–1308
Machanoff R, O'Neill J.P, Hsie AW (1981) Quantitative analysis of cytotoxicity and mutagenicity of benzo(a)pyrene in mammalian cells (CHO/HGPRT system). Chem Biol Interact 34:1–10
Manchester DK, Weston A, Choi JS, Trivers GE, Fennessey PV, Quitana E, Farmer PB, Mann DL, Harris CC (1988) Detection of benzo(a)pyrene diol epoxide-DNA adducts in human placenta. Proc Natl Acad Sci USA 85:9243–9247
Mandell JS, McLaughlin JK, Schlehofer B, Mellemgaard A, Helmert U, Lindblad P, McCredie M, Adami HO (1995) International renal-cell cancer study. IV. Occup Int J Cancer 61:601–605
Miller RL, Garfinkel R, Horton M, Camann D, Perera FP, Whyatt RM, Kinney PL (2004) Polycyclic aromatic hydrocarbons, environmental tobacco smoke, and respiratory symptoms in an inner-city birth cohort. Chest 126:1071–1078
Monroe JJ, Kort KL, Miller JE, Marino DR, Skopek TR (1998) A comparative study of in vivo mutation assays: analysis of hprt, lacI, and cII/cI as mutational targets for N-nitroso-N-methylurea and benzo(a)pyrene in big blue mice. Mutat Res 421:121–136
Moore CJ, Pruess-Schwartz D, Mauthe RJ, Gould MN, Baird WM (1987) Interspecies differences in the major DNA adducts formed from benzo(a)pyrene but not 7,12-dimethylbenz(a)anthracene in rat and human mammary cell cultures. Cancer Res 47:4402–4406
Muir C, Waterhouse J, Mack T, Powell J, Whelan S (1987) Cancer incidence in five continents, vol V. IARC scientific publication 88. International Agency for Research on Cancer, Lyons
Nair J, Barbin A, Guichard Y, Bartsch H (1995) 1,N^6-Ethenodeoxyadenosine and 3,N^4-ethenodeoxycytidine in liver DNA from humans and untreated rodents detected by immunoaffinity/32P-postlabelling. Carcinogenesis 16:613–617
Nath RG, Chung FL (1994) Detection of exocyclic 1,N^2-propanodeoxyguanosine adducts as common DNA lesions in rodents and humans. Proc Natl Acad Sci USA 91:7491–7495
Nowak D, Meyer A, Schmidt-Preuss U, Gatzemeier U, Magnussen H, Rüdiger HW (1992) Formation of benzo(a)pyrene-DNA adducts in blood monocytes from lung cancer patients with a familial history of lung cancer. J Cancer Res Clin Oncol 118:67–71
Oberly TJ, Huffman DM, Scheuring JC, Garriott ML (1993) An evaluation of 6 chromosomal mutagens in the AS52/XPRT mutation assay utilizing suspension culture and soft agar cloning. Mutat Res 319:179–187
OEHHA (1994) Benzo(a)pyrene as a toxic air contaminant. Prepared by the Staffs of the California Air Resources Board and the Office of Environmental Health Hazard Assessment
Okeanov AE, Sosnovskaya EY, Priatkina OP (2004) A national cancer registry to assess trends after the Chernobyl accident. Swiss Med Wkly 134:645–649
Ostro B (2004) Outdoor air pollution: assessing the environmental burden of disease at national and local levels. WHO environmental burden of disease series no 5. World Health Organization, Geneva
Pavanello S, Favretto D, Brugnone F, Mastrangelo G, Dal Pra G, Clonfero E (1999) HPLC/fluorescence determination of anti-BPDE-DNA adducts in mononuclear white blood cells from PAH-exposed humans. Carcinogenesis 20:431–435
Pavanello S, Pulliero A, Siwinska E, Mielzynska D, Clonfero E (2005) Reduced nucleotide excision repair and GSTM1-null genotypes influence anti-BPDE-DNA adduct levels in mononuclear white blood cells of highly PAH-exposed coke oven workers. Carcinogenesis 26:169–175
Petruzzelli S, Celi A, Pulera N, Baliva F, Vlegi G, Carrozzi L, Ciacchini G, Bottai M, Di Pede F, Paoletti P, Giuntini C (1998) Serum antibodies to benzo(a)pyrene diol epoxide-DNA adducts in the general population: effects of air pollution, tobacco smoking, and family history of lung diseases. Cancer Res 58:4122–4126
Pot F, Heinrich U (1990) Relative significance of different hydrocarbons for the carcinogenic potency of emissions from various incomplete combustion processes. IARC Sci Publ 104:288–297
Ramos KS, Moorthy B (2005) Bioactivation of polycyclic aromatic hydrocarbon carcinogens within the vascular wall: implications for human atherogenesis. Drug Metab Rev 37:595–610

Ray-Chaudhuri R, Kelley S, Iype PT (1980) Induction of sister chromatid exchanges by carcinogens mediated through cultured rat liver epithelial cells. Carcinogenesis 1:779–786
Roberts MJ, Wondrak GT, Laurean DC, Jacobson MK, Jacobson EL (2003) DNA damage by carbonyl stress in human skin cells. Mutat Res 522:45–56
Rogan EG, Devanesan PD, RamaKrishna NVS, Higginbotham S, Padmavathi NS, Chapman K, Cavalieri EL, Jeong H, Jankowiak R, Small GJ (1993) Identification and quantitation of benzo(a)pyrene-DNA adducts formed in mouse skin. Chem Res Toxicol 6:356–363
Roggeband R, Wolterbeek APM, Rutten AAJJL, Baan RA (1993) Comparative 32P-postlabeling analysis of benzo(a)pyrene-DNA adducts formed in vivo upon activation of benzo(a)pyrene by human, rabbit and rodent liver microsomes. Carcinogenesis 14:1945–1950
Roggeband R, Van Den Berg PTM, Van Der Wulp CJM, Baan RA (1994a) Detection of DNA adducts in basal and non-basal cells of the hamster trachea exposed to benzo(a)pyrene in organ culture. J Histochem Cytochem 42:1427–1434
Roggeband R, Wolterbeek APM, van den Berg PTM, Baan RA (1994b) DNA adducts in hamster and rat tracheas exposed to benzo(a)pyrene in vitro. Toxicol Lett 72:105–111
Rojas M, Alexandrov K, van Schooten FJ, Hillebrand M, Kriek E, Barsch H (1994) Validation of a new fluorometric assay for benzo(a)pyrene diolepoxide-DNA adducts in human white blood cells: comparisons with 32P-postlabeling and ELISA. Carcinogenesis 15:557–560
Ross J, Nelson G, Kligerman A, Erexson G, Bryant M, Earley K, Gupta R, Nesnow S (1990) Formation and persistence of novel benzo(a)pyrene adducts in rat lung, liver, and peripheral blood lymphocyte DNA. Cancer Res 50:5088–5094
Rothkamm K, Löbrich M (2003) Evidence for a lack of DNA double-strand break repair in human cells exposed to very low x-ray doses. Proc Natl Acad Sci USA 100:5057–5062
Rüdiger HW, Marxen J, Kohl FV, Melderis H, van Wichert P (1979) Metabolism and formation of DNA adducts of benzo(a)pyrene in human diploid fibroblasts. Cancer Res 39:1083–1088
Rüdiger HW, Nowak D, Hartmann K, Cerutti P (1985) Enhanced formation of benzo(a) pyrene:DNA adducts in monocytes of patients with a presumed predisposition to lung cancer. Cancer Res 45:5890–5894
Salama SA, Sierra-Torres CH, Oh HY, Hamada FA, Au WW (2001) Variant metabolizing gene alleles determine the genotoxicity of benzo(a)pyrene. Environ Mol Mutagen 37:17–26
Schuler D, Otteneder M, Saglsdorff P, Eder E, Gupta RC, Lutz WK (1997) Comparative analysis of 8-oxo-2′-deoxyguanosine in DNA by 32P- and 33P-postlabeling and electrochemical detection. Carcinogenesis 18:2367–2371
Shamsuddin AKM, Sinopoli NT, Hemminki K, Boesch RR, Harris CC (1985) Detection of benzo(a)pyrene:DNA adducts in human white blood cells. Cancer Res 45:66–68
Shane S, de Boer J, Watson DE, Haseman JK, Glickman BW, Tindall KR (2000) Lac I mutation spectra following benzo(a)pyrene treatment of Big Blue mice. Carcinogenesis 21:715–725
Shugart L (1985) Covalent binding of benzo(a)pyrene diol epoxide to DNA of mouse skin: in vivo persistence of adducts formation. J Toxicol Environ Health 15:255–263
Stoner GD, Daniel FB, Schenck KM, Schut HAJ, Goldblatt PJ, Sandwisch DW (1982) Metabolism and DNA binding of benzo(a)pyrene in cultured human bladder and bronchus. Carcinogenesis 3:195–201
Talaska G, Jaeger M, Reilman R, Collins T and Warshawsky D (1996) Chronic, topical exposure to benzo(a)pyrene induces relatively high steady-state levels of DNA adducts in target tissues and alters kinetics of adduct loss. Proc Natl Acad Sci USA 93:7789–7793
Theall G, Eisinger M, Grunberger D (1981) Metabolism of benzo(a)pyrene and DNA adduct formation in cultured human epidermal keratinocytes. Carcinogenesis 2:581–587
Tong C, Brat V, Williams GM (1981) Sister-chromatid exchange induction by polycyclic aromatic hydrocarbons in an intact cell system of adult rat-liver epithelial cells. Mutat Res 91:467–473
UNSCEAR (2000) Sources and effects of ionising radiation. United Nations Scientific Committee on the Effects of Atomic Radiation UNSCEAR 2000 report to the General Assembly, with scientific annexes. Volume II: Effects. http://www.unscear.org/reports/2000_2.html

Van Schooten FJ, Hillebrand MJX, Scherer E, den Engelse L, Kriek E (1991) Immunocytochemical visualization of DNA adducts in mouse tissues and human white blood cells following treatment with benzo(a)pyrene or its diol epoxide. A quantitative approach. Carcinogenesis 12:427–433

Van Schooten FJ, van Leeuwen FE, Hillebrand MJX, Rijke ME, Hart AAM, van Veen HG, Oosterink S, Kriek E (1990) Determination of benzo(a)pyrene-diol-epoxide-DNA adducts in white blood cell DNA from coke oven workers: the impact of smoking, J Natl Cancer Inst 82:927–933

Vineis P, Kogevinas M, Simonato L, Brennan P, Boffetta P (2000) Levelling-off of the risk of lung and bladder cancer in heavy smokers: an analysis based on multicentric case-control studies and a metabolic interpretation. Mutat Res 463:103–110

Walker VE, Fennel TR, Upton, B, MacNeela JP, Swenberg JA (1993) Molecular dosimetry of DNA and hemoglobin adducts in mice and rats exposed to ethylene oxide. Environ. Health Perspect 99:11-17

Weston A, Bowman ED (1991) Fluorescence detection of benzo(a)pyrene-DNA adducts in human lung. Carcinogenesis 12:1445–1449

Weston A, Rowe ML, Manchester DK, Farmer PB, Mann DL, and Harris CC (1989) Fluorescence and mass spectral evidence for the formation of benzo(a)pyrene anti-diol-epoxide-DNA and -hemoglobin adducts in humans. Carcinogenesis 10:251–257

Wiencke J, McDowell ML, Bodell WJ (1990) Molecular dosimetry of DNA adducts and sister chromatid exchanges in human lymphocytes treated with benzo(a)pyrene. Carcinogenesis 11:1497–1502

Wolff S (1998) The adaptive response in radiobiology: evolving insights and implications. Environ Health Perspect 106(Suppl 1):277–283

Wu KY, Ranasinghe A, Upton PB, Walker VE, Swenberg JA (1999) Molecular dosimetry of endogenous and ethylene oxide-induced N7-(2-hydroxyethyl)guanine formation in tissues of rodents. Carcinogenesis 20:1787–1792

9 Analysis of DNA Double-Strand Breaks by Means of γ-H2AX Foci

IRENA SZUMIEL

Abstract

DNA double strand break (DSB) quantification with the use of physicochemical methods is now often replaced by analysis of γ-H2AX foci in the nuclei of cells damaged with DSB-inducing agents. The H2A histone variant, H2AX, becomes phosphorylated by the phosphatidylinositol 3-kinase-like protein kinases, ATM and ATR, within 1–3 min after γ-irradiation. The phosphorylated form, called γ-H2AX, can be visualised as discrete foci with the use of specific antibodies with fluorescent tags and directly counted in the fluorescence microscope. More advanced techniques include flow cytometry or the use of confocal microscopy, image collection and computational analysis of distributions of foci size, fluorescence intensity and number per cell. Modification of the H2AX molecules is necessary to alter the local chromatin organisation in the surrounding of a DSB and thus to increase the accessibility of DNA to various repair factors which are colocalised in the foci. The foci are dispensable for the initial recruitment of proteins and for the early repair and signalling events. Afterwards, however, foci are needed to concentrate and retain the repair proteins in the vicinity of DSB. This seems to be important for repair of persistent lesions. The kinetics of DSB repair and that of formation and disappearance of foci do not correspond closely. Foci also are occasionally formed in S-phase cells as a result of broken replication forks. Nevertheless, γ-H2AX foci are now generally accepted as a DSB marker and are used to estimate radiation sensitivity.

9.1 Introduction

DNA double-strand breaks (DSB) are the most deleterious lesions directly or indirectly induced by DNA-damaging agents, and are occasionally formed during DNA replication owing to broken (collapsed) replication forks. Until recently, DSB quantification was time consuming and based on relations between physicochemical properties of isolated DNA or DNA in chromatin and the extent of DNA fragmentation under non-denaturing conditions. The DSB number was rather difficult to accurately estimate. The discovery that

Günter Obe and Vijayalaxmi (Eds.)
Chromosomal Alterations: Methods, Results and Importance in Human Health

discrete foci containing phosphorylated histone H2AX and some other DNA repair related proteins can be seen in the nuclei of irradiated cells and that there is an approximately 1:1 relation of foci to the calculated number of DSB (Sedelnikova et al. 2002; Rothkam and Loebrich 2003) created a basis for direct DSB estimation in single cells. Essentially, this is feasible; however, there are many confounding factors to be considered. Nevertheless, the method proved to be appropriate for detection of low damage levels and very useful in investigations of the DNA repair mechanisms; its application supplied much new information on the cellular response to DNA damage, as reviewed in this chapter.

9.2 Histone H2AX

Histones in vivo are subject to various modifications, such as phosphorylation, acetylation, methylation and ubiquitination. Phosphate moieties are transferred by nuclear enzymes on specific serine and threonine residues, usually located near the amino or carboxyl termini ("tails") of the histone molecules. Thus, their chemical properties are altered and this brings about changes in their interaction with DNA and with other proteins, and, in consequence, in the chromatin structure. Such changes are a necessary requirement for various functions of DNA performed in cells.

It has long been known that histones are heterogenous, but not all aspects of this heterogeneity are understood. An example of special interest for investigators of DNA damage is the family of H2A histones. It consists of three subfamilies present in varying proportions in various cell types: H2A1-H2A2, H2AZ and H2AX. In mammals, histone H2AX represents 2–25% of the H2A family (Redon et al. 2002). A specific function in DNA repair has been ascribed to the phosphorylated form of histone H2AX, called γ-H2AX (Rogakou et al. 1998). The current state of knowledge of the biological role of histone γ-H2AX is outlined in the following sections.

9.2.1 Histone H2AX Is Phosphorylated in Response to Induction of DNA Double-Strand Breaks

One effect of ionising radiation (or other agents that introduce DSB into DNA) in vivo and in vitro is induction of posttranslational modifications of proteins which can be detected by electrophoresis on two-dimensional gels on the basis of altered electrophoretic mobility and molecular mass. Labelling with ^{32}P can be used to indicate freshly phosphorylated molecules. In this way, phosphorylation of histone H2AX at serine 139 was identified (at the C-terminus of the molecule) and the modified form was named γ-H2AX (Rogakou et al. 1998).

The same modification of H2AX takes place in cells undergoing DNA fragmentation due to apoptosis (Rogakou et al. 2000). DSB are transient intermediates in genomic rearrangements within immunoglobulin and T-cell receptor genes (Chen et al. 2000; Petersen et al. 2001) and these processes also are accompanied by generation of γ-H2AX. Likewise, treatment with drugs that form DNA–protein cross-links or DNA adducts and present a steric hindrance for replication forks induces DSB and H2AX phosphorylation (Furuta et al. 2003; Meng et al. 2005). Altogether, at present, it is generally accepted that γ-H2AX serves as a marker of DSB that are both directly and indirectly generated. Detection of γ-H2AX in cells on microscope slides by immunofluorescence with the use of specific antibodies became a standard procedure in radiobiology and cytotoxicology.

When detected in nuclear extracts by western blotting, γ-H2AX molecules appear very soon after exposure of cells to ionising radiation; in some cells, half-maximal amounts are reached after 1 min and maximal amounts usually after 10–30 min. For example, γ-H2AX formation was studied in Chinese hamster ovary (CHO) cells (Rogakou et al. 1998), irradiated with 200 Gy of γ- rays on ice and then returned to the incubator for recovery at 37°C. As soon as 20 s after return to the incubator, γ-H2AX was detected, and the maximal level of γ-H2AX was reached after 10 min and then it gradually declined during the following 90 min. In the same paper, it was shown that γ-H2AX is evolutionarily conserved from yeast to human, indicating that its formation is part of a universal response to DSB.

9.2.2 γ-H2AX Is Present in the Nuclear Foci

When DSB are induced in DNA, a complex cellular response is triggered. One component of this response is accumulation of proteins involved in DNA repair or damage signalling. They can be seen as nuclear foci in the fluorescence microscope when specific antibodies with fluorescent tags are applied (Maser et al. 1997). With the use of an anti-γ-H2AX antibody with a fluorescent tag, a pattern of γ-H2AX foci can be revealed within 1–3 min after γ-irradiation (Paull et al. 2000) with doses 0.6–2 Gy, considerably lower than that used for the biochemical studies mentioned above (Rogakou et al. 1998). An example of γ-H2AX foci in non-irradiated and X-irradiated CHO cells is shown in Fig. 9.1.

9.2.3 The Phosphatidylinositol 3-Kinase-Like Protein Kinases Are the Main Enzymes that Phosphorylate Histone H2AX

The presence of DSB in nuclear DNA activates three related protein kinases, all belonging to the family of phosphatidylinositol-3-like kinases: DNA-dependent protein kinase (DNA-PK), ataxia telangiectasia mutated (ATM) – its function is impaired in the heritable hypersensitivity to ionising radiation) – and ATM and

A B

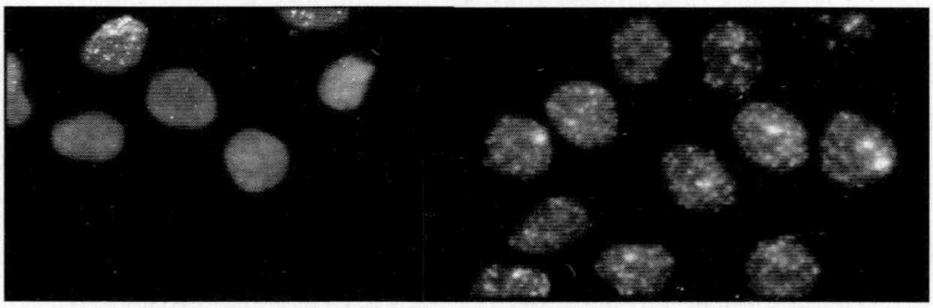

Fig. 9.1 γ-H2AX foci in CHO-K1 Chinese hamster ovary cells, **a** control and **b** X-irradiated (1 Gy, 20 min after exposure), as seen in the fluorescence microscope at × 400 magnification. (Courtesy of Maria Wojewódzka)

Rad3 related (ATR) – activated by arrest of DNA replication. These kinases form part of a monitoring system that detects DNA damage, generates alarm-signalling cascades and recruits repair enzymes to the damaged sites (reviewed in Szumiel 1998). The substrate specificities of these kinases in vitro are similar, but in vivo they have distinct substrates according to their functions.

All three kinases may carry out phosphorylation of H2AX in vivo in a redundant, overlapping manner (Stiff et al. 2004; Wang et al. 2005), but H2AX phosphorylation is practically absent in Atm–/– cells (devoid of both alleles of the Atm gene), whereas it is normal in DNA-PKcs–/– cells (lacking the catalytic subunit of the DNA-PK kinase) (Burma et al. 2001). Apparently, ATM is the main kinase that modifies H2AX, whereas DNA-PK is responsible for the minimal levels of H2AX phosphorylation in Atm–/– cells, since low doses of wortmannin, an inhibitor of phosphatidylinositol-3-like kinases, hence, also of DNA-PK, eliminate this trace amount of γ-H2AX. This observation is consistent with the finding that a fraction of nuclear ATM colocalises with γ-H2AX at sites of DSB in response to DNA damage (Andegeko et al. 2001). Foci with phosphorylated ATM (thus, with the kinase function activated) can be detected after irradiation; they appear and disappear with similar kinetics as γ-H2AX foci (Suzuki et al. 2006). Some studies implicate the contribution of additional kinases to H2AX phosphorylation (Wang et al. 2005).

When irradiated cells are treated with wortmannin, there is no effect on the formation of foci. In contrast, treatment of cells before irradiation completely prevents the accumulation of all components of the foci. These data show that H2AX phosphorylation takes place immediately after damage infliction and precedes sequential assembly of the studied repair proteins into repair foci (Paull et al. 2000). So, the likely scenario after irradiation is that a signal is produced to indicate that a DSB is present and must be repaired, ATM is activated at the site of the DSB and it immediately phosphorylates histone H2AX at the site of the break. This may then initiate the recruitment and assembly into foci of DNA repair, cell cycle checkpoint and damage-signalling factors (Burma et al. 2001; Fernandez-Capetillo et al. 2004).

9.2.4 Megabase Chromatin Domains Alter Conformation After H2AX Phosphorylation

As discussed by Rogakou et al. (1999) and Paull et al. (2000), modification of the H2AX molecules is necessary to alter the local chromatin organisation in the surrounding of a DSB and thus to increase the accessibility of DNA to various repair factors. Rogakou et al. (1998) calculated that approximately 1% of the H2AX becomes γ-phosphorylated per gray of ionising radiation. This radiation dose induces about 35 DSB per mammalian G_1 genome, which in 6×10^9 bp generates γ- H2AX distributed over 1% of the chromatin, i.e. about 0.03% of the H2AX histone appears to be modified per DSB, corresponding to about 2×10^6 bp of DNA per DSB. Several cell lines were examined which differed in the amount of the H2AX histone; nevertheless, the percentage of modified H2AX per DSB was approximately the same in all cells studied (0.03–0.06).

9.3 Repair and Checkpoint Factors Are Recruited into γ-H2AX Foci

After exposure to ionising radiation, γ-H2AX foci in the nuclei appear within minutes (Paull et al. 2000). Other components of the repair and checkpoint system (visualised via antibodies with different fluorescent tags) also can be detected by immunofluorescence as discrete foci that colocalise with γ-H2AX foci. Observation of the time course of localisation of various proteins in the foci indicates that they are assembled in a defined sequence. Mediator of DNA damage checkpoint protein 1 (MDC1) and p53 binding protein 1 (53BP1) colocalise very rapidly – 5–15 min after irradiation (Schultz et al. 2000; Stucki et al. 2005). MDC1 plays an important role in assembling other repair and checkpoint proteins (Xu and Stern 2003). Stucki and Jackson (2006) indicate that the interacting MDC1 and γ-H2AX molecules form an "affinity platform" on which other proteins are assembled and retained (Lukas et al. 2004; Stucki et al. 2005).

At a later time (about 1 h) γ-H2AX becomes colocalised with breast cancer 1 (BRCA1) followed by accumulation of RAD 50 or RAD 51 in different subsets of cells. RAD 51 is a component of the homologous recombination repair (HRR) system. RAD 50 is a part of the MRN complex (MRE11–RAD 50–NBS1, where NBS1 is Nijmengen breakage syndrome 1), which is a DSB sensor and hence is present at the site of the DSB immediately after damage generation (Lee and Paull 2005); however, it is also active in both HRR and non-homologous end joining (NHEJ). The components of the complex rapidly accumulate in the nuclear foci (Celeste et al. 2003a). The fork head associated (FHA) and the BRCA1 C-terminal (BRCT) domains of NBS1 are essential for physical interaction with γ-H2AX (Kobayashi et al. 2002) and thus for the localisation in the foci. Likewise, the BRCT repeat domain of MDC1 binds directly to γ-H2AX. Interestingly, RAD 50 and RAD 51 foci are not present in the same cell after damage induction (Maser et al. 1997), indicating that the two repair pathways, NHEJ and HRR, mutually exclude one another in individual cells.

9.4 Early DSB Rejoining Can Proceed Without Foci Formation

The recruitment of repair and checkpoint proteins to the sites of DSB is not dependent on the previous generation of the γ-H2AX foci, as concluded from the fact that X-irradiated H2AX-deficient cells are able to repair DSB and arrest progression through the cell cycle. Obviously, the foci are dispensable for the initial recruitment of proteins and for the early repair and signalling events. Afterwards, however, foci are needed to concentrate and retain the proteins in the vicinity of DSB (Celeste et al. 2003a). This seems to be important for repair of persistent lesions, since the H2AX-deficient cells are more radiosensitive than the wild-type cells, and mutant mice show defects in class switch recombination and spermatogenesis, decreased genomic stability and enhanced susceptibility to carcinogens and to ionising radiation (Celeste et al. 2003 a, b; Pilch et al. 2003). In agreement with these statements is the finding that the number of γ-H2AX foci gradually increases during the first 30–60 min after irradiation (Han et al. 2006; Wykes et al. 2006), whereas during that time a considerable fraction of DSB are repaired as judged from the results of the neutral comet assay or pulsed-field gel electrophoresis (PFGE).

9.5 Foci May Be Present Long After DSB Are Rejoined

Attempts were made to establish relations between the kinetics of DSB rejoining and foci numbers also in the late repair periods. In theory, repaired DSB should give a signal for dephoshorylation of γ-H2AX molecules and disintegration of the focus. Indeed, protein phosphatase 2A (PP2A) is involved in removing γ-H2AX foci. The PP2A catalytic subunit and γ-H2AX coimmunoprecipitate and colocalise in DNA damage foci and PP2A dephosphorylates the histone in vitro (Chowdhury et al. 2005). Hence, a decrease in foci frequency or in the amount of γ-H2AX detected by flow cytometry or western blotting should reflect the repair kinetics and correlate with survival; however, these relations are not so straightforward:

- DSB rejoining measured by PFGE or comet assay proceeds much faster than the decrease in foci frequency (Banath et al. 2004).
- Decrease in foci frequency is not associated with a decrease in the global γ-H2AX amount (estimated by flow cytometry or western blotting), suggesting that the modified histone is redistributed in the chromatin (Bouquet et al. 2006).
- DSB rejoining is not correlated with a decrease in the global γ-H2AX amount. For example, human melanoma cells with mutated ATM show a constant γ-H2AX level for 2 h after irradiation with 2 Gy X-rays, whereas the DSB rejoining proceeds normally (MacPhail et al. 2003a).
- When PP2A is inhibited or silenced by RNA interference, γ-H2AX foci persist, and the cells are hypersensitive to DNA damage.

Generally, it is assumed that persistent foci mark unrepaired or misrepaired DSB. Banath et al. (2004) pointed out that residual unrepaired DSB may be left undetected by the comet assay, which has a detection limit of 50 DSB per cell. Thus, the presence of only a few DSB per cell may keep the alarm system of the cell switched on and prevent the disassembly of the foci. Stucki and Jackson (2006) postulate that foci serve to amplify the signalling for cell cycle checkpoints in order to keep the cell safely arrested until all the residual damage has been repaired.

9.6 Practical Aspects of γ-H2AX Estimation

A description of the techniques for γ-H2AX histone detection can be found in Nakamura et al. (2006). Figure 9.2 shows an example of a procedure applied to obtain microscopic preparations suitable for foci scoring. The procedure must be modified according to the properties of the cell type studied, but the basic steps are the same: fixation, permeabilisation, blocking, incubation with the primary antibody followed by a secondary antibody. It should be added that since the early period of direct foci scoring with the use of an epifluorescence microscope, much more sophisticated approaches have become available: the use of confocal microscopy (especially important with less flat cells), image collection and computational analysis (Costes et al. 2006), including distributions of foci size, fluorescence intensity and number per cell. Figure 9.3 shows an example of foci analysis.

An alternative for foci scoring is flow cytometry. The detection of γ-H2AX is based on indirect immunofluorescence with the use of a secondary antibody tagged with fluorescein isothiocyanate (FITC), whereas DNA is counterstained with propidium iodide (PI). Then, bivariate analysis of the intensity of cellular green (FITC) and red (PI) fluorescence data can be used to assess the relation between the presence of γ-H2AX as a marker of DSB and the cell cycle phase (Huang and Darzynkiewicz 2006). Figure 9.4 shows examples of the flow cytometry estimation of γ-H2AX.

- Seed 1×10^5 cells into 6-well culture plate containing a glass cover slip in each well
- Treat with a DNA damaging agent
- Fix in 4% paraformaldehyde for 15 min
- Wash with PBS
- Permeabilize in 0.2% Triton X-100
- Block with blocking serum for 1.5 h
- Incubate with a mouse monoclonal anti-γH2AX antibody (1:1000) for 2 h
- Incubate with FITC-conjugated goat-anti-mouse secondary antibody (1:500) for 1h
- Add DAPI (4′, 6-diamidino-2-phenylindole) to the cells and incubate for 15 min
- Remove the cover slips from the plate and mount on to glass slides
- To prevent bias in selection of cells that display foci, count all cells in the field of vision (at least 50 cells)

Fig. 9.2 Example of the experimental procedure for preparation of microscope slides for examination of nuclear foci according to the description in Kaneko et al. (2005)

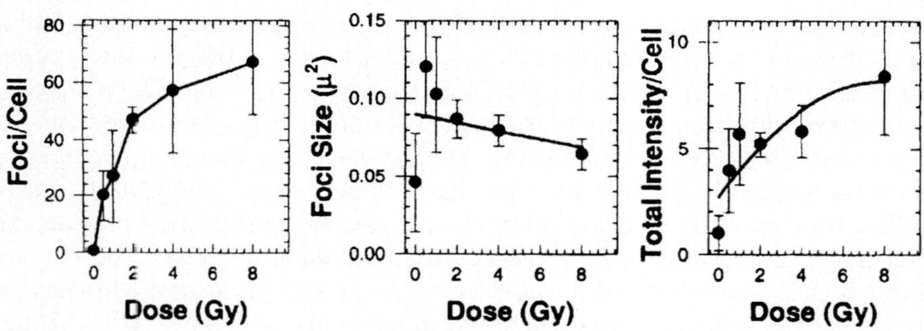

Fig. 9.3 Characterisation of γ-H2AX foci detected 30 min after various doses of X-rays in Chinese hamster V79 cells: average foci number per cell, average foci area per cell and average total cell fluorescence (mean and standard deviation, $n = 5$–10 cells). (Reproduced from MacPhail et al. 2003 with permission of Taylor and Francis Ltd.)

9.6.1 Foci Size and Phosphorylation Heterogeneity

As stated already, γ-H2AX serves as a generally acknowledged marker of DSB. Nevertheless, with the use of more advanced detection techniques, variation in size and differential distribution in cell cycle phases can be found even in

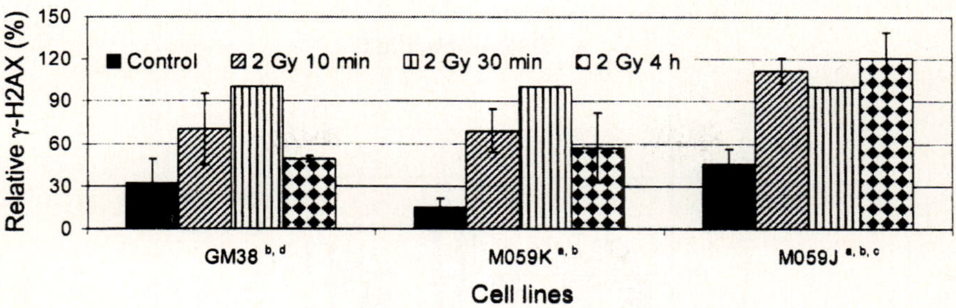

Fig. 9.4 Relative γ-H2AX phosphorylation levels determined after irradiation with 2-Gy γ-rays by flow cytometry and calculated from data obtained from three experiments on human fibroblasts GM38 and human glioma M059K and M059J cells, the latter with defective DNA dependent protein kinase dependent non-homologous end-joining repair. Values for the fluorescence signal intensity of cells stained with anti-γ-H2AX antibody were normalised to the value of the corresponding 30-min sample, arbitrarily set as 100%). Statistically significant differences: *a* $P<0.05$, control compared with 2 Gy, 10 min; *b* $P<0.05$, control compared with 2 Gy, 30 min; *c* $P<0.05$, control compared with 2 Gy, 4 h; *d* $P<0.05$, 2 Gy, 30 min compared with 2 Gy, 4 h; Student's *t* test. (Reproduced from Han et al. 2006 with permission of the Radiation Research Society)

cells not treated with DNA-damaging agents. Also the degree of phosphorylation considerably differs between individual foci.

Small foci are present in S-phase control cells and in all cell cycle phases early after low dose, low linear energy transfer (LET) irradiation (MacPhail et al. 2003b; Costes et al. 2006). It is assumed that replication forks occasionally collapse and the DSB thus generated evokes a normal cellular response aimed at repair. This accounts for the S-phase foci but not for those in other cell cycle phases. According to Han et al. (2006) in the analysed human cells, most of the DSB-unrelated foci (apart from those in early S-phase cells) have a volume smaller than 0.04 μm^3 and they persist after irradiation in unchanged numbers. Their origin and role are unknown. Figure 9.5 shows the numbers of DSB-related and DSB-unrelated foci with S-phase cells excluded from the analysis (although this exclusion does not change the general conclusions from these data; Han et al. 2006). It shows that the numbers of DSB-unrelated foci are very high and that without a careful analysis, the data on γ-H2AX foci can lead to erroneous conclusions.

Large foci can be found at a high incidence in non-irradiated senescent human cells and it is plausible to assume that they mark permanent DNA damage, since they also are found 2 h after high LET irradiation (Costes et al. 2006). In about 20–30% of foci, at higher magnification, with more sophisticated equipment it is possible to discern two to three spots within one large focus, seen as bright centres encircled with a dim halo. The number of spots better corresponds with the DSB number than that of the foci and may explain the often noted underestimation of foci number in relation to DSB (Han et al. 2006).

The extent of γ-H2AX phoshorylation and hence of fluorescence intensity of individual foci considerably differs, as shown in Fig. 9.6. It may be

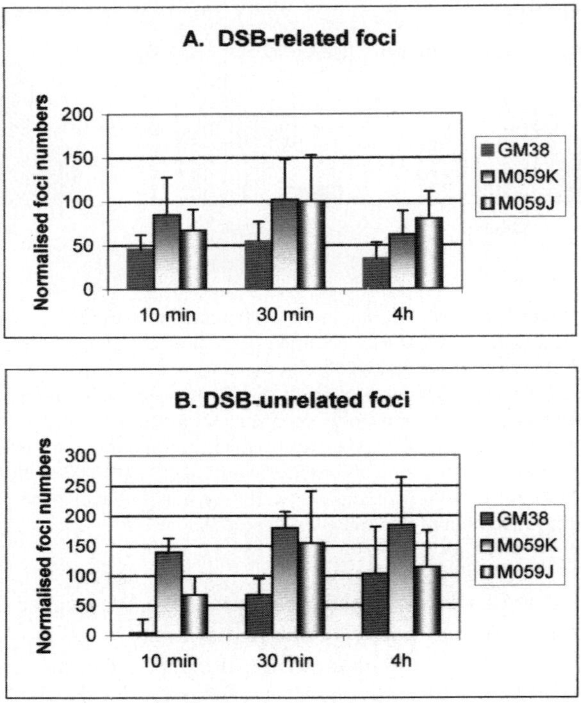

Fig. 9.5 Average numbers of double strand break (*DSB*) related **a** and DSB-unrelated **b** γ-H2AX foci after irradiation with 2 Gy γ-rays (mean ± standard deviation) with S-phase cells excluded from the analysis. Numerical data (numbers of foci normalised to the average nuclear volume at 30 min after irradiation) for human fibroblasts GM38 and human glioma M059K and M059J cells were taken from Han et al. (2006)

expected that as the foci are formed after damage infliction or persist for some time after DSB repair, the histone becomes gradually phosphorylated or dephosphorylated. The kinetics of these processes may depend on the complexity of the damage at individual sites.

Taking into account these confounding factors, it is advisable to set threshold levels of fluorescence intensity (e.g. one tenth of the maximal intensity) and of volume (e.g. 0.04 μm^3) (Han et al. 2006) in order to exclude DSB-unrelated foci from the analysis. The threshold values should be optimised for each cell type.

9.6.2 γ-H2AX Foci Frequency As a Measure of Radiosensitivity

The formation of foci is proportional to radiation dose and is similar in cells with differential radiosensitivity (Wykes et al. 2006). In contrast, the relation

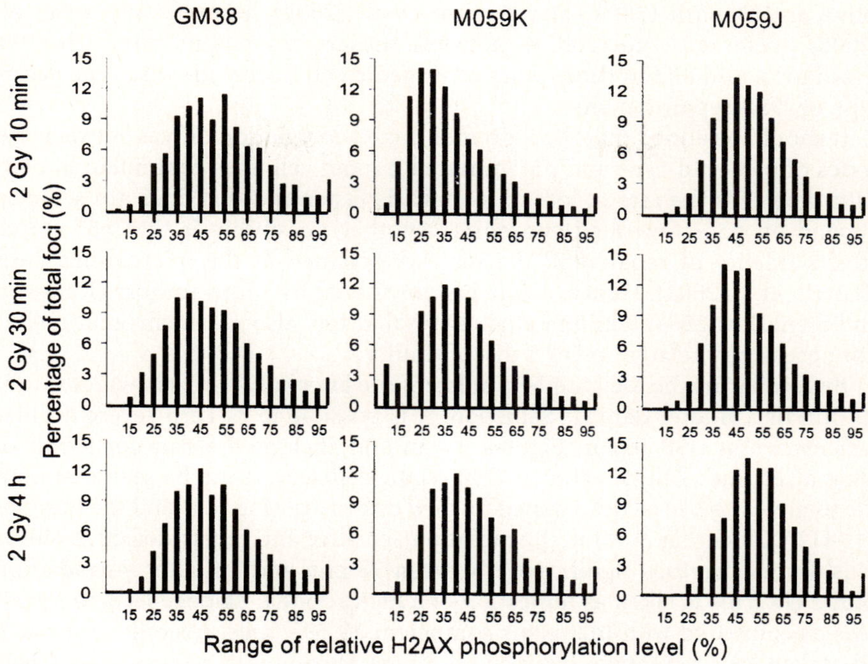

Fig. 9.6 Histogram of the relative H2AX phosphorylation level within DSB-related γ-H2AX foci in human fibroblasts GM38 and human glioma M059K and M059J cells. The *x*-axis represents the series of ranges of numerical value of the relative H2AX phosphorylation level (percentages of the phosphorylation level of the most highly phosphorylated foci in each individual cell) into which the data are sorted in a histogram analysis, and the *y*-axis represents the percentages of total DSB-related γ-H2AX foci within this range. (Reproduced from Han et al. 2006 with permission of the Radiation Research Society)

between the decrease in foci number and repair is by no means clear, as mentioned in Sect. 9.5. Then, what features of the cellular response to DSB are reflected by foci frequency at later times after DSB infliction?

Although the kinetics of focus formation and of the fast component of DSB repair (half-life 15–30 min) are not compatible, the slow component of DSB repair (1–4 h or more) would be consistent with the persistent accumulation of repair factors in foci. In this context, it is important that slowly repaired residual damage is better correlated with cell survival than the rate of fast repair (Badie et al. 1995). Also, correlations can be found between foci frequencies after 4 or 24 h, and radiation sensitivity expressed as surviving fraction after irradiation with 2 Gy (SF2). As discussed by Olive and Banath (2004), the latter parameter usually corresponds to the residual damage. Accordingly, such a correlation between the half-life of γ-H2AX loss and SF2 for a series of normal and tumour cell lines and murine tissues with a large variation in SF2 values was found by McPhail et al. (2003a),

Olive and Banath (2004) and Banath et al. (2004), whereas Wykes et al. (2006) presented a correlation between the mean focus number 4 h after irradiation and SF2 in three pairs of isogenic cell lines with wild type genotype or DSB repair defects.

It should be added that the aforementioned correlation differs between the series of p53-wild-type and p53-mutated human cell lines examined: in p53-wild-type cells the rate of loss of γ-H2AX is faster and this does not seem to depend on p53-controlled apoptosis. Another possibility is that p53 status affects fidelity of repair and the foci are retained at the misrejoined DSB (Banath et al. 2004). Hence, a conclusion, shared by some authors (Karlsson and Stenerlow 2004; Ichijima et al. 2005) that foci also mark chromatin sites not necessarily related to DNA discontinuity.

By comparing the foci numbers at later time intervals, 4 or 24 h, with those at 30 min or 1 h after damage infliction, it is possible to discern severe cellular deficiencies in DSB repair. Figures 9.4 and 9.5 include the data for a pair of glioma cell lines, one of which is DSB repair deficient. It can be seen that both the total γ-H2AX level determined by flow cytometry (Fig. 9.4) and the number of γ-H2AX foci (Fig. 9.5a) in the radiation-sensitive and repair-defective M059J cells do not decrease significantly between 30 min and 4 h after γ-irradiation with 2 Gy. This is taken as indication of a high residual damage level in M059J cells as compared with the repair competent M059K cells. A similar approach was taken to differentiate between wild-type murine cells (Atm+/+) and heterozygous cells (Atm+/–) with an intermediate radiosensitivity between normal and that of the homozygous recessive mutant (Atm–/–). No sharp distinction could be made after acute (2.5 Gy/min) irradiation, but 24-h exposure at 10 cGy/h allowed the radiosensitivity differences between the genotypes examined to be clearly distinguished, based on the foci frequencies (Kato et al. 2006).

Based on the same principle, an attempt was made to apply foci frequency as a predictor of radiosensitivity (Taneja et al. 2004). In a series of human tumour cell lines and xenograft tumours in athymic nude mice, γ-irradiation (3 Gy) induced γ-H2AX foci to a similar extent in all cell types, but the kinetics of focus removal were distinct and clearly dependent on radiosensitivity. Moreover, in a model for fractionated radiotherapy, accumulation of damage was estimated after a single dose or sequential doses of γ-radiation (a single 9-Gy dose or three fractions of 3 Gy at 24-h intervals. Divided doses enhanced the differences in the persistence of γ-H2AX foci between the head and neck cancer cell lines, radioresistant SQ-20B and radiosensitive SCC-61 cells: in SQ-20B cells, the foci disappeared with similar kinetics after each 3-Gy dose, whereas in SCC-61 cells, their number, size and fluorescence intensity increased. Western blot analysis of SQ-20B and SCC-61 cell extracts gave the same response pattern. Thus, γ-H2AX clearance may serve as a predictive test of tumour radiosensitivity to be applied in radiotherapy. Estimation of γ-H2AX accumulation by serial biopsy would help to establish "individually tailored therapy" by choosing the radiation dose and fractionation regimen according to radiosensitivity of the patient's tumour.

9.6.3 γ-H2AX Foci Size and Frequency Depending on Radiation Quality

In a sharp contrast to the uniform distribution of foci in cells exposed to X-rays, in the nuclei of cells irradiated with high LET radiation the foci are clustered (Goto et al. 2002). This provides direct support for other observations which indicate that high LET radiation produces highly clustered DNA damage, including complex lesions. In consequence, the repair proceeds at a slower rate than that of damage caused by low LET radiation (Pinto et al. 2005; Rydberg et al. 2005). Delayed disappearance of foci is, therefore, to be expected. Indeed, such observations were made by Karlsson and Stenerlow (2006) and Leatherbarrow et al. (2006): the relative numbers of residual foci correlated well with the residual damage estimated with the use of PFGE.

The formation of high and low LET radiation induced foci was studied by Costes et al. (2006). HCA2 human foreskin diploid fibroblasts in a confluent state were exposed to 30 cGy of 130 keV/μm nitrogen ions or X-rays (320 kV). The frequencies of the foci were initially similar for both kinds of radiation, but differences were seen both in the plateau level and in the time it was reached: 45 min after X-irradiation and 15 min after nitrogen ion irradiation. The frequency at the plateau was about 3.5 foci per 100 μm^2 for low LET radiation and about 2.0 foci per 100 μm^2 for high LET radiation. Another difference was in the focus size: in the 120 min after exposure, the average size increased from 0.20 to 0.45 μm^2 for high LET radiation and stayed in the same range from 0.22 to 0.30 μm^2 for low LET radiation. The size increase was accompanied by almost doubled fluorescence intensity, indicating a further phosphorylation of the histone during the 2-h period. Analysis of the size distribution allowed the authors to assign the role of high LET radiation to the very large (more than 0.7 μm^2) foci, the proportion of which rose from about 2% 5 min after exposure to about 20% 2 h after exposure to nitrogen ions.

9.7 Concluding Remarks

As outlined in this chapter, analysis of γ-H2AX foci in the nuclei of cells damaged with DSB-inducing agents allows us to identify the locality of the damage, its severity, repair and – when additional repair and checkpoint proteins are examined in overlapping foci – also to acquire new information on the molecular processes that take place at the damaged sites. Although γ-H2AX foci are now generally accepted as a DSB marker, there are background foci differing in size and number in various cell types and growth conditions (Banath et al. 2004) and this may complicate the analysis.

Recently, ATM foci have been proposed as an alternative method of DSB detection, more specific for ionising radiation induced damage (Suzuki et al. 2006). These foci rarely form in non-irradiated S-phase cells, are induced immediately after irradiation in a linear relation to dose in a dose range from 10 mGy to 1 Gy, and do not differ in the distribution of the mean focus size and the kinetics of disappearance in this dose range. With all these merits there is one disadvantage: for reasons unknown, in irradiated cells the size of S-phase foci is about half that of G1- or G2-phase foci, whereas the foci number is almost 2 times higher. So, for the use for exponentially growing cell cultures this difference must be taken into consideration. Clearly, the choice of marker for damage foci should be made depending on the aim of the experiment, not forgetting the confounding factors.

References

Andegeko Y, Moyal L, Mittelman L, Tsarfaty I, Shiloh Y, Rotman G (2001) Nuclear retention of ATM at sites of DNA double strand breaks. J Biol Chem 276:38224–3830

Badie C, Iliakis G, Foray N, Alsbeih G, Cedervall B, Chavaudra N, Pantelias G, Arlett C, Malaise EP (1995) Induction and rejoining of DNA double-strand breaks and interphase chromosome breaks after exposure to X rays in one normal and two hypersensitive human fibroblast cell lines. Radiat Res 144:26–35

Banath JP, MacPhail SH, Olive PL (2004) Radiation sensitivity, H2AX phosphorylation, and kinetics of repair of DNA strand breaks in irradiated cervical cancer cell lines. Cancer Res 64:7144–7149

Bouquet F, Muller C, Salles B (2006) The loss of gammaH2AX signal is a marker of DNA double strand breaks repair only at low levels of DNA damage. Cell Cycle 5:1116–1122

Burma S, Chen BP, Murphy M, Kurimasa A, Chen DJ (2001) ATM phosphorylates histone H2AX in response to DNA double-strand breaks. J Biol Chem 276:42462–42467

Celeste A, Fernandez-Capetillo O, Kruhlak MJ, Pilch DR, Staudt DW, Lee A, Bonner RF, Bonner WM, Nussenzweig A (2003a) Histone H2AX phosphorylation is dispensable for the initial recognition of DNA breaks. Nat Cell Biol 5:675–679

Celeste A, Difilippantonio S, Difilippantonio MJ, Fernandez-Capetillo O, Pilch DR, Sedelnikova OA, Eckhaus M, Ried T, Bonner WM, Nussenzweig A. (2003b) H2AX haploinsufficiency modifies genomic stability and tumor susceptibility. Cell 114:371–383

Chen HT, Bhandoola A, Difilippantonio MJ, Zhu J, Brown MJ, Tai X, Rogakou EP, Brotz TM, Bonner WM, Ried T, Nussenzweig A (2000) Response to RAG-mediated VDJ cleavage by NBS1 and gamma-H2AX. Science 29:1962–1965

Chowdhury D, Keogh MC, Ishii H, Peterson CL, Buratowski S, Lieberman J (2005) gamma-H2AX dephosphorylation by protein phosphatase 2A facilitates DNA double-strand break repair. Mol Cell 20:801–809

Costes SV, Boissiere A, Ravani S, Romano R, Parvin B, Barcellos-Hoff MH (2006) Imaging features that discriminate between foci induced by high- and low-LET radiation in human fibroblasts. Radiat Res 165:505–515

Fernandez-Capetillo O, Lee A, Nussenzweig M, Nussenzweig A (2004) H2AX: the histone guardian of the genome. DNA Repair 3:959–967

Furuta T, Takemura H, Liao ZY, Aune GJ, Redon C, Sedelnikova OA, Pilch DR, Rogakou EP, Celeste A, Chen HT, Nussenzweig A, Aladjem MI, Bonner WM, Pommier Y (2003) Phosphorylation of histone H2AX and activation of Mre11, Rad50, and Nbs1 in response to replication-dependent DNA double-strand breaks induced by mammalian DNA topoisomerase I cleavage complexes. J Biol Chem 278:20303–20312

Goto S, Watanabe M, Yatagai F (2002) Delayed cell cycle progression in human lymphoblastoid cells after exposure to high-LET radiation correlates with extremely localized DNA damage. Radiat Res 158:678–686

Han J, Hendzel MJ, Allalunis-Turner J (2006) Quantitative analysis reveals asynchronous and more than DSB-associated histone H2AX phosphorylation after exposure to ionizing radiation. Radiat Res 165:283–292

Huang X, Darzynkiewicz Z (2006) Cytometric assessment of histone H2AX phosphorylation: a reporter of DNA damage. Methods Mol Biol 314:73–80

Ichijima Y, Sakasai R, Okita N, Asahina K, Mizutani S, Teraoka H (2005) Phosphorylation of histone H2AX at M phase in human cells without DNA damage response. Biochem Biophys Res Commun 336:807–812

Kaneko H, Igarashi K, Kataoka K, Miura M (2005) Heat shock induces phosphorylation of histone H2AX in mammalian cells. Biochem Biophys Res Commun 328:1101–1106

Karlsson KH, Stenerlöw B (2004) Focus formation of DNA repair proteins in normal and repair-deficient cells irradiated with high-LET ions. Radiat Res 161:517–527

Kato TA, Nagasawa H, Weil MM, Genik PC, Little JB, Bedford JS (2006) γ-H2AX foci after low-dose-rate irradiation reveal *Atm* haploinsufficiency in mice. Radiat Res 166:47–54

Kobayashi J, Tauchi H, Sakamoto S, Nakamura A, Morishima K, Matsuura S, Kobayashi T, Tamai K, Tanimoto K, Komatsu K (2002) NBS1 localizes to gamma-H2AX foci through interaction with the FHA/BRCT domain. Curr Biol 12:1846–1851

Leatherbarrow EL, Harper JV, Cucinotta FA, O'Neill P (2006) Induction and quantification of gamma-H2AX foci following low and high LET-irradiation. Int J Radiat Biol 82:111–118

Lee JH, Paull TT (2005) ATM activation by DNA double-strand breaks through the Mre11-Rad50-Nbs1 complex. Science 308:551–554

Lukas C, Melander F, Stucki M, Falck J, Bekker-Jensen S, Goldberg M, Lerenthal Y, Jackson SP, Bartek J, Lukas J (2004) Mdc1 couples DNA double-strand break recognition by Nbs1 with its H2AX-dependent chromatin retention. EMBO J 23:2674–2683

MacPhail SH, Banath JP, Yu TY, Chu EH, Lambur H, Olive PL (2003a) Expression of phosphorylated histone H2AX in cultured cell lines following exposure to X-rays. Int J Radiat Biol 79:351–358

MacPhail SH, Banath JP, Yu Y, Chu E, Olive PL (2003b) Cell cycle-dependent expression of phosphorylated histone H2AX: reduced expression in unirradiated but not X-irradiated G1-phase cells. Radiat Res 159:759–767

Maser RS, Monsen KJ, Nelms BE, Petrini JH (1997) hMre11 and hRad50 nuclear foci are induced during the normal cellular response to DNA double-strand breaks. Mol Cell Biol 17:6087–6096

Meng LH, Kohlhagen G, Liao ZY, Antony S, Sausville E, Pommier Y (2005) DNA-protein cross-links and replication-dependent histone H2AX phosphorylation induced by aminoflavone (NSC 686288), a novel anticancer agent active against human breast cancer cells. Cancer Res 65:5337–5343

Nakamura A, Sedelnikova OA, Redon C, Pilch DR, Sinogeeva NI, Shroff R, Lichten M, Bonner WM (2006) Techniques for gamma-H2AX detection. Methods Enzymol 409:236–250

Olive PL, Banath JP (2004) Phosphorylation of histone H2AX as a measure of radiosensitivity. Int J Radiat Oncol Biol Phys 58:331–335

Paull TT, Rogakou EP, Yamazaki V, Kirchgessner CU, Gellert M, Bonner WM (2000) A critical role for histone H2AX in recruitment of repair factors to nuclear foci after DNA damage. Curr Biol 10:886–895

Petersen S, Casellas R, Reina-San-Martin B, Chen HT, Difilippantonio MJ, Wilson PC, Hanitsch L, Celeste A, Muramatsu M, Pilch DR, Redon C, Ried T, Bonner WM, Honjo T, Nussenzweig MC, Nussenzweig A (2001) AID is required to initiate Nbs1/gamma-H2AX focus formation and mutations at sites of class switching. Nature 414:660–665

Pilch DR, Sedelnikova OA, Redon C, Celeste A, Nussenzweig A, Bonner WM (2003) Characteristics of gamma-H2AX foci at DNA double-strand breaks sites. Biochem Cell Biol 81:123–129

Pinto M, Prise KM, Michael BD (2005) Evidence for complexity at the nanometer scale of radiation-induced DNA DSBs as a determinant of rejoining kinetics. Radiat Res 164:73–85

Redon C, Pilch D, Rogakou E, Sedelnikova O, Newrock K, Bonner W (2002) Histone H2A variants H2AX and H2AZ. Curr Opin Genet Dev 12:162–169

Rogakou EP, Pilch DR, Orr AH, Ivanova VS, Bonner WM (1998) DNA double-stranded breaks induce histone H2AX phosphorylation on serine 139. J Biol Chem 273:5858–5868

Rogakou EP, Boon C, Redon C, Bonner WM (1999) Megabase chromatin domains involved in DNA double-strand breaks in vivo. J Cell Biol 146:905–916

Rogakou EP, Nieves-Neira W, Boon C, Pommier Y, Bonner WM (2000) Initiation of DNA fragmentation during apoptosis induces phosphorylation of H2AX histone at serine 139. J Biol Chem 275:9390–9395

Rothkamm K, Lobrich M (2003) Evidence for a lack of DNA double-strand break repair in human cells exposed to very low x-ray doses. Proc Natl Acad Sci USA 100:5057–5062

Rydberg B, Cooper B, Cooper PK, Holley WR, Chatterjee A (2005) Dose-dependent misrejoining of radiation-induced DNA double-strand breaks in human fibroblasts: experimental and theoretical study for high- and low-LET radiation. Radiat Res 163:526–534

Schultz LB, Chehab NH, Malikzay A, Halazonetis TD (2000) p53 binding protein 1 (53BP1) is an early participant in the cellular response to DNA double-strand breaks. J Cell Biol 151:1381–1390

Sedelnikova OA, Rogakou EP, Panyutin IG, Bonner WM (2002) Quantitative detection of (125)IdU-induced DNA double-strand breaks with gamma-H2AX antibody. Radiat Res 158:486–492

Stiff T, O'Driscoll M, Rief N, Iwabuchi K, Lobrich M, Jeggo PA (2004) ATM and DNA-PK function redundantly to phosphorylate H2AX after exposure to ionizing radiation. Cancer Res 64:2390–2396

Stucki M, Jackson SP (2006) γH2AX and MDC1: Anchoring the DNA-damage-response machinery to broken chromosomes. DNA Repair 5:534–543

Stucki M, Clapperton JA, Mohammad D, Yaffe MB, Smerdon SJ, Jackson SP (2005) MDC1 directly binds phosphorylated histone H2AX to regulate cellular responses to DNA double-strand breaks. Cell 123:1213–1226

Suzuki K, Okada H, Yamauchi M, Oka Y, Kodama S, Watanabe M (2006) Qualitative and quantitative analysis of phosphorylated ATM foci induced by low-dose ionizing radiation. Radiat Res 165:499–504

Szumiel I (1998) Monitoring and signaling of radiation-induced damage in mammalian cells. Radiat Res 150:S92–101

Taneja N, Davis M, Choy JS, Beckett MA, Singh R, Kron SJ, Weichselbaum RR (2004) Histone H2AX phosphorylation as a predictor of radiosensitivity and target for radiotherapy. J Biol Chem 279:2273–2280

Wang H, Wang M, Wang H, Bocker W, Iliakis G (2005) Complex H2AX phosphorylation patterns by multiple kinases including ATM and DNA-PK in human cells exposed to ionizing radiation and treated with kinase inhibitors. J Cell Physiol 202:492–502

Wykes SM, Piasentin E, Joiner MC, Wilson GD, Marples B (2006) Low-dose hyper-radiosensitivity is not caused by a failure to recognize DNA double-strand breaks. Radiat Res 165:516–524

Xu X, Stern DF (2003) NFBD1/MDC1 regulates ionizing radiation-induced focus formation by DNA checkpoint signaling and repair factors. FASEB J 17:1842–1848

10 Comet Assay

Wolfgang-Ulrich Müller

Abstract

The "comet assay" was designed by Ostling and Johanson in 1984. Its name was derived from the shape of the nuclear DNA after some steps of manipulation. The comet assay can be applied to all cells, tissues and organs as long as it is possible to generate single cells. The concept of the assay is comparatively simple. In order to study DNA damage either directly after exposure to an agent or after repair has taken place, single cells are placed in a matrix on a microscope slide, all non-DNA material is removed, and an electrophoresis run in a weak electrical field is carried out. Depending on the experimental conditions, various types of DNA damage result in negatively charged loops and fragments which are drawn in the direction of the anode and either form the tail of the "comet" or, in the case of small fragments, migrate into the anode material. The comet assay in its various modifications definitely is a valuable tool for the assessment of DNA damage and is now used in many fields of research. A major problem is that many researchers underestimate the difficulties involved with the proper execution of this assay. The greatest danger for any seemingly "simple" assay is an uncritical use of the method resulting in a lot of contradictory results with the final result of abandoning the assay, because it is "unreliable". Hopefully, the comet assay will not meet that fate.

10.1 Introduction

The "comet assay" was designed by Ostling and Johanson (1984). Its name was derived from the shape of the nuclear DNA (Fig. 10.1) after some steps of manipulation, which will be outlined below. Meanwhile, the name "single cell gel electrophoresis" (SCGE) is used quite often, although this term is somewhat misleading as electrophoresis is not done with "single cells", but with the DNA of single cells.

The comet assay can be applied to all cells, tissues and organs as long as it is possible to generate single cells. The concept of the assay is comparatively simple. In order to study DNA damage either directly after exposure to an

Günter Obe and Vijayalaxmi (Eds.)
Chromosomal Alterations: Methods, Results and Importance in Human Health

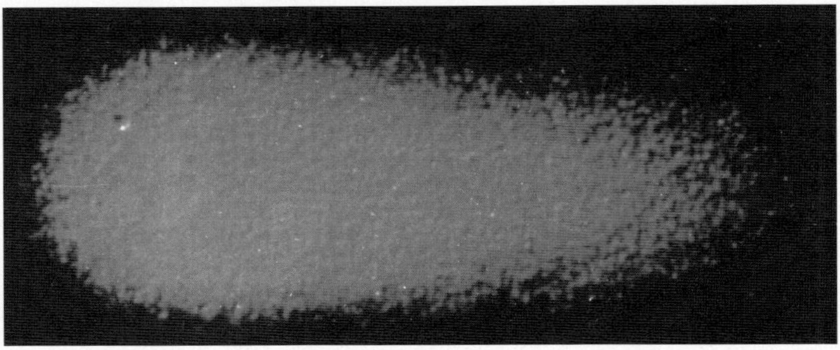

Fig. 10.1 Comet after exposure of lymphocytes to 10-Gy X-rays

agent or after repair has taken place, single cells are placed in a matrix on a microscope slide, all non-DNA material is removed (some nuclear-core proteins will persist) and an electrophoresis run in a weak electrical field is carried out. Depending on the experimental conditions, various types of DNA damage result in negatively charged loops and fragments which are drawn in the direction of the anode and either form the tail of the "comet" or, in the case of small fragments, migrate into the anode material. Thus, the release of DNA from a highly supercoiled DNA–protein complex is actually measured in the comet assay and there is a lot of evidence that the comet tail consists of DNA loops and not of DNA fragments (Collins et al. 1997b).

The comet assay is now used in many fields of research. In the following, the most important aspects of this assay will be outlined and some crucial issues that must be kept in mind, when running this assay, will be addressed.

10.2 Method

10.2.1 Basic Aspects

Many modifications of the method originally introduced by Ostling and Johanson (1984) have been suggested (Singh et al. 1988; Olive et al. 1990; Müller et al. 1994). Detailed guidelines with in-depth discussion of all the methodological aspects have been published (Rojas et al. 1999; Tice et al. 2000; Olive 2002; Collins 2004; Müller et al. 2004).

The principal steps of the procedure, however, have not changed:

- Use single cells
- Embed the cells into agarose on a microscope slide
- Remove all non-DNA material
- Unwind DNA
- Perform an electrophoresis run

- Stain the comets
- Evaluate

Basically, two approaches (with various modifications) are used: unwinding and electrophoresis are done either under more or less neutral or under alkaline conditions. Irrespective of the protocol chosen, it is of utmost importance to standardize the experimental conditions as far as possible. This is particularly important with regard to temperature, pH and voltage. An increase in voltage, which is usually in the range between 0.5 and 5 V/cm, can induce comet tails even in unexposed cells. In addition, one has to keep in mind that different cell types may need different assay conditions, which have to be identified before the experiments proper are started.

10.2.2 Scoring of Comets

Various methods are used for the quantitative assessment of the comets. Essentially, two different approaches have been introduced: visual scoring (Garcia et al. 2004) and image analysis (Böcker et al. 1997). In both cases, there are many different end points that are used to achieve a quantitative result. End points in visual scoring are, for example, the length of the comet tail or grouping into two to five categories according to the estimation by eye of the amount of DNA in the comet tail. In image analysis, fluorescence of the DNA in the comet is measured quantitatively and the results are expressed as, for example, tail-to-head ratio, percentage of DNA in the tail, or tail moment (i.e. the product of tail length and tail intensity). The last parameter is very popular, although, as Collins (2004) pointed out correctly, it is hard to understand why. In particular, the parameter is not linear with respect to dose, and the shape and, thus, the length of the comet tail very much depends on the experimental conditions chosen. A further disadvantage of the tail moment was addressed by Kumaraval (2006): tail moment is measured in arbitrary units and different image-analysis systems give different values, so comparisons are difficult to carry out.

Visual scoring has, of course, the big advantage that no expensive image-analysis hardware and software is required. On the other hand, the subjective factor is much higher in the case of visual scoring compared with image analysis. It is therefore particularly important to evaluate in a double-blind mode when visual scoring is applied. In addition, there are indications that the sensitivity of image-analysis systems is better (Ladeira et al. 2005).

10.2.3 Methods Aimed at the Detection of Specific Types of DNA Damage

A problem, and one which is consistently ignored by most newcomers to the field, is the fact that the comet assay does not detect a specific type of DNA damage, but reflects several types of damage. Among these are single-strand and double-strand breaks, base damage, incomplete excision-repair sites (in particular, after exposure to UV radiation), alkali-labile sites, cross-links and conformational changes. A lot of the modifications of the original method mentioned before aim at the detection of specific DNA lesions, like single-strand or double-strand breaks. One should keep in mind, however, that such simple allocations like "neutral version means detection of double-strand breaks" and "alkaline version means detection of single-strand breaks" are always wrong, because of the unavoidable presence of other modifications of the DNA.

What one can achieve, however, is a shift in the direction of a specific type, but one can never exclude all the other types completely. For example, the alkaline version will always include in addition to single-strand breaks (and some other types of damage) double-strand breaks. Owing to the much higher number of single-strand breaks compared with double-strand breaks, there is some justification to associate the alkaline version with the detection of single-strand breaks, but one should never forget that other types of damage contribute to the comet tail.

A useful categorization in terms of pH of unwinding and electrophoresis has been mentioned by Kumaravel (2006):

- Neutral (pH 7–8): double-strand breaks and cross-links
- Alkaline (pH 12.1–12.4): single-strand and double-strand breaks, incomplete excision-repair sites, cross-links
- Strongly alkaline (pH>12.6): alkali-labile sites in addition to all types listed above

One must be somewhat cautious with too simple classifications. The original method of Östling and Johanson (1984) was based on a pH of 9.5 and clearly picked up also single-strand breaks.

Descriptions of the methods aimed at the predominant detection of a specific type of damage can be obtained in the following publications:

- Single-strand breaks (Olive et al. 1990).
- Single strand breaks plus alkali-labile sites (Singh et al. 1988; McKelvey et al. 1990).
- Double-strand breaks (Olive et al. 1991).
- Base damage (Collins 2004). Enzymes are used that recognize specific damaged bases and introduce breaks at these sites. For example, endonuclease III (Endo III) recognizes oxidized pyrimidines and formamidopyrimidine DNA glycosylase (FPG) identifies 8-oxo-guanine and other damaged purines.
- Cross-links (Pfuhler and Wolf 1996; McKenna et al. 2003). Whereas the other types of DNA damage increase the amount of DNA in the comet tail, cross-links diminish it.
- Apoptosis (Olive et al. 1993b). DNA of apoptotic cells is assumed to show a very characteristic picture in the comet assay: extremely small or non-existent

heads and large, diffuse tails. This interpretation of the comet shape, however, has been doubted, because it can be shown that such comets disappear as a result of repair (which is not possible in apoptotic cells) and the DNA fragments generated during apoptosis are so small that they would get lost during lysis or electrophoresis (Collins 2004).

- Defects in specific DNA sequences (Santos et al. 1997; McKelvey-Martin et al. 1998; McKenna et al. 2003). A particularly interesting amplification of the comet assay is the application of the fluorescence in situ hybridization technique. This approach allows the detection of damage within defined DNA sequences and, thus, also in specific genes.

10.2.4 Problem of Standardization of Methods

Various suggestions have been made to standardize the method of the comet assay in order to be able to compare results more easily (Tice et al. 2000; Hartmann et al. 2003). On the one hand, however, the suggestions made for standardization are rather vague in most instances and, on the other hand, experience in many fields shows that it is almost impossible to convince scientists to change their methods. Thus, it will take quite a while before some sort of agreement might be reached, if it will be possible at all.

In any case, it is important that all methodological details are outlined precisely in the publications and that, at least, the methods are standardized in one and the same laboratory. This is particularly true for all temperatures, pH values and voltages.

10.2.5 Detection Limits

In particular in the field of radiation research, some information is available with regard to the lowest detection limit of the assay. One has to be very careful in this context to distinguish between results obtained with or without repair. Exclusion of repair is only possible in vitro, because you can prevent repair just by cooling the cells. Owing to the rapid repair of DNA damage, you will determine in vivo only the unrepaired fraction of DNA a couple of hours after the exposure. Even within minutes, the amount of damaged DNA will be markedly lower than immediately after exposure.

Modifications specifically intended to increase the sensitivity of the alkaline version of the assay made possible the detection of DNA damage in lymphocytes after X-ray doses in the range of 50 mGy (Singh et al. 1994). Effects of even lower doses down to 6 mGy after γ-ray exposure have been demonstrated (Malyapa et al. 1998). This corresponds to fewer than ten lesions per diploid genome (under the assumption that 1 Gy induces 1,160 strand breaks per diploid cell) (Moller 2006b). And finally, Singh et al. (1995)

reported 1.02 mGy as lowest detection level after external exposure of lymphocytes to technetium-99m.

The effects of such low doses cannot be detected under neutral conditions, simply because the number of damaged DNA sites that can be identified under neutral conditions is much lower (see before). Thus, at least, about 5 Gy is necessary to induce a detectable response in the neutral version of the assay that is related to double-strand breaks. Sometimes, an initial increase in the dose–response curve between 0 and 2 Gy is attributed to double-strand breaks; this increase, however, is due to a loss of DNA supercoiling by single-strand breaks (Olive 1999).

One should be careful with very high doses, because an extensive number of strand breaks will finally induce a lot of small fragments; and fragments smaller than about 50 kb may be lost during lysis and electrophoresis (Olive 1999).

There is another factor that has to be taken into consideration when proliferating cells are analysed: the number of cells in S phase of the cell cycle. The sensitivity of the neutral assay is much lower when cells in S phase are present (Olive et al. 1992), whereas there seems to be no effect of S-phase cells in the alkaline version (Olive et al. 1992; Malyapa et al. 1998).

10.3 Factors That Might Affect the Outcome of Comet Analyses

Several factors have been studied for possible effects on the comet assay. For an extensive review see Moller et al. (2000). It is very striking that many conflicting results have been published with regard to most of the factors mentioned in the following. This raises suspicion as to the quality of some of the published data.

10.3.1 Age

Moller (2006a) summarizing published data concluded that increased DNA migration is observable with increasing age.

10.3.2 Gender

Moller (2006a) concluded that no effect of gender on the comet assay is detectable, when one looks at all the data published. If at all, there is some indication that men show slightly higher amounts of DNA damage in the conventional alkaline version of the comet assay, whereas such a trend is not detectable for Endo III or FPG sensitive sites.

10.3.3 Smoking

In a meta-analysis, Hoffmann et al. (2005) summarized the data on the effect of smoking on the comet assay. Overall, they found an indication that the levels of DNA damage were higher in smokers compared with those in non-smokers. This result, however, is mainly due to a minority of studies not using image analysis, but arbitrary damage scores, because when the analysis was restricted to studies using image analysis, there was only a tendency for effects on the comet assay. Moller (2006a) concluded from a literature survey that no effect of smoking is detectable in the comet assay.

10.3.4 Alcohol

The comparatively few studies aimed at the role of alcohol in the comet assay are extremely contradictory. Definitely, additional studies are required.

10.3.5 Season

When running the comet assay one should be aware of the possibility that seasonal variation can occur (Moller 2006b). At least part of this variation is due to different sunlight exposure.

10.3.6 Physical Exercise

Extensive physical exercise has been reported to induce DNA damage that is detectable in the comet assay. This was shown for rickshaw pullers (Pandey et al. 2006), treadmill runners (Hartmann et al. 1994), short-distance triathletes (Hartmann et al. 1998) and marathon runners (Tsai et al. 2001). The results are somewhat conflicting, because Tsai et al. (2001) reported that oxidized pyrimidines (detected by Endo III) were responsible for most of the damage, whereas Hartmann et al. (1998) found more breaks and explicitly no increase in oxidized DNA bases (detected by FPG). (By the way, the author of this chapter did not find an effect in the comet assay after a marathon; but perhaps he did not run hard enough.)

10.3.7 Environmental Exposures

Air pollution and diet might have an impact on the results of the comet assay (Moller 2006b).

10.4 Advantages/Disadvantages

A number of advantages make the comet assay very attractive:

- It is possible to determine initial DNA damage and DNA repair. Initial damage can be monitored at low temperatures (e.g. on ice) in order to prevent repair processes; repair kinetics can be obtained after various times after exposure at 37°C.
- The assay is rapid. One can obtain results within a couple of hours after sampling of the cells.
- Few cells are necessary to obtain meaningful results. Many assays testing genotoxic effects need some tens or some hundreds of thousands cells; the comet assay can be done with ten to some hundred cells.
- Cell culturing is not required (although, of course, one can use cell cultures). When cells have to be cultured in order to run a specific assay, there always looms the danger that some cells will not survive under the culture conditions chosen. This can result in a serious selection bias.
- Cell proliferation is not necessary; the only prerequisite is the viability of the cells. In particular, several cytogenetic assays (dicentrics, reciprocal translocations, micronuclei) require mitosis. Thus, the result of these assays can be seriously affected by conditions that interfere with cell proliferation, for example by selecting against severely damaged cells, which are no longer able to reach mitosis.
- One can determine distributions not just means. Quite a number of assay systems are only able to give averages of populations of cells. In many cases, however, it is much more important to obtain an idea of the distribution of the characteristics of single cells. If one looks, for example, at two different tumours whose cells show on average the same repair capabilities, it makes a huge difference if the cells in one tumour all show more or less the same repair capacity and in the other tumour some cells with a very efficient repair are counterbalanced by a similar number of cells with poor repair. Those cells with the efficient repair may survive and result in a relapse of the tumour.
- The assay can be made very sensitive, so that the effects of comparatively low doses or concentrations can be assessed: This is a clear advantage over pulsed-field gel electrophoresis. For example, the effects of repair and the repair kinetics after non-lethal radiation doses can easily be detected in the comet assay; this is particularly important with respect to studies in the context of radiation therapy.
- With the exception of the image-analysis system and the software, the costs are low. When one uses visual scoring, one can avoid even the high costs of the image-analysis system. Visual scoring, however, has some drawbacks (see above).

One should not conceal that there are also some disadvantages.

- Various types of DNA damage are involved. This problem was discussed in detail in Sect. 10.2.3.
- The fidelity of the repair processes cannot be checked. In the comet assay the amount of DNA in the comet tail is taken as a measure of DNA damage. If the comet tail is reduced in the course of time after exposure to a damaging agent, this is seen as the result of repair processes. It is not possible, however, to draw conclusions as to the correct execution of repair.
- The assay is not possible with fixed material. This means that archived material cannot be used. It is, however, of course, possible to freeze and thaw cells as long as it is done without affecting the viability of the cells.

10.5 Applications

The comet assay has been introduced in many fields. One of the crucial questions still is which cell type is most suitable for the assay. Frequently, lymphocytes are used, because they offer some advantages: they can easily be obtained; they are single cells by nature; they are all in G_0 phase of the cell cycle. Some problems, however, should not be forgotten: there are several lymphocyte subpopulations with different sensitivities; it is not clear whether lymphocyte results are representative for other cell types; they are not target cells for cancer.

When proliferating cells are used in the assay, one has to keep in mind that cells in various cell cycle stages may behave differently in the comet assay, so the distribution of cells among the different stages may heavily affect the outcome of the assay.

There are several review articles on applications of the various types of the comet assay:

- Radiobiology (Olive 1999)
- Biomonitoring (Collins et al. 1997a; Kassie et al. 2000; Moller et al. 2000; Faust et al. 2004)
- Aquatic biomonitoring (Mitchelmore and Chipman 1998)
- Genotoxicity testing (Brendler-Schwaab et al. 2005; Moller 2005)
- Carcinogen monitoring (Anderson et al. 1998)
- Genetic ecotoxicology (Cotelle and Ferard 1999) (with the application of the comet assay in plants, worms, molluscs, fish, amphibians and mammalians)
- Various applications (Rojas et al. 1999; Collins 2004)

Some of these applications will be highlighted in the following sections.

10.5.1 Individual Radiosensitivity

People do not show identical or, at least, similar radiosensitivities. This is particularly true for some genetic diseases, but it also applies to the general population (ICRP 1999; NRPB 1999). Such differences play a crucial role after radiation accidents and in radiotherapy. DNA repair capacity of the individual is one important element that determines individual radiosensitivity, but differences in the initial DNA damage may also play a role. It is, therefore, no surprise that the comet assay has been used to predict individual radiosensitivity (Müller et al. 1999, 2001; Alapetite et al. 1999; McKeown et al. 2003).

10.5.2 Hypoxic Cells

In tumour radiotherapy the fraction of hypoxic cells plays an important role, because these cells are markedly less radiosensitive compared with euoxic cells. It is possible to identify hypoxic cells by the comet assay (Olive and Durand 1992; Olive et al. 1993a).

10.5.3 Genotoxicity of Chemicals and Identification of Carcinogenic Substances

A tremendous number of genotoxicity studies have been carried out (e.g. see Table 2 in Rojas et al. 1999, which lists almost 200 chemicals that had been tested in the comet assay until 1999). About 85% of the chemicals tested in the studies summarized by Rojas et al. showed a genotoxic effect; one must, however, keep in mind that most of these chemicals were chosen because there was already some information on genotoxic efficiency.

Numerous chemicals were studied in an in vivo comet assay by Sasaki et al. (2000). The major conclusion of the authors was that the comet assay used in various organs of mice and rats is a suitable technique to evaluate the in vivo genotoxicity of in vitro genotoxic agents. The assay had a high response ratio for rodent genotoxic carcinogens and a high negative response ratio for rodent genotoxic non-carcinogens.

10.5.4 Predictor of Cancer Risk

Baseline, benzopyrene diol epoxide and γ-radiation-induced DNA damage in peripheral blood lymphocytes was significantly higher in (untreated) patients with oesophageal cancer (Shao et al. 2005), bladder cancer (Schabath et al. 2003) or lung cancer (Zhang et al. 2000) compared with that in healthy individuals. The authors attributed these results to an increased genomic instability of the cancer cells. Similar results have been obtained for breast cancer cases

(Smith et al. 2003) and various cancer types (Sigurdson et al. 2005). In a very large study (397 women) Rajeswari et al. (2000) showed that lymphocytes of women with breast tumour showed the highest amount of DNA in the comet tails, followed by first-degree female relatives of breast cancer patients and the lowest values were shown by women without tumours without an enhanced family risk.

It remains to be elucidated, however, whether the comet assay has predictive power before cancer starts to develop or whether the observed effects are just a characteristic of already existing cancer cells. Thus, an extensive prospective study is required.

No effect on the comet assay was observed when lymphoblastoid cell lines with a heterozygous BRCA1 mutation were analysed (Trenz et al. 2005).

10.5.5 Biomonitoring

The possibilities to apply the comet assay in the environment for biomonitoring are virtually unlimited. Various possible genotoxic surroundings have been checked in the past (e.g. waste grounds, polluted air and water, areas contaminated with heavy metals) and publications with new fields of analysis appear all the time.

10.5.6 Assessment of Exposure to Ionizing Radiation in Vivo

There have been some attempts to estimate exposure to ionizing radiation of human beings. The dominant lesions after radiation exposure that are measured in the conventional comet assay are strand breaks. Owing to the comparatively rapid repair processes, already a couple of hours after radiation exposure only unrepaired DNA breaks (in addition to unrepaired DNA non-break lesions, which are identified by the conventional assay) can be detected by the comet assay. As one cannot prevent repair in vivo by, for example, cooling, it is not surprising that no effects could be demonstrated in people living in areas contaminated by the Chernobyl accident (Frenzilli et al. 1998; Garcia and Mandina 2005). Similarly, there was no indication that radiotherapy affects the outcome of the comet assay when lymphocytes were studied using the only slightly modified Ostling and Johanson technique (Müller et al. 2002).

There is, however, an indication that lymphocytes of people chronically exposed to ionizing radiation show a reduced repair capacity after an additional radiation exposure in vitro; this could be shown for patients subjected to radioiodine therapy and for children and adults living in the Chernobyl region (Plappert et al. 1997). In addition, Wojewodzka et al. (1998) reported that strand breakage was enhanced in workers exposed to chronic low-dose irradiation (unfortunately, without mentioning the doses received), but no increase in base damage was observed (Kruszewski et al. 1998). The authors mentioned, however, marked interindividual variability in the exposed as well as in the control group. Very high variability was also observed in a study on radon exposure (Hellman et al. 1999): there was a borderline significance ($P = 0.05$) for the relation between indoor radon exposure and tail moment in the alkaline comet assay for the whole group, but, for example, the white blood cells of the individual with the highest radon exposure had almost the lowest tail moment. This result was commented on by the authors in the following way: "Ionising radiation is obviously only one of many factors (environmental as well as constitutional) determining the individual background level of DNA damage".

10.5.7 UV Damage

UV radiation does not induce strand breaks. Nevertheless, the comet assay can be used to estimate damage induced by UV radiation, because base repair is carried out by excision of the affected DNA sites, thus introducing strand breaks. Therefore, in contrast to most other damaging agents, no comets can be detected directly after UV exposure, but an increase in the DNA in the comet tail is observed during the minutes and hours after the exposure owing to base repair. More detailed information can be obtained from the following publications: Gedik et al. (1992), Green et al. (1992) and Arlett et al. (1993).

10.6 Final Remark

The comet assay in its various modifications definitely is a valuable tool for the assessment of DNA damage. A major problem is that many researchers underestimate the difficulties involved with the proper execution of this assay. A lot of basic knowledge and a lot of experience is required to avoid mistakes ending up in seriously wrong conclusions. The greatest danger for any seemingly "simple" assay is an uncritical use of the method resulting in a lot of contradictory results with the final outcome of abandoning the assay, because it is "unreliable". Hopefully, the comet assay will not meet that fate.

References

Alapetite C, Thirion P, de la Rochefordière A, Cosset J-M, Moustacchi E (1999) Analysis by alkaline comet assay of cancer patients with severe reactions to radiotherapy: Defective rejoining of radioinduced DNA strand breaks in lymphocytes of breast cancer patients. Int J Cancer 83:83–90

Anderson D, Yu TW, McGregor DB (1998) Comet assay responses as indicators of carcinogen exposure. Mutagenesis 13:539–555

Arlett CF, Lowe JE, Harcourt SA, Waugh APW, Cole J, Roza L, Diffey BL, Mori T, Nikaido O, Green MHL (1993) Hypersensitivity of human lymphocytes to UV-B and solar irradiation. Cancer Res 53:609–614

Böcker W, Bauch T, Müller W-U, Streffer C (1997) Image analysis of comet assay measurements. Int J Radiat Biol 72:449–460

Brendler-Schwaab S, Hartmann A, Pfuhler S, Speit G (2005) The in vivo comet assay: use and status in genotoxicity testing. Mutagenesis 20:245–254

Collins A, Dusinska M, Franklin M, Somorovska M, Petrovska H, Duthie S, Fillion L, Panayiotidis M, Raslova K, Vaughan N (1997a) Comet assay in human biomonitoring studies: reliability, validation, and applications. Environ Mol Mutagen 30:139–146

Collins AR (2004) The comet assay for DNA damage and repair: principles, applications, and limitations. Mol Biotechnol 26:249–261

Collins AR, Dobson VL, Dusinska M, Kennedy G, Stetina R (1997b) The comet assay: what can it really tell us? Mutat Res 375:183–193

Cotelle S, Ferard JF (1999) Comet assay in genetic ecotoxicology: a review. Environ Mol Mutagen 34:246–255

Faust F, Kassie F, Knasmuller S, Boedecker RH, Mann M, Mersch-Sundermann V (2004) The use of the alkaline comet assay with lymphocytes in human biomonitoring studies. Mutat Res 566:209–229

Frenzilli G, Lori A, Panasiuk G, Ferdeghini M, Barale R (1998) Comet assay on children's leukocytes 8 years after the Chernobyl disaster. Mutat Res 415:151–158

Garcia O, Mandina T (2005) DNA damage evaluated by the comet assay in lymphocytes of children with 137Cs internal contamination caused by the Chernobyl accident. Mutat Res 565:191–197

Garcia O, Mandina T, Lamadrid AI, Diaz A, Remigio A, Gonzalez Y, Piloto J, Gonzalez JE, Alvarez A (2004) Sensitivity and variability of visual scoring in the comet assay. Results of an inter-laboratory scoring exercise with the use of silver staining. Mutat Res 556:25–34

Gedik CM, Ewen SWB, Collins AR (1992) Single-cell gel electrophoresis applied to the analysis of UV-C damage and its repair in human cells. Int J Radiat Biol 62:313–320

Green MHL, Lowe JE, Harcourt SA, Akinluyi P, Rowe T, Cole J, Anstey AV, Arlett CF (1992) UV-C sensitivity of unstimulated and stimulated human lymphocytes from normal and xeroderma pigmentosum donors in the comet assay: A potential diagnostic technique. Mutat Res 273:137–144

Hartmann A, Plappert U, Raddatz K, Grünert-Fuchs M, Speit G (1994) Does physical activity induce DNA damage? Mutagenesis 9:269–272

Hartmann A, Pfuhler S, Dennog C, Germadnik D, Pilger A, Speit G (1998) Exercise-induced DNA effects in human leukocytes are not accompanied by increased formation of 8-hydroxy-2′-deoxyguanosine or induction of micronuclei. Free Radic Biol Med 24:245–251

Hartmann A, Agurell E, Beevers C, Brendler-Schwaab S, Burlinson B, Clay P, Collins A, Smith A, Speit G, Thybaud V, Tice RR (2003) Recommendations for conducting the in vivo alkaline comet assay. 4th international comet assay workshop. Mutagenesis 18:45–51

Hellman B, Friis L, Vaghef H, Edling C (1999) Alkaline single cell gel electrophoresis and human biomonitoring for genotoxicity: a study on subjects with residential exposure to radon. Mutat Res 442:121–132

Hoffmann H, Hogel J, Speit G (2005) The effect of smoking on DNA effects in the comet assay: a meta-analysis. Mutagenesis 20:455–466

ICRP (1999) Genetic susceptibility to cancer. ICRP publication 79. Pergamon, Oxford

Kassie F, Parzefall W, Knasmuller S (2000) Single cell gel electrophoresis assay: a new technique for human biomonitoring studies. Mutat Res 463:13–31

Kruszewski M, Wojewodzka M, Iwanenko T, Collins AR, Szumiel I (1998) Application of the comet assay for monitoring DNA damage in workers exposed to chronic low-dose irradiation. II. Base damage. Mutat Res 416:37–57

Kumaravel TS, Jha AN (2006) Reliable Comet assay measurements for detecting DNA damage induced by ionising radiation and chemicals. Mutat Res 605:7–16

Ladeira MS, Rodrigues MA, Freire-Maia DV, Salvadori DM (2005) Use of Comet assay to assess DNA damage in patients infected by Helicobacter pylori: comparisons between visual and image analyses. Mutat Res 586:76–86

Malyapa RS, Bi C, Ahern EW, Roti Roti JL (1998) Detection of DNA damage by the alkaline comet assay after exposure to low-dose gamma radiation. Radiat Res 149:396–400

McKelvey VJ, Tice RR, Singh NP (1990) A technique for measuring DNA single strand breaks and alkali labile sites in individual cells. Br J Cancer 62:529–530

McKelvey-Martin VJ, Ho ET, McKeown SR, Johnston SR, McCarthy PJ, Rajab NF, Downes CS (1998) Emerging applications of the single cell gel electrophoresis (comet) assay. I. Management of invasive transitional cell human bladder carcinoma. II. Fluorescent in situ hybridization comets for the identification of damaged and repaired DNA sequences in individual cells. Mutagenesis 13:1–8

McKenna DJ, Gallus M, McKeown SR, Downes CS, McKelvey-Martin VJ (2003) Modification of the alkaline Comet assay to allow simultaneous evaluation of mitomycin C-induced DNA cross-link damage and repair of specific DNA sequences in RT4 cells. DNA Repair (Amst) 2:879–890

McKeown SR, Robson T, Price ME, Ho ET, Hirst DG, McKelvey-Martin VJ (2003) Potential use of the alkaline comet assay as a predictor of bladder tumour response to radiation. Br J Cancer 89:2264–2270

Mitchelmore CL, Chipman JK (1998) DNA strand breakage in aquatic organisms and the potential value of the comet assay in environmental monitoring. Mutat Res 399:135–147

Moller P (2005) Genotoxicity of environmental agents assessed by the alkaline comet assay. Basic Clin Pharmacol Toxicol 96(Suppl 1):1–42

Moller P (2006a) Assessment of reference values for DNA damage detected by the comet assay in human blood cell DNA. Mutat Res 612:84–104

Moller P (2006b) The alkaline comet assay: towards validation in biomonitoring of DNA damaging exposures. Basic Clin Pharmacol Toxicol 98:336–345

Moller P, Knudsen LE, Loft S, Wallin H (2000) The comet assay as a rapid test in biomonitoring occupational exposure to DNA-damaging agents and effect of confounding factors. Cancer Epidemiol Biomarkers Prev 9:1005–1015

Müller W-U, Bauch T, Streffer C, Niedereichholz F, Böcker W (1994) Comet assay studies of radiation-induced DNA damage and repair in various tumour cell lines. Int J Radiat Biol 65:315–319

Müller W-U, Bauch T, Streffer C (1999) The application of comet and micronucleus assay for the assessment of individual radiation risk. In: Baumstark-Khan C, Kozubek S, Horneck G (eds): Fundamentals for the assessment of risks from environmental radiation. Kluwer, Dordrecht, pp 127–134

Müller W-U, Bauch T, Stüben G, Sack H, Streffer C (2001) Radiation sensitivity of lymphocytes of healthy individuals and cancer patients as measured by the comet assay. Radiat Environ Biophys 40:83–89

Müller W-U, Bauch T, Streffer C, von Mallek D (2002) Does radiotherapy affect the outcome of the comet assay? Br J Radiol 75:608–614

Müller W-U, Ciborovius J, Bauch T, Johannes C, Schunck C, Mallek U, Böcker W, Obe G, Streffer C (2004) Analysis of the action of the restriction endonuclease AluI using three different comet assay protocols. Strahlenther Onkol 180:655–664

NRPB (1999) Genetic heterogeneity in the population and its implications for radiation risk. National Radiological Protection Board, Chilton

Olive PL (1999) DNA damage and repair in individual cells: applications of the comet assay in radiobiology. Int J Radiat Biol 75:395–405

Olive PL (2002) The comet assay. An overview of techniques. Methods Mol Biol 203:179–194

Olive PL, Durand RE (1992) Detection of hypoxic cells in a murine tumor with the use of the comet assay. J Nat Canc Inst 84:707–711

Olive PL, Banáth J, Durand RE (1990) Heterogeneity in radiation-induced DNA damage and repair in tumor and normal cells measured using the "comet" assay. Radiat Res 122:86–94

Olive PL, Wlodek D, Banáth JP (1991) DNA double-strand breaks measured in individual cells subjected to gel electrophoresis. Cancer Res 51:4671–4676

Olive PL, Wlodek D, Durand RE, Banáth JP (1992) Factors influencing DNA migration from individual cells subjected to gel electrophoresis. Exp Cell Res 198:259–267

Olive PL, Durand RE, Le Riche J, Olivotto IA, Jackson SM (1993a) Gel electrophoresis of individual cells to quantify hypoxic fraction in human breast cancers. Cancer Res 53:733–736

Olive PL, Frazer G, Banáth JP (1993b) Radiation-induced apoptosis measured in TK6 human B lymphoblast cells using the comet assay. Radiat Res 136:130–136

Ostling O, Johanson KJ (1984) Microelectrophoretic study of radiation-induced DNA damages in individual mammalian cells. Biochem Biophys Res Commun 123:291–298

Pandey AK, Bajpayee M, Parmar D, Rastogi SK, Mathur N, Seth PK, Dhawan A (2006) DNA damage in lymphocytes of Indian rickshaw pullers as measured by the alkaline comet assay. Environ Mol Mutagen 47:25–30

Pfuhler S, Wolf HU (1996) Detection of DNA-crosslinking agents with the alkaline comet assay. Environ Mol Mutagen 27:196–201

Plappert UG, Stocker B, Fender H, Fliedner TM (1997) Changes in the repair capacity of blood cells as a biomarker for chronic low-dose exposure to ionizing radiation. Environ Mol Mutagen 30:153–160

Rajeswari N, Ahuja YR, Malini U, Chandrashekar S, Balakrishna N, Rao KV, Khar A (2000) Risk assessment in first degree female relatives of breast cancer patients using the alkaline comet assay. Carcinogenesis 21:557–561

Rojas E, Lopez MC, Valverde M (1999) Single cell gel electrophoresis assay: methodology and applications. J Chromatogr B Biomed Sci Appl 722:225–254

Santos SJ, Singh NP, Natarajan AT (1997) Fluorescence in situ hybridization with comets. Exp Cell Res 232:407–411

Sasaki YF, Sekihashi K, Izumiyama F, Nishidate E, Saga A, Ishida K, Tsuda S (2000) The comet assay with multiple mouse organs: comparison of comet assay results and carcinogenicity with 208 chemicals selected from the IARC monographs and US NTP Carcinogenicity Database. Crit Rev Toxicol 30:629–799

Schabath MB, Spitz MR, Grossman HB, Zhang K, Dinney CP, Zheng PJ, Wu X (2003) Genetic instability in bladder cancer assessed by the comet assay. J Natl Cancer Inst 95:540–547

Shao L, Lin J, Huang M, Ajani JA, Wu X (2005) Predictors of esophageal cancer risk: assessment of susceptibility to DNA damage using comet assay. Genes Chromosomes Cancer 44:415–422

Sigurdson AJ, Hauptmann M, Alexander BH, Doody MM, Thomas CB, Struewing JP, Jones IM (2005) DNA damage among thyroid cancer and multiple cancer cases, controls, and long-lived individuals. Mutat Res 586:173–188

Singh NP, Graham MM, Singh V, Khan A (1995) Induction of DNA single-strand breaks in human lymphocytes by low doses of gamma-rays. Int J Radiat Biol 68:563–569

Singh NP, McCoy MT, Tice RR, Schneider EL (1988) A simple technique for quantitation of low levels of DNA damage in individual cells. Exp Cell Res 175:184–191

Singh NP, Stephens RE, Schneider EL (1994) Modifications of alkaline microgel electrophoresis for sensitive detection of DNA damage. Int J Radiat Biol 66:23–28

Smith TR, Miller MS, Lohman KK, Case LD, Hu JJ (2003) DNA damage and breast cancer risk. Carcinogenesis 24:883–889

Tice RR, Agurell E, Anderson D, Burlinson B, Hartmann A, Kobayashi H, Miyamae Y, Rojas E, Ryu JC, Sasaki YF (2000) Single cell gel/comet assay: guidelines for in vitro and in vivo genetic toxicology testing. Environ Mol Mutagen 35:206–221

Trenz K, Schutz P, Speit G (2005) Radiosensitivity of lymphoblastoid cell lines with a heterozygous BRCA1 mutation is not detected by the comet assay and pulsed field gel electrophoresis. Mutagenesis 20:131–137

Tsai K, Hsu TG, Hsu KM, Cheng H, Liu TY, Hsu CF, Kong CW (2001) Oxidative DNA damage in human peripheral leukocytes induced by massive aerobic exercise. Free Radic Biol Med 31:1465–1472

Wojewodzka M, Kruszewski M, Iwanenko T, Collins AR, Szumiel I (1998) Application of the comet assay for monitoring DNA damage in workers exposed to chronic low-dose irradiation. I. Strand breakage. Mutat Res 416:21–35

Zhang H, Buchholz TA, Hancock D, Spitz MR, Wu X (2000) Gamma-radiation-induced single cell DNA damage as a measure of susceptibility to lung cancer: a preliminary report. Int J Oncol 17:399–404

11 Origin of Chromosome Aberrations: Mechanisms

Peter E. Bryant

Abstract

After more than 60 years of observation and research, the molecular mechanisms underlying radiation-induced chromosome aberration formation are still controversial. Here I review the classical and recent models for the origin of aberrations and some of the current controversies. The experimental evidence to date strongly suggests that the mechanisms underlying the two aberration types, chromosome and chromatid aberrations, differ. It is also likely that two separate mechanisms exist for the formation of chromosome exchanges; both contributing to the overall yield. Recent evidence supports the previously postulated existence of an alternative pathway to the classical 'two-hit' mechanism for chromosome exchanges, involving interaction of two DNA double-strand breaks (DSB). This alternative pathway involves a cell's response to a single DSB, perhaps of a relatively complex nature, that interacts with DNA of an undamaged chromosome. Like chromosome aberrations, chromatid aberrations comprise both breaks and exchanges that could have different underlying mechanisms. Recent studies have focussed on the formation and origin of chromatid breaks. It is known that these breaks are formed from single DSB, but are unlikely to be formed (as has been postulated previously) simply by an expansion of the DSB. Here I argue that chromatid breaks are most likely to result from chromatin rearrangements, initiated by a DSB, but are formed by a mechanism involving misjoining of chromatin ends by topoisomerase II.

11.1 Introduction

Chromosome aberrations originate in somatic cells as a result of exposure to physical agents such as ionizing radiation or to genotoxic chemicals. In this review I will focus on the action of ionizing radiation, although the mechanisms discussed may apply also to the formation of spontaneous and some types of chemically induced aberrations, especially those produced by radiomimetic agents such as bleomycin and zeocin. Ionizing radiation and these chemical clastogens produce both chromosome-type and chromatid-type aberrations,

Günter Obe and Vijayalaxmi (Eds.)
Chromosomal Alterations: Methods, Results and Importance in Human Health

although the mechanisms underlying the two aberration types appear to be different.

Let us begin by defining what we mean by chromosome and chromatid aberrations. Chromosome aberrations are formed when cells are exposed to radiation whilst in either the G1 or the G0 (quiescent) phase when the chromosomes are unduplicated. As a consequence, aberrations are observed in the metaphase chromosomes after the cell has progressed from the G1 phase of the cell-cycle, through the S phase and eventually arrived at mitosis. In the case of G0 cells (e.g. human peripheral blood T lymphocytes) it is necessary to stimulate cells back into cell division with phytohaemagglutinin so that the cells enter the cell cycle and proceed to mitosis, where the metaphase chromosomes can be observed. Chromatid aberrations are formed in duplicated chromosomes as a result of exposure of cells to genotoxic agents in either the S phase or the G2 phase of the cell cycle. One or both chromatids may be involved in an aberration.

Chromosome aberrations observed at metaphase fall into two major groups: breaks, resulting in chromosome fragments, and 'exchanges' that have originated via the misjoining of two chromosomes or the two broken ends of the same chromosome. Since they lack a centromere, chromosome fragments often become excluded from the main nucleus as the nuclear envelope reforms, and then appear as 'micronuclei' in the cell cytoplasm (Fig. 11.1).

Chromosome exchanges fall into two categories; firstly the unstable type. These are formed by the misjoining of two broken chromosomes and result

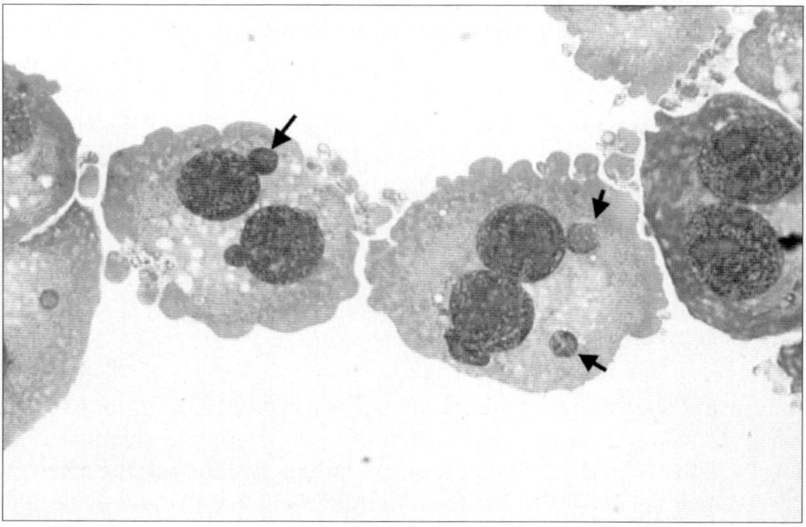

Fig. 11.1 Binucleate (Chinese hamster) cells showing multiple micronuclei (*arrows*) following irradiation and treatment with cytochalasin B

in linkage of two centromeres (dicentrics) or formation of a covalently linked ring within a single chromosome containing one centromere. In either of these cases the remaining acentric chromatin fragments also become linked, and are duplicated during the S phase. Unstable chromosome aberrations are usually lethal to a cell.

Stable chromosome rearrangments occur when two broken chromosomes misjoin and exchange material in a reciprocal fashion, allowing the cell to survive. The consequences of stable rearrangements may be the truncation and/or activation of DNA sequencies, e.g. an oncogene as in the case of the Philadelphia chromosome associated with chronic myeloid leukaemia. The mechanisms of chromosome aberrations are discussed in detail later.

Similarly, chromatid aberrations comprise both break and exchange types. However, in 'normal' human or rodent cells, following low doses of ionizing radiation, the single chromatid break (CB) predominates. At higher doses, and in certain mutant cell strains deficient in DNA repair, exchanges occur at higher frequencies. There are a variety of such exchanges, some intrachromosomal (e.g. sister-chromatid union) and some interchromosomal, i.e. formed between chromatids of two different chromosomes.

Chromosome and chromatid aberrations can be observed without the requirement of the cell to enter mitosis. The technique of premature chromosome condensation (Johnson and Rao 1970; Pandita and Hittleman 1992) uses the fusion of a mitotic cell with the test G1 or G0 cell, or more recently treatment of cells with calyculin A (Gotoh et al. 1999; Kawata et al. 2004) leading to the premature condensation of the interphase cell chromosomes. This technique, as we shall see, has resulted in the acquisition of valuable clues as to the mechanism of chromosome aberrations.

11.2 Primary DNA Damage

In reviewing the mechanisms of chromosome aberrations we have to first consider the initiating cellular damage that leads to their formation. Although it is known and accepted that ionizing radiation produces many types of chemical lesions in the DNA molecules of cells, such as strand breaks, base damage and cross-links, it is generally accepted that the DNA double-strand break (DSB) is the critical initiating lesion. The evidence for this comes primarily from the experiments of Natarajan et al. (1980), who used a poration technique employing Sendai virus to introduce *Neurospora* single-strand endonuclease into living cells to convert some of the induced single-strand breaks into DSB. It was found that this treatment led to a significant increase in the frequency of DSB in cells exposed to a prior dose of X-rays, and a corresponding increase in the number of chromosome aberrations was observed. These experiments were followed up by techniques in which various restriction endonucleases were introduced into dividing cells by various porating agents (Bryant 1984; Natarajan and Obe 1984; Winegar and Preston 1988). It was found that restriction endonuclease treatments (producing DSB with 3′-hydroxyl and 5′-phosphoryl end structures of various types) led to the formation of chromosome aberrations of all types. It was noted moreover that the 'blunt'-ended variety of DSB caused by enzymes such as *Pvu*II and *Eco*RV were much more efficient than the cohesive type produced by enzymes such as *Bam*H1, with a four-base 5′-overhang (Bryant 1984; Bryant and Christie 1989). It was deduced from these experiments that radiation induction of chromosome aberrations stemmed principally from the formation of DSB and that it was possible that DSB with overlapping ends might be rejoined by a different mechanism (e.g. single strand break rejoining) than that of blunt-ended DSB, which are known to be mainly rejoined by the non-homologous end-joining (NHEJ) pathway. Hence, it is likely that the reason why cohesive-ended DSB are much less effective at causing chromosome aberrations is because the rejoining rate for single-strand breaks is extremely fast (i.e. a lower probability of interactions leading to aberrations occurring). It was recognized that the end groups of DSB induced by restriction endonucleases were most probably in the majority of cases different from those induced by ionizing radiation. The reason for this is that the energy imparted by ionizing events in biological structures is many times greater than the energy of a covalent bond. Hence, following irradiation of a cell any bond within its DNA structure is susceptible to breakage, leaving untidy or 'dirty' ends of a DSB.

DSB are subject to rejoining by three known mechanisms: homologous recombinational repair (HRR), homologous single-strand annealing (SSA) and NHEJ (Thacker and Zdzienicka 2004). The formation of chromosome exchanges involves one or more of these mechanisms (Griffin and Thacker 2004; Griffin et al. 2005). Rejoining of DSB can be monitored by techniques such as low-voltage agarose-gel electrophoresis, and it is found that DSB rejoin with biphasic kinetics (Iliakis et al. 2004) (half-times of repair 10 min and 2h) which could possibly reflect the rejoining of different types of DSB. DSB can also be monitored using the disappearance of the γ-H2AX foci (Rotkamm and Lobrich 2003; MacPhail et al. 2003); these are large (mega-base-pair) domains of histone H2AX phosphorylation around the sites of DSB (Rogakou et al. 1999). Chromosome exchanges are reported to be formed rapidly following irradiation (Holmberg and Gumauskas 1986; Holmberg 1990); however, the process is reported to be dose-dependent – rapid formation at low doses and slower formation at higher doses (Darroudi et al. 1998)

The frequency of chromosome aberrations depends on the rate and extent of rejoining of DSB since DSB-rejoining deficient mutant strains (e.g. xrs) show elevated levels of chromosome aberrations when treated with either radiation or restriction endonucleases (Kemp and Jeggo 1986; Bryant et al. 1987). However, other cell lines, notably those derived from ataxia telangiectasia cases show elevated chromosome aberration frequencies against a background of normal (or faster) rejoining of DSB compared with normals (Mozdarani and

Bryant 1989b; Foray et al. 1995). Thus, the conversion of DSB into chromosome aberrations is a complex process (particularly for chromatid aberrations; see later) not always dependent on the rate of rejoining of DSB.

11.3 Models of Chromosome Aberration Formation

Essentially, three models have been proposed for the formation of chromosome aberrations; these are the classical 'breakage-first' or 'breakage-and-reunion' model (Sax 1940; Lea and Catcheside 1942; Bender et al. 1974), the molecular recombination model (Chadwick and Leenhouts 1978) and the more recent transcription-based model (Radford 2002). It should be noted that some authors have included Revell's 'exchange' model (Revell 1959) as an alternative to the breakage-first model. However, this model was originally proposed to account only for chromatid aberrations although some authors have taken it as applying also to chromosome aberrations (Neary 1965; Savage 1998). I have chosen to discuss the model only as it applies to chromatid aberrations.

11.3.1 The Breakage-First Model

The breakage-first model is generally applied to the formation of *both* chromosome and chromatid aberrations; however, as mentioned already, evidence indicates that the mechanism involved in the formation of chromosome aberrations (i.e. in G0/G1 cells) is different from that for chromatid aberrations formed in the G2 phase of the cell cycle. Therefore, I will discuss the mechanisms for chromosome and chromatid aberrations separately.

The breakage-first model, in its molecular form (Bender et al. 1974), maintains that the single DNA molecule has to be broken (i.e. by a DSB) before interaction can take place with another DSB, leading to a misjoined structure that can give rise to an 'exchange' aberration, which may be of an unstable type (i.e. likely to result in a missegregation of chromatin between daughter cells at mitosis) such as a dicentric or ring with an associated acentric fragment, or a stable type such as a reciprocal translocation, which could allow the cell to continue proliferating, perhaps with the overexpression of a novel truncated oncogene, leading to neoplastic transformation. An example of this is the Philadelphia chromosome, a chromosome translocation [t(9;22)(q34;q11)] found in the bone-marrow disease chronic myeloid leukaemia, where the translocation of the *ABL* protooncogene from its normal site on chromosome 9q34 to chromosome 22q11 where it forms a fusion hybrid gene with the breakage cluster region. The hybrid gene expresses a novel tyrosine kinase, leading to loss of cell-cycle control, promotion of cell growth and inhibition of DNA repair, leading to genomic instability. A similar case is Bukitt lymphoma, a B-cell malignancy characterized by a chromosome 8;14 translocation. In this case the translocation of the *MYC* protooncogene (normally located at 8q24), most often to 14q32, results in *MYC* activation by enhancers of the heavy chain immunoglobulin gene located at 14q32. Such specific translocations are a feature of many malignancies (Rabbitts 1994) and the progression of cells towards the neoplastic state and subsequent metastasis may involve more than one such translocation resulting from genomic instability in cells as they move towards the fully malignant state.

One of the problems facing proponents of the breakage-first model has been the finding that chromosomes in interphase nuclei occupy seemingly discrete 'domains', adherent to the inner surface of the nucleus (Cremer et al. 1996) and therefore it is difficult to visualize how chromosomes can interact over the relatively large distances of several microns, around or across the nucleus of a cell. Early work in *Drosophila* (Serebrovsky 1929) proposed that the chromosomes needed to be actually physically touching before interaction (misjoining) can take place, sometimes referred to as the 'contact-first' hypothesis, and it has been pointed out by Savage (2000) that preexisting proximity of chromosome segments, if not direct contact, is likely to be the reason for the occurrence of such typical translocations as the Philadelphia chromosome in chronic myeloid leukaemia (see above) and the RET/H4 intrachange in many radiation-induced thyroid tumours (Nikiforova et al. 2000).

Under the breakage-first model, the failure of a DSB to rejoin would, as the cell enters metaphase, give rise to two chromosome fragments (one centric and one acentric) which will have been, like exchange aberrations, duplicated during semiconservative replication. Owing to the absence of a centromere, acentric fragments, although comprising two chromatid fragments, cannot be segregated into the two daughter cells at mitosis, and are often excluded from the cell's nucleus during the process of envelope reformation. Such excluded fragments become enclosed in a smaller nuclear envelope and are termed 'micronuclei'. Occasionally, the centric fragment, or even a whole lagging chromosome, may become excluded from the nucleus.

In some of the earliest work on induction of chromosome and chromatid aberrations in *Tradescantia* microspores, Sax (1939, 1940) made the distinction between 'one-hit' and 'two-hit' aberrations; one-hit aberrations being classified as terminal deletions (unrejoined breaks) and two-hit aberrations as fusions (or exchanges) between two chromosomes. Sax reported that if the dose rate was kept constant, the yield of two-hit aberrations is proportional to the square of the radiation dose, whereas one-hit aberrations were linear with dose. The quadratic induction of (two-hit) exchanges is easily understood in molecular terms, since the probability of inducing two DSB in two chromosomes increases as the radiation dose increases. It is now generally recognized that, at least for mammalian cells, there is also a linear (single-hit) component of the dose response for exchanges, resulting from the fact that, even with X-rays, some clusters of ionizing events are large enough to induce two DSB in two chromosomes. For low linear energy transfer (LET) radiation such as X-rays this component is relatively small. However, for high LET

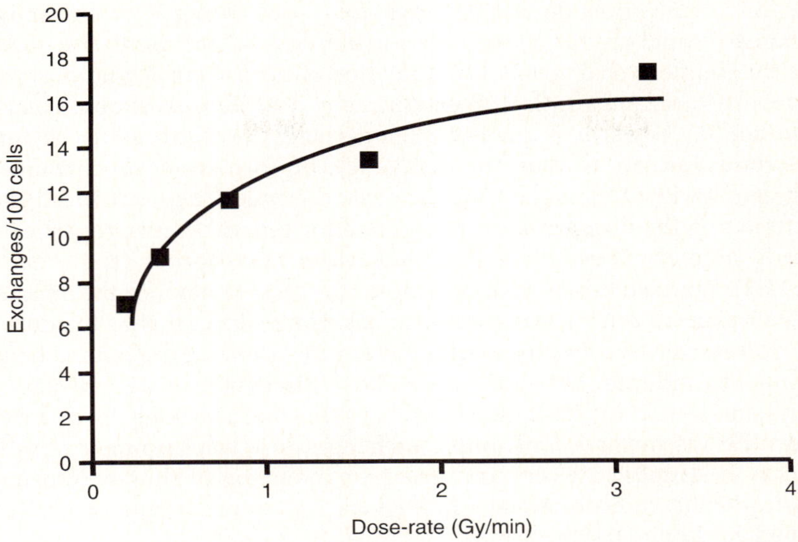

Fig. 11.2 The dose-rate effect for chromosome exchanges in *Tradescantia* microspores following a dose of 3.2 Gy. (Drawn from data of Sax 1939)

radiation (e.g. α-particles) the linear component of exchange induction dominates. Sax also recognized that by increasing the time of exposure to a given dose of radiation, the yield of two-hit aberrations would decrease (Fig. 11.2). This dose-rate effect is very relevant today in clarifying the mechanism of chromosome exchanges and I will return to this point later.

The breakage-first model was seriously challenged by the finding that very low energy X-rays (carbon-K X-rays that produce electrons with tracks of less 7 nm tracks) efficiently induce simple exchange-type aberrations and the dose–effect response for these aberrations is linear with dose (Virsik et al. 1980; Thacker et al. 1986; Griffin et al. 1998). The implication of this work is that exchanges can be induced by single ionizing events, and thus can arise from a single DSB. In support of the single DSB pathway to chromosome exchanges, Ludwików et al. (2002) used the property of the late replication of a region (Xq) of the Chinese hamster chromosome to label the bulk of chromosomes with ^{125}I while leaving Xq free of incorporated label. The frequency of exchanges between the unlabelled region and labelled chromosomes was found to be half that between different labelled chromosomes. The results support the existence of an alternative (single DSB) mechanism for the formation of exchanges.

Simple chromosome exchanges, i.e. those defined as being formed by interactions between two chromosomes carrying one DSB each (Savage 2002), should, according to the breakage-first model, be induced as a quadratic function of radiation dose. The single-hit–no-hit concept of chromosome exchanges would predict a completely linear dose–effect curve. However, the data for simple exchange low LET radiation is controversial, and dose–effect curves are linear–quadratic with exponents of dose for total simple exchanges around 1.35 (Edwards et al. 1999; Loucas and Cornforth 2001). A further important finding is that the frequency of formation of chromosome exchanges by low LET radiation is dose rate dependent and is reduced also by fractionating the dose with a time interval for repair. These properties were already noted by Sax (1939) in *Tradescantia* microspores (see above and Fig. 11.2). Sax used this as an argument for the requirement for the interaction of two lesions in one or two chromosomes. Sax realized that as the dose rate was decreased there was more time for repair of one of the lesions before a second was induced. Relevant to the above discussion of the breakage-first versus one-hit–no-hit formation of exchanges is that a striking dose-rate effect is seen for the formation of simple exchanges in human primary fibroblasts (Loucas et al. 2004). At very low dose rate, induction of simple exchanges is linear, possibly indicating that a limited fraction of the simple exchanges are formed via a one-DSB mechanism.

Other lines of evidence point to the requirement for two interacting DSB in the formation of chromosome exchanges. For example Cornforth (1990) elegantly demonstrated that by fusing cells together and thereby mixing irradiated chromosomes from one cell with unirradiated chromosomes from another, only a few interactions between the two sets of chromosomes occurred. Essentially only interactions leading to exchange aberrations occurred between the irradiated cell's chromosomes. Also, Wang et al. (1997) showed that exchanges (translocations) induced by the radiomimetic bleomycin in the *aprt* gene of Chinese hamster cells were almost all (six out of seven) reciprocal exchanges between the aprt sequence and an unrelated sequence on a different chromosome, consistent with a homology-independent (NHEJ) pathway and the breakage-first model. Support for the breakage-first model also came from the experiments of Richardson and Jasin (2000), who examined several recombinant embryonic stem (ES) cell clones in which recombination events had occurred between DSB in inserted sequences engineered to contain one or two I-*Sce*I sites, and cut by endogenous expression of the I-*Sce*I endonuclease. In support of the breakage-first model, only those recombinant clones resulting from interaction of DSB in cells with two I-*Sce*I sites (when incised by transiently expressed omeganuclease this yields a DSB with 3′-cohesive ends with four-base overlap) showed translocations, whereas those with only a single site did not. Analysis of the recombinational joins showed only a small degree of homology between the two sites, suggestive of SSA rather than homologous recombinational rejoining of the two sites. Whether such misjoining is representative of global misjoining of DSB in irradiated cells is not known, and it is possible that radiation-induced DSB, which are of a more complex nature (e.g. in the case of DSB induced by carbon-K X-rays, the electrons of which cause dense clusters of ionizations) may be misjoined in a different way from those produced by omeganuclease at I-*Sce*I sites.

Further recent evidence in support of the breakage-first model comes from experiments demonstrating movement together and apparent fusion of γ-H2AX foci within minutes of the passage of α-particle tracks or microbeams of ultrasoft X-rays (Aten et al. 2004; Hamada et al. 2006). This clustering of phosphorylated histone domains following DSB formation was shown to also implicate the MRE11 complex, which colocalizes with γ-H2AX clusters (Aten et al. 2004). These findings argue for a breakage-first mechanism of chromosome exchange formation and could possibly be an explanation for the formation of complex chromosome translocations involving interactions between breaks in several chromosomes as revealed by multicolour fluorescence in situ hybridization (FISH) (Griffin et al. 1995, 1996, 1998; Loucas et al. 2004; Anderson et al. 2002). However, this view of chromatin movement following induction of DSB has been challenged recently by Kruhlak et al. (2006), who demonstrated limited mobility and a simple expansion or diffusion of chromatin neighbouring a DSB following its induction by UV laser light. However, the question arises as to whether the difference in the types of damage induced by UV radiation (e.g. pyrimidine dimers which distort chromatin structure) as compared with α-particle or ultrasoft X-ray tracks (Aten et al. 2004; Hamada et al. 2006) could play a role in the modification of chromatin following exposure.

11.3.2 The Molecular Misrecombination Model

The misrecombination model for the formation of chromosome aberrations (Chadwick and Leenhouts 1978) invokes the Resnick (1976) model of homologous recombinational rejoining of DSB, which is prevalent in yeast but is thought to be of much lesser importance in mammalian cells (Essers et al. 2000; Wang et al. 2001; Iliakis et al. 2004). Essentially, the misrecombination model proposes that a DSB in one chromosome undergoing HRR with its homologue leads to misjoining with the homologous chromosome. At first glance this scheme would seem a likely explanation for single-hit exchanges; however, data do not appear to support the model, at least for normal cells. First of all, in normal (wild-type) cells, exchanges do not appear to be preferentially formed between homologous chromosome pairs (Savage 2002; Cornforth et al. 2002). For example, the frequency of exchanges between two human homologues in a hamster–human hybrid was the same as that in a hybrid containing only one human chromosome, i.e. there was no preferential interaction between the two homologues (Marcon et al. 2000). The frequency of chromosome exchanges between chromosomes 21 and 22 in Down syndrome (trisomy 21) as compared with normals (Grigorova and Natarajan 1998) was found to be similar, suggesting that homologous recombination does *not* play a role in the formation of exchanges in these cells. In contrast to these studies, Plan et al. (2005) reported an excess of homologous exchanges in human lymphocytes over that expected on the basis of random (non-homologous) interactions. Although the vast majority of dicentrics were heterologous, suggesting that homologous misrecombination is not the dominant mechanism, a role for this type of misjoining could not be ruled out entirely. Plan et al. suggest the excess could feasibly result from the relative proximity of homologues within the cell nucleus.

In *Xrcc2*$^{-/-}$ (a gene involved in the homologous recombination pathway) knockout mouse embryonic fibroblasts, the frequency of exchanges between homologues was increased dramatically, with cells showing a sevenfold increase in spontaneous homologous rearrangements (Deans et al. 2003). Also, in *Xrcc2*$^{-/-}$ ES cells the frequency of both spontaneous and radiation-induced exchanges was increased dramatically (Griffin et al. 2005). Thus, HRR is clearly important in the maintenance of genomic stability and in *preventing* formation of interchromosome rearrangements, so its proposed role in misrecombination leading to the formation of radiation-induced chromosome exchanges (Chadwick and Leenhouts 1978) is therefore doubtful. More plausible seems the proposal that misrecombination between chromosomes, leading to rearrangments, could arise via homologous SSA (Griffin and Thacker 2004) whereby minihomologies or microhomologies between heterologous chromosomes (e.g. satellite or Alu sequences which are common between all chromosomes) lead to misjoining. Such a process might conceivably occur, for example, during replication, or in the G1 phase during active gene expression where homologous sequences might be susceptible to strand invasion by a damaged strand and followed by strand-nicking to yield an interchromosomal exchange. Otherwise two DSB would be required to interact pairwise as in conventional SSA-mediated joining of DSB (Lin et al. 1984). Data of Griffin et al. (2005) show that deletion of the *Ercc1* gene (involved in SSA repair) results in elevated fragmentation of chromosomes after X-irradiation, whereas the frequency of simple and complex exchanges is only marginally increased, possibly indicating a role for SSA in the formation of exchanges.

11.3.3 The Transcription-Based Model

The idea that chromosomes, normally visualized by FISH during interphase as discrete concave domains attached to the nuclear matrix (Cremer and Cremer 2001), can interact over relatively large distances within the cell nucleus (Cremer et al. 1996; Savage 1998, 2000) raised the possibility that large loop domains may come together via channels within the nucleus to link transcription 'factories' (Iborra et al. 1996; Cook 1999). The idea of such interaction centres, corresponding to transcription factories, forms one component of the transcription-based model for the formation of chromosome aberrations (Radford 2002). The other component is the observation that some 80% of chromosome exchange aberrations occur in the light G-bands (R-bands) of X-irradiated normal human chromosomes (Holmberg and Jonasson 1973; Mattei et al. 1979; Nakagome et al. 1983; Radford et al. 1988) and in Chinese hamster chromosomes (Slijepcevic and Natarajan 1994;

Folle et al. 1998). These light bands are gene-rich and transcriptionally active regions of the chromosomes (Musio et al. 2002). The model proposes that a DSB induced in a loop from one chromosome becomes associated with a topoisomerase I (topo I) at a stalled RNA polymerase site. This trapped DNA–topo I then interacts with another topo I bound to a transcription unit on another chromosome. Radford proposes that the two DNA–topo I cleavages are then reversed, and in doing so misjoin the two chromosomes. A further cleavage would be required in order to resolve the structure to give a dicentric chromosome and a fragment or a reciprocal translocation.

As mentioned above, the idea of interactions between chromosomes at transcription factories and involving extruded loop domains has been proposed previously (Griffin et al. 1996; Savage 2000) and it seems feasible that regions of DNA undergoing active transcription at such factories, and hence having a more open structure, would be more vulnerable to cleavage and/or interaction via topoisomerases or indeed by homologous SSA (see before). Griffin et al. (2005) showed that deletion of the *Csb* gene ($Csb^{-/-}$ knockout in ES cells), a gene involved in transcription-coupled repair allowing more rapid nucleotide excision repair in the transcribed damaged DNA strand in active genes (van Hoffen et al. 2003), led to a significant reduction in complex exchanges relative to that in wild-type cells. The reduction in complex exchanges in $Csb^{-/-}$ cells was especially marked following α-particle irradiation. Griffin et al. (2005) suggest that the absence of Csb protein might hinder access of DSB to interactive transcription sites, or might simply block the multiple interactions, leading to complex exchanges.

Thus, while there is no direct evidence for the transcription-based model as such, DSB in gene-rich (transcribed) regions of chromosomes are more likely to undergo exchange. Furthermore, transcription sites in the cell are implicated in the formation of exchanges between DSB, possibly with unbroken strands, or between two or more DSB to form complex aberrations.

In summary, these studies indicate that although the major mechanism for formation of exchanges requires breakage first (i.e. two interacting DSB) it seems probable that at an alternative pathway may exist whereby some DSB, possibly those of a more complex nature (Griffin et al. 1998), can interact with a homologous sequence on an undamaged heterologous chromosome, leading to an exchange aberration.

11.4 Models of Chromatid Aberration Formation

11.4.1 Formation and Disappearance of Chromatid Breaks

Chromosome aberrations formed in the G2 phase of the cell cycle, i.e. where DNA has been duplicated, can occur in one or other or both of the sister chromatids. In human and other mammalian cell lines, the G2 phase is a highly radiosensitive phase and relatively small doses of radiation lead to significant blockade of the cell's progression into mitosis. Therefore, relatively small radiation doses are generally used (less than 1 Gy) and the frequency of exchanges in normal cells is very low. For example, in normal human lymphocytes the comparative frequencies were found to be 2.5 exchanges per 100 cells per gray compared with the maximum frequency of breaks 451 breaks per 100 cells per gray, giving a break-to-exchange ratio of 180 (MacLeod and Bryant 1992). Most recent studies of chromatid aberrations in mammalian cells have focussed on formation of chromatid breaks (CB). CB are induced linearly with dose, but the frequency of CB is not constant with time; the maximum frequency occurs at around 0.5 h following irradiation, with an exponential decline in the frequency thereafter (MacLeod and Bryant 1992; Parshad 1996) In contrast, the frequency of exchanges is more or less constant between 0.5 and 4h following irradiation.

The exponential decline in CB with time after irradiation has been variously interpreted in terms of repair or processing of DNA lesions such as base damage (Preston 1980) or DSB (Mozdarani and Bryant 1989a; MacLeod and Bryant 1992; Bryant and Slijepcevic 1993; Parshad 1996).

The induction of CB is linear with dose. This was demonstrated by Sax (1940) using *Tradescantia* microspores (Fig. 11.3), and when the maximum frequency of CB in human or rodent cells was taken as an end point, the induction of CB (Fig. 11.3) was also found to be linear (Bryant 1998; Bryant et al. 1998; Mozdarani et al. 2001). This view of induction and rejoining of CB has been challenged (Savage 2004) on the grounds that variability in the transit times of individual cells through the G2 phase will render measurements of breaks at various times after irradiation too imprecise to interpret them in this way. However, using the rapid premature chromosome condensation

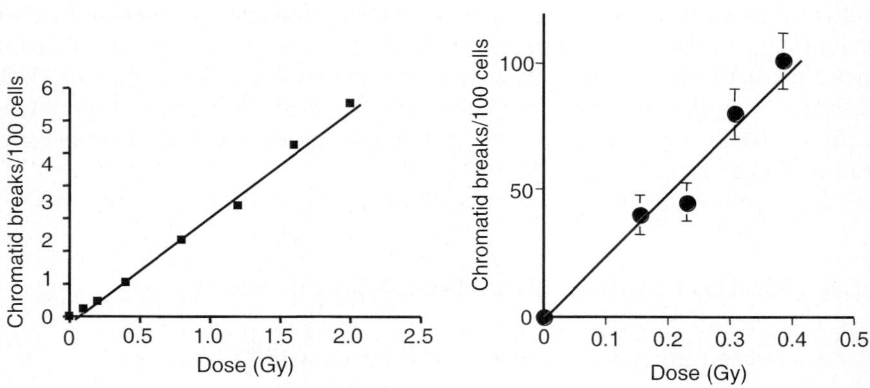

Fig. 11.3 *Left*: Frequency of chromatid breaks in *Tradescantia* microspores as a function of radiation dose. *Right*: Frequency of (G2) chromatid breaks in human lymphocytes. (*Left* drawn from data of Sax 1940; *right* from Bryant 1998)

induced by calyculin A (with which the induction of G2 CB can be studied without the putative complication of cell-cycle kinetics) the dose–effect relationship for CB was similarly shown to be linear (Gotoh et al. 1999; Kawata et al. 2004), confirming the single-event (DSB) origin of CB. Taken together, the decline in CB frequency with time during the G2 phase, both with a standard G2 assay (MacLeod and Bryant 1990, 1992; Parshad 1996) and with the premature chromosome condensation technique, which is independent of cell-cycle parameters (Pandita and Hittelman 1992; Gotoh et al. 1999; Kawata et al. 2000, 2004), suggests that for human and other mammalian cell lines, the kinetics of disappearance of CB does indeed represent a rejoining process. Moreover, the fact that treatment of cells with nucleoside analogues (arabinoside cytosine and adenine) following irradiation abrogates the rejoining or in some cases actually increases the frequencies of CB (Preston 1980; Mozdarani and Bryant 1989a, b; Antoccia et al. 1994; Parshad 1996) suggests a metabolically active rejoining process underlies the decline in CB with time.

11.4.2 Models of Chromatid Breakage

Four different models have been applied to account for CB. These are the breakage-first model (Sax 1940; Bender et al. 1974) the 'exchange' model (Revell 1959; Savage 1986; Harvey and Savage 1997) the misrecombinational model (Chadwick and Leenhouts 1978, 1998) and the signal model (Bryant 1998; Bryant et al. 2004).

The breakage-first model appears at first to fit much of the data for CB on the assumption that a CB can be equated with DSB (Bender et al. 1974; Parshad et al. 1990; Parshad 1996): linear induction of CB, and disappearance ('rejoining') of CB with time, as outlined already. However, several examples show that rejoining of CB does not accurately reflect the rejoining of DSB. For example, the rate of decline in frequency of CB with time in a DSB repair mutant (xrs5) was the same as in its wild-type (CHOK1) parental line, although the frequency of CB was more than 3 times that in CHOK1 (MacLeod and Bryant 1990; Bryant and Slijepcevic 1993). Also, in cell lines mutated in *Atm*, *Xrcc2* and *Xrcc8* genes, radiation-induced CB are significantly elevated (Taylor 1978; Mozdarani and Bryant 1989; Mozdarani et al. 2001; Bryant et al. 1993), while these cell lines all show normal or even faster rejoining of bulk DSB (Jones et al. 1990; Thacker and Ganesh 1990; Foray et al. 1995). A further point made previously (Bryant 1998; Bryant et al. 2004) is that of the huge difference in size between a DSB (where only a few base pairs are thought to be deleted;Pfeiffer et al. 2005) and a CB, which can be 5–20 Mbp (up to about 10% of the length of an average hamster chromosome). Two DSB might interact to produce such deletions, but this would yield a quadratic dose–effect relationship for CB instead of the observed linear induction (see before). Finally, it was shown that treatment of Chinese hamster ovary (CHO) cells with the nucleoside analogue arabinoside adenine at 100 μM abrogates the rejoining of CB without affecting the rejoining of DSB (Bryant et al. 2004).

The fact that *Xrcc2* (HRR pathway), *Xrcc5* (NHEJ pathway) and *Atm* gene mutations all lead to enhanced CB induction but do not affect rejoining indicates that these genes normally protect the cell from CB induction and are not actively involved in the production or processing of CB following irradiation with X-rays or γ-rays.

Thus, the breakage-first model does not satisfactorily account for the observed CB data. This conclusion is supported by evidence that indicates that the mechanism for CB is different from that for chromosome aberrations, where, as we have seen, breakage-first is the major mechanism. For example, Scott et al. (1999) showed a complete lack of correspondence between the frequencies of CB and micronuclei (a measure of chromosome fragment frequency – these arising by breaks and from unstable exchanges: rings and dicentrics) in peripheral blood lymphocytes of a group of breast cancer patients with variable individual chromosome radiosensitivity. Further evidence for this difference between the mechanisms of chromosome fragments and CB was provided by experiments of Vral et al. (2002) that showed the relative biological effectiveness (RBE) for neutrons for induction of micronuclei to be 5, whereas the RBE for CB was 1.0. These experiments suggest that the quality of DSB (i.e. the complexity) is an important determinant of chromosome aberrations, but does not appear to substantially affect the formation of CB.

11.4.3 The Revell 'Exchange' Model

The breakage-first model was challenged by Revell (1959), who proposed a model for chromatid aberrations (Fig. 11.4) under which the initiating lesions (unspecified) are those formed close together at the cross-over points of chromatin loops (Savage 1986). These lesions would interact, i.e. undergo subsequent breakage and misjoining, to give exchange-type aberrations, or if the rearrangements were incomplete they would lead to the various types of CB observed. Thus, in terms of Revell's model, CB (or terminal deletions) are a secondary product of an incomplete rearrangement. The model predicts that CB (i.e. a break in one chromatid, and not including isochromatid breaks) will be of five different types (Fig. 11.4).

Two of these types involve a rearrangement between sister chromatids and can be detected by a colour switch at the breakpoint in harlequin (fluorescence plus Giemsa) stained chromosomes (Fig. 11.5).

The frequency of such interarm rearrangements in irradiated and restriction endonuclease treated mammalian cells, in unirradiated cells and in omeganuclease-treated cells containing I-*Sce*I break sites (Rogers-Bald et al. 2000) is reported to be around 16%, and a constant ratio (Harvey and Savage 1997). This percentage is, of course, much lower than that originally predicted by Revell, who argued that all CB types would be produced at equal frequencies, thus yielding two fifths (or 40%) of the single CB appearing as

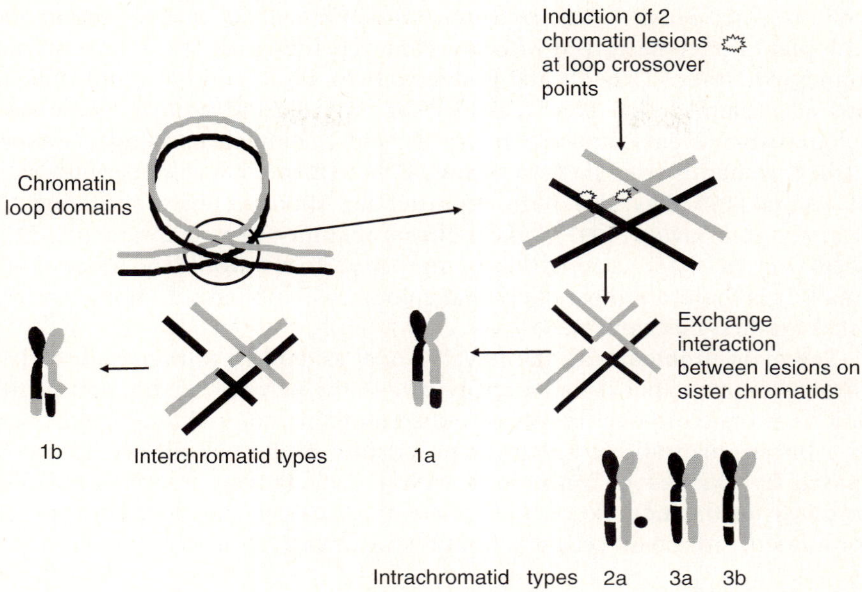

Fig. 11.4 Revell's exchange model. Note that the lesions at loop crossover points were not specified by Revell, but presumably would at some point have to be converted into double-strand breaks

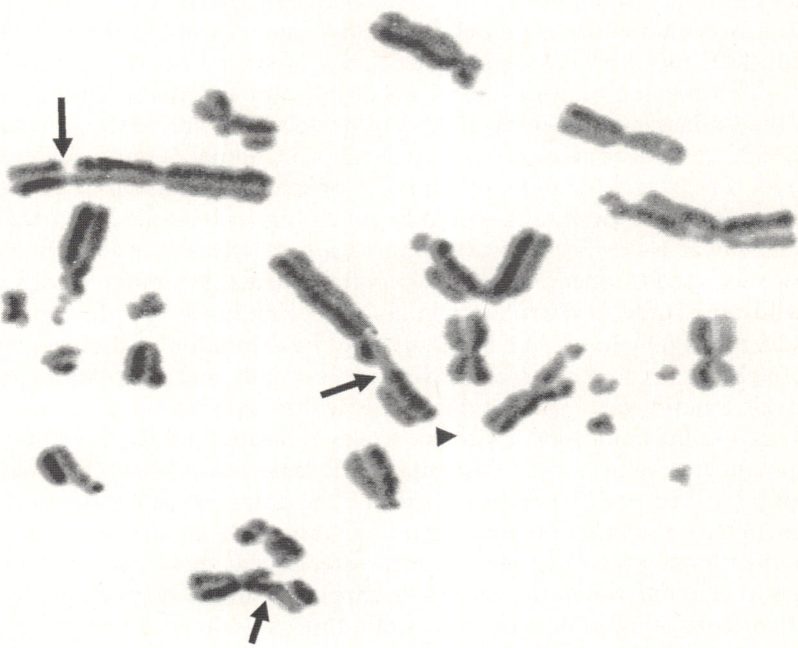

Fig. 11.5 Harlequin (fluorescence plus Giemsa) stained Chinese hamster chromosomes with interchromatid 'colour-switch' breaks (arrows). Other non-colour-switch breaks are considered to be formed by intrachromatid rearrangements. (From Bryant 1998)

'colour-switch breaks'. Although some measurements in earlier experiments with plant materials agreed with this ratio (Heddle et al. 1969), later studies found lower ratios (Heddle and Bodycote 1970; Wolff and Bodycote 1975). It was also claimed that the breakage-first model could accommodate such colour-switch events since the theory allowed incomplete exchanges between sister chromatids (Heddle 1994; Savage 1998). However, a colour-switch ratio of around 16% has also been reported for spontaneously occurring CB (Harvey and Savage 1997), where the probability of two interacting DSB occurring in sister chromatids at the same locus would be vanishingly small. It is therefore most likely that colour-switch breaks do not arise by a breakage-first mechanism.

The main problem with the Revell model as applied to human or rodent chromosomes is that it its basic premise is the pairwise interaction of two lesions at the cross-over points of looped domains, so predicting a quadratic or a linear–quadratic (assuming a proportion of the two-hit events will be caused by clusters of ionizations within one electron track) dose–effect response for the induction of CB. As mentioned above, the dose effect for CB for human and rodent cells was found to be linear.

11.4.4 The Misrecombination Model

Chadwick and Leenhouts (1978, 1998) included chromatid aberrations in their misrecombinational model. Under their model a single unrejoined DSB would lead to a non-colour-switch CB, and a stalled or incomplete inter-chromatid recombinational event, resulting from an attempt by the cell to use the undamaged sister chromatid to perform recombinational repair of the DSB, would lead to a colour-switch CB. This model seems unlikely to be correct, firstly because if it were the only or main mechanism of CB formation, the colour-switch ratio would be much higher than actually observed; possibly near 100%. Secondly, recombination-deficient cells are reported to show increased frequencies of colour-switch breaks as compared with those of wild-type controls (Mozdarani et al. 2001), indicating that the *Xrcc2* gene product normally has a role in preventing the formation of this type of CB; although the method of break scoring in this study has been challenged, in that chromatid gaps smaller than the width of a chromatid had been included in the colour-switch-break scores (Thacker and Griffin 2002). The arguments for including all clear gaps in CB scores have been rehearsed previously (see reply to Thacker and Griffin by Bryant et al. 2002a and in Bryant et al. 2002b). A further obvious point is that because colour switches (associated with colour-switch breaks) occur precisely at those gap sites that are equal to or *smaller* than the width of a chromatid in the same way as at gaps that are *larger* than the chromatid width, this shows that the narrower gaps are similarly the result of breakage and rearrangement of chromatin between the sister chromatids.

11.4.5 The Signal Model

The signal model (Bryant 1998; Bryant et al. 2004) was based on the loop domain idea of the Revell model (Revell 1959). It was proposed that radiation damage (a single DSB) occurring within a large mega-base-pair looped chromatin domain, portrayed as a simple single loop (Fig. 11.6) but probably much more complex and possibly equivalent to a transcription factory, generates a damage signal. Such a damage signal could be either *trans*- or *cis*-acting. One possible candidate for such a *trans*-acting damage signal is the rapid and prolonged Ser139 phosphorylation of histone H2AX by Atm protein over a megabase domain surrounding a DSB following irradiation (Rogakou et al. 1999; Burma et al. 2001). However, we have observed that the homozygous mutation of the *Atm* gene results in a twofold *elevation* of CB frequency (Mozdarani and Bryant 1989). H2AX-phosphorylation by ATM seems not be required for initial recruitment of DNA repair factors to DSB sites (Celeste et al. 2003) and more recently it has been proposed that DSB are initially recognized by 53BP1 protein owing to its property of interaction with constitutively methylated (at Lys79) histone H3, which may be exposed at DSB sites by local opening of the higher-order chromatin structure (Huyen et al. 2004; Zgheib et al. 2005). Local chromatin modification at DSB sites could feasibly be the signal reducing the fidelity of chromatid decatenation by topoisomerase II as cells progress towards mitosis.

Topoisomerase II is known to be subject to error-prone joining of cut DNA strands following treatment with agents inducing reactive oxygen species

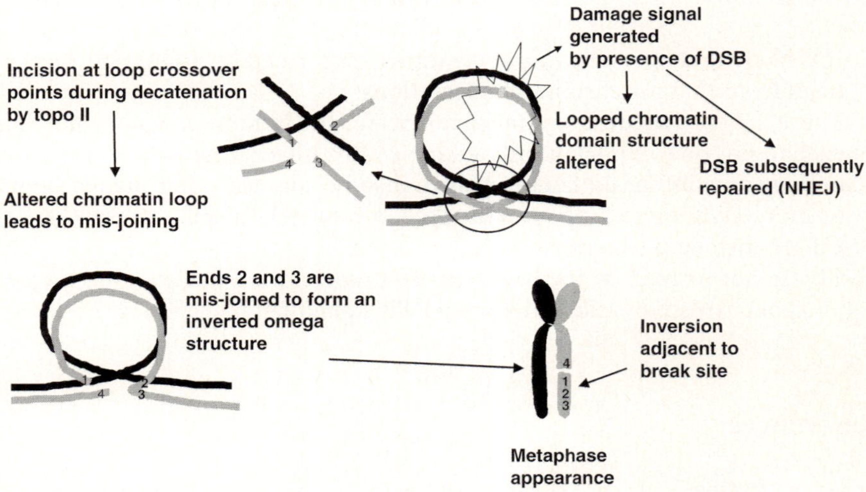

Fig. 11.6 A suggested scheme for the steps involved in the signal mechanism as applied to incomplete intrachromatid rearrangements, leading to chromatid breaks at metaphase. *DSB* double-strand breaks, *NHEJ* non-homologous end-joining, *topo II* topoisomerase II

(Li et al 1999). If incomplete, such misjoining of chromatin ends would lead to a CB; one of the several types proposed under the Revell model.

It is important to note that the signal model differs fundamentally from the Revell model in that it proposes only a *single* DSB is required to produce a CB, a proposal that is supported by data showing that a single (I-*Sce*I site) DSB in a wild-type Chinese hamster background is sufficient to induce a CB (Rogers-Bald et al. 2000). Also, CB are induced linearly with dose by carbon-K X-rays, where the track lengths of electrons produced are too short to 'hit' more than one DNA strand (Bryant et al. 2003).

The disappearance of CB with time (MacLeod and Bryant 1990, 1992) under the signal model would be explained by the sequential completion of the topoisomerase II misjoining and resultant incorporation of the chromatin rearrangement. Subsequently, the initiating DSB would be rejoined (independently of the rearrangement process) by one of the three known DSB rejoining pathways (HRR, SSA or NHEJ).

Thus, the signal model predicts the observed linear induction of CB, provides a possible framework with which to explain the lack of direct correspondence between the DSB rejoining and disappearance of CB with time after irradiation.

11.5 Conclusions

The mechanisms of formation of chromosome aberrations remain very controversial; however, some conclusions can be drawn from studies to date:

1. It is likely that chromosome aberrations are formed by a different mechanism from that for chromatid aberrations.
2. There may be two different mechanisms for the induction of chromosome exchanges; one requiring only a single DSB interacting with a microhomology on another undamaged chromosome, and the other involving two or more DSB interacting via a breakage-first mechanism.
3. CB are formed from single DSB.
4. CB are not formed by the breakage-first mechanism, but by a more complex mechanism possibly involving DSB signalling.

References

Anderson RM, Stevens DL, Goodhead DT (2002) M-FISH analysis shows that complex chromosome aberrations induced by alpha -particle tracks are cumulative products of localized rearrangements. Proc Natl Acad Sci USA 99:12167–12172

Antoccia A, Palitti F, Raggi T, Catena C, Tanzarella C (1994) Lack of effect of inhibitors of DNA synthesis/repair on the ionizing radiation-induced chromosomal damage in G2 stage of ataxia telangiectasia cells. Int J Radiat Biol 66:309–17

Aten JA, Stap J, Krawczyk PM, van Oven CH, Hoebe RA, Essers J, Kanaar R (2004) Science 303:92–95

Bender AM, Griggs HG, Bedford JS (1974) Mechanisms of chromosomal aberrations production III: Chemicals and ionizing radiation. Mutat Res 23:197–212

Bryant PE (1998) The signal model: a possible explanation for the conversion of DNA double-strand breaks into chromatid breaks Int J Radiat Biol 73:243–251

Bryant PE (1984) Enzymatic restriction of mammalian cell DNA using *Pvu II* and *Bam H1*: Evidence for double strand break origin of chromosome aberrations. Int J Radiat Biol 46:57–65

Bryant PE, Christie AF (1989) Induction of chromosomal aberration in CHO cells by restriction endonucleases: effects of blunt- and cohesive-ended double strand breaks in cells treated by "pellet" methods. Mutat Res 213:233–241

Bryant PE, Slijepcevic P (1993) G2 chromatid aberrations: kinetics and possible mechanisms. Environ Mol Mutagen 22:250–256

Bryant PE, Jeggo P, Birch D (1987) High chromosomal sensitivity of Chinese hamster xrs 5 cells to restriction endonuclease induced DNA double-strand breaks. Int J Radiat Biol 52:537–554

Bryant PE, Jones NJ, Liu N (1993) Radiosensitive Chinese hamster irs 2 cells show enhanced chromosomal sensitivity to ionising radiation and restriction endonuclease induced blunt-ended double-strand breaks. Mutagenesis 8:141–147

Bryant PE, Finnegan C, Swaffield L, Mozdarani H (1998) G2 chromatid breaks in murine scid cells. Mutagenesis 13:481–485

Bryant PE, Mozdarani H, Liu N, Jones NJ (2002a) Reply to the letter from Thacker and Griffin: Problems in scoring radiation-induced chromatid breaks. Int J Radiat Biol 78:947–948

Bryant PE, Gray L, Riches AC, Steel CM, Finnon P, Howe O, Kesterton I, Vral A, Curwen GB, Smart V, Tawn EJ, Whitehouse CA (2002b) The G2 assay: a technical report. Int J Radiat Biol 78:863–866

Bryant PE, Jones C, Armstrong G, Frankenberg-Schwager M, Frankenberg D (2003) Induction of chromatid breaks by carbon-K ultra-soft X-rays. Radiat Res 159:247–250

Bryant PE, Gray LJ, Peresse N (2004) Progress towards understanding the nature of chromatid breakage. Cytogenet Genome Res 104:65–71

Burma S, Chen BP, Murphy M, Kurimasa A, Chen D (2001) ATM phosphorylates histone H2AX in response to DNA double-strand breaks. J Biol Chem 276:42462–42467

Celeste A, Fernandez-Capetillo O, Kruhlak MJ, Pilch DR, Staudt DW, Lee A, Bonner RF, Bonner WM, Nussenzweig A (2003) Histone H2AX phosphorylation is dispensible for the initial recognition of DNA breaks. Nat Cell Biol 5:675–679

Chadwick KH, Leenhouts HP (1978) The rejoining of DNA double-strand breaks and a model for the formation of chromosomal rearrangements. Int J Radiat Biol 33:517–529

Chadwick KH, Leenhouts HP (1998) Radiation induced chromosome aberrations: some biophysical considerations. Mutat Res 404:113–117

Cook PR (1999) The organization of replication and transcription. Science 284:1790–1795

Cornforth MN (1990) Testing the notion of the one-hit exchange. Radiat Res 121:21–27

Cornforth MN, Greulich-Bode KM, Loucas BD, Arsuaga J, Vazquez M, Sachs RK, Bruckner M, Molls M, Hahnfelt P, Hlatky L, Brenner DJ (2002) Chromosomes are predominantly located randomly with respect to each other in interphase human cells. J Cell Biol 159:237–244

Cremer T, Cremer C (2001) Chromosome territories, nuclear architecture and gene regulation in mammalian cells. Nat Rev Genet 2:292–301

Cremer C, Munkel C, Granzow M, Jauch A, Dietzel S, Eils R, Guan XY, Meltzer RS, Trent JM, Langowski J, Cremer T (1996) Nuclear architecture and the induction of chromosomal aberrations. Mutat Res 366:97–116

Darroudi F, Fomina J, Meijers M, Natarajan AT (1998) Kinetics of the formation of chromosome aberrations in X-irradiated human lymphocytes, using PCC and FISH. Mutat Res 404:55–66

Deans B, Griffin CS, O'Regan P, Jasin M, Thacker J (2003) Homologous recombination deficiency leads to profound genetic instability in cells derived from *Xrcc2*-knockout mice. Cancer Res 63:8181–8187

Edwards AA, Hone PA, Moquet JE, Lloyd DC (1999) Simple chromosome exchanges are not linear with dose. Int J Radiat Biol 75:1113–1117

Essers J, van Steeg H, de Wit J, Swagemakers SM, Vermeij M, Hoeijmakers JH, Kanaar R (2000) Homologous and non-homologous recombination differentially affect DNA damage repair in mice. EMBO J 19:1703–1710

Folle GA, Marinez-Lopez, Boccardo E, Obe G (1998) Localization of chromosome breakpoints: implication of the chromatin structure and nuclear architecture. Mutat Res 404:17–26

Foray N, Arlett CF, Malaise EP (1995) Dose-rate effect on induction and repair of radiation-induced DNA double-strand breaks in a normal and an ataxia-telangiectasia human fibroblast cell line. Biochemie 77:900–905

Gotoh E, Kawata T, Durante M (1999) Chromatid break rejoining and exchange aberration formation following γ-ray exposure: analysis in G2 human fibroblasts by chemical-induced premature chromosome condensation. Int J Radiat Biol 75:1129–1135

Griffin C, Thacker J (2004) The role of homologous recombination repair in the formation of chromosome aberrations. Cytogenet Genome Res 104:21–27

Griffin C, Marsden SJ, Stevens DL, Simpson P, Savage JRK (1995) Frequencies of complex chromosome exchange aberrations induced by 238Pu α-particles and detected by fluorescence in situ hybridization using single chromosome-specific probes. Int J Radiat Biol 67:431–439

Griffin C, Stevens DL, Savage JRK (1996) Ultrasoft 1.5 keV Aluminium K X-rays are efficient producers of complex chromosome exchange aberrations as revealed by fluorescent in situ hybridization. Radiat Res 146:144–150

Griffin C, Hill MA, Papworth DG, Townsend KMS, Savage JRK, Goodhead DT (1998) Effectiveness of 0.28keV carbon K ultrasoft X-rays at producing simple and complex chromosome exchanges in human firbroblasts in vitro detected using FISH. Int J Radiat Biol 73:591–598

Griffin C, de Waard H, Deans B, Thacker J (2005) The involvement of key DNA repair pathways in the formation of chromosome rearrangements in embryonic stem cells. DNA Repair 4:1019–1027

Grigorova M, Natarajan AT (1998) Relative involvement of chromosome #21 in radiation induced exchange aberrations in lymphocytes of Down syndrome patients. Mutat Res 404:67–75

Hamada N, Schettino G, Kashino G, Vaid M, Suzuki K, Kodama S, Vojnovic B, Folkard M, Watanabe M, Michael BD, Prise KM (2006) Histone H2AX phosphorylation in normal human cells irradiated with focused ultrasoft X-rays: evidence for chromatin movement during repair. Radiat Res 166:31–38

Harvey AN, Savage JRK (1997) Investigating the nature of chromatid breaks produced by restriction endonucleases. Int J Radiat Biol 71:21–28

Heddle J (1994) Revelling in cytogenetics. Environ Mol Mutagen 23:(Suppl 24):35–38

Heddle J, Bodycote DJ (1970) On the formation of chromosomal aberrations. Mutat Res 9:117–126

Heddle J, Whissell, D, Bodycote DJ (1969) Changes in chromosome structure induced by radiations: a test of the two chief hypotheses. Nature 221:1158–1160

Holmberg M (1990) On the time course of the interactions between DNA breaks in the production of a radiation-induced chromosome exchange aberration. Mutat Res 232:267–272

Holmberg M, Gumauskas E (1986) The role of short-lived DNA lesions in the production of chromosome exchange aberrations. Mutat Res 160:221–229

Holmberg M, Jonasson J (1973) Preferential location of X-ray induced chromosome breakage in the R-bands of human chromosomes. Hereditas 73:57–67

Huyen Y, Zgheib O, DiTullio RA Jr, Gorgoulis VG, Zacaratos P, Petty TJ, Sheston EA, Mellert HS, Stavridi ES, Halazonetis TD (2004) Methylated lysine 79 of histone H3 targets 53BP1 to DNA double-strand breaks. Nature 432:406–411

Iborra FJ, Pombo A, Jackson DA, Cook PR (1996) Active RNA polymerases are localised within discrete transcription 'factories' in human nuclei. J Cell Sci 109:1–14.

Iliakis G, Wang H, Perrault AR, Boecker W, Rosidi B, Windhofer F, Wu W, Guan J, Terzoudi G, Pantelias G (2004) Mechanisms of DNA double-strand break repair and chromosome aberration formation. Cytogenet Genome Res 104:14–20

Johnson RT, Rao PN (1970) Mammalian cell fusion: induction of premature chromosome condensation in interphase nuclei. Nature 226:717–722

Jones NK, Stewart SA, Thompson LH (1990) Biochemical and genetic analysis of the Chinese hamster mutants irs 1 and irs 2 and their comparison to cultured ataxia telangiectasia cells. Mutagenesis 5:15–23

Kawata T, Gotoh E, Durante M, Wu H, George K, Furusawa Y, Cucinotta FA (2000) High-LET radiation-induced aberrations in prematurely condensed G2 chromosomes of human fibroblasts. Int J Radiat Biol 76:929–937

Kawata T, Ito H, Uno T, Saito M, Yamamoto S, Furusawa Y, Durante M, George K, Wu H, Cucinotta FA (2004) G2 chromatid damage and repair kinetics in normal human fibroblast cells exposed to low- or high-LET radiation. Cytogenet Genom Res 104:211–215

Kemp LM, Jeggo PA (1986) Radiation-induced chromosome damage in X-ray sensitive mutants (xrs) of the Chinese hamster ovary cell line. Mutat Res 166:255–263

Kruhlak MJ, Celeste A, Dellaire G, Fernandez-Capetillo O, Müller WG, McNally JG, Bazett-Jones DP, Nussenzweig A (2006) Changes in chromatin structure and mobility in living cells at sites of DNA double-strand breaks. J Cell Biol 172:823–834

Lea DE, Catcheside DG (1942) The mechanism of the induction by radiation of chromosome aberrations in *Tradescantia*. J Genet 44:216–222

Li T, Chen AY, Mao Y, Wang H, Liu LF (1999) Activation of topoisomerase II-mediated excision of chromosomal DNA loops during oxidatie stress. Genes Dev 13:1553–1560

Lin F-L, Sperle K, Sternberg N (1984) Model for homologous recombination during transfer of DNA into mouse L cells: role for DNA ends in the recombination process. Mol Cell Biol 4:1020–1034

Loucas BD, Cornforth MN (2001) Complex chromosome exchanges induced by gamma rays in human lymphocytes: an mFISH study. Radiat Res 155:660–671

Loucas BD, Eberle R, Bailey SM, Cornforth MN (2004) Influence of dose-rate on the induction of simple and complex chromosome exchanges by gamma-rays. Radiat Res 162:339–349

Ludwików G, Xiao Y, Hoebe RA, Franken NAP, Darroudi F, Stap J, van Oven CH, van Noorden CJF, Aten JA (2002) Induction of chromosome aberrations in unirradiated chromatin after partial irradidiation of a cell nucleus. Int J Radiat Biol 78:239–247

Macleod RAF, Bryant PE (1990) Similar kinetics of chromatid aberrations in X-irradiated xrs 5 and wild-type Chinese hamster cells. Mutagenesis 5:407–410

Macleod RAF, Bryant PE (1992) Effects of adenine arabinoside and coformycin of the kinetics of G2 chromatid aberrations in X-irradiated human lymphocytes. Mutagenesis 7:285–290

MacPhail SH, Banath JP, Yu TY, Chu EHM, Lambur H, Olive PL (2003) Expression of phosphorylated histone H2AX in cultured cell lines following exposure to X-rays. Int J Radiat Biol 79:1–8

Marcon F, Boei JJWA, Natarajan AT (2000) Recombination between two homologous chromosomes does not play a dominant role in the formation of chromosome aberrations. Int J Radiat Biol 76:1343–1348

Mattei MG, Ayme S, Mattei JF, Aurran Y, Giraud AF (1979) Distribution of spontaneous chromosome breaks in man. Cytogenet Cell Genet 23:95–102

Mozdarani H, Bryant PE (1989) Kinetics of chromatid aberrations in G2 ataxia telangiectasia cells exposed to X-rays and ara A. Int J Radiat Biol 55:71–84

Mozdarani H, Liu N, Jones NJ, Bryant PE (2001) The *XRCC2* human repair gene influences recombinational rearrangments leading to chromatid breaks. Int J Radiat Biol 77:859–865

Musio A, Mariani T, Vezzoni P, Frattini A (2002) Heterogenous gene distribution reflects human genome complexity as detected at the cytogenetic level. Cancer Genet Cytogenet 134:168–171

Nakagome Y, Matsubara T, Fujita H (1983) Distribution of break-points in human structural rearrangements. Am J Hum Genet 35:288–300

Natarajan AT, Obe G (1984) Molecular mechanisms involved in the production of chromosomal aberrations III. Restriction endonucleases. Chromosoma 90:120–127

Natarajan AT, Obe G, van Zeeland AA, Palitti F, Meijers M, Verdegaal-Immerzell EAM (1980) Molecular mechanisms involved in the production of chromosomal aberrations; II Utilization of Neurospora endonuclease for the study of aberration production by X-rays in G1 and G2 stages of the cell cycle. Mutation Res 69:293–305

Neary GJ (1965) Chromosome aberrations and the theory of RBE: 1. General considerations. Int J Radiat Biol 9:477–502

Nikiforova MN, Stringer JR, Blough MRM, Fagin JA, Nikiforova YE (2000) Proximity of chromosomal loci that participate in radiation induced rearrangements in human cells. Science 290:138–141

Pandita TK, Hittelman WN (1992) Initial chromosome damage but not DNA damage is greater in ataxia telangiectasia cells. Radiat Res 130:94–103

Parshad R (1996) Deficient DNA repair capacity, a predisposing factor in breast cancer. Br J Cancer 74:1–5

Parshad R, Sanford KK, Jones GM (1990) Chromatid damage after G2 phase X-irradiation of cells from cancer prone individuals implicates deficiency in DNA repair. Proc Natl Acad Sci USA 80:5612–5616

Pfeiffer P, Feldmann E, Odersky A, Kuhfittig-Kulle S, Goedecke W (2005) Analysis of DNA double strand break repair by nonhomologous end joining in cell-free extracts from mammalian cells. Methods Mol Biol 291:351–71

Plan Y, Hlatky L, Hahnfeldt P, Sachs R, Loucas B, Cornforth M (2005) Full-color painting reveals an excess of radiation-induced dicentrics involving homologous chromosomes. Int J Radiat Biol 81:613–620

Preston RJ (1980) The effect of cytosine arabinofuranoside on the frequency of X-ray induced chromosome aberrations in normal human leukocytes, Mutat Res 69:71–79

Rabbitts TH (1994) Chromosomal translocations in human cancer. Nature 372:143–149

Radford IR (2002) Transcription-based model for the induction of interchromosomal exchange events by ionizing irradiation in mammalian cell lines that undergo necrosis. Int J Radiat Biol 78:1081–1093

Radford IR, Hodgson GS, Matthews JP (1988) Critical DNA target size model of ionizing radiation-induced mammalian cell death. Int J Radiat Biol 54:63–79

Resnick MA (1976) The repair of double-strand breaks in DNA: A model involving recombination. J Theor Biol 59:97–106

Revell SH (1959) The accurate estimation of chromatid breakage, and its relevance to a new interpretation of chromatid aberrations by ionizing radiations. Proc R Soc Lond Ser B 150:563–589

Richardson C, Jasin M (2000) Frequent chromosome translocations induced by DNA double-strand breaks. Nature 405:697–700

Rogakou EP, Boon C, Redon C, Bonner WM (1999) Megabase chromatin domains involved in DNA double-strand breaks in vivo. J Cell Biol 146:905–916

Rogers-Bald M, Sargent RG, Bryant PE (2000) Production of chromatid breaks by single dsb: evidence supporting the signal model. Int J Radiat Biol 76:23–29

Rotkamm K, Lobrich M (2003) Evidence for a lack of DNA double-strand break repair in human cells exposed to very low X-ray doses. Proc Natl Acad Sci USA 100:5057–5062

Savage JRK (1986) The 'Revell loop'. Clin Cytogenet Bull 1:192–196

Savage JRK (1998) A brief survey of aberration origin theories. Mutat Res 404:139–147

Savage JRK (2000) Proximity matters. Science 290:62–63

Savage JRK (2002) Reflections and meditations upon complex chromosomal exchanges. Mutat Res 512:93–109

Savage JRK (2004) On the nature of visible chromosomal gaps and breaks. Cytogenet Genome Res 104:46–55

Sax K (1939) The time factor in X-ray production of chromosome aberrations. Proc Nat Acad Sci USA 25:225–233

Sax K (1940) An analysis of X-ray induced chromosomal aberrations in *Tradescantia*. Genetics 25:41–68

Scott D, Barber JBP, Spreadborough AR, Burrill W, Roberts SA (1999) Increased radiosensitivity in breast cancer patients: a comparison of two assays. Int J Radiat Biol 75:1–10

Serebrovsky AS (1929) A general scheme for the origin of mutations. Am Nat 63:374–378

Slijepcevic P, Natarajan AT (1994) Distribution of radiation-induced G1 exchange and terminal deletion breakpoints in Chinese hamster chromosomes as detected by G banding. Int J Radiat Biol 66:747–755

Taylor AMR (1978) Unrepaired DNA strand breaks in irradiated ataxia telangiectasia lymphocytes suggested from cytogenetic observations. Mutat Res 50:407–418

Thacker J, Ganesh AN (1990) DNA break repair, radioresistance of DNA synthesis. and camptoghecin sensitivity in the radiation-sensitive irs mutants: Comparisons to ataxia-telangiectasia cells. Mutat Res 235:49–58

Thacker J, Griffin CS (2002) Problems in scoring radiation-induced chromatid breaks. Comments on the paper: The XRCC2 human repair gene influences recombinational rearrangements leading to chromatid breaks. Int J Radiat Biol 78:945–952

Thacker J, Wilkinson RE, Goodhead DT (1986) The induction of chromosome exchange aberrations by carbon ultrasoft X-rays in V79 hamster cells. Int J Radiat Biol 49:645–656

Thacker J, Zdzienicka MZ (2004) The *XRCC* genes: expanding roles in DNA double-strand break repair. DNA Repair 3:1081–1090

van Hoffen A, Balajee AS, van Zeeland AA, Mullenders LH (2003) Nucleotide excision repair and its interplay with transcription. Toxicology 193:79–90

Virsik RP, Schafer C, Harder D, Goodhead DT, Cox R, Thacker J (1980) Chromosome aberrations induced in human lymphocytes by ultrasoft Al (K) and C (K) X-rays. Int J Radiat Biol 38:545–557

Vral A, Thierens H, Bryant PE, De Ridder L (2001) A higher micronucleus yield in B- versus T-cells after low-dose gamma-irradiation is not linked with defective Ku86 protein. Int J Radiat Biol 77:329–339

Wang H, Zeng Z-C, Bui T-A, Sonada E, Takata M, Takeda S, Iliakis G (2001) Efficient rejoining of radiation-induced DNA double-strand breaks in vertebrate cells deficient in genes of the RAD52 epistasis group. Oncogene 20:2212–2224

Wang P, Zhou R-H, Zou Y, Jackson-Cook CK, Povirk L (1997) Highly conservative reciprocal translocations formed by apparent joining of exchanged DNA double-strand break ends. Proc Natl Acad Sci USA 94:12018–12023

Winegar RA, Preston JR (1988) The induction of chromosome aberrations by restriction endonucleases that produce blunt-end or cohesive-end double-strand breaks. Mutat Res 197:141–149

Wolff S, Bodycote J (1975) The induction of chromatid deletions in accord with the breakage-and-reunion hypothesis. Mutat Res 29:85–91

Zgheib O, Huyen Y, DiTullo RA Jr, Snyder A, Venere M, Stavridi ES, Halazonetis TD (2005) ATM signaling and 53BP1. Radiother Oncol 76:119–122

12 Chromatin Structure and the Formation of Chromosomal Alterations

Gregor Kreth and Christoph Cremer

Abstract

Current experimental studies on the induction of chromosomal aberrations suggest the influence of different structural features of chromosome organization which are responsible for deviations from the predicted aberration frequencies when only the DNA content of the chromosomes involved is considered. The nonrandom neighbourhood relationship resulting from the experimentally observed radial positioning of chromosome territories in lymphocyte nuclei was reported as one important factor. As an additional, second contribution the compaction level of a chromosome territory also has a major impact on aberration frequencies. Different compaction levels were observed, e.g. for the active and inactive X-chromosome territories in amniotic fluid or fibroblast nuclei. A recent study of Branco and Pombo (*PLOS Biol.* 4:e138, 2006) revealed, however, also for other chromosome territories of similar DNA content, remarkable differences in the volumes between gene-rich and gene-poor chromosome territories. In the present study we summarize computer simulations of nuclear structure where we quantify the effects of these nuclear features on aberration frequencies for the example of the active and the inactive X chromosome.

12.1 Introduction

Current experimental studies on the induction of chromosomal aberrations suggest the influence of different structural features of chromosome organization which are responsible for deviations from the predicted aberration frequencies when only the DNA content of the chromosomes involved is considered. Two geometrical constraints of interphase chromosome organization might be responsible and are discussed in this chapter: the nonrandom neighbourhood relationship of chromosome territories (CTs) derived from the experimentally observed radial arrangement of gene-dense and gene-poor CTs (Cremer et al. 2001; Boyle et al. 2001; Tanabe et al. 2002), and the presence of different compaction levels of CTs according to their gene densities/transcriptional activities (Branco and Pombo 2006). Arsuaga et al.

Günter Obe and Vijayalaxmi (Eds.)
Chromosomal Alterations: Methods, Results and Importance in Human Health

(2004) reported that the aberration frequencies of at least two clusters of chromosomes show a statistically significant deviation from a random chromosome association. These clusters comprise the nucleolus-organizing CTs {13, 14, 15, 21, 22} and a cluster of gene-rich chromosomes {1, 16, 17, 19, 22} which are located in the interior of lymphocyte nuclei. We quantified such effects by computer simulations where we regarded explicitly the gene density correlated radial arrangement of CTs in simulated nuclei (Kreth et al. 2004, 2007). For these calculations, a direct correlation between DNA content and volume was assumed and no specific assumptions about a more decondensed compaction of gene-rich chromosomes were made.

This is, however, in contrast to the recent study of Branco and Pombo (2006). They presented evidence for a gene density/transcriptional activity dependent compaction level of CTs. The experimentally observed remarkable difference in the aberration frequencies between (1;21) and (1;22) [the frequency of (1;22) is nearly 3 times higher than that of (1;21)] could be explained by an increased volume of CT 22. Both chromosomes possess nearly the same DNA content but CT 22 revealed nearly twice the volume of CT 21. To quantify the effect of such different compaction levels, in the present study we performed quantitative computer simulations of the active and the inactive X-chromosome territories, taking into account the different experimentally observed compaction levels and also different localizations for the inactive X-chromosome territory in the nuclear volume.

12.2 Modelling of Virtual Radiation Events

To estimate the effects of the higher-order spatial chromosome organization on aberration frequencies, we used an earlier version of the already published spherical 1-Mbp chromatin domain computer model of nuclear structure (Kreth et al. 2004). Again, each chromosome was modelled by a linear chain of spherical domains of 1 Mbp (with a diameter of 500 nm each). To account for the observed dynamic variability in the distances between the domains, zigzag random walk linkers (e.g. four linker segments of 300-nm length) were used. The territory compaction was modelled with the assumption of an elastic "tensegrity force" $\mathbf{F}_a$ acting on the 1-Mbp domains and directed towards the bary centre ("centre of gravity") of the entire territory. Between spatially interacting domains, a hard-sphere potential was assumed, allowing a minimal distance of a half domain radius (i.e. 125 nm). To achieve a higher compaction level for the inactive X chromosome (Fig. 12.1) in comparison with that of the active X chromosome, $\mathbf{F}_a$ was adjusted in such a way that after the virtual imaging and the segmentation process (Kreth et al. 2002) the experimentally obtained volume and enveloping surface ratios were obtained (see also Sect. 12.3).

Under the assumption of a random distribution of double-strand breaks (DSBs) within the DNA for low linear energy transfer (LET) radiation, the

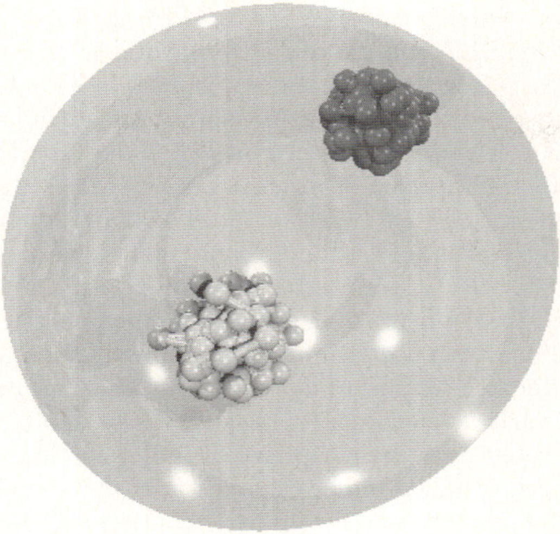

Fig. 12.1 Surface rendering of a simulated nucleus according to the earlier version of the spherical 1-Mbp chromatin domain model (Kreth et al. 2002). For the visualization, only the active (the *light grey* chromosome) and the inactive (at the top of the nucleus, the *dark grey* chromosome) were chosen. The compaction level of the inactive X-chromosome territory is increased in comparison with that of the active X-chromosome territory to correspond to the experimental volume and enveloping surface ratios

probability of a break occurring within a certain 1-Mbp domain was modelled using Poisson distribution mathematics. The yield of DSBs was adapted to ensure a mean number of 50 DSBs per gray per nucleus according to experimental observations (referred to in Cornforth et al. 2002). The DSBs within the 1-Mbp domains were placed into randomly chosen locations. To determine an exchange (interchange/intrachange) between two DSBs in two 1-Mbp domains, only those 1-Mbp domains containing DSBs were considered which revealed a distance (the distance between the midpoints of both 1-Mbp domains) smaller than or equal to 500 nm. An exchange event depending on the distance between the two DSBs in the two domains was counted according to a normalized probability function (Kreth et al. 2004, 2007).

12.3 Modelling Results

The influence of the geometry of the spatial organization of interphase chromosomes on the yield of aberration frequencies can be estimated in a first step assuming a proportional relationship yield (contact surface) between two adjacent CTs. This implies also that the yield is proportional to the cross-section, which is proportional to r^2, of a single CT. With the assumption that the DNA content of a chromosome is direct proportional to its volume, and therefore the DNA content is proportional to r^3, we obtained the previously reported relation that the yield is proportional the two-thirds power of the DNA content (Kreth et al. 1998, 2004; Cornforth et al. 2002). When a so-called statistical arrangement of the CTs is assumed to exist, this relation holds true for all CTs. (Kreth et al. 1998, 2004, 2007). In mammalian cells, however, preferential radial arrangements were reported which suggest a gene density correlated arrangement of CTs in lymphocyte nuclei (Boyle et al. 2001) and a size-dependent arrangement in fibroblast nuclei (Bolzer et al. 2005). Such a nonrandom positioning has the consequence that CTs which are located preferentially near the nuclear periphery have a reduced contact surface to other CTs, because the part of the surface of such a CT which is located near the periphery cannot participate in a misrejoining event. The second important contribution is the so-called compaction level (ratio of DNA content to CT volume). CTs revealed obviously different compactions depending on their gene density or transcriptional activity. In such cases the linear relationship between DNA content and volume no longer holds true (Branco and Pombo 2006). One important example where different volumes were observed for the same DNA content is the experimentally observed difference between the active and inactive X-chromosome territories in female fibroblast and amniotic fluid cells (Fig. 12.2). Eils et al. (1996) determined the mean ratios between active and inactive X-chromosome territories in these cell nuclei and found for the enveloping surface (A_{env})

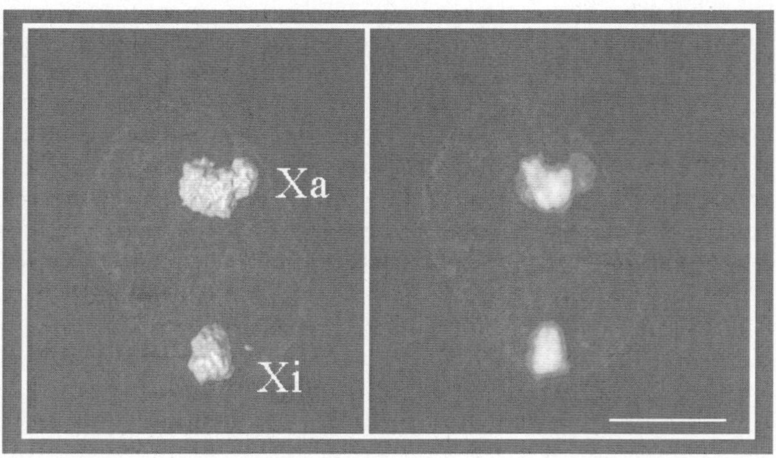

Fig. 12.2 Reconstruction of the active and inactive X-chromosome territories of an amniotic fluid cell of a confocal data stack. *Left*: Volume rendering. *Right*: Optical section. Cells were kindly provided by I. Solovei, University of Munich, Germany. *Xa* active X-chromosome territory, *Xi* inactive X-chromosome territory. Bar denotes 10μm

the ratio $A_{env}(Xa)/A_{env}(Xi) \approx 1.4$ and for the volume enclosed by the enveloping surface the ratio $V(Xa)/V(Xi) \approx 1.2$, where Xa and Xi refer to the active and inactive CTs, respectively. For the segmentation process, however, they chose the same threshold for both X-chromosome territories; this resulted in a slight underestimation of the surface of the active X-chromosome territory in contrast to that of the inactive X-chromosome territory. The reason for this underestimation of the surface of the active X-chromosome territory and the better conservation of the surface of the inactive X-chromosome territory surface is the sharper increase of the intensity profile of the inactive X-chromosome territory. In this way, lower intensities of the brighter-intensity profile of the active X-chromosome territory are lost, while the higher intensities of the inactive X-chromosome territory are conserved, when the same threshold is applied for both X-chromosome territories. This underestimation was quantified by model calculations. In this model, each CT was described by a linear chain of spherical 500-nm domains with a DNA content of 1 Mbp. The number of domains corresponds directly to the DNA content of each chromosome. The morphological difference of both X-chromosome territories was adjusted in this model to correspond to the experimental volume and enveloping surface ratios when the same threshold and segmentation process for the virtually convolved simulated data stacks were used (Kreth et al. 2002). Because of the underestimation, the real volume and the enveloping surfaces revealed slightly higher differences: $A_{env\ true}(Xa)/A_{env\ true}(Xi)=1.77$ and $V_{true}(Xa)/V_{true}(Xi)=1.67$. With use of these "true" values for the creation of virtual confocal microscopy images evaluated by image-analysis algorithms used for the experimental data, good agreement with the latter was obtained.

Taking into account the different morphologies for the active and the inactive X-chromosome in the computer model (Kreth et al. 2002), we created a random DSB pattern in a simulated nucleus according to low-LET γ-radiation at 3 Gy. In all, 400 statistically independent cell configurations were calculated and 32 different DSB distributions per nucleus were determined to end up with a sample of 12,800 configurations. For this study, the autosomes and the active X-chromosome were arranged in a statistical manner in the simulated nucleus. No preferential radial arrangements like in lymphocyte or fibroblast nuclei (Kreth et al. 2004) were assumed. For the inactive X-chromosome, however, different assumptions were tested: (1) a localization in the centre of the nuclear volume, (2) a statistical arrangement and (3) a localization at the nuclear envelope. The one-chromosome yields (an aberration event was assigned to each chromosome involved in an interchange aberration) were normalized in such a way that the sum of the one-chromosome yields for all chromosomes together amounted to 1. The resulting one-chromosome yields were plotted against the DNA content of each chromosome (Fig. 12.3). As mentioned before, the aberration yields for the statistically distributed autosomes and the active X-chromosome are close to the exponential fitting curve with exponent 2/3. The reduced enveloping surface and the different assumed

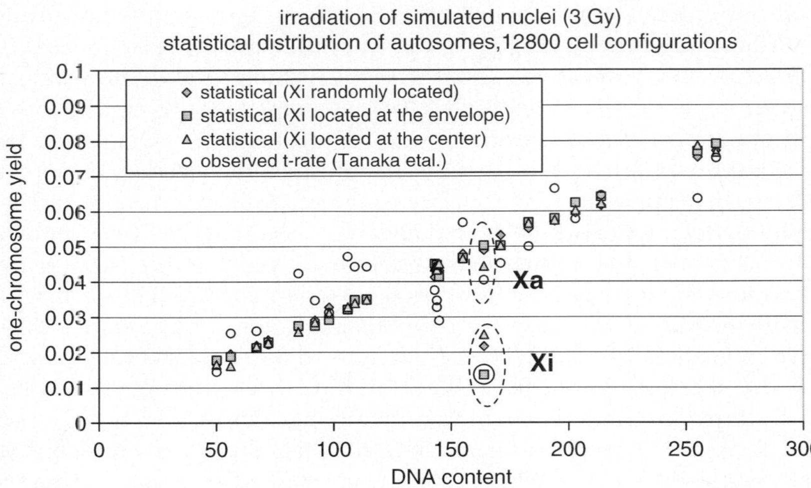

Fig. 12.3 Comparison between the calculated one-chromosome yields and the observed translocation rate (*t-rate*) of Hiroshima A-bomb survivors (Tanaka et al. 1983) of female cells. For the simulated data, the experimentally observed difference in the morphology was explicitly considered. Additionally different positions in the nuclear volume were tested for the inactive X-chromosome. All other simulated chromosome territories were placed randomly. For the comparison, the calculated one-chromosome yields and the observed translocation rate were normalized to 1

positionings for the inactive X-chromosome territory, however, show a clear effect on the aberration yields.

For the three different positionings we observed the following:

1. For the central positioning of the inactive X-chromosome territory (the active X-chromosome territory was located statistically), the aberration yield of the inactive X-chromosome territory shows the highest value because of the permanent accessibility of the complete surface of the chromosome; the aberration frequency ratio between the active and the inactive X-chromosome territories is $f(\mathrm{Xa})/f(\mathrm{Xi}) = 1.78$.
2. For the statistical arrangement (active X-chromosome territory and inactive X-chromosome territory located statistically) the yield of the inactive X-chromosome territory was reduced because of the probability of it being localized also at the nuclear periphery. Consequently the aberration frequency ratio was increased to $f(\mathrm{Xa})/f(\mathrm{Xi}) = 2.26$.
3. For the localization of the inactive X-chromosome territory at the nuclear envelope, however, the yield was drastically reduced; the reduced contact surface had a significant effect on the aberration yields. Here, the aberration frequency ratio between the statistically distributed active X-chromosome territory and the inactive X-chromosome territory was increased and $f(\mathrm{Xa})/f(\mathrm{Xi}) = 3.71$ was obtained.

The comparison with an observed translocation rate (again the sum of the one-chromosome yields for all chromosomes together was normalized to 1) from Hiroshima A-bomb survivors (Tanaka et al. 1983) revealed complete agreement with case 3 for the inactive X-chromosome territory. The scattering of the data points of the Hiroshima A-bomb survivors around the exponential fitting curve might mirror insufficient statistics but could also be the result of a nonrandom positioning of the other CTs. This possibility is suggested by simulations where a nonstatistical arrangement was assumed, e.g. for lymphocyte cells (Kreth et al. 2004, 2007).

In another approach, also for other CTs a close relationship between the contact surfaces, or an equivalent measure the intermingling volume, was reported. Branco and Pombo (2006) developed a novel fluorescence in situ hybridization (FISH) procedure for ultrathin cryosections which allows the use of wide-field microscopy with no reduction in the axial resolution. This allows a more precise evaluation of the volume of each CT and also of the intermingling volumes between CTs. Remarkably, they found that the volumes of most CTs were not only correlated with the DNA content but also with its gene density/transcriptional activity, which obviously corresponds to the example of the active and inactive X-chromosome territories. One prominent example is the observed volumes for CTs 21 and 22. Despite their similar DNA contents of 49 Mbp for CT 22 and 46 Mbp for CT 21, CT 22, which is a very gene rich chromosome, showed nearly twice of the volume of CT 21. They compared these results with aberration frequencies obtained by multicolour FISH (Arsuaga et al. 2004) and found a close relationship between the intermingling volume of two CTs and the evaluated aberration frequencies of Arsuaga et al. This means, for example, that the aberration yield for (1;22) is 3 times higher than that for (1;21).

12.4 Conclusions

To understand the influence of geometrical constraints of the nuclear architecture on aberration frequencies, we used a computer model which explicitly describes the polymer structure of the diploid human female genome, taking into account additionally the different morphologies of the active and inactive X-chromosome territories. As an additional constraint, we considered also different positionings in the nuclear volume for the inactive X-chromosome. Based on calculations of low-LET γ-radiation at 3 Gy, one chromosome yields were calculated for the different positionings. The expected results were confirmed: a central positioning of the inactive X-chromosome territory revealed the highest value for the aberration frequency because of the permanent accessibility of the complete contact surface. In the case of a random localization of the inactive X-chromosome territory, the yield was reduced slightly because of the possibility of it being located also at the periphery with a certain probability. The positioning of the inactive X-chromosome territory at the periphery revealed a drastic reduction of the inactive X-chromosome aberration frequency, which reflects the presence of only a part of the permanent reduction of a part of the surface of the territory. Remarkably this yield is in complete agreement with the observed yield for the inactive X-chromosome from survivors of the Hiroshima A-bomb. The experimentally observed peripheral positioning of the inactive chromosome in fibroblast and amniotic fluid cells would confirm this result. In addition to our previous studies where the effect of the non-random positioning of all CTs on aberration frequencies was investigated (Kreth et al. 2004), in the present study we tested for a single CT, the inactive X-chromosome territory, and the resulting aberration yields, when different locations for this chromosome are assumed, and additionally the different compaction level of the inactive X-chromosome territory in comparison with that of the active X-chromosome territory was considered.

Acknowledgements: This work was supported by the EC (contract no. F16R-CT-2003-508842, RISC-RAD). We also acknowledge Jan Boei and Thomas Cremer for stimulating discussions.

References

Arsuaga J, Greulich-Bode KM, Vazquez M, Bruckner M, Hahnfeldt P, Brenner DJ, Sachs R, Hlatky L (2004) Chromosome spatial clustering inferred from radiogenic aberrations. Int J Radiat Biol 80:507–515

Bolzer A, Kreth G, Solovei I, Koehler D, Saracoglu K, Fauth C, Müller S, Eils R, Cremer C, Speicher M, Cremer T (2005) Three-dimensional maps of all chromosomes in human male fibroblast nuclei and prometaphase rosettes. PLoS Biol 3:e157

Boyle S, Gilchrist S, Bridger JM, Mahy NL, Ellis JA, Bickmore WA (2001) The spatial organization of human chromosomes within the nuclei of normal and emerin-mutant cells. Hum Mol Genet 10:211–219

Branco MR, Pombo A (2006) Intermingling of chromosome territories in interphase suggests role in translocations and transcription-dependent associations. PLOS Biol 4:e138

Cornforth MN, Greulich-Bode KM, Loucas BD, Arsuaga J, Vazquez M, Sachs RK, Bruckner M, Molls M, Hahnfeldt P, Hlatky L, Brenner DJ (2002) Chromosomes are predominantly located randomly with respect to each other in interphase human cells. J Cell Biol 159: 237–244

Cremer M, von Hase J, Volm T, Brero A, Kreth G, Walter J, Fischer C, Solovei I, Cremer C, Cremer T (2001) Non-random radial higher-order chromatin arrangements in nuclei of diploid human cells. Chromosome Res 9:541–567

Eils R, Dietzel S, Bertin E, Granzow M, Schröck E, Speicher MR, Volm T, Ried T, Robert-Nicoud M, Cremer C, Cremer T (1996) Three-dimensional reconstruction of painted human interphase chromosomes: active and inactive X-chromosome territories have similar volumes but differ in shape and surface structure. J Cell Biol 135:1427–1440

Kreth G, Münkel C, Langowski J, Cremer T, Cremer C (1998) Chromatin structure and chromosome aberrations: modeling of damage induced by isotropic and localized irradiation. Mutat Res 404:77–88

Kreth G, Edelmann P, Münkel C, Langowski J, Cremer C (2002) Translocation frequencies for X and Y chromosomes predicted by computer simulations of nuclear structure. In: Sobti RC,

Obe G, Athwal RS (eds) Some aspects of chromosome structure and functions. Narosa, New Delhi, pp 57–71

Kreth G, Finsterle J, Cremer C (2004) Virtual radiation biophysics: implications of nuclear structure. Cytogenet Genome Res 104:157–161

Kreth G, Pazhanisamy SK, Hausmann M, Cremer C (2007) Cell type specific quantitative predictions of radiation-induced chromosome aberrations: a computer mode approach. Radiat Res (in press)

Tanabe H, Müller S, Neusser M, von Hase J, Calcagno E, Cremer M, Solovei I, Cremer C, Cremer T (2002) Evolutionary conservation of chromosome territory arrangements in cell nuclei from higher primates. Proc Natl Acad Sci USA 99:4424–4429

Tanaka K, Kamada N, Ohkita T, Kuramoto A (1983) Nonrandom distribution of chromosome breaks in lymphocytes of atomic bomb survivors. J Radiat Res 24:291–304

13 Role of Chromatin Structure and Activity in the Induction of Chromosomal Aberrations

Wilner Martínez-López, María Vittoria Di Tomaso, Leticia Méndez-Acuña, and María Mühlmann

Abstract

It is assumed that unrepaired or misrepaired DNA double-strand breaks lead to chromosomal aberrations. However, little information is available about the role of chromatin structure and remodelling in the formation of double-strand breaks and their processing into chromosome damage. Open chromatin conformation, recognized as euchromatin, possesses transcriptional activity, while tighter chromatin conformation, named heterochromatin, remains transcriptionally inactive. A higher order of differential chromatin structure can be observed at the chromosome level as banding patterns. In order to study the role of chromatin organization and activity in the induction of chromosomal aberrations, different approaches were employed: (1) analysis of the distribution of chromosome breakpoints induced by ionizing radiation (neutrons or γ-rays) in relation to the CHO G-banding pattern; (2) correlation of the CHO chromosome H4 acetylation pattern with radiation-induced breakpoint localization; (3) study of the influence of chromatin remodelling processes related to DNA replication and cell cycle progression analysing etoposide-induced and methyl methanesulfonate induced breakpoints distributed along Xp (euchromatic) and Xq (heterochromatic) CHO X chromosome arms; (4) localization of γ-ray-induced breakpoints to euchromatin and heterochromatin by exposing human fibroblast lines with different numbers of X chromosomes. Several factors, such as chromatin activity, chromatin remodelling processes associated with DNA synthesis or cell cycle progression as well as chromatin compartmentalization in euchromatin, heterochromatin and G-bands, have been found associated with the non-random distribution of chromosome damage.

13.1 Introduction

Almost every process occurring within the nuclei of eukaryotic cells is affected by chromatin structure (Jenuwein and Allis 2001). During the last few years, several studies have been carried out to understand how DNA replication, DNA repair and gene transcription can be regulated by chromatin

Günter Obe and Vijayalaxmi (Eds.)
Chromosomal Alterations: Methods, Results and Importance in Human Health

structure (Felsenfeld and Groudine 2003; Morrison et al. 2004; Polo and Almouzni 2005). Histone modifications (acetylation, phosphorylation and methylation) establish the so-called epigenetic histone code (Strahl and Allis 2000). Histone acetylation/deacetylation (controlled by acetyltransferases and deacetylases) is the most frequent epigenetic process, determining either gene activation or inactivation, respectively. Histone acetylation allows chromatin accessibility by reducing positive charges of lysines located in histone amino tails, diminishing the attractive force between the nucleosome core and negatively charged DNA (Jeppesen 1997; Turner 2000). The dynamic chromatin organization controls the access to DNA of many proteins, such as transcription factors, and influences DNA damage induction and processing (Green and Almouzni 2002; Rubbi and Milner 2003; Akhtar and Cavalli 2005; Chai et al. 2005).

DNA double-strand breaks (DSB) are the ultimate lesions for the formation of chromosomal aberrations (Obe et al. 2002), but the role of chromatin structure in the formation of DSB and their processing into chromosome damage is not very well defined.

Two major types of chromatin occur in all eukaryotic cells, namely, euchromatin and heterochromatin. Euchromatin is characterized by an open chromatin conformation, transcriptional activity and sensitivity to DNase I. Heterochromatin is held in a tighter chromatin conformation and is transcriptionally inactive as well as insensitive to DNase I (Mühlmann-Díaz and Bedford 1994). A further order of chromatin organization can be observed at microscope level as different banding patterns. A Giemsa banding pattern can be obtained, for example, after exposing fixed metaphase chromosomes to trypsin followed by Giemsa staining (G-banding, G-bands). G-bands are darkly stained bands (G-dark bands) at fixed position in the chromosomes with unstained areas between them (G-light bands). On the other hand, treatments of chromosome spreads with hot salt reveal the presence of a reverse staining pattern (R-banding, R-bands). G-light bands or R-bands correspond to early replicating GC-rich regions that contain housekeeping genes and short interspersed repeated elements. G-dark bands contain tissue-specific genes, late replicating AT-rich DNA and long interspersed repeated elements (Korenberg and Rykowsky 1988; Goldman et al. 1984). Constitutively heterochromatic C-bands, which mainly occur in the centromere regions, contain tandem non-coding highly repeated satellite DNA sequences. AT-rich DNA stretches of scaffold-associated regions mainly bind to the chromosome scaffold in G-dark band regions (Saitoh and Laemmli 1994). During S phase, G- and C-bands transiently decondense and recondense after DNA synthesis (Manuelidis and Chen 1990). Eukaryotic type II DNA topoisomerase is very active during replication and cell cycle related chromatin remodelling by transitorily cutting DNA strands to alleviate torsional stress (Champoux 2001). The clastogenic anti-topoisomerase II agents, such as etoposide, trap the cleavable complex of topoisomerase II and hinder the resealing of DNA strand breaks introduced by the enzyme (Palitti 1993; Palitti et al. 1994; Degrassi et al. 2004).

We analysed the role of chromatin structure and activity in the induction of chromosomal aberrations using different approaches:

1. Treating Chinese hamster ovary (CHO) cells with ionizing radiation (neutrons or γ-rays) to assign induced chromosome breakpoints to G-light and G-dark bands.
2. Immunollabelling of CHO chromosomes with antibodies against acetylated H4 histones to correlate radiation-induced breakpoints with the histone H4 acetylation pattern.
3. Studying the influence of chromatin remodelling processes related to DNA replication and cell cycle progression on chromosome damage by the analysis of the distribution of breakpoints induced by etoposide and the monofunctional alkylating agent methyl methanesulfonate (MMS) on the euchromatic Xp and heterochromatic Xq arms of the X chromosome of CHO cells during early and late S or G2 phases of the cell cycle.
4. Exposing lines of human fibroblasts with different numbers of X chromosomes to γ-rays and localizing induced breakpoints to euchromatin and heterochromatin.

13.2 Methodological Aspects

13.2.1 Localization of Radiation-Induced Breakpoints on CHO G-banded Chromosomes

CHO cells were irradiated either with 2 Gy γ-rays or 2 Gy fission neutrons. Chromosome slides were exposed to a 0.025% trypsin solution to obtain G-bands. Three hundred breakpoints for each type of irradiation were scored. Breakpoints sites were assigned to G-light or G-dark bands on 262-band resolution G-band idiograms drawn according to relative chromosome and band length.

13.2.2 Immunolabelling of CHO Acetylated H4 Histones and Densitometric Analysis

CHO metaphases were prepared for immunolabelling by cytocentrifugation and exposed to rabbit antiserum R5/12, which recognizes acetylated H4 histone ($H4^{+a}$) at Lys12 and binds predominantly to diacetylated, triacetylated and tetraacetylated H4 (Turner and Fellows 1989). Indirect immunofluorescent labelling with rabbit antisera was carried out as described by Jeppesen and Turner (1993). Slides were counterstained with Hoechst 33258.

Images of chromosomes were captured with a Zeiss Axioskop fluorescence microscope fitted with a Chroma 83000 triple band pass filter set and a Photometrics CH250 CCD camera, using Digital Scientific capture and analysis extensions to IPLab Spectrum v3.1. The chromosome images were used to construct densitometric curves.

13.2.3 Etoposide and MMS Treatments on CHO Cells: Breakpoint Mapping on the CHO X Chromosome

Exponentially growing cells were exposed for 40 min with either etoposide (10 or 20 μM) or MMS (10 or 20 mM) (Sigma) and were simultaneously pulse-labelled with 30 μM 5-bromo-2′-deoxyuridine (BrdUrd) (Sigma). BrdUrd incorporation was detected by immunolabelling.

The distribution of 1,200 breakpoints in Xp and Xq was analysed in damaged BrdUrd immunostained CHO metaphases, treated with etoposide in early and late S and G2 phases or with MMS in early and late S phases.

The numbers of breakpoints per aberration were determined by Obe and Winkel (1985) as follows: chromatid breaks 1; isochromatid breaks 2; quadriradials and triradials (chromatid interchanges) 2; interstitial deletions 2; duplication deletions 3; intrachanges 2; dicentrics 2; translocations 2; and rings 2. Xp and Xq expected breakpoint frequencies were estimated according to arm length.

13.2.4 Radiation Exposure of Human Cell Lines with Different Numbers of X Chromosomes

The following human fibroblast cell lines were used: AG1521 (46,XY) and AG1522 (46,XY); GM2149 (46,XX); GM4626 (47,XXX); GM00157 (49, XXXXY). Confluent fibroblasts were exposed to γ-rays (^{137}Cs) and first division cells were analysed.

13.2.4.1 Fluorescence in Situ Hybridization

Metaphase spreads were "painted" by the method of fluorescence in situ hybridization (FISH) using chromosome-specific DNA probe libraries for X chromosomes following the method of Pinkel et al. (1988) and Lucas et al. (1992).

Aberrations in the X chromosomes were scored in bicoloured chromosomes with one (translocations) or two (dicentric chromosomes) centromeres. In addition, acentric fragments were scored.

13.2.4.2 BrdUrd Incorporation and Immunolabelling

To ascertain late replicating patterns of X chromosomes, BrdUrd was added to cultures 3 h before addition of colcemid, and mitotic cells were collected and fixed 2 h later. The X chromosomes containing BrdUrd were detected by immunolabelling.

FISH was used to recognize X chromosomes with a biotinylated X-specific centromere probe which was postlabelled with fluorescein isothiocyanate (FITC)–avidin.

13.2.4.3 In Situ Nick Translation

DNase I sensitivity of fixed metaphase chromosomes from cell lines with multiple X chromosomes was determined by in situ nick translation according to the method of Kerem et al. (1983). Chromosome spreads were incubated in a nick translation buffer containing a mixture of DNase I/polymerase I, deoxyribonucleoside triphosphates and biotin–2′-deoxyuridine-5′-triphosphates. Slides were treated with a biotinylated X-specific centromere probe and detected by avidin conjugated with FITC. Propidium iodide was used for counterstaining.

13.3 Distribution of Chromosome Breakpoints in CHO Cells

13.3.1 Mapping of Radiation-Induced Breakpoints

Considerable differences have been reported with respect to the pattern of non-random chromosomal distribution of radiation-induced breakpoints in human cells analysed with G-banding techniques (Cooke et al. 1975; Buckton 1976; Dubos et al. 1978; Bauchinger and Götz 1979; Lee and Kamra 1981; Tanaka et al. 1983; Barrios et al. 1989; Johnson et al. 1999). Although non-random localizations of radiation-induced breakpoints were confirmed with chromosome painting techniques, contradictory results were reported with respect to the involvement of specific chromosomes and chromosome regions in aberrations (Tucker and Senft 1994; Dominguez et al. 1996; Grigorova et al. 1998; Griffin et al. 1995; Luomahaara et al. 1999; Cigarran et al. 1998; Kiuru et al. 2000; Knehr et al. 1994, 1996; Xiao and Natarajan 1999; Schleiermacher et al. 2003).

In CHO cells we found that breakpoints induced by γ-rays or neutrons are non-randomly distributed with respect to chromosome length and G-bands in a similar manner (Martínez-López et al. 1998). Breakpoints occurred preferentially in G-light bands and 22% clustered in 7% of the total chromosome length (Folle et al. 1998).

Similar results were found in hamster cells following exposure to X-rays (Slijepcevic and Natarajan 1994a, b) and to endonucleases (Folle et al. 1997, 1998), suggesting that chromatin conformation and nuclear structure play a role in the distribution of breakpoints.

G-light band regions containing active genes can be expected to have a more open conformation, making DNA more accessible to endonucleases and oxygen radicals induced by radiation (Löbrich et al. 1996; Forrester and Radford 2004). The width of the minor groove of DNA could also have an influence on the induction of lesions by influencing the accessibility of hydrogen atoms in sugar moieties to OH radicals (Sy et al. 1997).

The open conformation of transcribed chromatin might allow greater mobility of broken DNA ends, which may increase the chance of rejoining wrong partners, leading to chromosomal aberrations. In contrast, inactive chromatin is so tightly packaged that the movement of broken DNA ends is very constrained, so exchanges are unlikely to occur and restitution is the rule (Savage 1998).

Differential repair of randomly induced primary lesions could also explain the non-random distribution of induced breakpoints (Pellicia et al. 1985; Pantelias 1994; Sanders et al. 2004). More experimental evidence is needed to understand the possible role of repair mechanisms in chromosome breakpoint distribution.

13.3.2 Histone Acetylation Pattern and Clustering of Breakpoints

Transcriptionally active chromatin has a more accessible structure than transcriptionally inactive chromatin, as indicated by its increased sensitivity to nuclease digestion (Hebbes et al. 1994). The loosening of nucleosome structure caused by histone acetylation may facilitate the binding of initiating factors and RNA polymerase transcription complex to DNA. Immunolabelling of human metaphase chromosomes with fluorescent antibodies against acetylated histone H4 results in hyperacetylated (R-bands), relatively underacetylated (G-bands) and hypoacetylated (heterochromatin) regions (Jeppesen 1997; Brinkman et al. 2006). Underacetylation of inactive X chromosomes and constitutive heterochromatic chromosome regions was shown by Jeppesen and Turner (1993). Thus, the pattern of H4 hyperacetylation ($H4^{+a}$) can be considered a chromosomal marker for gene expression.

In CHO chromosomes we confirmed that the distribution of $H4^{+a}$ corresponds to an R-band pattern. The distribution of $H4^{+a}$ in CHO chromosomes and the underacetylated heterochromatic long arm of the X chromosome are shown in Fig. 13.1 (Martínez-López et al. 2001).

Radiation-induced breakpoints are clustered in highly acetylated chromosome regions (Fig. 13.1), confirming that chromosomal aberrations occur more frequently in active than in inactive chromatin (Martínez-López et al. 2001). Similar to human chromosomes (Jeppesen and Turner 1993), active genes and induced breakpoints seem to concentrate in the subtelomeric and telomeric regions in most CHO chromosomes (Folle et al. 1997, 1998; Martínez-López et al. 1998).

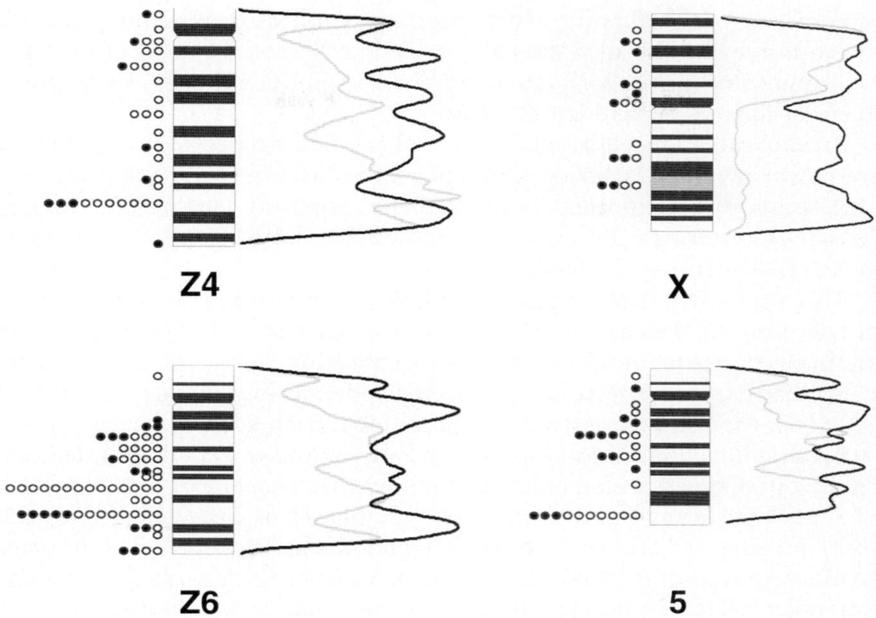

Fig. 13.1 Ideograms of Chinese hamster ovary chromosomes 5, Z4, Z6 and X employed in the analysis of the intrachromosomal localization of ionizing radiation induced breakpoints. On the *right side of the ideograms* representative densitometric tracings of histone H4 acetylation pattern (*grey*) and Hoechst 33258 (*black*) fluorescence of these chromosomes are presented. *Empty circles* correspond to breakpoints induced by neutrons and *filled circles* correspond to γ-rays. In chromosomes 5 and Z6, both breakage-prone bands map to the same G-light segments since the two terminal G-dark bands from Z6 originate in a reciprocal translocation involving the distal segments of 1p and 5q. The secondary constriction of the X chromosome is indicated by a *shaded area*

13.3.3 Distribution of Breakpoints Induced by Etoposide and MMS in the X Chromosome of CHO Cells

The advantage of inducing DSB by trapping the cleavable complex formed by topoisomerase II, make the anti-topoisomerase II compound etoposide a useful tool to study the influence of chromatin remodelling processes on the distribution of chromosome damage (Martínez-López et al. 2004).

The CHO X chromosome has a short euchromatic arm (Xp) and a long heterochromatic arm (Xq), which are early- and late-replicating, respectively; as revealed by BrdUrd pulse-labelling. The differential euchromatin/heterochromatin replication timing of Xp and Xq cells provides a valuable model to analyse the influence of DNA synthesis and chromatin remodelling on the distribution of induced breakpoints.

Xq is more sensitive to etoposide during late S phase (Martínez-López et al. 2004). To assess the influence of DNA replication and chromatin remodelling on breakpoint distribution, we studied breakpoints produced by etoposide and the alkylating agent MMS in Xp and Xq arms of the CHO X chromosome during early and late S phase. Additionally cells were exposed to etoposide in early and late G2 phase.

Breakpoints induced by etoposide and MMS in Xp or Xq during S phase are shown in Fig. 13.2. During early S phase (replication of Xp) etoposide- and MMS-induced breakpoints clustered in Xp. Similarly, during late S phase (when Xq replicates) breakpoints produced by both agents concentrated in Xq (Di Tomaso et al. 2006).

This shows that DNA replication plays an important role in breakpoint distribution. In the case of MMS, increased DNA accessibility due to chromatin decondensation associated with replication may explain the breakpoint distribution. The results with etoposide are explainable by the high activity of topoisomerase II and its inhibition during DNA synthesis.

Breakpoints induced by etoposide in early G2 phase were distributed randomly but mapped preferentially to Xp in late G2 phase (Fig. 13.2).

Studies performed in different cell cycle phases of CHO cells (G1 or G2) with ionizing radiation or restriction endonucleases revealed conflicting results with respect to localization of breakpoints in Xp or Xq (Slijepcevic and Natarajan 1994a, b; Folle et al. 1997; Grigorova et al. 1998; Xiao and Natarajan 1999). Contradictory findings could be due to subtle variations in cell cycle stages at the time of treatment.

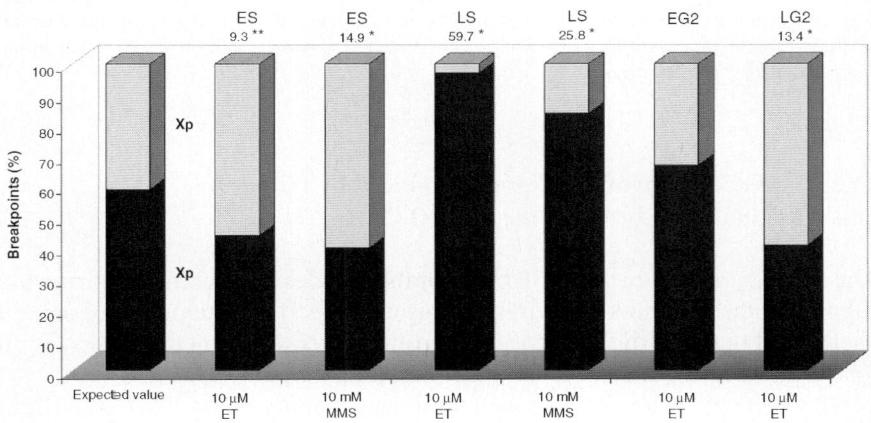

Fig. 13.2 The distribution (as percentages) of chromosome breakpoints induced by etoposide (*ET*; 10 μM) and methyl methanesulfonate (*MMS*; 10 mM) along Xp (*grey*) and Xq (*black*) in early S phase (*ES*) and late S phase (*LS*) as well as induced only by etoposide (10 μM) during early G2 phase (*EG2*) or late G2 phase (*LG2*). The *bar on the left* indicates expected frequencies of breakpoints according to Xp and Xq relative length. χ^2 values are indicated on the *top of the bars*. *p0.0005; **p<0.005

13.4 Chromosome Damage in Human Cell Lines with Multiple X Chromosomes

13.4.1 Replication Timing of Active and Inactive Human X Chromosomes

Studies on replication timing during S phase showed that each pair of human chromosomes has the same replication pattern, except for the X chromosomes in female cells. The inactive X chromosome is late-replicating as are extra X chromosomes in cells with multiple inactive X chromosomes; therefore, replication patterns seen after BrdUrd incorporation (replication bands) allow us to decide in which stage of the S phase the cell was when BrdUrd was added (Drouin et al. 1990).

To determine late labelling patterns of X chromosomes in cells with multiple X chromosomes, an X-centromere specific probe in combination with BrdUrd immunolabelling was applied. Male cells had only a small late-replicating pericentromeric region in their X chromosome. Normal female cells have one entirely late replicating X chromosome. The 47,XXX and 49,XXXXY cells exhibited two and three late-replicating X chromosomes, respectively.

13.4.2 Nuclease Sensitivity of Human Metaphase Chromosomes

It is well known that active or potentially active genes have a preferential DNase I sensitivity probably due to a less condensed chromatin conformation at the transcribed site (Weintraub and Groudine 1976). By in situ nick translation it is possible to distinguish between active and inactive X chromosomes on fixed metaphases (Kerem et al. 1983).

In order to determine whether the chromatin structure on both active and inactive X chromosomes presents uneven sensitivities to DNase I, in situ nick translation was performed on metaphase spreads from human cell lines containing different numbers of X chromosomes. The late-replicating supernumerary X chromosomes in 47,XXX and 49,XXXXY cells appeared to be very resistant to DNase I digestion.

It is worth mentioning that digestion of metaphase chromosomes with *Alu*I, *Bam*HI or DNase I followed by Giemsa staining produced G-bands in CHO cells (Folle and Obe 1995). A differential sensitivity of heterochromatin and euchromatin to endonuclease digestion was also observed by Chan et al. (2000). In CHO chromosomes the heterochromatic long arm of the X chromosome remained almost unaffected by enzyme treatment, showing that heterochromatin is less accessible to nucleases (Folle et al. 1997; Folle and Obe 1995).

Table 13.1 Total frequency of aberrations induced by γ-rays in cell lines containing different numbers of X chromosomes as well as the frequency of aberrations involving only X chromosomes

Cell line	Karyotype	X status	Total aberrations scored per cell (average 1.754)	Observed X aberrations per cell	Normalized X aberrations per cell[a]	Expected X aberrations per cell[b]
AG 1521	46,XY	1 (Xa)	1.89	0.18	0.17	0.17
AG 1522	46,XY	1 (Xa)	1.77	0.16	0.16	0.17
GM2149	46,XX	2 (1 Xa+1 Xi)	1.86	0.28	0.26	0.34
GM04626	47,XXX	3 (1 Xa+2 Xi)	1.66	0.34	0.36	0.51
GM00157	49,XXXXY	4 (1 Xa+3 Xi)	1.59	0.22	0.24	0.68

Xa active X chromosome, *Xi* inactive X chromosome

[a]Normalization was done to correct for small differences in cellular radiosensitivity among cell lines. The corrected or normalized values were estimated as follows:

$$\frac{\text{average of the total aberrations}}{\text{total aberrations per cell}} \times \text{aberrations in X chromosomes per cell.}$$

[b]Expected X aberrations per cell were estimated based on average aberrations involving active X chromosomes.

13.4.3 Chromosomal Aberrations Induced by γ-rays in Human X Chromosomes

Seabright (1973), employing G-banding, reported fewer X-ray induced aberrations in female human X chromosomes than expected from the length.

The frequencies of γ-ray induced aberrations involving the X chromosome in various cell lines with an active X chromosome and different numbers of inactive X chromosomes were measured by employing a specific chromosome painting probe. No increase in the yield of aberrations in the X chromosomes was found in 47,XXX or 49,XXXXY cells compared with 46,XX cells (Table 13.1), indicating that the chromatin structure of extra X chromosomes or their activity status affects the induction of chromosomal aberrations (Mühlmann-Díaz and Bedford 1994).

13.5 Conclusion

Chromosomal aberrations are the result of different cellular mechanisms interacting with DNA lesions. Most of these lesions are repaired in such a way that they do not lead to chromosome damage. Only a few DNA lesions (unrepaired or misrepaired) are transformed into chromosomal aberrations as observed by light microscopy.

Complex interplay of factors such as DNA conformation, transcription and replication can lead to a non-random interchromosomal and intrachromosomal distribution of chromosome breakpoints. Chromatin activity and remodelling processes associated with DNA replication and cell cycle progression are also involved in the topology of breakpoints. Differential distribution of euchromatin and heterochromatin along chromosomes as well as the compartmentalization of chromatin structure in G-light and G-dark bands can also affect the localization of breakpoints.

Acknowledgements: This work was supported in part by PDT/BID-DINACYT (project 91/29) and PEDECIBA from Uruguay and by CONICET and SECYT from Argentina (project PICTR 2002-00121).

References

Akhtar A, Cavalli G (2005) The epigenome network of excellence. PLoS Biol 3:762–763

Barrios L, Miró R, Caballín MR, Fuster C, Guedea F, Subias A, Egozcue J (1989) Cytogenetic effects of radiotherapy breakpoint distribution in induced chromosome aberrations. Cancer Genet Cytogenet 41:61–70

Bauchinger M, Götz G (1979) Distribution of radiation induced lesions in human chromosomes and dose-effect relation analysed with G-banding. Radiat Environ Biophys 16:335–366

Brinkman AB, Roelofsen T, Pennings SW, Martens JH, Jenuwein T, Stunnenberg HG (2006) Histone modification patterns associated with the human X chromosome. EMBO Rep 7:628–634

Buckton KE (1976) Identification with G and R banding of the position of breakage points induced in human chromosomes by in vitro X-irradiation. Int J Radiat Biol 29:475–488

Chai B, Huang J, Cairns BR and Laurent BC (2005) Distinct roles for the RSC and Swi/Snf ATP-dependent chromatin remodellers in DNA double-strand break repair. Genes Dev 19: 1656–1661

Champoux JJ (2001) DNA topoisomerases: structure, function, and mechanism. Ann Rev Biochem 70:369–413

Chan JK, Park PC, De Boni U (2000). Association of DNAse sensitive chromatin domains with the nuclear periphery in 3T3 cells in vitro. Biochem Cell Biol 78:67–78

Cigarran S, Barrios L, Barquinero JF, Caballin MR, Ribas M, Egozcue J (1998) Relationship between the DNA content of human chromosomes and involvement in radiation-induced structural aberrations, analysed by painting. Int J Radiat Biol 74:449–455

Cooke P, Seabright M, Wheeler M (1975) The differential distribution of X-rays induced chromosome lesions in trypsin-banded preparations from human subjects. Humangenetik 28:221–231

Degrassi F, Fiore M, Palitti F (2004) Chromosomal aberrations and genomic instability induced by topoisomerase-targeted antitumor drugs. Curr Med Chem Anti-Cancer Agents 4:317–325

Di Tomaso MV, Martínez-López W, Folle GA, Palitti F (2006) Modulation of chromosome damage localisation by DNA replication timing. Int J Radiat Biol 82:877–886

Dominguez I, Boei JJWA, Balajee AS, Natarajan AT (1996) Analysis of radiation-induced chromosome aberrations in Chinese hamster cells by FISH using chromosome-specific DNA libraries. Int J Radiat Biol 70:199–208

Drouin R, Lemieux N, Richer CL (1990) Analysis of DNA replication during S-phase by means of dynamic chromosome banding at high resolution. Chromosoma 99:273–280

Dubos C, Pequignot EV, Dutrillaux B (1978) Localisation of gamma-rays induced chromatid breaks using a three consecutive staining techniques. Mutat Res 49:127–131

Felsenfeld G, Groudine M (2003) Controlling the double helix. Nature 421:448–453

Folle GA, Obe G (1995) Localization of chromosome breakpoints induced by AluI and BamHI in Chinese hamster ovary (CHO) cells treated in the G1 phase of the cell cycle. Int J Radiat Biol 68:437–445

Folle GA, Boccardo E, Obe G (1997) Localization of chromosome breakpoints induced by DNase I in Chinese hamster ovary (CHO) cells. Chromosoma 106:391–399

Folle GA, Martínez-López W, Boccardo E, Obe G (1998) Localization of chromosome breakpoints: implication of the chromatin structure and nuclear architecture. Mutat Rese 404:17–26

Forrester HB, Radford IR (2004) Ionizing radiation-induced chromosomal rearrangements occur in transcriptionally active regions of the genome. Int J Radiat Biol 80:757–767

Goldman MA, Holmquist GP, Gray MC, Caston LA, Nag A (1984) Replication timing of mammalian genes and middle repetitive sequences. Science 224:686–692

Green CM, Almouzni G (2002) When repair meets chromatin. EMBO Rep 31:28–33

Griffin CS, Marsden SJ, Stevens DL, Simpson P, Savage JR (1995) Frequencies of complex chromosome exchange aberrations induced by 238Pu alpha-particles and detected by fluorescence in situ hybridization using single chromosome-specific probes, Int J Radiat Biol 67:431–439

Grigorova M, Brand R, Xiao Y, Natarajan AT (1998) Frequencies and types of exchange aberrations induced by X-rays and neutrons in Chinese hamster splenocytes detected by FISH using chromosome-specific DNA libraries. Int J Radiat Biol 74:297–314

Hebbes TR, Clayton AL, Thorne AW, Crane-Robinson C (1994) Core histone hyperacetylation co-maps with generalized DNase I sensitivity in the chicken beta-globin chromosomal domain. EMBO J 13:1823–1830

Jenuwein T, Allis CD (2001) Translating the histone code. Science 293:1074–1080

Jeppesen P (1997) Histone acetylation: a possible mechanism for the inheritance of cell memory at mitosis. BioEssays 19:67–74

Jeppesen J, Turner BM (1993) The inactive X chromosome in female mammals is distinguished by a lack of histone H4 acetylation, a cytogenetic marker for gene expression. Cell 74:281–289

Johnson KL, Brenner DJ, Nath J, Tucker JD, Geard CR (1999) Radiation-induced breakpoint misrejoining in human chromosomes: random or non-random?. Int J Radiat Biol 75:131–141

Kerem BS, Goitein R, Richler C, Marcus M, Cedar H (1983) In situ nick translation distinguishes between active and inactive X chromosomes. Nature 304:88–90

Kiuru A, Lindholm C, Auvinen A, Salomaa S (2000) Localization of radiation-induced chromosomal breakpoints along human chromosome 1 using a combination of G-banding and FISH. Int J Radiat Biol 76:667–672

Knehr S, Zitzelsberger H, Braselmann H, Bauchinger M (1994) Analysis for DNA-proportional distribution of radiation-induced chromosome aberrations in various triple combinations of human chromosomes using fluorescence in situ hybridization. Int J Radiat Biol 65:683–690

Knehr S, Zitzelsberger H, Braselmann U, Nahrstedt U, Bauchinger M (1996) Chromosome analysis by fluorescence in situ hybridization: further indications for a non-DNA-proportional involvement of single chromosomes in radiation-induced structural aberrations. Int J Radiat Biol 70:385–392

Korenberg JR, Rykowski MC (1988) Human genome organization: Alu, Lines, and the molecular structure of metaphase chromosome bands. Cell 53:391–400

Lee C, Kamra O (1981) The pattern of radiation induced transmissible aberrations in a human cell culture. Hum Genet 57:380–384

Löbrich M, Cooper PD, Rydberg B (1996) Non-random distribution of DNA double-strand breaks induced by particle irradiation. Int J Radiat Biol 70:493–503

Lucas JN, Awa A, Straume T, Poggensee M, Kodama Y Nakano M, Ohtaki K, Weier HU Pinkel D, Gray J (1992) Rapid translocation frequency analysis in humans decades after exposure to ionizing radiation. Int J Radiat Biol 62:53–63

Luomahaara S, Lindholm C, Mustonen R, Salomaa S (1999) Distribution of radiation-induced exchange aberrations in human chromosomes 1, 2 and 4. Int J Radiat Biol 75:1551–1556

Manuelidis L, Chen TL (1990) A unified model of eukaryotic chromosome. Cytometry 11:8–25

Martínez-López W, Boccardo E, Folle GA, Porro V, Obe G (1998) Intrachromosomal localization of aberration breakpoints induced by neutrons and γ-rays in Chinese hamster ovary (CHO) cells. Radiat Res 150:585–592

Martínez-López W, Folle GA, Obe G, Jeppesen P (2001) Chromosome regions enriched in hyperacetylated histone H4 are preferred sites for endonuclease- and radiation-induced breakpoints. Chromosome Res 9:69–75

Martínez-López W, Folle GA, Cassina G, Méndez-Acuña L, Di Tomaso MV, Obe G, Palitti F (2004) Distribution of breakpoints induced by etoposide and X-rays along the CHO X chromosome. Cytogenet Genome Res 104:182–187

Morrison A, Highland J, Krogan NJ, Arbel-Eden A, Greenblatt JF, Haber JE, Shen X (2004) INO80 and gamma-H2AX interaction links ATP-dependent chromatin remodelling to DNA damage repair. Cell 119:767–775

Mühlmann-Díaz MC, Bedford JS (1994) A comparison of radiation-induced aberrations in human cells involving early and late replication X-chromosomes. In: Obe G, Natarajan AT (eds) Chromosomal alterations. Origin and significance. Springer, Berlin, pp 125–131

Obe G, Winkel EU (1985) The chromosome breaking activity of the restriction endonuclease AluI in CHO cells is independent of the S-phase of the cell cycle. Mutat Res 152:25–29

Obe G, Savage JRK, Johannes C, Goedecke W, Jeppesen P, Natarajan AT, Martínez-Lopez W, Folle GA, Drets ME (2002) Chromosomal aberrations: formation, identification and distribution. Mutat Res 504:17–36

Palitti F (1993) Mechanism of induction of chromosomal aberrations by inhibitors of DNA topoisomerases. Environ Mol Mutagen 22:275–277

Palitti F, Mosesso P, Di Chiara D, Schinoppi A, Fiore M, Bassi L (1994) Use of antitopoisomerase drugs to study the mechanisms of induction of chromosomal damage. In: Obe G, Natarajan AT (eds) Chromosomal alterations. Origin and significance. Springer, Berlin, pp 103–115

Pantelias GE (1994) Factors determining the yields of radiation-induced chromosomal aberrations as visualized by means of premature chromosome condensation in interphase cells. In: Obe G, Natarajan AT (eds) Chromosomal alterations. Origin and significance. Springer, Berlin, pp 140–149

Pellicia F, Micheli A, Olivieri G (1985) Inter- and intra-chromosomal distribution of chromatid breaks induced by X-rays during G2 in human lymphocytes. Mutat Res 150:293–298

Pinkel D, Landegent J, Collins C, Fuscoe J, Segraves R, Lucas J, Gray J (1988) Fluorescence in situ hybridization with human chromosome specific libraries: Detection of trisomy 21 and translocations of chromosome 4. Proc Natl Aacd Sci USA 85:9138–9142

Polo SE, Almouzni G (2005) Histone metabolic pathways and chromatin assembly factors as proliferation markers. Cancer Lett 220:1–9

Rubbi CP, Milner J (2003) p53 is a chromatin accessibility factor for nucleotide excision repair of DNA damage. EMBO J 22:975–986

Saitoh Y, Laemmli UK (1994) Metaphase chromosome structure: bands arise from a differential folding path of the highly AT-rich scaffold. Cell 76:609–622

Sanders MH, Bates SE, Wilbur BS, Holmquist GP (2004) Repair rates of R-band, G-band and C-band DNA in murine and human cultured cells. Cytogenet Genome Res 104:1–4

Savage JRK (1998) A brief survey of aberration origin theories. Mutat Rese 404:139–147

Schleiermacher G, Janoueix-Lerosey I, Combaret V, Derre J, Couturier J, Aurias A, Delattre O (2003) Combined 24-color karyotyping and comparative genomic hybridization analysis indicates predominant rearrangements of early replicating chromosome regions in neuroblastoma. Cancer Genet Cytogenet 141:32–42

Seabright M (1973) Noninvolvement of the human X chromosome in X-ray-induced exchanges. Cytogenet Cell Genet 12:342–356

Slijepcevic P, Natarajan AT (1994a) Distribution of X-ray-induced G2 chromatid damage among Chinese hamster chromosomes: influence of chromatin conformation. Mutat Res 323:113–119

Slijepcevic P, Natarajan AT (1994b) Distribution of radiation-induced G1 exchange and terminal deletion breakpoints in Chinese hamster chromosomes as detected by G banding. Int J Radiat Biol 66:747–755

Strahl BD, Allis CD (2000) The language of covalent histone modifications. Nature 403:41–45

Sy D, Savoye C, Begusova M, Michalik V, Charlier M, Spotheim-Maurizot M (1997) Sequence-dependent variations of DNA structure modulate radiation-induced strand breakage. Int J Radiat Biol 72:147–155

Tanaka K, Kamada N, Ohkita T, Kuramoto, A (1983) Non-random distribution of chromosome breaks in lymphocytes of atomic bomb survivors. J Radiat Rese 24:291–304

Tucker JD, Senft JR (1994) Analysis of naturally occurring and radiation-induced breakpoint locations in human chromosomes 1, 2 and 4. Radiat Rese 140:31–36

Turner BM (2000) Histone acetylation and an epigenetic code. BioEssays 22:836–845

Turner BM, Fellows G (1989) Specific antibodies reveal ordered and cell-cycle-related use of histone-H4 acetylation sites in mammalian cells. Eur J Biochem 179:131–139

Weintraub H, Groudine M (1976) Chromosomal subunits in active genes have an altered conformation. Science 193:848–856

Xiao Y, Natarajan AT (1999) Non-proportional involvement of Chinese hamster chromosomes 3, 4, 8 and 9 in X-rays-induced chromosomal aberrations. Int J Radiat Biol 75:943–951

14 A Biomedical Perspective of Telomere Structure and Function

Máximo E. Drets and Federico F. Santiñaque

Abstract

Telomeres are specialized nucleoproteic terminal structures of eukaryotic chromosomes that play an essential role in the stability of the genetic material within the cell, preventing its exonucleolytic degradation, end-to-end fusions and other chromosome-lethal events. Interest in this multifunctional chromosome segment has increased considerably since it intervenes in cell senescence, gene transcriptional regulation, premature aging syndromes and cancer, which are areas of great biomedical importance. Numerous "cryptic" microdeletions and rearrangements occurring in the subtelomeric region have been detected by fluorescent in situ hybridization and are related to mental retardation, congenital malformations and other severe clinical human conditions. Microphotometric scanning of T-banded subtelomeric segments of normal human and Chinese hamster ovary (CHO) chromosomes produces highly detailed graphic images allowing the detection of a differential distribution of high-density chromatin patterns between sister chromatids, namely, of similar size in both sister chromatids, predominating in one of the chromatids or concentrated in only one chromatid; these observations have been confirmed by scanning endoreduplicated CHO chromosomes. Furthermore, minute sister-chromatid exchanges were detected in normal human and CHO chromosomes. This precise analytical method additionally revealed that the highest chromatin densities of each chromosome arm were located in opposite chromatids in the majority of the chromosomes analyzed. The significance of the major aspects of function and molecular composition of the telomeric/subtelomeric segment and their relationship to biomedical problems recently reported, as well as some microphotometrical observations on the human and CHO chromatin structure are briefly reviewed.

14.1 Introduction

Telomeres are specialized DNA/protein structures located at the end of eukaryotic chromosomes that play an essential role in the stable maintenance of the genetic material within the cell. By capping chromosome ends, telomeres

Günter Obe and Vijayalaxmi (Eds.)
Chromosomal Alterations: Methods, Results and Importance in Human Health

prevent its exonucleolytic degradation, end-to-end fusions and other potential chromosome-lethal events (Blackburn 2001; Zakian 1995).

The study of telomeric area is increasing its biomedical importance since this multifunctional chromosome segment intervenes not only in keeping constant the number and structure of the chromosomes but it is also involved in cell senescence, gene transcriptional regulation, premature aging syndromes and cancer (Blackburn and Greider 1995; Blasco 2005; Baur et al. 2001; Greider 1998). It has also been shown that microdeletions and subtle rearrangements occurring in the subtelomeric region can cause mental retardation, congenital abnormalities and other severe clinical conditions.

In this chapter, we describe major aspects of the molecular composition and function of the telomeric/subtelomeric segment considering its possible relationship to some microscopic observations made on the structure of these areas in human and Chinese hamster ovary (CHO) chromosomes. Finally, we briefly discuss some aspects not yet well clarified on this critical chromosome segment.

14.2 The Complex Molecular Structure of the Terminal Chromosome Region

14.2.1 Telomeric DNA and Telomerase: Cell Senescence and Cancer

Mammalian telomeres are composed of tandem arrays of TTAGGG sequences (Blackburn and Gall 1978) and of several associated telomeric proteins, including some related to DNA repair/recombination (Hande 2004). The physical end of telomeric DNA presents a long single-stranded overhang of the G-rich strand that is found in all chromosome ends, irrespective of whether they were replicated by the lagging or the leading DNA synthesis (Makarov et al. 1997). This particular configuration of the telomeric DNA is conserved in eukaryotic cells and is involved in the stabilization of a specialized telomeric structure called a t-loop that is observable by electron microscopy in telomeric chromatin fibres (Griffith et al. 1999; Nikitina and Woodcock 2004). The t-loop model postulates that the terminal DNA forms a large loop capping the 3′ single-stranded overhang in the telomeric double-stranded DNA.

The length of telomeric DNA is established during zygotic development (Graakjaer et al. 2004) but it degrades during each cycle of DNA synthesis owing to incomplete telomere replication by conventional DNA polymerases and owing to the action of nucleases (Hug and Lingner 2006). The ribo nucleoprotein telomerase, a specific reverse transcriptase that extends the terminal repeated DNA, employs its specific RNA as a template for the addition of TTAGGG sequences to chromosome ends (Lingner et al. 1997); therefore, telomerase plays a key role in telomere length maintenance, compensating the progressive shortening of chromosome ends. However, it has been proposed that the progressive shortening of telomeric DNA produced in the absence of telomerase functions as a "biological clock" regulating the time of cell senescence or apoptosis (Harley 1995). Telomere shortening is observed in normal somatic cells of adult organisms but not in germinal cells or during the first stages of the embryonic development where it is compensated by the action of telomerase (Cong et al. 2002). The telomerase activity is detected in 90% of cancer cells, enabling an unlimited proliferation of neoplastic cells (Hiyama and Hiyama 2002). Since short telomeres lead to cell senescence or chromosome alterations, the length of telomeric DNA is a critical factor for the stability of telomeres, cell aging and cancer (Rodier et al. 2005). Recent findings indicate that chronic stress, cigarette smoking and obesity are significantly associated with oxidative damage and accelerated shortening of telomeres in women (Epel et al. 2004; Valdes et al. 2005).

14.2.2 Telomeric Chromatin

The telomeric chromatin contains structural proteins that bind sequence-specifically to the telomeric double-stranded DNA (TRF1 and TRF2) or to the G-rich single-stranded DNA (POT1) (Chong et al. 1995; van Steensel and de Lange 1997; Baumann and Cech 2001). These proteins interact with other telomeric proteins, including TIN2, Rap1 and Tpp1, to form a specific high-order chromatin structure that is associated with telomeres throughout the cell cycle (Smith and de Lange 1997; Liu et al. 2004). This protein complex that preserves telomeres formed by the subunits TRF1, TRF2, TIN2, Rap1, Tpp1 and POT1 has been collectively named "shelterin" (de Lange 2005).

Telomeric proteins form specialized structures with an elevated high-order complexity in the telomeric DNA. The protein TRF2 is capable of forming large duplex loops in the telomeric DNA (t-loops) (Stansel et al. 2001; Yoshimura et al. 2004). Furthermore, the main component of the telomeric chromatin, TRF1, binds to the telomeric DNA with a high spatial flexibility which could bend the telomeric DNA, to form a loop or, with an elevated tendency, pairing of telomeric duplexes (Bianchi et al. 1997, 1999; Griffith et al. 1998). The intratelomeric pairing of (TTAGGG)*n* repeats induced by TRF1 is stimulated when the TIN2 protein associates with the homodimeric complex (Kim et al. 2003). Griffith et al. (1998) suggested that the intratelomeric pairing produced by TRF1 is responsible for the telomeric DNA coiled configuration that is observed by electron microscopy in mitotic chromosomes (Steinmüller et al. 1993).

Telomerase accessibility to telomeric DNA is highly regulated by the telomeric proteins and by the level of chromatin condensation (Fang and Cech 1995; Blasco 2004). However, the most important function of the telomeric proteins appears to be telomere protection. Mutations in the protein TRF2 produce end-to-end fusions, nucleolytic degradation and other aberrant telomere phenotypes, suggesting that this protein plays an important role in the protective activity of telomeres (van Steensel et al. 1998). Griffith et al. (1999) proposed that TRF2 switches telomeres into a non-recombinogenic state through the formation of the t-loops. Besides, mice with a targeted deletion of TRF1 or deficiency for TIN2 show embryonic lethality (Karlseder et al. 2003; Chiang et al. 2004). Several recent findings suggested that an altered expression or function of the telomeric proteins is present in some human premature aging syndromes and tumours (Munoz et al. 2006; Garcia-Aranda et al. 2006; Oh et al. 2005).

14.3 Cryptic Subtelomeric Aberrations: Their Importance in Human Health

The subtelomeric region is composed of a mixture of unique DNA sequences together with different families of polymorphic middle repetitive DNA (Henderson 1995; Riethman 2005). This area possesses the highest gene concentration of the human genome (Saccone et al. 1992). The region close to the chromosome termini also contains highly polymorphic repetitive DNA sequences which are extraordinarily variable, dynamic and of rapid evolution (Der-Sarkissian et al. 2002). Besides, the subtelomeric region is enriched 25-fold in short internal (TTAGGG)n sequences in relation to the rest of the genome, which could play a potential role in subtelomeric recombination and telomere healing (Riethman et al. 2004).

It has been reported that the subtelomeric area is the site of a high number of interchanges that are able to produce phenotypic variability and embryonic developmental alterations (Badge et al. 2000). Using fluorescent in situ hybridization (FISH), Knight and Flint (2000) reported a considerable number of minute aberrations in the subtelomeric region in patients with mental retardation and congenital malformation syndromes.

The chapter of this syndromatic association started being clarified when Lejeune et al. (1959) discovered trisomy 21 and Sutherland (1977) detected the X chromosome fragility. However, a considerable number of congenital conditions, presenting mental retardation and malformations were obscure and not diagnosed at all by that time, being considered "idiopathic". However, the use of FISH made it possible to detect in the subtelomeric region minute aberrations called "cryptic" and believed to be responsible for 7.5% of congenital malformation and mental retardation (Sutton et al. 2002). Moreover, cryptic aberrations have also been found in other severe medical conditions such as in haemato-oncology, cancerology, gynaecology and psychiatry, showing the practical importance of these studies for clinical medicine and prognosis. The chromosome picture of these cryptic chromosome syndromes observed up to now in different areas of pathology is so complex

Table 14.1 Examples of complex cryptic aberrations detected in the subtelomeric region of human chromosomes associated with syndromes of mental retardation and congenital malformations

der(16)t(16;19)(p13.3;p13.3); Brown et al. (2000)	del(3)(p25.3p26.2); Cargile et al. (2002)
4p, dup(4)(p16.3p15.3); Cotter et al. (2001)	t(8;20)(p23;p13); de Vries et al. (2001)
t(2;6)(q37.3;q26); Giardino et al. (2003)	t(3;16)(q29;p13.3); Holisnki-Feder et al. (2000)
t(18;21)(q2.11q21.3); Horn et al. (2003)	46,XX,der(18)t(2;18)(p25;p11.2); Horsley et al. (1998)
dup(17p11/15q11); Keller et al. (2003)	(46,XX,der(13)t(8;13)(q24;q34); Kleeftra et al. (2000)
der(14)t(9;14)(q34.3;q32.33); Leube et al. (2003)	46,XX,inv(6)(q22.1q27); Lorda-Sanchez et al. (2000)
del(9)(q34.3)dup19(13.3); Quigley et al. (2004)	del(5q35.3); Rauch et al. (2003)
der(18)t(5;18)pat; Reddy and Fugate (1999)	t(11;18)(q25;q23); Schultz et al. (2001)
17p [46,XX,der(6)t(6;17)(q25.3;p13)]; Souter et al. (2003)	t(2,7);(q37;q35)/7q35→qter; Speleman et al. (2000)
t(4q35;10q26); van Deutekom et al. (1996)	46,XX,der(18)t(17;18)(q25;p11.32); Velagaleti et al. (2003)
t(17p13.3)(20q13.33); Walter et al. (2004)	46,XX,der(16)t(16;1); Warburton et al. (2000)

that their description escapes the possibility to describe them in detail in this chapter (Table 14.1). Therefore, the reader is referred to general reports published on this topic, particularly the extensive study of Raynan et al. (2006) that covers 11,688 cases.

14.4 Light Microscope Methods for the Study of Telomere Structure

14.4.1 T-banding

The telomere/subtelomeric chromatin structure and its DNA GC richness confer to the terminal chromosome region a high resistance to heat denaturation, being selectively stained with the T-banding method (Dutrillaux 1973). This technique is based upon an incubation of chromosomes in a hot buffer followed by Giemsa staining, leading to the appearance of T-bands as darkly stained segments in a lightly stained chromosome. T-bands are considered to be a subset of GC-rich R-bands highly resistant to digestion with the restriction endonuclease *Mse*I (Ludeña et al. 1991). Scanning electron microscopy shows that this region corresponds to areas of highly aggregated fibres (Jack et al. 1986).

14.4.2 Telomere Microdensitometric Scanning

A microphotometer system composed of a Zeiss (Oberkochen, Germany) microscope system processor MSP65 and a fast-scanning stage (0.25-μm step) coupled to a MicroVAX 3300 computer (Digital Equipment Corporation, Maynard, MA, USA) and a Netstation 4211 graphics terminal (Tektronix, Beaverton, OR, USA) was used for the study of the telomere structure.

Each scanning of T-banded telomeres covered an object area of 100 × 100 stage steps. One hundred densitometric measurements were made per scanning step, totalling one million measurements per microscope object. The data obtained per step were averaged in order to reduce random noise produced by the electronic components. The instrument setting resulted from a compromise between scanning speed and noise disturbance (object measurement to noise ratio greater than 10), which increased the reliability and precision of the system in all image-analysis procedures. For scan conversion, absorbances were represented as two-dimensional image units (pixels) in the discrete pixel image plane of the graphics terminal. The picture element size corresponded approximately to the resolving power of the light microscope (1 pixel being approximately 0.2 μm). Screen coordinates and pixel size assignment resulted from the magnification of the photomicrograph, the number of scanning steps and the final size of the telomere image displayed on the colour graphics terminal screen.

14.4.3 Computer Graphics Image Processing

Changes of contrast and brightness allowed the pixeled images to be displayed on a 0–100 grey scale corresponding to the absorbance measurement limits of the Zeiss module. For image discrimination and threshold densities, T-banded telomeres were sorted into 12 classes ranging from 0 to 100 absorbance units. Absorbance lower than 5 units corresponded to the background by default. Arbitrary colours were used to differentiate sorted classes, allowing a visual enhancement of the chromosome regional densities displayed. Binary filtering (shrink and expand operation) reduced unwanted pixels displayed in the background surface resulting from cytoplasm debris or stain precipitates. Data interpolation was used to obtain an image resolution of 600 × 600 compared with the original scanning of the 100 × 100 measurement matrix, enabling a more detailed image of chromosomal segments, since the staircase effect was eliminated. The number of pixels of the subtelomeric T-banded chromosome region corresponding to each sorted rank per chromatid

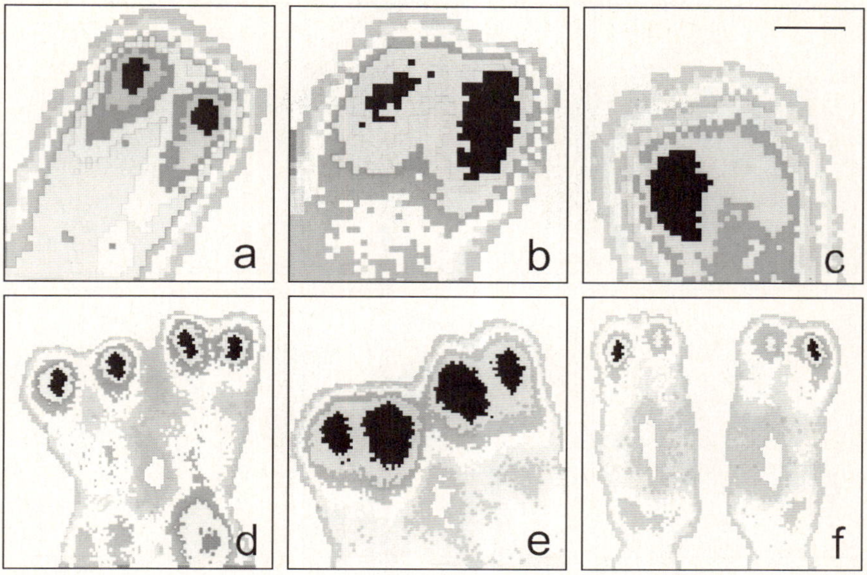

Fig. 14.1 Pixeled graphic images of high-density patterns detected in normal Chinese hamster ovary (CHO) telomeres (**a–c**) and in endoreduplicated CHO chromosomes (**d–f**). Note the similarity of the three basic high-density patterns found in normal and endoreduplicated chromosomes, indicating that only preexisting structures could replicate in the same maner. High-density areas are denoted in *black*. *Bar* 1 μm

was counted by drawing a line on the displayed image to define the chromatid borders using the graphic input rubberbanding feature of the Netstation 4211 graphics terminal. This feature allowed very precise quantitative measurements of the size of the high-density subtelomeric areas analysed to be obtained. Reprocessing of pixeled or interpolated colour graphic images allowed black-and-white illustrations to be obtained (Figs. 14.1–14.3) (see Drets et al. 1995a for a detailed description of the procedures).

14.5 Microdensitometric Analysis of T-banded Human and CHO Chromosomes

14.5.1 Differential Distribution of the Highest Chromatin Densities Between Sister Telomeres

The microphotometric analytical method developed in our laboratory was used for the study of the telomere structure of T-banded human and CHO chromosomes. This method allowed the detection of a differential distribution of high-density chromatin between sister telomeres (Drets et al. 1992)

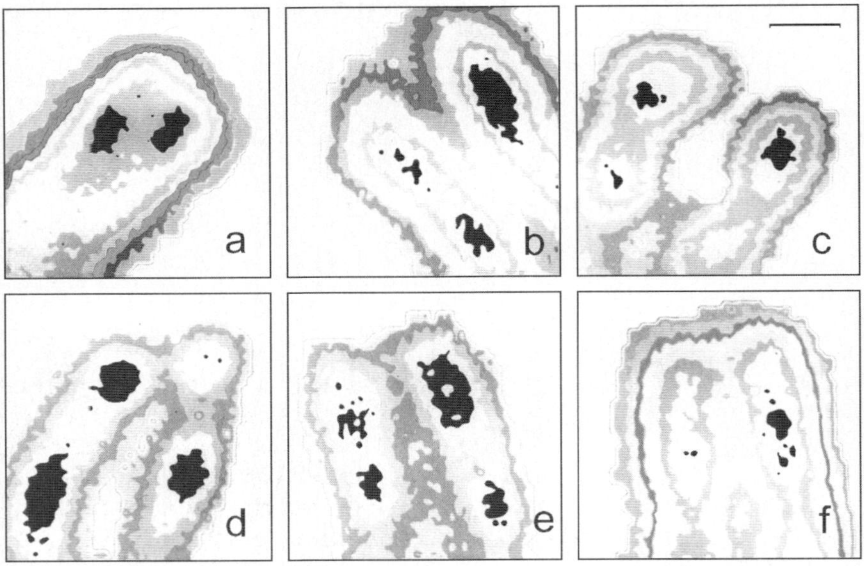

Fig. 14.2 Different examples of minute sister-chromatid exchanges of high-density segments detected in CHO (**a–c**) and human (**d–f**) chromosomes by means of microphotometrical scanning. Graphic images were obtained by interpolating microphotometric data. Exchanged blocks of high-density chromatin appear in **a** and **d.** Very tiny exchanges are seen in **b**, **e** and **f.** *Bar* 1 μm

that was arbitrarily classified into three types of density patterns, namely type I, high-density segments of similar size in both sister chromatids (Fig. 14.1a); type II, predominating in one of the chromatids (Fig. 14.1b); and type III, concentrated in only one chromatid (Fig. 14.1c). This unexpected distribution of high-density chromatin was confirmed by scanning T-banded endoreduplicated CHO chromosomes, since similar types of density patterns appeared in both sister chromosomes (Drets and Mendizábal 1998; Fig. 14.1d–f).

Interestingly, a prolonged incubation of chromosomes in the T-banding buffer induces the appearance of tiny holes in subtelomeric regions of sister chromatids. These holes were observed in normal and aberrant CHO chromosomes and were located in areas comparable to those of high-density telomeric areas, suggesting that a specific removal of high-density chromatin was produced (Drets et al. 1995b).

Recently, we studied the distribution of high-density chromatin at both ends of T-banded CHO chromosomes using our microdensitometric method and found that in the majority of the chromosomes that exhibited, in both ends, an asymmetrical distribution of high-density chromatin between sister telomeres (clearly noticeable in cases of the high-density pattern type III; Fig. 14.1c),

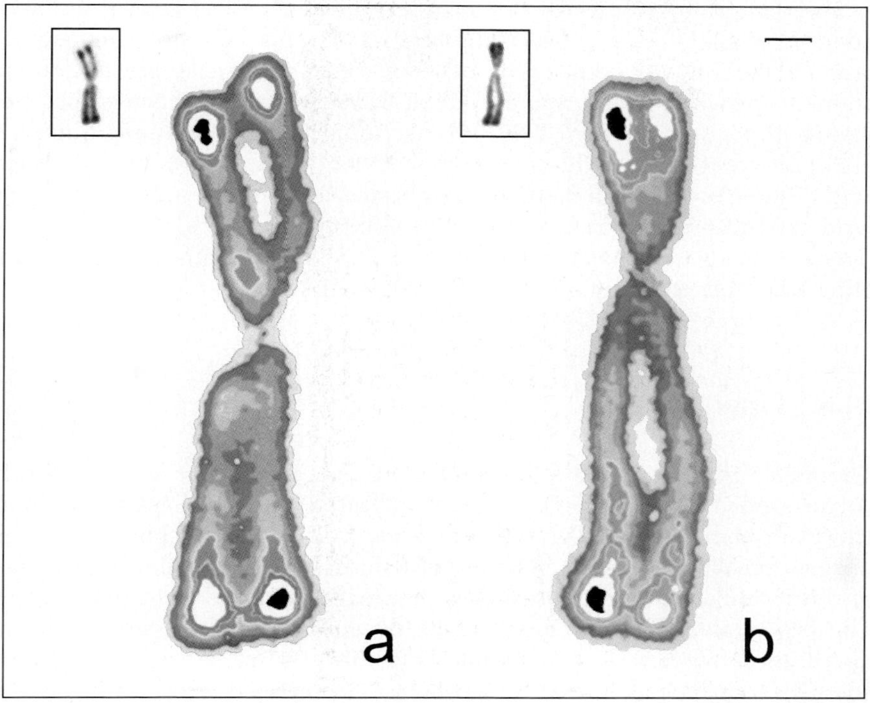

Fig. 14.3 Distribution of the high-density segments at opposite ends of T-banded CHO chromosomes. **a** High-density blocks are seen located in opposite chromatids (*trans* configuration). **b** High-density segments appearing in the same chromatids (*cis* configuration). The *insets* illustrate scanned chromosomes. *Bar* 1 μm

the highest-density segments of each arm were located in opposite chromatids. Preliminary observations showed that this configuration (*trans*) was detected in approximately 73% of the chromosomes analysed, and in a lower number of chromosomes (27%), high-density segments were found in the same chromatid (*cis* configuration). Figure 14.3 illustrates an example of both types of high-density segmentary configurations as detected by microdensitometric scanning of entire T-banded CHO chromosomes.

14.5.2 Minute Telomeric Sister-Chromatid Exchanges

Microphotometrical scanning of T-banded chromosomes also allowed the detection of a differential distribution of high-density areas between sister chromatids which looked like minute sister-chromatid exchanges that we named t-SCEs (Drets et al. 1992; Fig 14.2).

Several authors (Bailey et al. 2004; Bechter et al. 2004; Laud et al. 2005; Londoño-Vallejo et al. 2004; Wang et al. 2005), employing chromosome-orientation FISH, also detected minute sister-chromatid exchanges in telomere segments which are quite similar to the t-SCEs that we described earlier, but these authors used the term T-SCE to describing their observations. Since the use of different terms to describe similar, if not identical, structures detected with different cytological methods is undesirable, we proposed the use of the term ter-SCEs in order to standardize the nomenclature of these events (Drets et al. 2006). This new nomenclature would avoid any confusion with other terms generally used in human cytogenetics.

14.6 Final Remarks

The different types of graphic images obtained with our system closely corresponded to the T-banded segments observed with the light microscope. However, the computer graphic method used produced highly detailed images that allowed a more enhanced visualization of the chromatin structure of the terminal chromosome region than with the usual light microscope. This procedure revealed a complex high-density chromatin organization and distribution between sister chromatids in the terminal segments of human and CHO chromosomes. This finding was confirmed by scanning endo-reduplicated CHO chromosomes, since sister chromosomes exhibited similar patterns of high-density chromatin distribution. Besides, the detection of a "diagonal" distribution of high-density areas at opposite ends of T-banded CHO chromosomes by microphotometry suggests that these observations represented a telomeric non-random phenomenon.

It is of interest to mention that Bekaert et al. (2002) using three-dimensional confocal microscopy and quantitative FISH found significant differences in telomere signal intensity between sister chromatids of human lymphocyte chromosomes. These observations are comparable to the density patterns found by us. This view may be additionally supported by the observation that high-density areas colocalize with telomere probe signals in human metaphase chromosomes (Drets 2004), which are reportedly embedded in the subterminal region of the chromatids (Moyzis et al. 1988; Steinmüller et al. 1993).

Pardue and DeBaryshe (1999) stressed the point that several findings suggest that eukaryotic telomeres may have functions other than chromosome end protection and recognition as intact chromosomes. Evidence supporting the high functional activity of the terminal region is the detection of minute sister chromatid exchanges (t-SCEs) in T-banded human and CHO chromosomes. Owing to the fact that with our method, 5-bromo-2′-deoxyuridine is not used for detecting t-SCEs, and hence the DNA composition is intact in these minute SCEs, they may represent in some way an expression of the diverse functionality prevailing in the region (Obe et al. 2002; Drets 2004).

In this connection, Jin and Ikushima (2004) considered that the frequent occurrence of the tiny t-SCEs observed in the T-banded terminal segments of CHO and human chromosomes is a visualization of telomere recombination, a view that supports the fact that many rearrangements occur more frequently in subtelomeric domains than in other regions of the human genome and undergo high rates of spontaneous recombination as reported by Cornforth and Eberle (2001).

The differential distribution of high-density chromatin at telomeres is revealed by the T-banding procedure and it is not clear up to now how the underlying chromosomal organization responds to the banding phenomena that allow the detection of the complex high-density segments. It is a well-established fact that the main banding procedures (C-, G-, R- and T-banding) reveal specific chromosome structures related to local variations of the DNA composition, but all of them depended on a controlled chromosome denaturation and on the complex chemical composition of the Giemsa stain. In this connection, Sumner (1990) pointed out that in chromosome banding dye accessibility may be modified either by proteins or by variation of the DNA base pair composition (or both) along the genome. An interesting approach was reported by Saitoh and Laemmli (1994), who described that the bands arise from a differential folding path of the chromosome scaffold.

Even though considerable information on the molecular analysis of the telomeric/subtelomeric DNA and associated protein complexes is available nowadays, a clear picture at the high level of organization of this metaphase chromosome segment is still missing. In this respect, we believe that a combined molecular and detailed microscope study on this complex chromosome segment may lead to a better understanding of its structural role and function within the nucleus. Besides, it is expected that a deeper insight into aspects of the telomeric/subtelomeric region obtained through new molecular procedures or improved microscope analytical methods will also produce a considerable impact in medicine since it will contribute to clarifying many still obscure pathological conditions, permitting a more precise diagnosis, treatment and prevention of illnesses severely affecting human populations.

Acknowledgement: This work was supported in part by PEDECIBA (Uruguay).

References

Badge RM, Yardley J, Jeffreys AJ, Armour JA (2000) Crossover breakpoint mapping identifies a subtelomeric hotspot for male meiotic recombination. Hum Mol Genet 9:1239–1244

Bailey SM, Brenneman MA, Goodwin EH (2004) Frequent recombination in telomeric DNA may extend the proliferative life of telomerase-negative cells. Nucleic Acids Res 32:3743–3751

Baumann P, Cech TR (2001) Pot1, the putative telomere end-binding protein in fission yeast and humans. Science 292:1171–1175

Baur JA, Zou Y, Shay JW, Wright WE (2001) Telomere position effect in human cells. Science 292:2075–2077
Bechter OE, Zou Y, Walker W, Wright WE, Shay JW (2004) Telomeric recombination in mismatch repair deficient human colon cancer cells after telomerase inhibition. Cancer Res 64:3444–3451
Bekaert S, Koll S, Thas O, Van Oostveldt P (2002) Comparing telomere length of sister chromatids in human lymphocytes using three-dimensional confocal microscopy. Cytometry 48:34–44
Bianchi A, Smith S, Chong L, Elias P, de Lange T (1997) TRF1 is a dimer and bends telomeric DNA. EMBO J 16:1785–1794
Bianchi A, Stansel RM, Fairall L, Griffith JD, Rhodes D, de Lange T (1999) TRF1 binds a bipartite telomeric site with extreme spatial flexibility. EMBO J 18:5735–5744
Blackburn EH (2001) Switching and signaling at the telomere. Cell 106:661–673
Blackburn EH, Gall JG (1978) A tandemly repeated sequence at the termini of the extrachromosomal ribosomal RNA genes in Tetrahymena. J Mol Biol 120:33–53
Blackburn EH, Greider CW (1995) Telomeres. Cold Spring Harbor Laboratory Press, Cold Spring Harbor
Blasco MA (2004) Telomere epigenetics: a high-order control of telomere length in mammalian cells. Carcinogenesis 25:1083–1087
Blasco MA (2005) Telomeres and human disease: ageing, cancer and beyond. Nat Rev Genet 6:611–622
Brown J, Horsley SW, Jung C, Saracoglu K, Janssen B, Brough M, Daschner M, Beedgen B, Kerkhoffs G, Eils R, Harris PC, Jauch A, Kearney L (2000) Identification of a subtle t(16;19)(p13.3;p13.3) in an infant with multiple congenital abnormalities using a 12-colour multiplex FISH telomere assay, M-TEL. Eur J Hum Genet 8:903–910
Cargile CB, Goh DL, Goodman BK, Chen XN, Korenberg JR, Semenza GL, Thomas GH (2002) Molecular cytogenetic characterization of a subtle interstitial del(3)(p25.3p26.2) in a patient with deletion 3p syndrome. Am J Med Genet 109:133–138
Chiang YJ, Kim SH, Tessarolo L, Campisi J, Hodes RJ (2004) Telomere-associated protein TIN2 is essential for early embryonic development through a telomerase-independent pathway. Mol Cell Biol 24:6631–6634
Chong L, van Steensel B, Broccoli D, Erdjument-Bromage H, Hanish, J, Tempst P, de Lange T (1995) A human telomeric protein. Science 270:1663–1667
Cong Y, Wright WE, Shay JW (2002) Human telomerase and its regulation. Microbiol Mol Biol Rev 66:407–425
Cornforth MN, Eberle RL (2001) Termini of human chromosomes display elevated rates of mitotic recombination. Mutagenesis 16:85–89
Cotter PD, Kaffe S, Li L, Gershin IF, Hirschhorn K (2001) Loss of subtelomeric sequence associated with a terminal inversion duplication of the short arm of chromosome 4. Am J Med Genet 102:76–80
de Lange T (2005) Shelterin: the protein complex that shapes and safeguards human telomeres Genes Dev 19:2100–2110
de Vries BB, Lees M, Knight SJ, Regan R, Corney D, Flint J, Barnicoat A, Winter RM (2001) Submicroscopic 8pter deletion, mild mental retardation, and behavioral problems caused by a familial t(8;20)(p23;p13) Am J Med Genet 99:314–319
Der-Sarkissian H, Vergnaud G, Borde YM, Thomas G, Londoño-Vallejo JA (2002) Segmental polymorphism in the proterminal regions of a subset of human chromosomes. Genome Res 12:1673–1678
Drets ME (2004) Cytological indications on the complex structure of the subtelomeric region. Cytogenet Genome Res 104:137–141
Drets ME, Mendizábal M (1998) The underlying structure of the subtelomeric segments detected by microphotometrical scanning and graphic image analysis. Mutat Res 404:13–16
Drets ME, Obe G, Monteverde FJ, Folle GA, Medina II, De Galvez MG, Duarte JE, Mechoso BH (1992) Computerized graphic and light microscope analyses of T-banded chromosome segments of Chinese hamster ovary cells and human lymphocytes. Biol Zentralbl 111:204–214

Drets ME, Drets GA, Queirolo PJ, Monteverde FJ (1995a) Computer graphics as a tool in cytogenetic research and education. Comput Appl Biosci 11:463–468

Drets ME, Mendizábal M, Boccardo EM, Bonomi R (1995b) Further analyses of subtelomeric and paracentric holes induced in human and Chinese hamster ovary cell chromosomes. Biol Zentralbl 114:329–338

Drets ME, Santiñaque FF, Obe G (2006) A proposal of a standardised nomenclature for terminal minute sister chromatid exchanges. Genet Mol Biol 29:446–447

Dutrillaux B (1973) Nouveau système de marquage chromosomique: Les bandes T. Chromosoma 41:395–402

Epel ES, Blackburn EH, Lin J, Dhabhar FS, Adler NE, Morrow JD, Cawthon RM (2004) Accelerated telomeres shortening in response to life stress. Proc Natl Acad Sci USA 101:17312–17315

Fang G, Cech TR (1995) Telomere proteins. In: Blackburn EH, Greider CW (eds) Telomeres. Cold Spring Harbor Laboratory Press, Cold Spring Harbor, pp 69–105

Garcia-Aranda C, de Juan C, Diaz-Lopez A, Sanchez-Pernaute A, Torres AJ, Diaz-Rubio E, Balibrea JL, Benito M, Iniesta P (2006) Correlations of telomeres length, telomerase activity, and telomeric-repeat binding factor 1 expression in colorectal carcinoma. Cancer 106:541–551

Giardino D, Finelli P, Gottardi G, De Canal G, Della Monica M, Lonardo F, Scarano G, Larizza L (2003) Narrowing the candidate region of Albright hereditary osteodystrophy syndrome by deletion mapping in a patient with unbalanced cryptic translocation t(2;6)(q37.3;q26). Am J Med Genet 122:261–265

Graakjaer J, Pascoe L, Der-Sarkisian H, Thomas G, Kolvraa S, Christensen K, Londoño-Vallejo JA (2004) The relative lengths of individual telomeres are defined in the zigote and strictly maintained during life. Aging Cell 3:97–102

Greider CW (1998) Telomerase activity, cell proliferation, and cancer. Proc Natl Acad Sci USA 95:90–92

Griffith J, Bianchi A, de Lange T (1998) TRF1 promotes parallel pairing of telomeric tracts in vitro. J Mol Biol 278:79–88

Griffith JD, Comeau L, Rosenfield S, Stansel RM, Bianchi A, Moss H, de Lange T (1999) Mammalian telomeres end in a large duplex loop. Cell 97:503–514

Hande MP (2004) DNA repair factors and telomere-chromosome integrity in mammalian cells. Cytogenet Genome Res 104:116–122

Harley CB (1995) Telomeres and aging. In: Blackburn EH, Greider CW (eds) Telomeres. Cold Spring Harbor Laboratory Press, Cold Spring Harbor, pp 247–263

Hiyama E, Hiyama K (2002) Clinical utility of telomerase in cancer. Oncogene 21:643–649

Henderson E (1995) Telomere DNA structure. In: Blackburn EH, Greider CW (eds) Telomeres. Cold Spring Harbor Laboratory Press, Cold Spring Harbor, pp 11–34

Holinski-Feder E, Reyniers E, Uhrig S, Golla A, Wauters J, Kroisel P, Bossuyt P, Rost I, Jedele K, Zierler H, Schwab S, Wildenauer D, Speicher MR, Willems PJ, Meitinger T, Kooy RF (2000) Familial mental retardation syndrome ATR-16 due to an inherited cryptic subtelomeric translocation, t(3;16)(q29;p13.3). Am J Hum Genet 66:16–25

Horn D, Neitzel H, Tonnies H, Kalscheuer V, Kunze J, Hinkel GK, Bartsch O (2003) Familial MCA/MR syndrome due to inherited submicroscopic translocation t(18;21)(q2.11q21.3) with breakpoint at the Down syndrome critical region. Am J Med Genet 117:236–244

Horsley SW, Knight SJ, Nixon J, Huson S, Fitchett M, Boone RA, Hilton-Jones D, Flint J, Kearney L (1998) Del(18p) shown to be a cryptic translocation using a multiprobe FISH assay for subtelomeric chromosome rearrangements. J Med Genet 35:722–726

Hug N, Lingner J (2006) Telomere length homeostasis. Chromosoma 115:413–425

Jack EM, Harrison CJ, Allen TD, Harris R (1986) A structural basis for R- and T-banding: a scanning electron microscopy study. Chromosoma 94:395–402

Jin G, Ikushima T (2004) Frequent occurrence of UVB-induced sister chromatid exchanges in telomere regions and its implication to telomere maintenance. Cytogenet Genome Res 104:310–314

Karlseder J, Kachatrian L, Takai H, Mercer K, Hingorani S, Jacks T, de Lange T (2003) Targeted deletion reveals an essential function for the telomere length regulator Trf1. Mol Cell Biol 23:6533–6541

Keller K, Williams C, Wharton P, Paulk M, Bent-Williams A, Gray B, Ward A, Stalker H, Wallace M, Carter R, Zori R (2003) Routine cytogenetic and FISH studies for 17p11/15q11 duplications and subtelomeric rearrangements studies in children with autism spectrum disorders. Am J Med Genet 117:105–111

Kleefstra T, van de Zande G, Merkx G, Mieloo H, Hoovers JM, Smeets D (2000) Identification of an unbalanced cryptic translocation between the chromosomes 8 and 13 in two sisters with mild mental retardation accompanied by mild dysmorphic features. Eur J Hum Genet 8:637–640

Kim S, Han S, You Y, Chen DJ, Campisi J (2003) The human telomere-associated protein TIN2 stimulates interactions between telomeric DNA tracts *in vitro*. EMBO Rep 4:685–691

Knight SJL, Flint J (2000) Perfect endings: a review of subtelomeric probes and their use in clinical diagnosis. J Med Genet 37:401–409

Laud PR, Multani AS, Bailey SM, Wu L, Ma J, Kingsley C, Lebel M, Pathak S, DePinho RA, Chang S (2005) Elevated telomere-telomere recombination in WRN-deficient, telomere dysfunctional cells promotes escape from senescence and engagement of the ALT pathway. Genes Dev 19:2560–2570

Lejeune J, Gauthier M, Turpin R (1959) Étude des chromosomes somatiques de neuf enfants mongoliens. C R Acad Sci Paris 248:1721–1722

Leube B, Majewski F, Drechsler M, Royer-Pokora B (2003) Unbalanced cryptic translocation der(14)t(9;14)(q34.3;q32.33) identified by subtelomeric FISH. Clin Dysmorphol 12: 261–265

Lingner J, Hughes TR, Shevchenko A, Mann M, Lundblad V, Cech TR (1997) Reverse transcriptase motifs in the catalytic subunit of telomerase. Science 276:561–567

Liu D, O'Connor MS, Qin J, Songyang Z (2004) Telosome, a mammalian telomere-associated complex formed by multiple telomeric proteins. J Biol Chem 279:51338–51342

Londoño-Vallejo JA, Der-Sarkissian H, Cazes L, Bacchetti S, Reddel RR (2004) Alternative lengthening of telomeres is characterized by high rates of telomeric exchange. Cancer Res 64:2324–2327

Lorda-Sanchez I, Lopez-Pajares I, Roche MC, Sanz R, Rodriguez De Alba M, Gonzalez-Gonzalez MC, Ibanez A, Ramos C, Ayuso C (2000) Cryptic 6q subtelomeric deletion associated with a paracentric inversion in a mildly retarded child. Am J Med Genet 95:336–338

Ludeña P, Sentis C, De Cabo F, Velazquez M, Fernandez-Piqueras J (1991) Visualization of R-bands in human metaphase chromosomes by the restriction endonuclease MseI. Cytogenet Cell Genet 57:82–86

Makarov VL, Hirose Y, Langmore JP (1997) Long G tails at both ends of human chromosomes suggest a C strand degradation mechanism for telomere shortening. Cell 88:657–666

Moyzis RK, Buckinham JM, Cram LS, Dani M, Deaven LL, Jones MD, Meyne J, Ratliff RL, Wu JR (1988) A highly conserved repetitive DNA sequence (TTAGGG)n present at the telomeres of human chromosomes. Proc Natl Acad Sci USA 85:6622–6626

Munoz P, Blanco R, Blasco MA (2006) Role of the TRF2 telomeric protein in cancer and ageing. Cell Cycle 5:718–721

Nikitina T, Woodcock CL (2004) Closed chromatin loops at the ends of chromosomes. J Cell Biol 166:161–165

Obe G, Pfeiffer P, Savage JRK, Johannes C, Goedecke W, Jeppesen P, Natarajan AT, Martínez-López W, Folle GA, Drets ME (2002) Chromosomal aberrations: formation, identification and distribution. Mutat Res 504:17–36

Oh BK, Kim YJ, Park C, Park YN (2005) Up-regulation of telomere-binding proteins, TRF1, TRF2, and TIN2 is related to telomere shortening during human multistep hepatocarcinogenesis. Am J Pathol 166:73–80

Pardue ML, DeBaryshe PG (1999) Telomeres and telomerase: more than the end of the line. Chromosoma 108:73–82

Quigley DI, Kaiser-Rogers K, Aylsworth AS, Rao KW (2004) Submicroscopic deletion 9(q34.3) and duplication 19(p13.3): identified by subtelomere FISH probes. Am J Med Genet 125:67–72

Rauch A, Beese M, Mayatepek E, Dorr HG, Wenzel D, Reis A, Trautmann U (2003) A novel 5q35.3 subtelomeric deletion syndrome. Am J Med Genet 121:1–8
Raynan JB, Tepperberg JH, Papenhausen P, Lamb AN, Hedrick J, Eash D, Ledbetter DH, Martin CL (2006) Subtelomere FISH analysis of 11,688 cases: an evaluation of the frequency and pattern of subtelomere rearrangements in individuals with developmental disabilities. J Med Genet 43:478–489
Reddy KS, Fugate JK (1999) A half cryptic derivative der(18)t(5;18)pat identified by M-FISH and subtelomere probes: clinical findings and review of subtelomeric rearrangements. Clin Genet 56:328–332
Riethman H (2005) Human subtelomere structure and variation. Chromosome Res 13:505–515
Riethman H, Ambrosini A, Castaneda C, Finklestein J, Hu XL, Mudunuri U, Paul S, Wei J (2004) Mapping and initial analysis of human subtelomeric sequence assemblies. Genome Res 14:18–28
Rodier F, Kim SH, Nijjar T, Yaswen P, Campisi J (2005) Cancer and aging: the importance of telomeres in genome maintenance. Int J Biochem Cell Biol 37:977–990
Saccone S, De Sario A, Della Valle G, Bernardi G (1992) The highest gene concentrations in the human genome are in T bands of metaphase chromosomes. Proc Natl Acad Sci USA 89:4913–4917
Saitoh Y, Laemmli UK (1994) Metaphase chromosome structure: bands arise from a differential folding path of the highly AT-rich scaffold. Cell 76:609–622
Schultz LN, Schmidt P, Tabor A, Bryndorf T, Christesen B, Ludsteen C (2001) Cryptic familial t(11;18)(q25;q23) incidentally detected by interphase FISH. Clin Genet 59:279–283
Smith S, de Lange T (1997) TRF1, a mammalian telomeric protein. Trends Genet 13:21–26
Souter VL, Glass IA, Chapman DB, Raff ML, Parisi MA, Opheim KE, Disteche CM (2003) Multiple fetal anomalies associated with subtle subtelomeric chromosomal rearrangements. Ultrasound Obstet Gynecol 21:609–615
Speleman F, Callens B, Logghe K, Van Roy N, Horsley SW, Jauch A, Verschraegen-Spac MR, Leroy JG (2000) Subtelomeric familial translocation t(2;7)(q37;q35) leading to partial trisomy 7q35→qter: molecular cytogenetic analysis and clinical phenotype in two generations. Am J Med Genet 93:349–354
Stansel R, de Lange T, Griffith JD (2001) T-loop assembly *in vitro* involves binding of TRF2 near the 3′ telomeric overhang. EMBO J 20: 5532–5540
Steinmüller J, Schleiermacher E, Scherthan H (1993) Direct detection of repetitive whole chromosome paint and telomere DNA probes by immunogold electron microscopy. Chromosome Res 1:45–51
Sumner AT (1990) Chromosome banding. Unwin Hyman, London
Sutherland GR (1977) Fragile sites on human chromosomes: demonstration of their dependence on the type of tissue culture medium. Science 197:265–266
Sutton VR, Coveler KJ, Lalani SR, Kashork CD, Shaffer LG (2002) Subtelomeric FISH uncovers trisomy 14p32: lessons for imprinted regions, cryptic rearrangements and variant acrocentric short arms. Am J Med Genet 112:23–27
Valdes AM, Andrew T, Gardner JP, Kimura M, Oelsner E, Cherkas LF, Aviv A, Spector TD (2005) Obesity, cigarette smoking, and telomeres length in women. Lancet 366:662–664
van Deutekom JC, Bakker E, Lemmers RJ, van der Wielen MJR, Bik E, Hofker MH, Padberg GW, Frants RR (1996) Evidence for subtelomeric exchange of 3.3 kb tandemly repeated units between chromosomes 4q35 and 10q26: implications for genetic counselling and etiology of FSHD1. Hum Mol Genet 5:1997–2003
van Steensel B, de Lange T (1997) Control of telomere length by the human telomeric protein TRF1. Nature 385:740–743
van Steensel B, Smogorzewska A, de Lange T (1998) TRF2 protects human telomeres from end-to-end fusions. Cell 92:401–413
Velagaleti GV, Jalal SM, Michaelis RC, Rowe TF, Nichols JR, Lockhart LH (2003) Molecular cytogenetic characterization of a *de novo* unbalanced translocation leading to trisomy

17q25→qter and monosomy 18p11.3→pter in a girl with dysmorphic features. Clin Dysmorphol 12:29–33

Walter S, Sandig K, Hinkel GK, Mitulla B, Ounap K, Sims G, Sitska M, Utermann B, Viertel P, Kalscheuer V, Bartsch O (2004) Subtelomere FISH in 50 children with mental retardation and minor anomalies, identified by a checklist, detects 10 rearrangements including a *de novo* balanced translocation of chromosomes 17p13.3 and 20q13.33. Am J Med Genet 128:364–373

Wang RC, Smogorzewska A, de Lange T (2004) Homologous recombination generates T-loop-sized deletions at human telomeres. Cell 119:355–368

Warburton P, Mohammed S, Ogilvie CM (2000) Detection of submicroscopic subtelomeric chromosome translocations: a new case study. Am J Med Genet 91:51–55

Yoshimura SH, Maruyama H, Ishikawa F, Ohki R, Takeyasu K (2004) Molecular mechanisms of DNA end-loop formation by TRF2. Genes Cells 9:205–218

Zakian VA (1995) Telomeres: beginning to understand the end. Science 270:1601–1607

15 Cytokinesis-Block Micronucleus Assay: a Comprehensive "Cytome" Approach for Measuring Chromosomal Instability, Mitotic Dysfunction and Cell Death Simultaneously in One Assay

Michael Fenech

Abstract

The cytokinesis-block micronucleus (CBMN) assay was originally developed as an ideal system for measuring micronuclei; however, it can also be used to measure nucleoplasmic bridges (NPBs), nuclear buds, cell death (necrosis or apoptosis) and nuclear division rate. Current evidence suggests that (1) NPBs originate from dicentric chromosomes in which the centromeres have been pulled to the opposite poles of the cell at anaphase and are therefore indicative of DNA misrepair, chromosome rearrangement or telomere end-fusions, (2) NPBs may break to form micronuclei, (3) the nuclear budding process is the mechanism by which cells remove amplified and/or excess DNA and is therefore a marker of gene amplification and/or altered gene dose, (4) cell cycle checkpoint defects result in micronucleus formation, and (5) hypomethylation of DNA, induced nutritionally or by inhibition of DNA methyltransferase can lead to micronucleus formation either via chromosome loss or via chromosome breakage. The strong correlation between micronucleus formation, nuclear budding and NPBs ($r = 0.75$–0.77, $P<0.001$) induced by either folic acid deficiency or exposure to ionising radiation is supportive of the hypothesis that folic acid deficiency and/or ionising radiation cause genomic instability and gene amplification by the initiation of breakage–fusion–bridge cycles. In its comprehensive mode the CBMN assay measures all cells, including necrotic and apoptotic cells, as well as number of nuclei per cell to provide a measure of cytotoxicity and mitotic activity. The CBMN assay has in fact evolved into a "cytome" method for measuring comprehensively chromosomal instability phenotype and cellular viability caused by genetic defects and/or nutritional deficiencies and/or exogenous genotoxins, opening up an exciting future for the use of this methodology in the emerging fields of nutrigenomics and toxicogenomics and their combinations.

15.1 Introduction

Micronuclei originate from chromosome fragments or whole chromosomes that lag behind at anaphase during nuclear division (Fenech 2000a). The micronucleus index in human cells has become one of the standard cytogenetic tests for genetic toxicology testing. The cytokinesis-block micronucleus

Günter Obe and Vijayalaxmi (Eds.)
Chromosomal Alterations: Methods, Results and Importance in Human Health

(CBMN) assay is the preferred method for measuring micronuclei in cultured human and/or mammalian cells because scoring is specifically restricted to once-divided cells. These cells are recognised by their binucleated appearance after inhibition of cytokinesis by cytochalasin B (Fenech 2000a). Restricting scoring of micronuclei in binucleated cells prevents confounding effects caused by suboptimal cell division kinetics, which is a major variable in this assay (Fenech 2000a). Over the past 17 years the CBMN assay has evolved into a comprehensive method for measuring chromosome breakage, chromosome loss, non-disjunction, necrosis, apoptosis and cytostasis (Fenech 2000a; Fig. 15.1). In combination with the use of centromeric probes or kinetochore antibodies it is also possible to distinguish between chromosome breakage and loss as the mechanism of micronucleus formation and to measure malsegregation of chromosomes amongst the two nuclei in binucleated cells. However, new developments now indicate that this method can

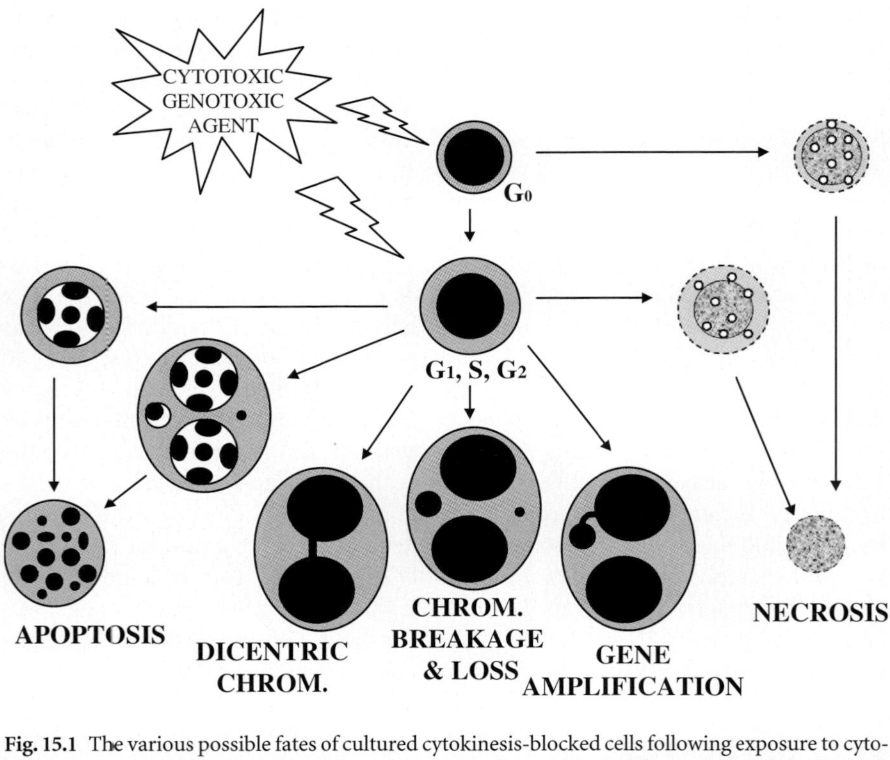

Fig. 15.1 The various possible fates of cultured cytokinesis-blocked cells following exposure to cytotoxic/genotoxic agents. Using these biomarkers within the cytokinesis-block micronucleus (CBMN) assay it is possible to measure the frequency of chromosome breakage (micronucleus), chromosome loss (micronucleus), chromosome rearrangement, e.g. dicentric chromosomes (nucleoplasmic bridges), gene amplification (nuclear buds), necrosis and apoptosis. In addition, cytostatic effects are readily estimated from the ratio of mononucleated, binucleated and multinucleated cells. Chromosome loss can be distinguished from chromosome breakage using pancentromeric probes or anti-kinetochore antibodies. In addition, non-disjunction (malsegregation of chromosomes) can also be measured in binucleated cells using chromosome-specific centromeric probes

also be used to measure nucleoplasmic bridges (NPBs), a biomarker of dicentric chromosomes resulting from telomere end-fusions or DNA misrepair, and to measure nuclear buds, a biomarker to detect gene amplification and altered gene dose events (Fenech 2000a; Umegaki and Fenech 2000; Haaf et al. 1999). The significance of these developments and the concept of the CBMN assay as a "cytome" assay of chromosomal instability are elaborated in the following sections. The "cytome" concept implies that every cell in the system studied is scored cytologically for its viability status (necrosis, apoptosis), its mitotic status (mononucleated, metaphase, anaphase, binucleated, multinucleated) and its chromosomal instability or damage status (presence of micronuclei, NPBs, nuclear buds and the number of centromere probe signals amongst nuclei of binucleated cells if such molecular tools are used in combination with the assay) (Fig. 15.2).

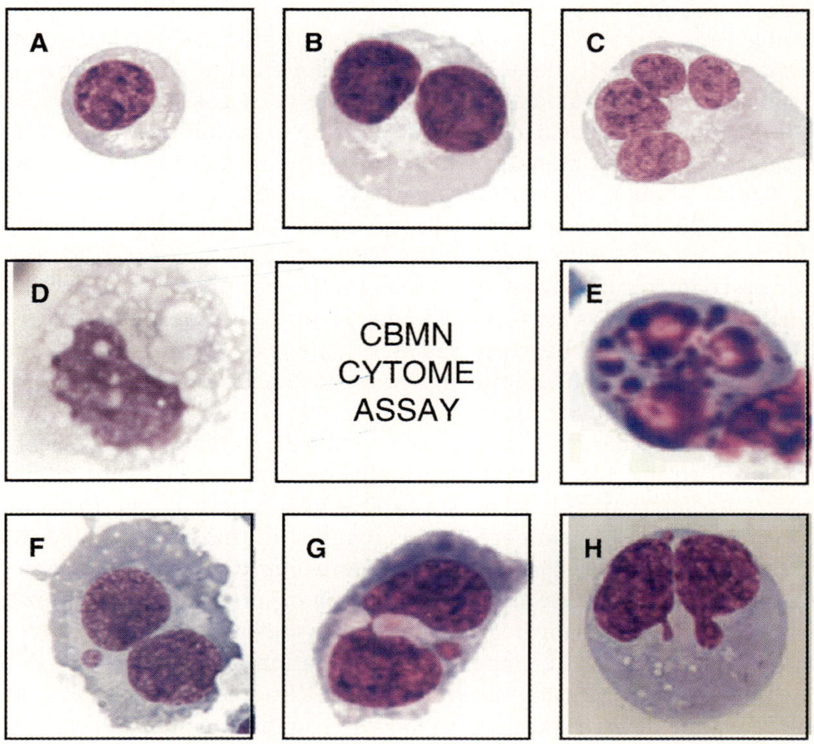

Fig. 15.2 Photomicrographs of the cells scored in the CBMN "cytome" assay. **a** Mononucleated cell, **b** binucleated cell, **c** multinucleated cell, **d** necrotic cell, **e** apoptotic cell, **f** binucleated cell containing one or more micronuclei, **g** binucleated containing a nucleoplasmic bridge and a micronucleus, **h** binucleated cell containing nuclear buds. The ratios of mononucleated, binucleated, multinucleated, necrotic and apoptotic cells are used to determine the mitotic division rate or nuclear division index (a measure of cytostasis) and cell death (cytotoxicity). The ratios of binucleated cells with micronuclei, nucleoplasmic bridges or nuclear buds relative to the total number of binucleated cells provides a measure of genome damage and/or chromosomal instability. For more details of the scoring criteria refer to Fenech et al. (2003)

15.2 Nucleoplasmic Bridges

We have proposed that NPBs between nuclei in binucleated cells should also be scored in the CBMN assay because they provide a measure of chromosome rearrangement which is otherwise not measured in this assay if only micronuclei are scored (Umegaki and Fenech 2000). The NPBs are assumed to occur when the centromeres of dicentric chromosomes are pulled to opposite poles of the cell at anaphase. It is rarely possible to observe dicentric anaphase bridges before the nuclear membrane is formed, because cells proceed through anaphase and telophase rapidly, completing cytokinesis and ultimately breakage of the NPB when the daughter cells separate. However, in the CBMN assay binucleated cells with NPBs are allowed to accumulate because cytokinesis is inhibited and the nuclear membrane is eventually formed around the chromosomes, allowing a NPB to be observed.

We have recently validated the use of NPBs as a biomarker of DNA damage in human WIL2-NS cells treated with hydrogen peroxide, superoxide or after coincubation with activated human neutrophils (Umegaki and Fenech 2000). NPB frequency in binucleated cells increased up to 20-fold relative to control cells in a dose-related manner with these agents that are known to induce DNA strand breaks. NPBs were positively correlated with micronucleus frequency in the same binucleated cells ($r > 0.82$, $P < 0.0001$). More than 60% of the binucleated cells with one or more NPBs were also observed to contain one or more micronuclei, which fits with the fact that acentric chromosome fragments (expressed as micronuclei) and dicentric chromosomes (expressed as NPBs) are known to be expressed simultaneously within a cell following exposure to agents such as ionising radiation that induce double-stranded DNA breaks (IAEA 1986).

The various mechanisms that could lead to NPB formation following DNA misrepair of strand breaks in DNA are illustrated and explained in detail in Fig. 15.3. An alternative mechanism for NPB formation is due to telomere end-fusions caused by telomere shortening, but in this case the NPB is not necessarily accompanied by the presence of an acentric chromosome fragment. The formation of anaphase and NPBs has been observed in models of rodent and human intestinal cancer in vivo and was shown to correlate positively with telomere length, indicating that NPB formation can also be used as a surrogate measure of critically short telomeres (Rudolph et al. 2001). More recent real-time in vitro studies have focussed on the fate of anaphase bridges and showed that during cytokinesis the bridges may be broken in more than one region, leading to the formation of a micronucleus (Hoffelder et al. 2004; Shimizu et al. 2005). Therefore, some micronuclei may also originate from anaphase bridges, although whether this actually happens in cytokinesis-blocked cells remains unclear.

The importance of scoring NPBs should not be underestimated because it provides direct evidence of genome damage resulting from misrepaired DNA

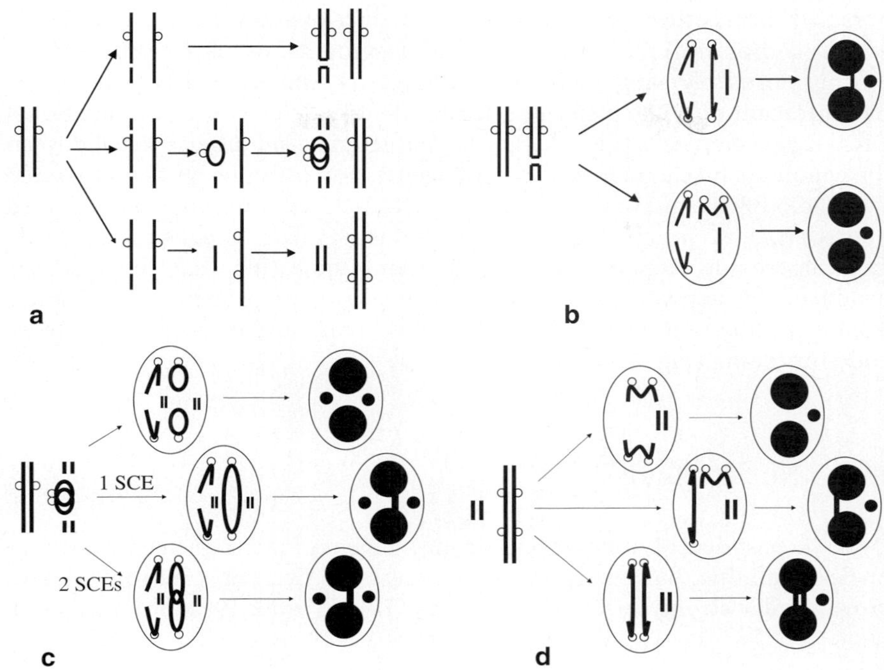

Fig. 15.3 **a** Initiating events in the formation of a nucleoplasmic bridge. *Top*: Double-strand DNA break is induced in one chromosome but remains unrepaired until after replication. Misjoining of the broken ends after replication leads to the formation of a dicentric chromatid and an acentric fragment. *Middle*: Two double-strand breaks on either side of the centromere of one chromosome are induced. The broken ends are misrepaired, producing a ring chromatid and two acentric fragments which are subsequently replicated. *Bottom*: Double-strand breaks are induced in two homologous or non-homologous chromosomes. Misjoining of the broken ends leads to the formation of a dicentric chromatid and acentric fragment which are subsequently replicated. **b** *Top*: Centromeres of a dicentric chromatid are pulled to opposite poles of the cell at anaphase to form a nucleoplasmic bridge and the acentric chromatid fragment lags behind to form a micronucleus. *Bottom*: Centromeres of a dicentric chromatid are both pulled to the same pole of the cell and no nucleoplasmic bridge is formed; however, the lagging acentric chromatid fragment results in the formation of a micronucleus. **c** *Top*: The ring chromatids are segregated normally but the acentric fragments lag at anaphase to form two micronuclei. *Middle*: The two ring chromatids, after completing one sister-chromatid exchange (*SCE*), are transformed into one large dicentric ring chromatid, which leads to the formation of a nucleoplasmic bridge when the centromeres are pulled to the opposite poles of the cell at anaphase. The accompanying acentric fragments lag at anaphase to form two micronuclei. *Bottom*: The two ring chromatids, after completing two SCEs, become concatenated, which leads to the formation of a nucleoplasmic bridge when the centromeres are pulled to the opposite poles of the cell at anaphase. The accompanying acentric fragments lag at anaphase to form two micronuclei. **d** *Top*: Centromeres of dicentric chromatids travel to the same pole of the cell and no nucleoplasmic bridges are formed. *Middle*: Centromeres of one of the dicentric chromatids are pulled to opposite poles of the cell, leading to the formation of one nucleoplasmic bridge. *Bottom*: Centromeres of both dicentric chromatids are pulled to opposite poles of the cell, leading to the formation of two nucleoplasmic bridges. In each of the above cases one micronucleus is produced from the lagging acentric chromosome fragment accompanying the dicentric chromosome

breaks which is otherwise not possible to deduce by scoring micronuclei only which could originate from acentric fragments of chromosome loss. Furthermore, following radiation-challenge or oxidant-challenge tests, the measurement of NPBs could be used to identify defects in recombinational DNA-repair pathways (Fig. 15.4) or to characterise and differentiate between the genotoxic mechanism of different agents. For example, we demonstrated that the NPB-to-micronucleus ratio in WIL2-NS cells induced by activated neutrophils is much higher than the ratios for hydrogen peroxide and superoxide, which were not different, indicating that the oxidants produced by activated neutrophils have a greater propensity to induce lesions leading to misrepair of DNA strand breaks and or possibly telomere end-shortening (Fig. 15.5).

15.3 Nuclear Buds

Over the past decade another unique mechanism of micronucleus, known as nuclear budding, has emerged. This process has been observed in cultures grown under strong selective conditions (Toledo et al. 1992; Shimura et al.

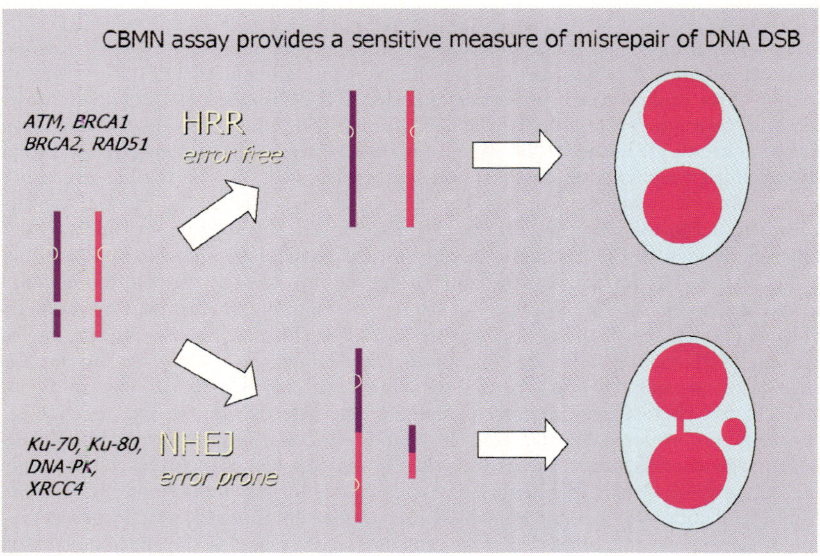

Fig. 15.4 Defects in genes involved in homologous recombinational repair (*HRR*) genes (e.g. *ATM*, *BRCA1*, *BRCA2*, *RAD51*) could be detected using the CBMN assay because under these conditions it may be anticipated that the cell would resort to the error-prone non-homologous end-joining (*NHEJ*) repair mechanism that is likely to lead to misrepair of double-strand breaks in DNA, formation of dicentric chromosomes and ultimately nucleoplasmic bridges which are detectable in the CBMN assay

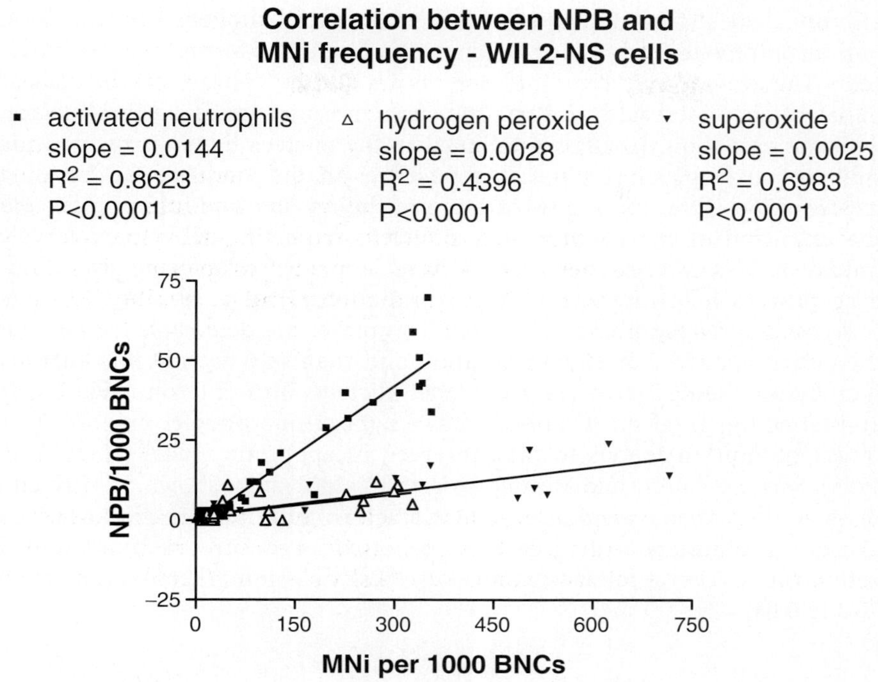

Fig. 15.5 Correlation between nucleoplasmic bridges and micronuclei in WIL2-NS cultures exposed to increasing doses of hydrogen peroxide or superoxide or activated neutrophils. (Data derived from experiments described in Umegaki and Fenech 2000). *MNi* micronuclei, *NPB* nucleoplasmic bridge, *BNC* binucleated cell

1999; Ma et al. 1993) which induce gene amplification. Shimizu et al. (1998, 2000) showed, using in vitro experiments with mammalian cells, that amplified DNA is selectively localised to specific sites at the periphery of the nucleus and eliminated via nuclear budding to form micronuclei during S phase of mitosis. Furthermore, DNA synthesis inhibitors such as hydroxyurea were shown to increase the rate of elimination of amplified DNA via this process. Amplified DNA may be eliminated through recombination between homologous regions within amplified sequences forming minicircles of acentric and atelomeric DNA (double minutes) which localise to distinct regions within the nucleus or through the excision of amplified sequences after segregation to distinct regions of the nucleus. The above suggests the hypothesis that the nucleus may have a capacity to sense excess DNA that does not fit well within the nuclear matrix, indicating a higher-order DNA repair or nuclear housekeeping process. Shimizu et al. (1998, 2000) have suggested that the nucleus eliminates excess amplified DNA by an active process that concentrates the amplified DNA to a peripheral point in the nucleus, following which this surplus DNA is budded out to form a micronucleus and eventually excluded from the cell altogether by extrusion of the micronucleus from the cytoplasm, leading to the formation of a "minicell". The process of nuclear budding occurs during S phase and the nuclear buds are characterised by having the same morphology as a micronucleus, with the exception that they are linked to the nucleus by a narrow or wide stalk of nucleoplasmic material depending on the stage of the budding process (Figs. 15.1, 15.2). The duration of the nuclear budding process and the extrusion of the resulting micronucleus from the cell remain largely unknown. It is also reported that micronuclei may also be formed by a budding process following exposure to γ-radiation (Haaf et al. 1999). In this process, Rad51-recombination protein complexes are detectable throughout the entire nucleus 3 h after irradiation and then concentrate into distinct foci before being extruded from the nucleus into micronuclei. Using end-labelling, Haaf et al. (1999) showed these micronuclei exhibit DNA fragmentation analogous to that observed in apoptotic nuclei (Haaf et al. 1999). Serrano-Garcia and Montero-Montaya (2001) have shown, in vitro and in vivo in human lymphocytes, that nuclear buds and micronuclei are expressed simultaneously following genotoxic exposure in human cells with positive correlation factors between these two biomarkers ranging from 0.73 to 0.84.

15.4 Micronuclei, NPBs and Nuclear Buds Induced by Folic Acid Deficiency

In a series of studies on folic acid deficiency in long-term primary human lymphocyte cultures we carefully quantified the interrelationship between micronuclei, NPBs and nuclear buds in an attempt to validate the use of these biomarkers and to determine more comprehensively the impact of folic acid deficiency on various aspects of genomic stability (Crott et al. 2001a, b; Kimura et al. 2004; Beetstra et al. 2006). Briefly, lymphocytes were cultured in duplicate in RPMI-1640 medium containing 5% dialysed foetal calf serum, 10 U/ml interleukin-2 and 12, 24, 60 or 120 nM folic acid for 9 days. The medium was refreshed at days 3 and 6, cytokinesis was inhibited on day 8 and cells were transferred to microscope slides 24 h later (for the detailed protocol see Crott et al. 2001a, b). We analysed separately the pooled data from this study and verified that folic acid concentration correlated significantly ($P<0.0001$) and negatively ($r = -0.63$ to -0.74) with all these markers of chromosome damage, which were minimised at 60–120 nM folic acid, the latter being greater than the concentration of folate normally observed in plasma (10–30n M). However, even more interestingly, we observed that the frequency of NPBs and nuclear buds correlated significantly and negatively with folic acid dose, suggesting that chromosome rearrangement and gene amplification are induced by folic acid deficiency. The strong cross-correlation between micronucleus, NPB and nuclear bud frequency ($r = 0.65$–0.77, $P<0.001$) suggests a common mechanism initiated by folic acid deficiency induced DNA breaks, although coincidence of effects with other DNA damage events (e.g. folic acid deficiency induced CpG hypomethylation) cannot be excluded.

15.5 Breakage–Fusion–Bridge Cycles Explain Micronuclei, NPBs and Nuclear Buds Induced by Folic Acid Deficiency

Because folic acid deficiency is known to cause gene amplification and chromosome damage such as double-stranded breaks (Blount et al. 1997; Fenech 2001; Melnyk et al. 1999), it is likely that these events contribute to the formation of micronuclei, NPBs and nuclear buds in this system. Gene amplification is thought to be a key event in cellular resistance to drugs like methotrexate (Biedler and Spengler 1976) and in tumour progression (Brison 1993). Several plausible models of gene amplification have been proposed (reviewed in Windle and Wahl 1992; Stark 1993); however, the presence of NPBs in the cytokinesis-blocked cells provides support for the "breakage–fusion–bridge (BFB) cycle" model described in the seminal work of McClintock (1942) in maize. According to this theory, sister chromatids which have both undergone double-stranded breakage fuse at a distal position (possibly telomeric) forming a dicentric chromosome which has two copies of homologous genes positioned between the two centromeres. During anaphase these dicentric chromosomes are drawn towards both poles and form (nucleoplasmic) bridges. During cytokinesis these dicentric chromosomes, which span both daughter nuclei, are thought to break unevenly and may form a chromosome with two copies of one or more genes and a chromosome (fragment) with no copies of these genes. The chromatids with multiple copy number of these genes may fuse again during interphase forming a dicentric chromosome (doubling again the gene copy number within the chromosome), which is then replicated during the next nuclear division leading to the next bridge–breakage–fusion cycle and further gene amplification. The BFB cycle mechanism is illustrated in Fig. 15.6.

The BFB cycle model has been validated as a mechanism for gene amplification in various systems, including (1) coformycin-induced amplification of the adenylate deaminase gene in Chinese hamster cells (Toledo et al. 1992), (2) *N*-(phosphonyacetyl)-L-aspartate induced amplification of the CAD gene in human fibrosarcoma cells which is enhanced by the expression of *Vpr*, a HIV accessory gene (Shimura et al. 1999), (3) methotrexate-induced

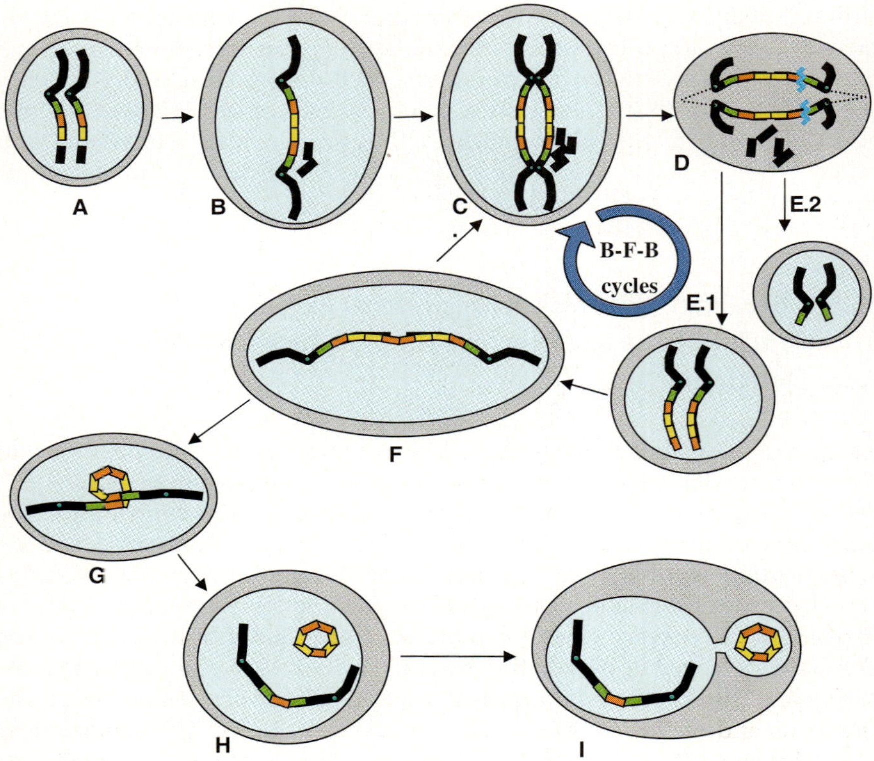

Fig. 15.6 Gene amplification via breakage–fusion–bridge cycles (*B-F-B*). Broken sister chromosomes **(A)** fuse at the "sticky" ends to form a dicentric chromosome **(B)**. The dicentric chromosome is replicated in S phase **(C)** and drawn to the cell's poles during anaphase (bridge) **(D)**. In this example, the dicentric chromosomes break in a non-central region and one daughter cell will receive chromosomes that contain two copies of the *yellow* and *red* genes **(E.1)**, while the other cell's chromosomes have a deletion in these genes **(E.2)**. The chromosomes with multiple copy number may fuse again **(F)** to propagate the amplification cycle (back to step C). Alternatively, recombination may occur between homologous sequences **(G)** and result in the "looping-out" of circular acentric DNA fragments or double minutes **(H)** which are subsequently extruded from the nucleus via budding **(I)**

amplification of the dihydrofolate reductase gene in Chinese hamster cells (Ma et al. 1993) and (4) actinomycin D induced amplification of the multidrug resistance 1 gene in Chinese hamster cells through the induction of fragile sites which determined the initiation and size of amplicons (Coquelle et al. 1997). Folic acid deficiency induced fragile site expression and DNA hypomethylation may have also contributed to the promotion of gene amplification and resulted in elimination of this DNA by nuclear budding in our system. For example, the induction of hypomethylation by 5-azacytidine has been reported to enhance *N*-(phosphonylacetyl)-L-aspartate induced amplification of the CAD gene in Syrian hamster kidney cells (Perry et al. 1992).

15.6 Micronucleus Formation Caused by Hypomethylation of Heterochromatin and Silencing of Cell Cycle Checkpoint Genes

The DNA methylation inibitor 5-azacytidine also induces distinct undercondensation of the heterochromatin regions of chromosomes 1, 9, 15, 16 and Y, and the specific loss of these chromosomes as micronuclei in human lymphocytes in vitro (Guttenbach and Schmid 1994). Similarly it has been shown that 5-azacytidine causes decondensation in the homogenously staining region in the highly methylated heteromorphic chromosome 15 and loss of this genetic material as micronuclei in human diploid lung fibroblast cell line TIG-7 (Satoh et al. 2004). The immunodeficiency–centromeric instability–facial anomalies (ICF) syndrome, which is caused by mutation in the DNA methyltransferase gene, is characterised by despiralisation of heterochromatin of chromosomes 1, 9 and 16 and loss of this chromatin into micronuclei and nuclear blebs (Xu et al. 1999). These events are likely to be relevant to the ageing process because Suzuki et al. (2002) demonstrated that in vitro ageing of normal human fibroblasts results in concomitant demethylation of satellite 2 and satellite 3 DNA which is abundant in the juxtacentromeric DNA of chromosomes 1, 9 and 16 and the increasing frequency of micronuclei that specifically contained these sequences. We have also recently demonstrated that folate deficiency caused a significant 26–35% increment in frequency of aneuploidy of chromosome 17 ($P = 0.0017$) and aneupoidy of chromosome 21 ($P = 0.0008$) relative to 120 nM 5-methyltetrahydrofolate or folic acid. The pattern of aneuploidy in binucleated cells was significantly correlated with that observed in mononucleated cells ($R = 0.51$–0.75, $P<0.0004$) and was consistent with a model based on chromosome loss or partial aneusomy rescue as the cause rather than non-disjunction, although the latter mechanism could not be excluded (Wang et al. 2004). Figure 15.7 illustrates possible mechanisms for the relationship between genome methylation and micronucleus formation.

It is evident that micronucleus expression could be increased as a result of hypomethylation of satellite DNA; however, it is also possible that increased genome damage may be caused by hypermethylation of CpG islands within or adjacent to the promoter regions of housekeeping genes involved in cell cycle checkpoints and DNA repair. For example CpG island hypermethylation of the mitotic spindle checkpoint genes such as *APC*, *BUB1* and *HCDC4* could reduce their expression and therefore increase the possibility of chromosome malsegregation leading to micronucleus formation (Rajagopalan et al. 2004; Rao et al. 2005).

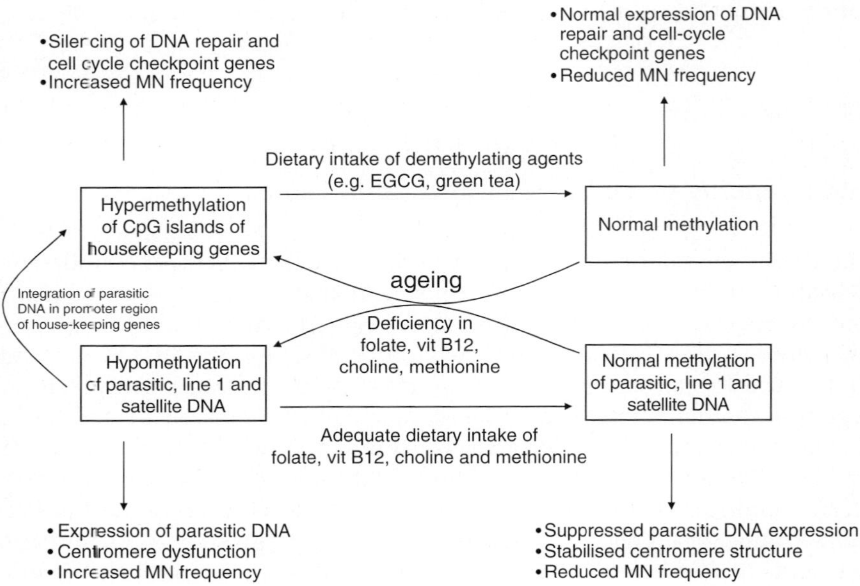

Fig. 15.7 The possible effects of dietary factors on methylation of CpG islands, parasitic DNA, gene expression, centromere function and micronucleus (*MN*) formation. *EGCG* epigallocatechin gallate. For further explanation refer to Fenech (2001, 2003) and Kimura et al. (2004)

15.7 Conclusion

In conclusion, the CBMN method has evolved into an efficient "cytome" assay of chromosomal instability, cell death and cytostasis enabling direct and/or indirect measurement of various aspects of cellular and nuclear dysfunction such as:

- Chromosomal instability phenotype
- BFB cycles
- DNA misrepair, chromosome breaks and asymmetrical rearrangement
- Telomere end-fusions
- Cell cycle checkpoint malfunction
- Malsegregation of chromosomes (when used in combination with pancentromeric or chromosome-specific probes)
- Gene amplification and nuclear elimination of excess DNA
- DNA hypomethylation (if combined with centromeric probes for chromosomes 1, 9 and 16)
- Altered mitotic activity and/or cytostasis
- Cell death by necrosis or apoptosis

The Micronucleus Assay Working Group of the International Workshop on Genotoxicity Test Procedures concluded that demonstration of cell proliferation in both control and treated cells is compulsory for the acceptance of the test because micronucleus expression is compromised when the proportion of dividing cells is reduced by cytostatic or poor tissue culture conditions (Kirsch-Volders et al. 2000, 2003; Fenech 2000b). It is evident that the best way to eliminate any possible confounding effects of altered cell division kinetics is to use the cytokinesis-block method regardless of the cell type used by scoring micronuclei in binucleated cells, which in addition allows measurement of NPBs as well as measurement of nuclear buds as well as cytostatic and cytotoxic effects. The choice not to use cytochalasin-B is only justifiable in those rare situations when cytokinesis defects are being specifically investigated.

The CBMN assay has already been proven to be an effective tool for the study of cellular and nuclear dysfunction caused by in vitro or in vivo ageing, micronutrient deficiency or excess, genotoxin exposure and genetic defects in genome maintenance. More recent studies indicate that this method is likely to also prove fruitful in the emerging fields of nutrigenomics and toxicogenomics and their combinations as it becomes increasingly clear that nutrient status also impacts on sensitivity to exogenous genotoxins (Beetstra et al. 2005). The ultimate test of the CBMN cytome assay is its ability to predict disease outcomes prospectively. These anticipated results, which were presented at the EEMS 2005 Kos Meeting and the HUMN Workshop at the 9th ICEM in San Francisco in September 2005, are indicative that the micronucleus index in lymphocytes is indeed predictive of cancer risk. Whether the use of the additional biomarkers in the CBMN assay (e.g. NPBs, nuclear buds) improves the predictive capacity of the assay remains to be determined, but a study by El-Zein et al. (2006) shows that the micronucleus, NPB and nuclear bud indices are sensitive indicators of lung cancer risk in smokers. As continuous variables, these measures were associated with 2-, 29- and 6-fold increases in the risk for spontaneous levels of the micronucleus, NPB and nuclear bud biomarkers and 2.3-, 45.5- and 10-fold increases in cancer risk for NNK-induced micronuclei, NPBs and nuclear buds, respectively [NNK is nicotine-derived nitrosamine 4-(methylnitrosamino)-1-(3-pyridyl)-1-butanone, a potent inducer of lung adenocarcinoma] (El-Zein et al. 2006).

References

Beetstra S, Thomas P, Salisbury C, Turner J, Fenech M (2005) Folic acid deficiency increases chromosomal instability, chromosome 21 aneuploidy and sensitivity to radiation-induced micronuclei. Mutat Res 578:317–326

Beetstra S, Salisbury C, Turner J, Altree M, McKinnon R, Suthers G, Fenech M (2006) Lymphocytes of BRCA1 and BRCA2 germ-line mutation carriers, with or without breast cancer, are not abnormally sensitive to the chromosome damaging effect of moderate folate deficiency. Carcinogenesis 27:517–524

Biedler JL, Spengler BA (1976) Metaphase chromosome anomaly: association with drug resistance and cell-specific products. Science 191:185-187

Blount BC, Mack MM, Wehr CM, MacGregor JT, Hiatt RA, Wang G, Wickramasinghe SN, Everson RB, Ames BN (1997) Folic acid deficiency causes uracil misincorporation into human DNA and chromosome breakage: implications for cancer and neuronal damage. Proc Natl Acad Sci USA 94:3290–3295

Brison O (1993) Gene amplification and tumor progression. Biochim Biophys Acta 1155:25–41

Coquelle A, Pipiras E, Toledo F, Buttin G, Debattisse M (1997) Expression of fragile sites triggers intrachromosomal mammalian gene amplification and sets boundaries to early amplicons. Cell 89:215–225

Crott JW, Mashiyama ST, Ames BN, Fenech M (2001a) Folic acid deficiency increases chromosome breakage and rearrangement, gene amplification and DNA-uracil content in human lymphocytes in vitro: effect of MTHFR C677T polymorphism. Cancer Epidemiol Biomarkers Prev 10:1089–1096

Crott JW, Mashiyama ST, Ames BN, Fenech M (2001b) Methylenetetrahydrofolate reductase C677T polymorphism does not alter folic acid deficiency-induced uracil incorporation into primary human lymphocyte DNA in vitro. Carcinogenesis 22:1019–1025

El-Zein RA, Schabath MB, Etzel CJ, Lopez MS, Franklin JD, Spitz MR (2006) Cytokinesis-blocked micronucleus assay as a novel biomarker for lung cancer risk. Cancer Res 66:6449–6456

Fenech M (2000a) The in vitro micronucleus technique. Mutat Res 455:81–95

Fenech M (2000b) A mathematical model of the in vitro micronucleus assay predicts false negative results if micronuclei are not specifically scored in binucleated cells or in cells that have completed one nuclear division. Mutagenesis 15:329–336

Fenech M (2001) The role of folic acid and vitamin B12 in genomic stability of human cells. Mutat Res 475:57–68

Fenech M (2003) Nutritional treatment of genome instability: a paradigm shift in disease prevention and in the setting of recommended dietary allowances. Nutr Res Rev 16:109–122

Fenech M, Chang WP, Kirsch-Volders M, Holland N, Bonassi S, Zeiger E (2003) HUMN project: detailed description of the scoring criteria for the cytokinesis-block micronucleus assay using isolated human lymphocyte cultures. Mutat Res 534:65–75

Guttenbach M, Schmid M (1994) Exclusion of specific human chromosomes into micronuclei by 5-azacytidine treatment of lymphocyte ceultures. Exp Cell Res 211:127–132

Haaf T, Raderschall E, Reddy G, Ward DC, Radding CM, Golub EI (1999) Sequestration of mammalian Rad51-recombination protein into micronuclei. J Cell Biol 144:11–20

Hoffelder DR, Luo L, Burke NA, Watkins SC, Gollin SM, Saunders WS (2004) Resolution of anaphase bridges in cancer cells. Chromosoma 112:389–397

IAEA (1986) Biological dosimetry: chromosomal aberrations analysis for dose assessment. Technical reports series no 260. International Atomic Energy Agency, Vienna, pp 13–27

Kirsch-Volders M, Sofuni T, Aardema M, Albertini S, Eastmond D, Fenech M, Ishidate M Jr, Kirchner S, Lorge E, Morita T, Norppa H, Surralles, Vanhauwaert A, Wakata A (2003) Report from the in vitro micronucleus assay working group. Mutat Res 540:153–163

Kimura M, Umegaki K, Higuchi H, Thomas P, Fenech M (2004) MTHFR C677T polymorphism, folic acid and riboflavin are important determinants of genome stability in cultured human lymphocytes. J Nutr 134:48–56

Kirsch-Volders M, Sofuni T, Aardema M, Albertini S, Eastmond D, Fenech M, Ishidate M, Lorge E, Norppa H, Surrales J, von der Hude W, Wakata A (2000) Report from the In Vitro Micronucleus Assay Working Group. Environ Mol Mutagen 35:167–172

Ma C, Martin S, Trask B, Hamlin JL (1993) Sister chromatid fusion initiates amplification of the dihydrofolate reductase gene in Chinese hamster cells. Genes Dev 7:605–620

McClintock B (1942) The fusion of broken ends of chromosomes following nuclear fusion. Proc Natl Acad Sci USA 28:458–463

Melnyk S, Pogribna M, Miller BJ, Basnakian AG, Pogribny IP, James SJ (1999) Uracil misincorporation, DNA strand breaks, and gene amplification are associated with tumorigenic cell

transformation in folic acid deficient/repleted Chinese hamster ovary cells. Cancer Lett 146:35–44
Perry ME, Rolfe M, McIntyre P, Commane M, Stark GR (1992) Induction of gene amplification by 5-aza-2′-deoxycytidine. Mutat Res 276:189–197
Rajagopalan H, Jallepalli PV, Rago C, Velculescu VE, Kinzler KW, Vogelstein B, Lengauer C (2004) Inactivation of hCDC4 can cause chromosomal instability. Nature 428:77–81
Rao CV, Yang YM, Swamy MV, Liu T, Fang Y, Mahmood R, Jhanwar-Uniyal M, Dai W. Colonic tumorigenesis in BubR1+/–ApcMin/+ compound mutant mice is linked to premature separation of sister chromatids and enhanced genomic instability. Proc Natl Acad Sci USA 102:4365–4370
Rudolph KL, Millard M, Bosenberg MW, DePinho RA (2001) Telomere dysfunction and evolution of intestinal carcinoma in mice and humans. Nat Genet 28:155–159
Satoh T, Yamamoto K, Miura KF, Sofuni T (2004) Region-specific chromatin decondensation and micronucleus formation induced by 5-azacytidine in human TIG-7 cells. Cytogenet Genome Res 104:289–294
Serrano-Garcia L, Montero-Montaya R (2001) Micronuclei and chromatid buds are the result of related genotoxic events. Env Mol Mutagenesis 38:38–45
Shimizu N, Itoh N, Utiyama H, Wahl GM (1998) Selective entrapment of extrachromosomally amplified DNA by nuclear budding and micronucleation during S phase. J Cell Biol 140: 1307–1320
Shimizu N, Shimuara T, Tanaka T (2000) Selective elimination of acentric double minutes from cancer cells through the extrusion of MNi. Mutat Res 448:81–90
Shimizu N, Shingaki K, Kaneko-Sasaguri Y, Hashizume T, Kanda T (2005) When, where and how the bridge breaks: anaphase bridge breakage plays a crucial role in gene amplification and HSR generation. Exp Cell Res 302:233–243
Shimura M, Onozuka Y, Yamaguchi T, Hatake K, Takaku F, Ishizaka Y (1999) MNi formation with chromosome breaks and gene amplification caused by Vpr, an accessory gene of human immunodeficiency virus. Cancer Res 59:2259–2264
Stark GR (1993) Regulation and mechanisms of mammalian gene amplification. Adv Cancer Res 61:87–113
Suzuki T, Fujii M, Ayusawa D (2002) Demethylation of classical satellite 2 and 3 DNA with chromosomal instability in senescent human fibroblasts. Exp Gerontol 37:1005–1014
Toledo F, Le Roscouet D, Buttin G, Debatisse M (1992) Co-amplified markers alternate in megabase long inverted repeats and cluster independently in interphase nuclei at early steps of mammalian gene amplification. EMBO J 11:2665–2673
Umegaki K, Fenech M (2000) Cytokinesis-block micronucleus assay in WIL2-NS cells: a sensitive system to detect chromosomal damage induced by reactive oxygen species and activated human neutrophils. Mutagenesis 15:261–269
Wang X, Thomas P, Xue J, Fenech M (2004) Folate deficiency induces aneuploidy in human lymphocytes in vitro-evidence using cytokinesis-blocked cells and probes specific for chromosomes 17 and 21. Mutat Res 551:167–180
Windle BE, Wahl GM (1992) Molecular dissection of mammalian gene amplification: new mechanistic insights revealed by analyses of very early events. Mutat Res 276:199–224
Xu GL, Bestor TH, Bourc'his D, Hsieh CL, Tommerup N, Bugge M, Hulten M, Qu X, Russo JJ, Veigas-Peguinot E (1999) Chromosome instability and immunodeficiency syndrome caused by mutations in a DNA methyltransferase gene. Nature 402:187–191

16 In Vivo Rodent Micronucleus Assay

Makoto Hayashi

Abstract

Genotoxicity plays an important role for the safety evaluation of chemicals. Chromosomal aberration is one of two major end points of genotoxicity. The rodent haematopoietic cell micronucleus assay is most widely used as an in vivo test to evaluate structural and numerical chromosomal aberrations. The historical aspects of the development of the in vivo micronucleus test, the mechanism of micronucleus formation, and the characteristics of the test are reviewed. One of the limitations of the micronucleus assay is the use haematopoietic cells and it is desired to develop the genotoxic method targeted at cells other than haematopoietic cells. The micronucleus assay using cells other than bone marrow cells, for example using liver, skin, testis, and colon cells, has been developed and validated successfully. The validation studies have been organized as collaborative studies of the Mammalian Mutagenicity Study Group belonging to the Japanese Environmental Mutagen Society and have been published. The other recent development of the micronucleus test is automated evaluation systems. Image analysis and flow cytometry were introduced to the test system with great success. These automated methods are being validated for regulatory purposes. The in vivo multiple end point assay system, which is more efficient, is also discussed here. When we use transgenic mutagenicity model animals in the micronucleus assay, we can obtain information on both gene mutation and chromosomal aberrations concomitantly, and possibly on DNA damage when the micronucleus assay is combined with the comet assay. Such a multiple end point genotoxicity assay system should be highly appreciated from the viewpoint of animal welfare.

16.1 Introduction

Genotoxicity plays an important role for the safety evaluation of chemicals. It is well known that there are in vitro and in vivo assay systems for the evaluation of chemical genotoxicity on different end points. The bacterial gene mutation test and the chromosomal aberrations test using mammalian cultured cells

Günter Obe and Vijayalaxmi (Eds.)
Chromosomal Alterations: Methods, Results and Importance in Human Health

are representative examples. It can be said that the genotoxicology field has been developed on the basis of in vitro assay systems. It is, however, apparent that there are limitations of in vitro assay systems for chemical safety evaluation and risk assessment for human health, and in vivo assay systems are becoming more important from the viewpoint of weight of evidence and risk assessment. Several in vivo assay systems have been developed and used for different end points. Among these, the rodent micronucleus test using haematopoietic cells has been most widely and frequently used to detect induction of chromosomal aberration. The historical aspects of the development of the in vivo micronucleus test and the characteristics of the test will be reviewed.

In addition, the current development of the micronucleus test and future perspectives of the assay will be summarized and discussed. As previously mentioned, the in vivo rodent micronucleus test is widely used to assess the clastogenic and possibly the aneugenic potential of chemicals for regulatory purposes. Haematopoietic cells, however, have been mainly targeted to evaluate cytogenetic effects and immature bone marrow erythrocytes or reticulocytes in peripheral blood cells have been observed. Bone marrow cells are effectively exposed by the test chemical after treatment by intraperitoneal injection, gavage, and intravenous injection because the blood–bone marrow barrier is not effective. Some chemicals, however, are metabolized in liver to active metabolites; some of those metabolites have a very short lifespan and cannot reach bone marrow cells. This is one of the limitations of the micronucleus assay using haematopoietic cells. It is desired to develop the genotoxic method to target organs other than haematopoietic organs. Several test systems, e.g. transgenic animal model, comet assay, and also the micronucleus test, have been developed, but only a few systems have been validated to be used for assessment of chemical safety.

Micronucleus assay using cells other than bone marrow cells, e.g. using liver, skin, testis, and colon cells, have been developed and validated successfully. The validation studies have been organized as collaborative studies of the Mammalian Mutagenicity Study Group (MMS) belonging to the Japanese Environmental Mutagen Society (JEMS) and some of them have been published. The technical issues of the micronucleus assay using colon and skin were established and the methods were evaluated by some model colon and skin carcinogens, respectively. The methods could detect the model chemicals as micronucleus inducers. The assay systems using testis and liver cells have almost been established and validated. For the testis assay, we introduced double staining with 4′,6-diamidino-2-phenylindole and acridine orange. The staining made identification of spermatocyte from other cells at the different spermiogenesis stages easy. For the liver micronucleus assay, the partial hepatectomy method has been used, but we do not know the precise influence of hepatectomy. We developed the assay using young rat hepatocytes without partial hepatectomy and other pretreatment. We evaluated the system using some model liver carcinogens and the method could detect these chemicals as micronucleus inducers.

The other recent development of the micronucleus test is automated evaluation systems. Image analysis and flow cytometry were introduced to the test system with great success. Now these automated methods are being validated for regulatory purposes. The spleen function of many mammalian species, including rat and human, eliminates the majority of micronucleated erythrocytes from the peripheral blood and the peripheral blood of such species was thought not to be suitable for the micronucleus test. Because flow cytometry can analyse a large number of cells and detect small increases of micronucleated erythrocytes, the peripheral blood of rats and humans became usable in the micronucleus test.

Finally, I will propose the in vivo multiple end point assay system. When we use transgenic mutagenicity models in the micronucleus assay, we can obtain information on both gene mutation and chromosomal aberrations concomitantly, and possibly DNA damage by the comet assay. Such a multiple end point genotoxicity assay system should be highly appreciated.

16.2 Historical Overview

Since the nineteenth century, the micronucleus has been known as the Howell–Jolly body in peripheral blood of mammals. In the 1970s, the micronucleus assay was established as an alternative method to chromosomal aberration assessment by metaphase analysis. Now, the assay is most widely used as an in vivo assay and is a member of a standard core battery of genotoxicity assays. The major historical events in connection with the micronucleus are as follows:

Nineteenth century Howell, Jolly; small body in erythrocytes of malignant anaemia cat and rat.

1959 *Vicia faba* roots (fast neutrons and γ-rays); chromosomal aberration evaluation by the induction of micronuclei (Evans et al. 1959).

1970–1975 Chinese hamster, mouse (Trenimon and other strong alkylating agents); chromosomal aberration evaluation test system: nomenclature “Microkern test”; establishment of rodent micronucleus test (Boller und Schmid 1970; Heddle 1973; Schmid 1975).

1976 Primary culture of human peripheral lymphocytes; human material/in vitro (Countryman and Heddle 1976).

1979 Transplacental application; embryonic micronucleus assay (Cole et al. 1979; King and Wild 1979).

1980 Mouse peripheral blood micronucleus assay; spleen function of rat and other species (MacGregor et al. 1980).

1980 Aneugens can be detected by the micronucleus test by size difference/centromere staining (Yamamoto and Kikuchi 1980; Högstedt and Karlsson 1985; Degrassi and Tanzarella 1988; Thomson and Perry 1988; Miller et al. 1991).

1980 Guidelines included the micronucleus test as a member of the standard battery.
1981–1983 Male germ cell micronucleus test (Lähdetie and Parvinen 1981; Tates et al. 1983).
1982- Automation of micronucleus analysis by flow cytometry/image analysis (Hutter and Stöhr 1982; Hayashi et al. 1992a, b; Romagna and Staniforth 1989; Tates et al. 1990; Asano et al. 1998; Dertinger et al. 1996).
1983 First report of the Gene Tox Program; IPCS project on evaluation of short-term test (Heddle et al. 1983).
1983 Introduction of fluorescent staining; acridine orange/pyronin Y and Hoechst 33258 (Hayashi et al. 1983; MacGregor et al. 1983).
1984 JEMS·MMS started the Collaborative Study Group for the Micronucleus Test (CSGMT) (reviewed in Sutou 1996).
1985 Human lymphocytes and other nucleated cells; introduction of cytochalasin B (Fenech and Morley 1985).
1990 Acridine orange supravital staining method for mouse peripheral blood micronucleus assay (Hayashi et al. 1990).
1993 First IWGT in Melbourne, Australia (Hayashi et al. 1994).
1997 Revision of OECD test guideline 474 (OECD 1997).
1999 Second IWGT in Washington, DC, USA (Hayashi et al. 2000).
2005 Fourth IWGT in San Francisco, CA, USA (Hayashi et al. 2007).

16.3 The Rodent Micronucleus Assay for Evaluation of Chromosomal Aberration Induction of Agents

Evaluation of micronucleus frequency in rodents, the micronucleus assay, is the assay in a battery of genotoxicity tests which is primarily selected as an in vivo assay. It is, however, applicable to any experimental cell systems that undergo the cell cycle in vitro and in vivo, e.g. rodent haematopoietic cells, in vitro cultured mammalian cells, plants. The micronucleus assay is not only an alternative method to the chromosomal aberration assay analysing metaphase chromosomes but also has its own characteristics as an assay to evaluate genotoxicity of agents. The characteristics of the assay are summarized as follows:

- Any dividing cell population with any karyotype can be used.
- Accurate data can be obtained because of the simple end point.
- Wider window for detection can be expected.
- Spindle poisons can be also detected.
- Background levels of micronucleated erythrocytes are low and stable.
- No additional treatment, e.g. colcemid or 5-bromo-2′-deoxyuridine, is required.

But:

- Types of chromosomal aberration cannot be classified.
- Pseudomicronuclei are observed under some circumstances.

In this chapter, only the in vivo rodent erythrocyte micronucleus is focused upon. For the assay using other materials, the reader is referred to the references.

The mechanism of formation of micronuclei from structural chromosomal aberrations and the correlation between frequencies of micronucleated erythrocytes and cells with chromosomal aberrations have been reported (Hayashi et al. 1984). The acentric fragments of chromatid/chromosome resulting from chromosomal aberration form micronuclei efficiently. The flow of micronucleus formation and the relationship between chromosomal aberration and the micronucleus is shown schematically in Fig. 16.1. The *k* value in the figure shows the relationship between frequencies of micronucleated cells and cells with chromosomal aberrations. The micronuclei are also the result of the interference of the mitotic spindle and whole chromosomes remaining in the cytoplasm when new daughter nuclei were formed form the micronucleus beside the main (daughter) nuclei (Fig. 16.2).

16.3.1 Rodent Bone Marrow Micronucleus Assay

Femoral marrow cells are usually suspended in foetal bovine serum and smeared on clean glass slides. Then these cells are fixed, stained and the frequency of micronucleated polychromatic erythrocytes is evaluated. To improve the identification of young erythrocytes, acridine orange and Hoechst 33258 plus pyronin Y fluorescent staining methods have been introduced to identify RNA containing erythrocytes as young erythrocytes (Hayashi et al. 1983, MacGregor et al. 1983). The acridine orange method has been reevaluated for its usefulness in the micronucleus assay (Tinwell and Ashby 1989). The precise protocol of this method was defined by the guidelines (e.g. OECD 1997) after extensive discussion among experts at the IWGT meetings (Hayashi et al. 1994, 2000). These guidelines and references should be referred to for more detail methodological aspects.

16.3.2 Rodent Peripheral Blood Micronucleus Assay

The usefulness of mouse peripheral blood erythrocytes as experimental material for the micronucleus assay has been reported (MacGregor et al. 1980). Bone marrow cells, however, are still the major target cells for studying micronucleus induction by clastogens. If peripheral blood erythrocytes could be used instead of bone marrow cells without any reduction in the sensitivity

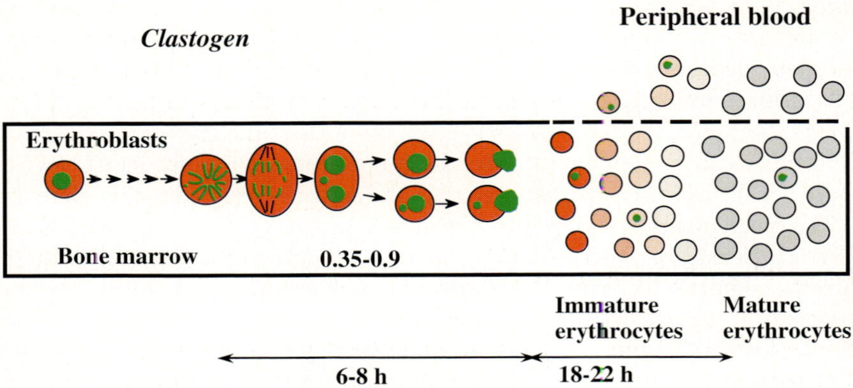

Fig. 16.1 Model of micronucleated erythrocyte formation in bone marrow by clastogens and the relationship between chromosomal aberration and the micronucleus. The *k* value shows the relationship between frequencies of micronucleated cells and cells with chromosomal aberrations

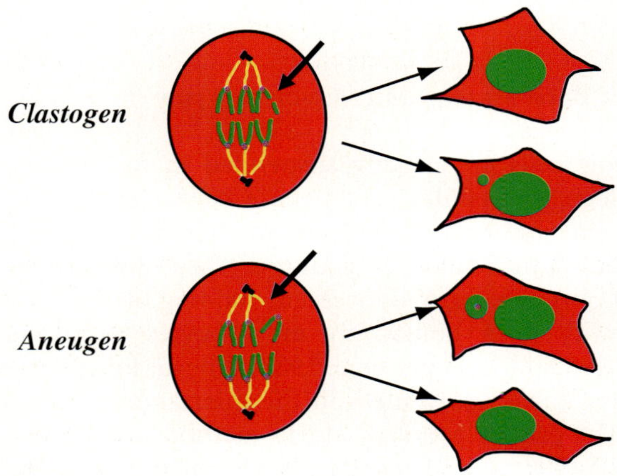

Fig. 16.2 Model of micronucleus induction by clastogens and aneugens. The micronuclei induced by aneugens are larger in size than those induced by clastogens and also have a centromere

of the test to identify the clastogenic effects of test chemicals, there should be several advantages. Firstly, the cell population is more uniform and, therefore, easier to score; therefore, there is a big possibility to automate the scoring process by image analysis and flow cytometry. Secondly, a small number of blood cells could repeatedly be taken from the tail of mice without killing the exposed animals; therefore, it would be possible to make preparations at the optimal sampling time after treatment(s) by periodic sampling from the same individual. Thus, no pilot experiments to establish optimal sampling times will be required. It is also possible to take a very small amount of blood from other toxicity experiments without interfering with the study.

16.4 The Micronucleus Assay Using Tissues Other Than Bone Marrow

As previously described, the micronucleus assay can apply any cell populations when they are dividing. Haematopoietic cells, i.e. bone marrow cells and peripheral blood erythrocytes, have been routinely used as target tissue/cells of this assay. Because the haematopoietic cell population in bone marrow is always dividing, researchers in the cytogenetic field preferred to use this cell population for their study historically. Recently, however, for assessment of human health the tendency is more for risk assessment than hazard identification. If this is the case, not only bone marrow and peripheral blood cells but also the site of contact of the tissues/cells and accumulation of the test chemical and also the target site of carcinogenicity should be evaluated to understand the mechanism of action.

16.4.1 Liver

It has been pointed out that several compounds which are metabolized to a biologically active form give a negative response in the bone marrow micronucleus assay (Morita et al. 1997). Some active metabolites have a very short lifespan and do not reach bone marrow at sufficient concentrations to induce micronuclei. In fact, some rodent liver carcinogens, including dialkyl nitrosamines, nitro aromatic compounds, and azo derivatives, gave negative results in a bone marrow assay (Angelosanto 1995). It is worthwhile therefore to consider the use of liver for detecting liver carcinogens.

Owing to low mitotic activity in adult animals, procedures to mitogenize liver cells were introduced. Tates et al. (1980) reported a method incorporating partial hepatectomy before or after chemical treatment in the liver micronucleus assay. The division of hepatocytes is stimulated by partial hepatectomy, and positive responses were observed in induction of micronuclei in liver cells after treatment with liver carcinogens that gave negative results in bone marrow. There are two shortcomings in the partial hepatectomy method: the first one is that it is not easy to perform hepatectomy on all the animals used in the assay, and the second is the activity level of enzymes such as cytochrome P450, styrene monooxygenase, epoxide hydrolase, and glutathione *S*-epoxide transferase 12 h after partial hepatectomy (Rossi et al.

1987). Another method used a treatment with mitogens for liver cells, e.g. 4-acetyl aminofluorene (Braithwaite and Ashby 1988).

Use of proliferating tissue is a prerequisite for the micronucleus assay and the hepatocytes of young rat (i.e. 4-week-old rats) are still proliferating, and the percentages of S-phase cells are more than 40 times greater than in adult rat liver (Parton and Garriott 1997). Sipes and Gandolfi (1993) reported that P450 levels reach a maximum for rats at 4 weeks of age. For example, diethylnitrosamine (DEN), which is a well-known rodent liver carcinogen, was used as a model compound. A negative response for DEN was obtained in the bone marrow micronucleus assay (Morita et al. 1997), but there was a positive response in the liver micronucleus assay using the partial hepatectomy method (Tates et al. 1980). The method using young rat liver showed an unambiguous positive response in induction of micronuclei.

16.4.2 Colon

For the colon micronucleus test, classic paraffin sections have been used to evaluate induction of micronuclei or nuclear anomalies in epithelial cells of intact colon (Wargovich et al. 1983; Migliore et al. 1989). To overcome the disadvantages of the section method, e.g. the improper position of sections which might miss micronuclei, steps were taken to develop a method using single cell suspensions of cells from mouse colon (Blakey et al. 1985; Ohyama and Tokumitsu 1996; Zhurkov et al. 1996). One of the methods developed by Tokumitsu et al. modified the method for the chromosomal aberration test (Tokumitsu 1976; Ohyama et al. 1991, 1992, 1993). This method was easier to perform and gave more accurate and reproducible data. The most probable reason was the analysis of intact cells using fluorescent staining (Hayashi et al. 1983). The method was evaluated by the collaborative study of JEMS/MMS CSGMT and showed good performance and interlaboratory reproducibility (Ohyama et al. 2002).

16.4.3 Skin

The skin is a major interface between the body and the environment; therefore, many mutagens and carcinogens are exposed directly. Accordingly, an in vivo skin micronucleus assay should be a useful tool for assessing genotoxic risk in certain regulatory situations, e.g. for the assessment of cosmetic ingredients. The assay using rodent skin has been evaluated and a validation study organized by CSGMT is being undertaken. The assay was characterized by four improvements: (1) cold-treating the epidermis with trypsin, (2) peeling it off with fine forceps, (3) treating it in hypotonic solution, and (4) staining it with acridine orange (Nishikawa et al. 1999, 2002, 2005). The data showed that skin epidermal basal cells (which are termed “basal + suprabasal” cells) could be used to evaluate genotoxicity following epidermal application of chemicals.

16.4.4 Testes

The assessment of the cytogenetic effect of agents is one of the major points for human health assessment in terms of genotoxicity. The use of a micronucleus assay in mammalian germ cells was proposed in the early 1980s by Lähdetie and Parvinen (1981) and by Tates et al. (1983). Lähdetie and Parvinen's method is referred to as the "dissection method". It takes advantage of the organization of seminiferous tubules, where different spermatogenic stages are specifically associated. On the other hand, the method proposed by Tates et al. (1983) was based on the preparation of a cell suspension from testicular tubules. The number of papers reporting data obtained by the spermatid micronucleus assay is still limited, compared with the large amount of information available from assays on somatic cells. A summary of the most recent data (from 1994) obtained with the spermatid micronucleus assay in rats and mice is presented in Table 2 of the IWGT report (Hayashi et al. 2000).

16.5 Automation of the Micronucleus Assay

The end point of the micronucleus assay is clearly defined and easily recognized; therefore, there is a possibility to easily automate it by image analysis and also by flow cytometry. Manual microscopic evaluation of micronuclei is still tedious and time-consuming and requires well-trained laboratory personnel, although it is faster and easier than metaphase analysis. To obtain more objective data quickly, automation techniques have been developed both with computerized image analysis (Tates et al. 1990; Romagna and Staniforth 1989; Parton et al. 1996; Asano et al. 1998) and with flow cytometry (Hutter and Stöehr 1982; Hayashi et al. 1992a, b; Krishna et al. 1998; Tometsko et al. 1993; Dertinger et al. 1996, 2003). Romagna and Staniforth (1989) emphasized the importance of careful pretreatment of samples when using bone marrow and May–Grünwalt Giemsa staining for the micronucleus assay. The use of peripheral blood and acridine orange supravital staining eliminates the need for pretreatment, but high-quality preparation is important for automated analysis. Adequate colour intensity and well-spread blood cells are necessary to distinguish the micronucleated reticulocytes accurately. The image-analysis system was applied to the peripheral blood micronucleus assay using the acridine orange coated glass slide method (Asano et al. 1998).

Unlike microscopy-based image analysis, flow cytometry analyses cells in aqueous suspension with high speed. The early systems were not capable of differentiating micronucleated immature erythrocytes accurately from micronucleated mature erythrocytes. A significant advance in this area came when Grawé et al. (1992, 1997a, b, 1998) described a flow cytometry method which was capable of scoring micronuclei in the immature erythrocyte population. Dertinger et al. (1996, 2003) developed a procedure which is based on RNase treatment in conjunction with propidium iodide and anti-CD71-fluorescein isothiocyanate staining. This protocol is capable of resolving all four erythrocyte populations of interest: mature and immature erythrocytes, with and without micronuclei. An advantage of flow cytometry scoring over manual scoring and image analysis is its extremely high analysis rates. Modern flow cytometers are able to evaluate tens of thousands of immature and hundreds of thousands of mature erythrocytes for micronuclei in several minutes. The increment of the number of cells analysed increases the sensitivity of the micronucleus assay from a statistical viewpoint (Abramsson-Zetterber et al. 1995; Abramsson-Zetterber 2003; Torous et al. 2006; Asano et al. 2006). The methods are almost ready for the assay to be used for regulatory purposes.

16.6 Multiple End Point Assays

Genotoxicity plays an important role for the safety evaluation of chemicals. When the carcinogenicity of a chemical evident, the threshold can be estimated only when the genotoxic mechanism does not operate for carcinogenesis, otherwise a threshold cannot be set. Without a genotoxic mechanism – non-genotoxic carcinogen – the threshold can be estimated but with the genotoxic mechanism – genotoxic carcinogen – it cannot be estimated. It is well known that there are in vitro and in vivo assay systems for the evaluation of chemical genotoxicity on different end points. The bacterial gene mutation test and the chromosomal aberrations test using mammalian cultured cells are representative examples. It can be said that the genotoxicology field has been developed on the basis of in vitro assay systems. It is, however, apparent that there are limitations of in vitro assay systems for chemical safety evaluation and risk assessment for human health and in vivo assay systems are becoming more important from the viewpoint of weight of evidence. Several in vivo assay systems have been developed and used for different end points. Among these, the rodent micronucleus test using haematopoietic cells has been most widely and frequently used to detect induction of chromosomal aberration. However, global assessment of genotoxic risk is desired for human health risk evaluation. For this purpose, in vivo assays came into the spotlight because of concern over weight of evidence and also mode of action. On the other hand, however, the animal welfare movement and the 3R (replace, reduce, refine) concern are affecting the increase of animal experimentation. The peripheral blood micronucleus assay does not require killing animals and can be integrated into other toxicological studies (Hayashi et al. 2000). If the transgenic animals are used for the general toxicological study, then we can incorporate the comet assay to assess DNA damage, gene mutation assessment, the micronucleus assay to evaluate cytogenetic effect together with the general toxicity assessment (Suzuki et al. 1997; Kohara et al. 2001,

2002a, b). Such a multiple end point assay system will be important and it can provide information of more global aspects, including genotoxicity. Moreover, the technique can be incorporated into the general toxicological study using rats and provide much information about the toxicity of the test chemical for global assessment (Hayashi et al. 2000; Wakata 1998; Hamada et al. 2001). The combination of the toxicological studies will reduce the need for experimental animals.

References

Abramsson-Zetterberg L (2003) The dose-response relationship at very low doses of acrylamide is linear in the flow cytometer-based mouse micronucleus assay. Mutat Res 535:215–222

Abramsson-Zetterberg L, Grawé J, Zetterberg G (1995) Flow cytometric analysis of micronucleus induction in mice by internal exposure to 137 Cs at very low dose rates. Int J Radiat Biol 67:29–36

Angelosanto FA (1995) Tissues other than bone marrow that can be used for cytogenetic analyses. Environ Mol Mutagen 25:338–343

Asano N, Katsuma Y, Tamura H, Higashikuni N, Hayashi M (1998) An automated new technique for scoring the rodent micronucleus assay: computerized image analysis of acridine orange supravitally stained peripheral blood cells. Mutat Res 404:149–154

Asano N, Torous DK, Tometsko CR, Dertinger SD, Morita T, Hayashi M (2006) Practical threshold for micronucleated reticulocyte induction observed for low doses of mitomycin C, Ara-C and colchicine. Mutagenesis 21:15–20

Blakey DH, Duncan AMV, Wargovich MJ, Goldverg MT, Bruce WR, Heddle JA (1985) Detection of nuclear anomalies in colonic epithelium of the mouse. Cancer Res 45:242–249

Boller K, Schmid W (1970) Chemische Mutagenese beim Sauger. Das Knochemark des Chinesischen Hasmters als in vivo-Testsystem. Haematologische Befunde nach Behandlung mit Trenimon. Humangenetik 11:35–54

Braithwaite I, Ashby J (1988) A non-invasive micronucleus assay in the rat liver. Mutat Res 203:23–32

Cole RJ, Taylor NA, Cole J, Arlette CF (1979) Transplacental effects of chemical mutagens detected by the micronucleus test. Nature 227:317–318

Countryman PI, Heddle JA (1976) The production of micronuclei from chromosome aberrations in irradiated cultures of human lymphocytes. Mutat Res 41:321–332

Degrassi F, Tanzarella C (1988) Immunofluorescent staining of kinetochores in micronuclei: a new assay for the detection of aneuploidy. Mutat Res 203:339–345

Dertinger SD, Chen Y, Miller RK, Brewer KJ, Smudzin T, Torous DK, Hall NE, Olvany KA, Murante FG, Tometsko CR (2003) Micronucleated CD71-positive reticulocytes: a blood-based endpoint of cytogenetic damage in humans. Mutat Res 542:77–87

Dertinger SD, Torous DK, Tometsko KR (1996) Simple and reliable enumeration of micronucleated reticulocytes with a single-laser flow cytometer. Mutat Res 371:283–292

Evans HJ, Neary GJ, Williamson FS (1959) The relative biological efficiency of single doses of fast neutrons and gamma-rays in *Vicia faba* roots and the effect of oxygen: Part II. Chromosome damage: the production of micronuclei. Int J Radiat Biol 3:216–229

Fenech M, Morley AA (1985) Measurement of micronuclei in lymphocytes. Mutat Res 147:35–54

Grawé J, Zetterberg G, Amneus H (1992) Flow-cytometric enumeration of micronucleated polychromatic erythrocytes in mouse peripheral blood. Cytometry 13:750–758

Grawé J, Adler I-D, Nüsse M (1997a) Quantitative and qualitative studies of micronucleus induction in mouse erythrocytes using flow cytometry. II. Analysis of micronuclei of aneugenic

and clastogenic origin by dual color FISH on populations of bone marrow PCEs flow sorted on the basis of their relative DNA content. Mutagenesis 12:9–15
Grawé J, Nüsse M, Adler I-D (1997b) Quantitative and qualitative studies of micronucleus induction in mouse erythrocytes using flow cytometry: I. Measurement of micronucleus induction in peripheral blood polychromatic erythrocytes by chemicals with known and suspected genotoxicity. Mutagenesis 12:1–8
Grawé J, Abramsson-Zetterberg L, Zetterberg G (1998) Low dose effects of chemicals as assessed by the flow cytometric in vivo micronucleus assay. Mutat Res 405:199–208
Hamada S, Sutou S, Morita T, Wakata A, Asanami S, Hosoya S, Ozawa S, Kondo K, Nakajima M, Shimada H, Osawa K, Kondo Y, Asano N, Sato S, Tamura H, Yajima N, Marshall R, Moore C, Blakey DH, Schechtman LM, Weaver JL, Torous DK, Proudlock R, Ito S, Namiki C, Hayashi M (2001) Evaluation of the rodent micronucleus assay by a 28-day-treatment protocol: summary of the 13th collaborative study by CSGMT/JEMS·MMS. Environ Mol Mutagen 37:93–110
Hayashi M, Sofuni T, Ishidate M Jr (1983) An application of acridine orange fluorescent staining to the micronucleus test. Mutat Res 120:241–247
Hayashi M, Sofuni T, Ishidate M Jr (1984) Kinetics of micronucleus formation in relation to chromosomal aberrations in mouse bone marrow. Mutat Res 127:129–137
Hayashi M, Morita T, Kodama Y, Sofuni T, Ishidate M Jr (1990) The micronucleus assay with mouse peripheral blood reticulocytes using acridine orange coated slides. Mutat Res 245:245–249
Hayashi M, Norppa H, Sofuni T, Ishidate M Jr (1992a) Mouse bone marrow micronucleus test using flow cytometry. Mutagenesis 7:251–256
Hayashi M, Norppa H, Sofuni T, Ishidate M Jr (1992b) Flow cytometric micronucleus test with mouse peripheral erythrocytes. Mutagenesis 7:257–264
Hayashi M, Tice RR, MacGregor JT, Anderson D, Blakey DH, Kirsch-Volders M, Oleson FB Jr, Pacchierotti F, Romagna F, Shimada H, Sutou S, Vannier B (1994) *In vivo* rodent erythrocyte micronucleus assay. Mutat Res 312:293–304
Hayashi M, MacGregor JT, Gatehouse DG, Adler I-D, Blakey DH, Dertinger SD, Krishna G, Morita T, Russo A, Sutou S (2000) *In vivo* rodent erythrocyte micronucleus assay. II. Some aspects of protocol design including repeated treatments, integration with toxicity testing, and automated scoring. Environ Mol Mutagen 35:234–252
Hayashi M, MacGregor JT, Gatehouse DG, Blakey DH, Dertinger SD, Abramsson-Zetterberg L, Krishna G, Morita T, Russo A, Asano N, Suzuki H, Ohyama W, Gibson D (2007) In vivo erythrocyte micronucleus assay: III. Validation and regulatory acceptance of automated scoring and the use of rat peripheral blood reticulocytes, with discussion of non-hematopoietic target cells and a single dose-level limit test. Mutat Res 627:10–30
Heddle JA (1973) A rapid in vivo test for chromosomal damage. Mutat Res 18:187–190
Heddle JA, Hite M, Kirkhart B, Mavourmin K, MacGregor JT, Newell GW, Salamone MF (1983) The induction of micronuclei as a measure of genotoxicity: A report of the U.S. Environmental Protection Agency Gene-Tox program. Mutat Res 123:527–549
Hutter K-J, Stöehr M (1982) Rapid detection of mutagen induced micronucleated erythrocytes by flow cytometry. Histochemistry 75:353–362
Högstedt B, Karlssson (1985) A The size of micronuclei in human lymphocytes varies according to inducing agent used. Mutat Res 156:229–232
King MT, Wild D (1979) Transplacental mutagenesis: The micronucleus test on fetal mouse blood. Hum Genet 51:183–194
Kohara A, Suzuki T, Honma M, Hirano N, Ohsawa K, Ohwada T, Hayashi M (2001) Mutation spectrum of o-aminoazotoluene in the cII gene of lambda/*lacZ* transgenic mice (Muta™ mouse). Mutat Res 491:211–220
Kohara A, Suzuki T, Honma M, Oomori T, Ohwada T, Hayashi M (2002a) Dinitropyrenes induce gene mutations in multiple organs of the lambda/*lacZ* transgenic mouse (Muta™ mouse). Mutat Res 515:73–83
Kohara A, Suzuki T, Honma M, Ohwada T, Hayashi M (2002b) Mutagenicity of aristolochic acid in the lambda/*lacZ* transgenic mouse (Muta™ mouse). Mutat Res 515:63–72

Krishna G, Urda G, Theiss J (1998) Principles and practices of integrating genotoxicity evaluation into routine toxicology studies: A pharmaceutical industry perspective. Environ Mol Mutagen 32:115–120

Lähdetie J, Parvinen M (1981) Meiotic micronuclei induced by X-ray in early spermatids of the rat. Mutat Res 81:103–115

MacGregor JT, Wehr CM, Gould DH (1980) Clastogen-induced micronuclei in peripheral blood erythrocytes: the basis of a improve micronucleus test. Environ Mutagen 2:509–514

MacGregor JT, Wehr CM, Langlois RG (1983) A simple fluorescent staining procedure for micronuclei and RNA in erythrocytes using Hoechst 33258 and pyronin Y. Mutat Res 120:269–275

Migliore L, Ventura L, Barale R, Loprieno N, Castellino S, Pulci R (1989) Micronuclei and nuclear anomalies induced in the gastro-intestinal epithelium of rat treated with formaldehyde. Mutagenesis 4:327–334

Miller BM, Zitzelsberger HF, Weier HU, Adler I-D (1991) Classification of micronuclei in mouse erythrocytes: immunofluorescent staining using CREST antibodies compared to in situ hybridization with biotinylated gamma satellite DNA. Mutagenesis 6:297–302

Morita T, Asano N, Awogi T, Sasaki YF, Sato S, Shimada H, Sutou S, Suzuki T, Wakata A, Sofuni T, Hayashi M (1997) Evaluation of the rodent micronucleus assay in the screening of IARC carcinogens (groups 1, 2A, and 2B) The summary report of the 6th collaborative study by CSGMT/JEMS·MMS. Mutat Res 389:3–122

Nishikawa T, Haresaku M, Adachi K, Masuda M, Hayashi M (1999) Study of a rat skin in vivo micronucleus test: data generated by mitomycin C and methyl methanesulfonate. Mutat Res 444:159–166

Nishikawa T, Hasekura M, Fukushima A, Nakamura T, Adachi K, Masuda M, Hayashi M (2002) Further evaluation of an in vivo micronucleus test on rat and mouse skin: results with five skin carcinogens. Mutat Res 513:93–102

Nishikawa T, Nakamura T, Fukushima A, Takagi Y (2005) Further evaluation of the skin micronucleus test: Results obtained using 10 polycyclic aromatic hydrocarbons. Mutat Res 588:58–63

OECD (1997) Guideline for the testing of chemicals. mammalian erythrocyte micronucleus test. Guideline 474, July 1997

Ohyama W, Tokumitsu T (1996) An in vivo micronucleus test using colonic epithelial cells of mice. Environ Mutat Res Commun 17:265–269

Ohyama W, Urakawa M, Tokumistu T (1991) Chromosome observations of mouse intestinal epithelial cells. Mutat Res 252:258

Ohyama W, Urakawa M, Tokumistu T (1992) Chromosome aberrations induced by MMC in the intestinal epithelial cells of mice. Mutat Res 272:278

Ohyama W, Urakawa M, Tokumistu T (1993) Chromosome aberrations in the intestinal epithelial cells of mice treated with 1,2-dimethylhydrazine. Mutat Res 292: 287

Ohyama W, Gonda M, Miyajima H, Kondo K, Noguchi T, Yoshida J, Hatkeyama S, Watabe E, Ueno Y, Hayashi M, Tokumitsu T (2002) Collaborative validation study of the in vivo micronucleus test using mouse colonic epithelial cells. Mutat Res 518:39–45

Parton JW, Garriott ML (1997) An evaluation of micronucleus induction in bone marrow and in hepatocytes isolated from collagenase perfused liver or from formalin-fixed liver using four-week-old rats treated with known clastogens. Environ Mol Mutagen 29:379–385

Parton JW, Hoffman WP, Garriott ML (1996) Validation of an automated image analysis micronucleus scoring system. Mutat Res 370:65–73

Romagna F, Staniforth CD (1989) The automated bone marrow micronucleus test. Mutat Res 213:91–104

Rossi AM, Romano M, Zaccaro L, Pulci R, Salmona M (1987) DNA synthesis, mitotic index, drug-metabolising systems and cytogenetic analysis in regenerating rat liver. Mutat Res 182:75–82

Schmid W (1975) The micronucleus test. Mutat Res 31:9–15

Sipes GI, Gandolfi AJ (1993) Biotransformation of toxicants. In: Amdur MO, Doull J, Klassen CD (eds) Casarett and Doull's toxicology, the basic science of poisons. McGraw-Hill, New York, pp 88–126

Sutou S (1996) Achievements by CSGMT/JEMS·MMS: the Collaborative Study Group for the Micronucleus Test in the Mammalian Mutagenesis Study Group of the Environmental Mutagen Society of Japan. Mutat Res 340:151–174

Suzuki T, Hayashi M, Wang X, Yamamoto K, Ono T, Myhr BC, Sofuni T (1997) A comparison of the genotoxicity of ethylnitrosourea and ethyl methanesulfonate in *lacZ* transgenic mice (Muta™ mouse). Mutat Res 395:75–82

Tates AD, Neuteboom I, Hofker M, den Englse L (1980) A micronucleus technique for detecting clastogenic effects of mutagens/carcinogens (DEN, DMN) in hepatocytes of rat liver in vivo. Mutat Res 74:1–20

Tates AD, Dietrich JJ, de Vogel N, Newteboom I, Bos A (1983) A micronucleus method for detection of meiotic micronuclei in male germ cells of mammals. Mutat Res 121:131–138

Tates AD, van Welie MT, Ploem JS (1990) The present state of the automated micronucleus test for lymphocytes. Int J Radiat Biol 58:813–825

Thomson EJ, Perry PE (1988) The identification of micronucleated chromosomes: a possible assay for aneuploidy. Mutagenesis 3:415–418

Tinwell H, Ashby J (1989) Comparison of acridine orange and Giemsa stains in several mouse bone marrow micronucleus assay – including a triple dose study. Mutagenesis 4:476–481

Tokumitsu T (1976) Chromosomal study on mouse intestinal mucosal cells, with a note on aberrations induced by a carcinogen, MNNG. Proc Jpn Acad 52:513–516

Tometsko AM, Torous DK, Dertinger SD (1993) Analysis of micronucleated cells by flow cytometry. 3. Advanced technology for detecting clastogenic activity. Mutat Res 292:145–153

Torous D, Asano N, Tometsuko C, Sugunan S, Dertinger S, Morita T, Hayashi M (2006) Performance of flow cytometric analysis for the micronucleus assay – a reconstruction model using serial dilutions of malaria-infected cells with normal mouse peripheral blood. Mutagenesis 21:11–13

Wakata A, Miyamae Y, Sato S, Suzuki T, Morita T, Asano N, Awogi T, Kondo K, Hayashi M (1998) Evaluation of the rat micronucleus test with bone marrow and peripheral blood: The summary of the 9th collaborative study by CSGMT/MMS·JEMS. Environ Mol Mutagen 32:84–100

Wargovich MJ, Goldberg MT, Newmark HL, Bruce WR (1983) Nuclear aberrations as a short-term test for genotoxicity to the colon: Evaluation of nineteen agents in mice. J Natl Cancer Inst 71:133–137

Yamamoto KI, Kikuchi Y (1980) A comparison of diameters of micronuclei induced by clastogens and spindle poisons. Mutat Res 71:127–131

Zhurkov VS, Sycheva LP, Salamatova O, Vyskubenko IF, Feldt EG, Sherenesheva NI (1996) Selective induction of micronuclei in the rat/mouse colon and liver by 1,2-dimethylhydrazine: a seven-tissue comparative study. Mutat Res 368:115–120

17 Sister-Chromatid Exchanges

Andrzej Wojcik and Günter Obe

Abstract

The sister chromatid exchange (SCE) test is a relatively simple assay, sensitive to most mutagenic compounds. Its major disadvantage is that in order to detect SCEs cells must be either treated with the label for one round of replication prior to treatment with the mutagen studied or treated simultaneously, but fixed and harvested following two rounds of replication. Thus, one can never be sure if the label (usually 5-bromo-2′-deoxyuridine, BrdU) used does not interact with the mutagen studied or if the scoring of the second posttreatment mitoses is not a source of bias in the estimation of the SCE frequency. A further problem is the necessity to allow the cells to divide twice before mitotic cells are harvested. Given the fact that cells in culture do not grow in a synchronized manner, perturbations of the cell cycle kinetics provide a potential source of error in estimating the true SCE frequency. The aim of this chapter is to discuss some problems of the SCE assay, especially those related to the use of BrdU.

17.1 Introduction

Sister-chromatid exchanges (SCEs) reflect an interchange mechanism between sister chromatids of mitotic chromosomes (Latt 1981). The visualization of SCEs in linear chromosomes requires that the sister chromatids are stained differentially (Fig. 17.1). The most widely applied method with which this can be accomplished involves the use of the halogenated thymidine analogue 5-bromo-2′-deoxyuridine (BrdU) (Latt 1973; Perry and Wolff 1974).

In ring chromosomes, SCEs can be detected without differential staining of the chromatids, as first described in 1938 by Barbara McClintock (1938) in *Zea mais*. She noticed that sister chromatids of ring-shaped chromosomes may recombine to form a single, double-sized ring. The only explanation for this was to assume that an interchange occurred between the chromatids of the original ring chromosome. In 1957, Taylor et al. (1957) used autoradiography to visualize the incorporation of tritiated thymidine into the DNA of plant cells. Cells were labelled for one cell cycle with tritiated thymidine, the

Günter Obe and Vijayalaxmi (Eds.)
Chromosomal Alterations: Methods, Results and Importance in Human Health

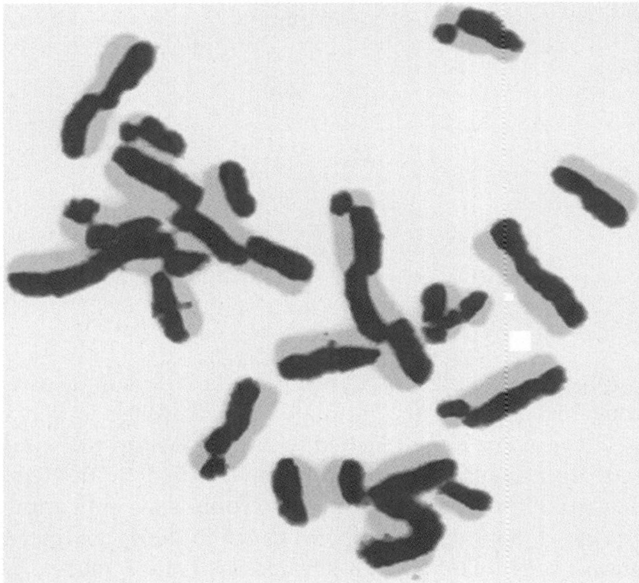

Fig. 17.1 A Chinese hamster ovary cell differentially labelled with 5-bromo-2′-deoxyuridine (BrdU) and stained by the fluorescence plus Giemsa technique with a fluorochrome-labelled monoclonal antibody

ensuing mitoses were suppressed with colchizine and during the next cell cycle cells were grown in the presence of thymidine. Following autoradiography, the authors analysed tetraploid metaphases in which the first cell cycle was exposed to tritiated thymidine and the second to thymidine. The chromatids of the chromosomes in tetraploid metaphases were tritiated thymidine–thymidine and thymidine–thymidine. In autoradiographs tritiated thymidine–thymidine but not thymidine–thymidine chromatids were covered with silver grains. Taylor et al. found SCEs between labelled and unlabelled chromatids. Some SCEs occurred at the same sites in two sister chromosomes of the tetraploid metaphases (twin SCEs) and in single chromosomes (single SCEs). One twin SCE corresponds to two single SCEs, indicating that twin SCEs result from replication of a SCE which occurred in S phase of the first cell cycle, and single SCEs occurred in S phase of the second cell cycle. Similar results were obtained with mammalian cells labelled with BrdU (Wolff and Perry 1975; Bruckmann et al. 1999b). The analysis of SCEs in autoradiographs is difficult because the differentiation of small SCEs is often impossible; therefore, the SCE test was not widely used. This situation changed after the development of the fluorescence plus Giemsa (FPG) method by Perry and Wolff (1974).

Since the introduction of FPG staining, numerous studies have provided evidence that SCEs are induced by a great majority of mutagens in cells that can incorporate labels such as BrdU for one replication cycle and then undergo a second replication, in which the presence of BrdU is optional (Latt 1981; Natarajan and Mullenders 1987). SCEs are quite easy to score and, thanks to the fact that their spontaneous number is around five per metaphase (Sutou 1981; Natarajan et al. 1986; Bender et al. 1992), only a low number of cells (usually between 30 and 50) need to be scored in order to detect a statistically significant enhancement of the SCE frequency. This makes the SCE system a favourite end point in mutagenesis studies.

However, in order to detect SCEs, cells must be either treated with BrdU for one round of replication prior to treatment with the mutagen studied or treated simultaneously, but fixed and harvested following two rounds of replication (Fig. 17.2). Thus, when compounds are tested for their mutagenic potential, one can never exclude the possibility that an enhanced SCE frequency is due to the interaction of the compound with BrdU. A further problem is the necessity to allow the cells to divide twice before mitotic cells are harvested. Given the fact that cells in culture do not grow in a synchronized manner, perturbations of the cell cycle kinetics provide a potential source of error in estimating the true SCE frequency.

The aim of this chapter is to discuss some problems of the SCE assay, especially those related to the use of BrdU.

17.2 The Influence of BrdU on the Spontaneous SCE Level

The mechanisms of differential staining by the BrdU–FPG method have been described in detail (Goto et al. 1978; Holmquist 1987). The impact of BrdU on the spontaneous level of SCE was originally studied by employing very low concentrations of BrdU. With the FPG technique, optimal quality of differential staining is achieved when the concentration of BrdU in the medium is about 20 μM (Goto et al. 1978). Much lower concentrations of BrdU can be applied when a monoclonal antibody against BrdU is used. Pinkel et al. (1985) showed for the first time that SCEs can be detected in Chinese hamster ovary (CHO) cells when the DNA is substituted with 5-bromouracil (BrU) at a level of 0.6%. These authors found that the baseline level of SCEs was about six per cell and that this frequency did not change up to a substitution level of 20%, but increased with higher substitution levels. Somewhat different results were reported by Natarajan et al. (1986), who incubated CHO cells in medium containing a mixture of 10 μM BrdU, 10 μM thymidine, 1 μM 5-fluorodeoxyuridine and 100 μM deoxycytidine. The frequency of SCEs obtained at the lowest level of BrdU substitution (1%) was about four per cell and increased steadily with increasing level of BrdU, reaching about seven

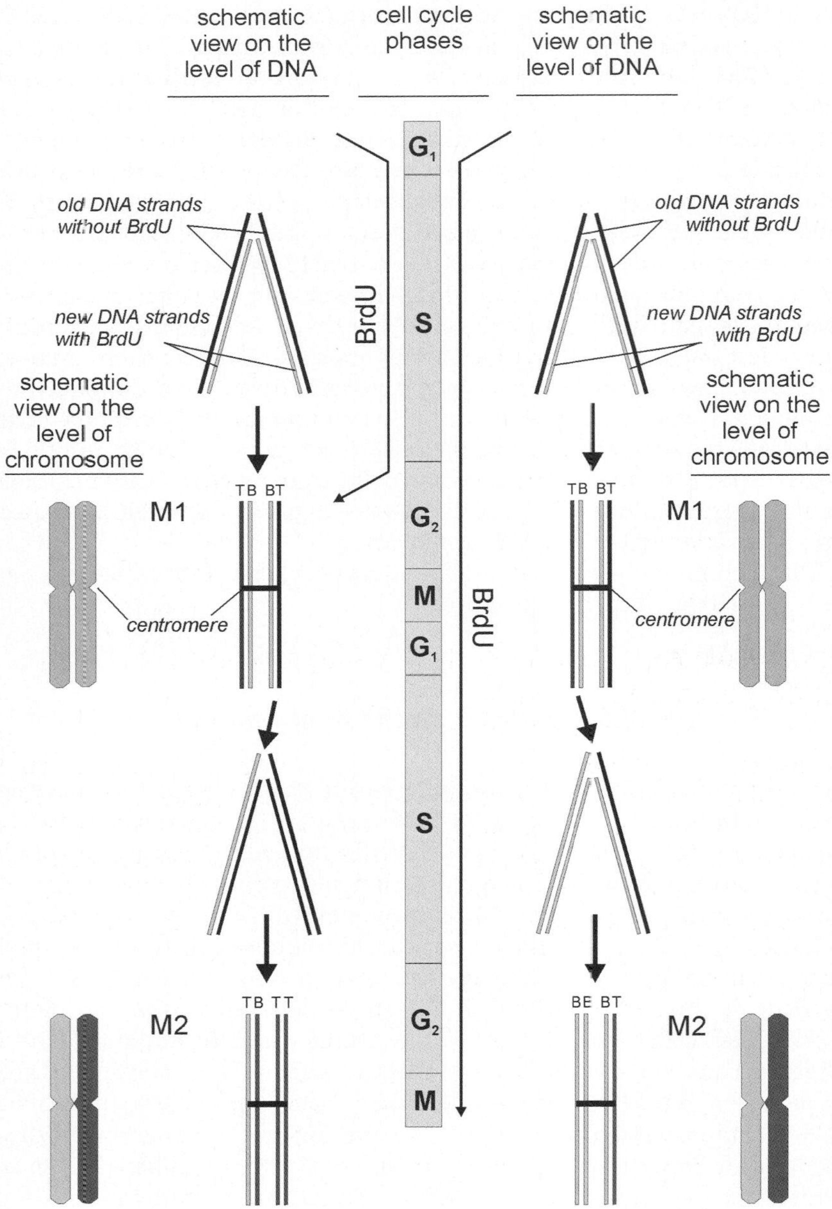

Fig. 17.2 Labelling scheme with BrdU of a cell which leads to differential staining of sister chromatids. *Left*: Labelling for one round of replication. *Right*: Labelling for two rounds of replication

per cell at a substitution level of 20%. A value of four SCEs was also calculated by Sutou (1981), who analysed ring chromosomes in CHO cells not substituted with BrdU. It has been suggested that the spontaneous value of about five SCEs per cell is an indirect manifestation of endogenous double-strand breaks (DSBs) in cellular DNA which are repaired by homologous recombination during replication (Vilenchik and Knudson 2003).

An alternative method of differential staining was developed by Bruckmann et al. (1999a, b). These authors electroporated CHO cells in the presence of biotin–2′-deoxyuridine-5′-triphosphate (dUTP), which was detected on microscope slides by immunological methods. The chemical structure of biotin-dUTP suggests that, unlike BrdU, it does not give rise to radicals. The SCE frequency of about six per cell was determined in all chromosomes, excluding X, Z10 and the small chromosomes Z12 and Z13, which were generally not labelled over their whole length. Since the omitted chromosomes comprise approximately 12% of the total chromosome length, the corrected, total genomic frequency of SCE was about seven. The higher level of SCE found with the biotin-dUTP method compared with the SCE level obtained with the lowest concentration of BrdU may indicate that biotin-dUTP induces a few additional SCEs, possibly by steric hindrance during DNA replication (Bruckmann et al. 1999a).

17.3 BrdU As a Source of DNA Damage Leading to SCE Formation

As discussed in the previous section a disadvantage of using BrdU for differential staining is that it induces SCEs by itself. The source of these SCEs is DNA damage, which arises owing to spontaneous or induced formation of uracilyl radicals and bromine atoms (Morris 1991; Fig. 17.3). One factor that causes the disintegration of BrdU is UV light (von Sonntag 2005). This is the reason why cell cultures should be kept in the dark following the addition of BrdU. An interesting question concerns the type of DNA damage which is responsible for SCEs in cells prelabelled with BrdU. The major photochemical

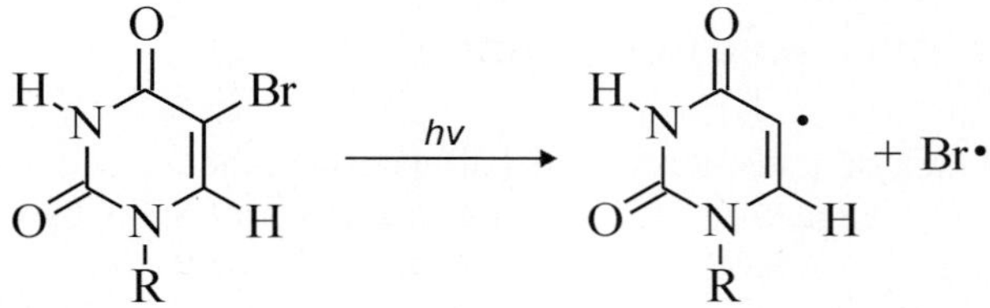

Fig. 17.3 Disintegration of 5-bromouracil leading to the formation of a uracilyl radical and a bromine atom

lesion in BrdU-substituted DNA is a single-strand break (SSB) (Hutchinson 1973); however, studies with the alkaline comet assay indicate that SSBs are repaired very quickly (Olive et al. 1990; Wojcik et al. 1996) and it was shown that UV radiation induces SCEs when cells prelabelled with BrdU are exposed in G1 phase of the cell cycle (Wojcik et al. 2003). Since SCEs are formed when DNA replication is stalled at a damaged template (Wolff et al. 1974; Painter 1980; Cortes et al. 1993; Wojcik et al. 2004), it appears unlikely that SSBs induced in G1 phase prevail in a non-repaired form until S phase, where they would give rise to SCEs. Furthermore, it has been shown that ionizing radiation, which gives rise to many SSBs, is a very poor inducer of SCEs (Littlefield et al. 1979). In fact it was demonstrated that following irradiation of CHO cells with X-rays, no SCEs are formed in cells not prelabelled with BrdU (Bruckmann et al. 1999b). This shows that it is the radiation damage to BrdU which is responsible for SCE formation. Considering the above, it appears unlikely that SSBs are the major lesions responsible for SCE formation.

Chemical agents that form DNA interstrand cross-links (ICLs) are among the most potent inducers of SCEs (Latt 1981). Wojcik et al. (2003) have shown that BrdU-induced damage is responsible for more than 80% of the SCEs formed in UV-irradiated cells unifilarily labelled with BrdU. Moreover, the authors pointed out that in DNA unifilarily labelled with BrdU and exposed to UV the bromine atom can, with a given likelihood, oxidize a purine base of the opposing DNA strand. This will yield two radicals at opposing strands (uracilyl radical and the oxidized purine) in the very neighbourhood, which is the precondition for an ICL (Wojcik et al. 2003; von Sonntag 2005). Indeed, the results of a recent study performed with the modified comet assay of Merk and Speit (1999) demonstrate that ICLs are formed in CHO cells unifilarily labelled with BrdU and exposed to UV radiation (Wojcik et al. 2006). This result was substantiated by the observation that CL-V4B cells exposed to BrdU and UV-C, but not to UV-C alone showed a very high level of chromosomal damage (Wojcik et al. 2006). CL-V4B cells have a deficient Rad51C paralogue that renders them extremely sensitive towards ICLs (Godthelp et al. 2002). Thus, it appears that ICLs induced by the disintegration of BrU are the source of SCEs.

17.4 The Influence of BrdU on the SCE Level Induced by Mutagenic Compounds

Owing to the lack of a straightforward alternative label to BrdU, only limited data exist on the impact of BrdU on the induction of SCEs by chemical and physical mutagens. Several authors reported that BrdU influences the SCE frequency induced by different mutagens (Popescu et al. 1980; Ockey 1981; Schvartzman and Tice 1982; Morgan and Wolff 1984; Zhao et al. 1992; Gonzalez-Beltran and Morales-Ramirez 2003). Given the fact that BrdU also inhibits the repair of DNA damage (Lawrence et al. 1995; Wang et al. 1994a, b) it may be difficult to judge whether its presence modifies the level of induced damage or the fidelity of repair.

From the considerations outlined in the previous section it can be assumed that BrdU will strongly enhance the action of agents which induce free-radical formation and promote the disintegration of BrU. As already mentioned, BrdU strongly enhances the SCE frequency induced by UV radiation (Wojcik et al. 2003). With respect to ionizing radiation, SCEs are only induced when cells prelabelled with BrdU are exposed (Bruckmann et al. 1999b). Stoilov et al. (2002) have applied the biotin-dUTP labelling technique to study the role of BrdU in the induction of SCEs by DNase I, Alu I and mitomycin C (MMC). CHO cells were treated in G1 phase of the cell cycle. Following treatment with Alu I, similar frequencies of SCEs were observed in cells prelabelled with BrdU and biotin-dUTP. DNase I induced significantly more SCEs in cells prelabelled with BrdU than with biotin-dUTP. MMC induced slightly more SCEs in cells labelled with biotin-dUTP, but the difference was not significant. The authors give two possible explanations for the observation that DNase I induced significantly more SCEs in cells prelabelled with BrdU than with biotin-dUTP. Firstly, it can be assumed that biotin-dUTP hinders the DNA incision activity of DNase I. The second explanation relies on the observations that DNase I induces SSBs which, when close together, lead to the formation of DSBs (Suck and Oefner 1986) and that the disintegration of BrU also may lead to the formation of SSBs (Morris 1991). If a SSB induced by DNase I lies in close proximity to a SSB induced by BrU, a DSB will be formed. In this sense, BrdU and DNase I would act synergistically. This assumes that DSB may be involved in SCE formation when formed in G1 phase of the cell cycle. The fact that restriction enzymes, neocarcinostatin and bleomycin effectively induce SCEs appears to support this (Natarajan et al. 1985; Obe et al. 1994; Schunck and Obe 1995). The mechanisms by which this occurs are not known, although newer data suggest that these SCEs are "false", in that they result from exchange-type chromosomal aberrations (Sayed Aly et al. 2002). The problem of "false" SCEs following treatment with S-phase-independent agents is discussed in the next section. In any case, no synergistic effect of BrdU would be expected with the DSB-inducing agent Alu I and the intercalating agent MMC, and this was confirmed experimentally (Stoilov et al. 2002).

17.5 The Problem of "False" and "True" SCE

Chemical mutagens, especially those capable of alkylating or cross-linking DNA, induce SCEs very effectively (Latt 1981; Perry and Thomson 1984). Such mutagens do not directly induce DSBs in the DNA and are described as "S-phase-dependent". The reason for this is that before DSBs are formed (which eventually leads to chromosomal aberrations), a cell must pass through an S phase. The chromosomal aberrations induced by S-phase-dependent agents are of chromatid-type (Savage 1975). In contrast, S-phase-independent agents are capable of directly inducing DSBs and they are a source of chromosome-type aberrations when cells are treated in G1 phase of the cell cycle. S-phase-independent agents are ionizing radiation, endonucleases such as DNase I, benzene nuclease, restriction endonucleases (RE) and a few antibiotics such as bleomycin or neocarcinostatin..

The observation that endonucleases induce SCEs was first reported by Natarajan et al. (1985) and Stoilov et al. (1986) for RE and Obe et al. (1994) for DNase I and benzene nuclease. Endonucleases are only effective in inducing SCEs when cells unifilarily labelled with BrdU are treated in G1 phase of the cell cycle (Morgan et al. 1989; Folle et al. 1992). The nature of DNA lesions responsible for SCE induction by endonucleases and other S-phase-independent agents has been a matter of debate because it is difficult to envisage how DSBs induced in G1 phase can lead to SCEs in S phase. Folle et al. (1992) proposed that apart from DSBs, RE could induce SSBs in canonical structures of chromosomal DNA or by cleavage of DNA–messenger RNA hybrids in G1 phase and that some of these SSBs could persist until S phase, where they would give rise to SCEs. If this were true, endonucleases should induce chromatid-type aberrations in cells treated in G1 phase; however, this is very rarely the case (Obe et al. 1987; Sayed Aly et al. 2002). In addition, Morgan et al. (1989) and Obe et al. (1994) have shown that when cells are treated with endonucleases in S phase, chromatid-type aberrations are induced but not SCEs.

The above considerations indicate that no other DNA lesions but the DSBs are the initial events leading to production of SCEs by endonucleases. Because an unrepaired DSB is a lethal event (Dikomey et al. 1998; Obe et al. 1992), cells possess efficient signalling and repair mechanisms preventing them from entering S phase before DSBs are repaired (Szumiel and Sochanowicz 1998; Pfeiffer et al. 2000; Obe et al. 2002). This clearly points towards G1 as the phase of the cell cycle in which SCE are produced by DSBs. However, how can DSBs lead to SCEs in G1 phase? A simple explanation is the assumption that the SCEs result from chromosomal exchange-type aberrations, and are therefore "false" SCEs (Fig. 17.4), as opposed to "true" SCEs, which are formed as a result of DNA recombination events during S phase. The existence of "false" SCEs was first postulated by Wolff et al. (1974) and Mühlmann-Diaz and Bedford (1995) for SCEs induced by X-rays. Bruckmann et al. (1999b) confirmed that only "false" SCEs are induced by X-rays when cells prelabelled with biotin-dUTP (but not BrdU: here true SCEs are formed owing to the disintegration of BrU – see Sect. 4) are exposed in G1 phase of the cell cycle. Sayed Aly et al. (2002) analysed the relationship between chromosomal aberrations and SCEs in CHO cells treated in G1 phase with DNase I and observed a very good, linear correlation between aberrations and SCEs. At the same time, no clear dose–response curve for SCEs was seen in cells without any chromosomal aberrations, suggesting that the SCEs induced by the endonuclease are all "false".

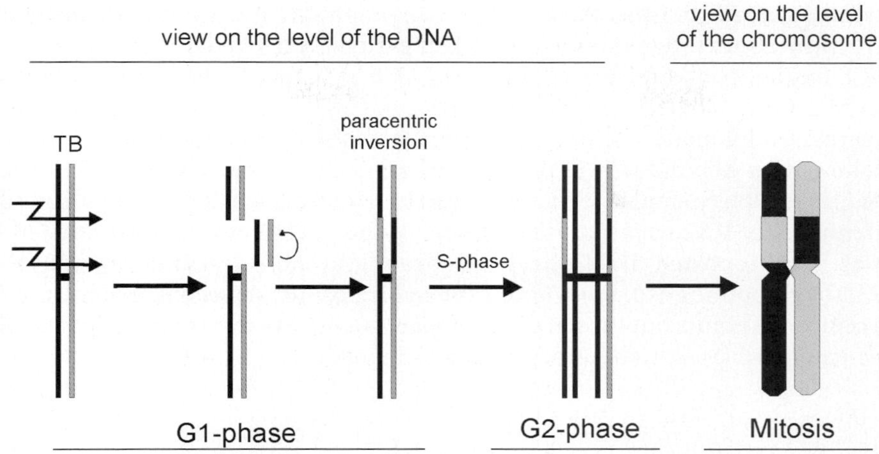

Fig. 17.4 How a paracentric inversion induced in G1 phase of the cell cycle can lead to the formation of a "false" sister-chromatid exchange

17.6 The Problem of Scoring SCEs in the Second Posttreatment Mitoses

SCEs can be visualized by growing cells for either two rounds of replication in the presence of BrdU or for one round with BrdU and the next without (Fig. 17.2). Both experimental schemes have advantages and disadvantages. If one wishes to analyse the first posttreatment mitoses, cells unifilarily labelled with BrdU must be treated. The disadvantage of this scheme is that, as discussed already, BrdU can influence the SCE-inducing capacity of the mutagen tested.

If BrdU is added simultaneously, or after treating cells with a DNA-damaging agent, the effect on SCEs can only be analysed in the second post-treatment mitoses. Two main problems are associated with this scheme. One problem is that most mutagens induce both SCEs and chromosomal aberrations and this may lead to mitotic death of cells when the mutagen is added together with BrdU (Carrano and Heddle 1973; Kirkland 1992). Thus, by scoring second posttreatment metaphases, one cannot be sure as to the extent of the elimination of damaged cells before they can be analysed for the presence of SCEs. This can be a source of bias in the estimation of the frequency of induced SCEs.

Another problem is related to the observation that DNA lesions induced by mutagens can persist for a number of cell cycles, giving rise to an enhanced SCE frequency during several posttreatment mitoses. This effect was studied extensively by Kaina and Aurich (1985), who pulse-treated V79 cells with different alkylating agents and scored SCEs in the first (M1), second (M2) and third (M3) posttreatment mitoses. The authors observed that following treatment with the methylating agents *N*-methyl-*N*-nitrosourea, *N*-methyl-*N'*-nitro-*N*-nitrosoguanidine and 1,3-dimethyl-3-phenyl-1-nitrosourea high SCE frequencies were detected in M2 and M3 at doses which induced no SCEs in M1. Ethyl methanesulphonate (EMS), *N*-ethyl-*N*-nitrosourea and methyl methanesulphonate (MMS) induced the highest SCE frequency in M2, followed by M3 and M1. MMC induced similar SCE frequencies in M1 and M2, a result very similar to that observed by Wojcik et al. (2005). An enhanced frequency of SCEs over two cell cycles following treatment with MMS and EMS was also reported by Gonzalez-Beltran and Morales-Ramirez (2003). Although nothing is known about the nature of the persistent lesions, their occurrence points out the necessity of using a uniform experimental protocol when planning interlaboratory comparison studies.

17.7 Conclusions

The SCE test is a relatively simple assay, sensitive to most mutagenic compounds. Its major disadvantage is that in order to detect SCEs, cells must be either treated with the label for one round of replication prior to treatment with the mutagen studied or treated simultaneously, but fixed and harvested following two rounds of replication. Thus, one can never be sure if the label used (usually BrdU) does not interact with the mutagen studied, or if the scoring of the second posttreatment mitoses is not a source of bias in the estimation of the SCE frequency. These factors must always be considered in the interpretation of the results obtained.

References

Bender MA, Preston RJ, Leonard RC, Pyatt BE, Gooch PC (1992) On the distribution of spontaneous SCE in human peripheral blood lymphocytes. Mutat Res 281:227–232

Bruckmann E, Wojcik A, Obe G (1999a) Sister chromatid differentiation with biotin-dUTP. Chromosome Res 7:185–189

Bruckmann E, Wojcik A, Obe G (1999b) X-irradiation of G1 CHO cells induces SCE which are both true and false in BrdU-substituted cells but only false in biotin-dUTP-substituted cells. Chromosome Res 7:277–288

Carrano AV, Heddle JA (1973) The fate of chromosome aberrations. J Theor Biol 38:289–304

Cortes F, Pinero J, Ortiz T (1993) Importance of replication fork progression for the induction of chromosome damage and SCE by inhibitors of topoisomerases. Chromosome Res 4:540–544

Dikomey E, Dahm-Daphi J, Brammer I, Martensen R, Kaina B (1998) Correlation between cellular radiosensitivity and non-repaired double-strand breaks studied in nine mammalian cell lines. Int J Radiat Biol 73:269–278

Folle GA, Johannes C, Mechoso BH, Medina II, Obe G (1992) The restriction endonuclease AluI induces sister chromatid exchange in Chinese hamster ovary cells. Mutagenesis 7:291–294

Godthelp BC, Wiegant WW, van Duijn-Goedhart A, Schaerer OD, van Buul PPW, Kanaar R, Zdzienicka MZ (2002) Mammalian Rad51C contributes to DNA cross-link resistance, sister chromatid cohesion and genomic stability. Nucleic Acids Res 30:2172–2182

Gonzalez-Beltran F, Morales-Ramirez P (2003) Repairability during G1 of lesions eliciting sister chromatid exchanges induced by methylmethanesulfonate or ethylmethanesulfonate in bromodeoxyuridine-substituted and unsubstituted DNA strands. Mutagenesis 18:13–17

Goto K, Maeda S, Kano Y, Sugiyama T (1978) Factors involved in differential giemsa-staining of sister chromatids. Chromosoma 66:351–359

Holmquist GE (1987) The magic of cytogenetic staining. In: Obe G, Basler A (eds) Cytogenetics. Springer, Berlin, pp 30–47

Hutchinson F (1973) The lesions produced by ultraviolet light in DNA containing 5-bromouracil. Q Rev Biophys 6:201–246

Kaina B, Aurich O (1985) Dependency of the yield of sister-chromaid exchanges induced by alylating agents on fixation time. Possible involvement of secondary lesions in sister-chromatid exchange induction. Mutat Res 149:451–461

Kirkland D. (1992) Chromosomal aberration tests in vitro: problems with protocol design and interpretation of results. Mutagenesis 7:95–106

Latt S (1973) Microfluorometric detection of deoxyribonucleic acid replication in human metaphase chromosomes. Proc Natl Acad Sci USA 70:3395–3399

Latt S (1981) Sister chromatid exchange formation. Annu Rev Genet 15:11–55

Lawrence TS, Davis MA, Normolle DP (1995) Effect of bromodeoxyuridine on radiation-induced DNA damage and repair based on DNA fragment size using pulsed-field gel electrophoresis. Radiat Res 144:282–287

Littlefield LG, Colyer SP, Joiner EJ, DuFrain RJ (1979) Sister chromatid exchanges in human lymphocytes exposed to ionizing radiation during G_0. Radiat Res 78:514–521

McClintock B (1938) The production of homozygous deficient tissues with mutant characteristics by means of aberrant mitotic behaviour of ring-shaped chromosomes. Genetics 23:315–376

Merk O, Speit G (1999) Detection of crosslinks with the comet assay in relationship to genotoxicity and cytotoxicity. Environ Mol Mutagen 33:167–172

Morgan WF, Wolff S (1984) Effect of 5-bromodeoxyuridine substitution on sister chromatid exchange induction by chemicals. Chromosoma 89:285–289

Morgan WF, Chung HW, Phillips JW, Winegar RA (1989) Restriction endonucleases do not induce sister-chromatid exchanges in Chinese hamster ovary cells. Mutat Res 226:203–209

Morris SM (1991) The genetic toxicology of 5-bromodeoxyuridine in mammalian cells. Mutat Res 258:161–188

Mühlmann-Diaz MC, Bedford JS (1995) Comparison of gamma-induced chromosome ring and inversion frequencies. Radiat Res 143:175–180

Natarajan AT, Mullenders LHF (1987) Sister chromatid exchanges. In: Obe G, Basler A (eds) Cytogenetics. Springer, Berlin, pp 339–344

Natarajan AT, Mullenders LHF, Meijers M, Mukherjee M (1985) Induction of sister-chromatid exchanges by restriction endonucleases. Mutat Res 144:33–39

Natarajan AT, Rotteveel AHM, van Pieterson J, Schliermann MG (1986) Influence of incorporated 5-bromodeoxyuridine on the frequencies of spontaneous and induced sister-chromatid exchanges, detected by immunological methods. Mutat Res 163:51–55

Obe G, Vasudev V, Johannes C (1987) Chromosome aberrations induced by restriction endonucleases. In: Obe G, Basler A (eds) Cytogenetics. Springer, Berlin, pp 300–314

Obe G, Johannes C, Schulte-Frohlinde D (1992) DNA double-strand breaks induced by sparsely ionizing radiation and endonucleases as critical lesions for cell death, chromosomal aberrations, mutations and oncogenic transformation. Mutagenesis 7:3–12

Obe G, Schunck C, Johannes C (1994) Induction of sister-chromatid exchanges by ALU I, DNase I, benzon nuclease and bleomycin in Chinese hamster ovary (CHO) cells. Mutat Res 307:315–321

Obe G, Pfeiffer P, Savage JRK, Johannes C, Goedecke W, Jeppesen P, Natarajan AT, Martinez-Lopez W, Folle GA, Drets M (2002) Chromosomal aberrations: formation, identification and distribution. Mutat Res 504:17–36

Ockey CH (1981) Methyl methane-sulfonate (MMS) induced SCEs are reduced by the BrdU used to visualize them. Chromosoma 84:243–256

Olive PL, Banath JP, Durand RE (1990) Heterogeneity in radiation-induced DNA damage and repair in tumor and normal cells measured using the "comet" assay. Radiat Res 122:86–94

Painter RB (1980) A replication model for sister-chromatid exchange. Mutat Res 70:337–341

Perry PE, Thomson EJ (1984) The methodology of sister chromatid exchanges. In: Kilbey BJ, Legator M, Nichols W, Ramel C (eds), Handbook of mutagenicity test procedures. Elsevier, Amsterdam, pp 495–529

Perry P, Wolff S (1974) New Giemsa method for differential staining of sister chromatids. Nature 251:156–157

Pfeiffer P, Goedecke W, Obe G (2000) Mechanisms of DNA double-strand break repair and their potential to induce chromosomal aberrations. Mutagenesis 15:289–302

Pinkel D, Thompson LH, Gray JW, Vanderlaan M (1985) Measurement of sister chromatid exchanges at very low bromodeoxyuridine substitution levels using a monoclonal antibody in Chinese hamster ovary cells. Cancer Res 45:5795–5798

Popescu NC, Amnnbaugh SA, DiPaolo JA (1980) Reduced N-methyl-N-nitro-N-nitrosoguanidine sister chromatid exchange induction in Chinese hamster V79 cells pre-exposed to 5-bromodeoxyuridine. Chromosoma 76:329–338

Savage JRK (1975) Classification and relationship of induced chromosomal structural changes. J Med Genet 12:103–122

Sayed Aly M, Wojcik A, Schunck C, Obe G (2002) Correlation of chromosomal aberrations and sister chromatid exchanges in individual CHO cells pre-labelled with BrdU and treated with DNaseI or X-rays. Int J Radiat Biol 78:1037–1044

Schunck C, Obe G (1995) Neocarzinostatin induces chromosomal aberrations and sister chromatid exchanges in Chinese hamster ovary cells. Mutagenesis 10:37–42

Schvartzman JB, Tice RR (1982) 5-Bromodeoxyuridine and its role in the production of sister chromatid exchange. In: Sandberg AA (ed) Sister chromatid exchange. Liss, New York, pp 123–134

Stoilov L, Mullenders LHF, Natarajan AT (1986) Influence of bromodeoxyuridine substitution of thymidine on sister-chromatid exchanges and chromosomal aberrations induced by restriction endonucleases. Mutat Res 174:295–301

Stoilov L, Wojcik A, Giri A, Obe G (2002) SCE formation after exposure of CHO cells pre-labelled with BrdU or biotin-dUTP to various DNA-damaging agents. Mutagenesis 17:399–403

Suck D, Oefner C (1986) Structure of DNase I at 2.0 A resolution suggets a mechanism for binding to and cutting DNA. Nature 321:620–625

Sutou S (1981) Spontaneous sister-chromatid exchanges in Chinese hamster cells in vivo and in vitro. Mutat Res 82:331–341

Szumiel I, Sochanowicz B (1998) Signal transfer in the cellular response to ionizing radiation. Nukleonika 43:133–146

Taylor JH, Woods PS, Hughes ML (1957) The organisation and duplication of chromosomes as revealed by autoradiographic studies using tritium-labeled thymidine. Proc Natl Acad Sci USA 43:122–128

Vilenchik MM, Knudson AG (2003) Endogenous DNA double-strand breaks: production, fidelity of repair, and induction of cancer. Proc Natl Acad Sci USA 100:12871–12876

von Sonntag C (2005) Free-radical-induced DNA damage and its repair. A chemical perspective. Springer, Heidelberg

Wang Y, Pantelias GE, Iliakis G (1994a) Mechanism of radiosensitization by halogenated pyrimidines: the contribution of excess DNA and chromosome damage in BrdU radiosensitization may be minimal in plateau-phase cells. Int J Radiat Biol 66:133–142

Wang Y, Pantelias GE, Iliakis G (1994b) The effect of scavenging of eaq- by acetone on the survival of CHO cells prelabelled with BrdU and exposed to X-rays. Radiat Res 66:133–142

Wojcik A, Sauer K, Zölzer F, Bauch T, Müller W-U (1996) Analysis of DNA damage recovery processes in the adaptive response to ionizing radiation in human lymphocytes. Mutagenesis 11:291–297

Wojcik A, von Sonntag C, Obe G (2003) Application of the biotin-dUTP chromosome labelling technique to study the role of 5-bromo-2′-deoxyuridine in the formation of UV-induced sister chromatid exchanges in CHO cells. J Photochem Photobiol B 69:139–144

Wojcik A, Bruckmann E, Obe G (2004) Insights into the mechanisms of sister chromatid exchange formation. Cytogenet Genome Res 104:304–309

Wojcik A, Stoilov L, Szumiel I, Legerski RJ, Obe G (2005) Rad51C-deficient CL-V4B cells exhibit normal levels of mitomycin C-induced SCEs but reduced levels of UVC-induced SCEs. Biochem Biophys Res Commun 326:805–810

Wojcik A, Bochenek A, Lankoff A, Lisowska H, Padjas A, Szumiel I, von Sonntag C, Obe G (2006) DNA interstrand crosslinks are induced in cells prelabelled with 5-bromo-2-deoxyuridine and exposed to UVC radiation. J Photochem Photobiol B 84:15–20

Wolff S, Perry P (1975) Insights on chromosome structure from sister chromatid exchange ratios and the lack of both isolabelling and heterolabelling as determined by the FPG technique. Exp Cell Res 93:23–30

Wolff S, Bodycote J, Painter RB (1974) Sister chromatid exchanges induced in Chinese hamster cells by UV irradiation of different stages of the cell cycle: the necessity for cells to pass through S. Mutat Res 25:73–81

Zhao JH, Tohda H, Oikawa A (1992) Camptothecin-induced sister chromatid exchange is dependent of the presence of bromodeoxyuridine and the phase of the cell-cycle. Mutat Res 282:49–54

18 Fluorescence in Situ Hybridisation

ILSE CHUDOBA

Abstract

The method of fluorescence in situ hybridisation (FISH) is based on the fundamental property of DNA strands to anneal to one another in a complementary fashion. However, there are different classes of DNA probes available for FISH investigations, but not all of them are suitable for all targets. Locus-specific probes are derived from single-copy DNA and produce a specific signal. In contrast, painting probes are heterogeneous DNA libraries, composed of different DNA classes, namely single copy sequences as well as repetitive sequences. Furthermore, DNA probes constituted of repetitive sequences only target specific structures of the chromosomes, i.e. the centromeres and the telomeres. All these probes have in common that they are directly or indirectly labelled and after hybridisation are visualised under an epifluorescence microscope. Comparative genomic hybridisation (CGH), a method to determine genetic imbalances with a resolution between 2 and 10 Mb, uses normal metaphase spreads as the target and the fluorescently labelled test DNA is hybridised together with differently labelled normal reference DNA. The resolution of CGH has been improved tremendously by the introduction of array-CGH, where the target chromosomes are replaced by DNA from, for example, bacterial artificial chromosomes with a spacing of 1 Mb and beyond. FISH has developed into a valuable tool for genetic investigations in clinical as well as tumour cytogenetics for detection of microdeletions, amplifications, specific fusion genes and others.

18.1 Introduction

Fluorescence in situ hybridisation (FISH) was introduced in 1986 (Pinkel et al. 1986). In general, FISH is applied to slides prepared from standard cytogenetic cell suspensions, but it can be carried out on tissue sections, blood or bone marrow smears as well. The method of FISH is based on the fundamental property of DNA strands to anneal to one another in a complementary fashion (complementary base pairing). Owing to this basic feature of single-stranded DNA, a specific DNA probe will hybridise to its complementary target DNA

Günter Obe and Vijayalaxmi (Eds.)
Chromosomal Alterations: Methods, Results and Importance in Human Health

sequence, no matter whether it is present in a metaphase chromosome, an interphase nucleus or as naked filamentary DNA (fibre-FISH).

There are different classes of DNA probes available for FISH investigations, but not all of them are suitable for all targets. For example, a whole chromosome painting probe is constituted of a library of different DNA sequences from the same chromosome, but specific for a variety of different loci on the respective chromosome. Such a probe is informative on metaphase chromosomes, but will contribute only little to no information on a DNA fibre.

FISH is carried out for different purposes. By using locus-specific probes, the presence or absence of a specific DNA sequence can be determined, or specific DNA probes can be assigned to particular chromosomal sites. In clinical cytogenetics, FISH investigations provide clarification of aberrations that were observed using standard cytogenetic techniques (GTG-banding). Furthermore, FISH is able to detect cryptic or submicroscopic chromosomal aberrations (e.g. microdeletions).

18.2 The Different Classes of DNA Probes

The human DNA is composed of different DNA classes (Table 18.1), which are of variable abundance. The basic nature of these DNA classes is of importance for the technical aspect of FISH experiments. A single copy sequence is present only once in the human haploid genome and, thus, will produce a specific signal.

In contrast, painting probes are heterogeneous DNA libraries and include all DNA classes. In order to avoid cross-hybridisation with repetitive DNA

Table 18.1 The different DNA classes of the human genome

Class	Copy number	Distribution	Percentage of the human genome
Single copy sequence	1	Unique copies throughout the whole genome	50
Low copy number repeat	2–20	Mainly throughout the euchromatin	10
Moderately repeated DNA	50–5,000	Scattered throughout the entire genome	25
Highly repeated DNA	10,000–500,000	Clustered in heterochromatin, like satellite DNA but also interspersed throughout the genome	15

and to improve the specificity of hybridisation, a so-called preannealing step using Cot-1 DNA has to be performed (Cremer et al. 1990).

Cot-1 DNA represents highly repetitive sequences among the nuclear DNA. It is obtained from total genomic DNA by selecting for rapidly reassociating DNA sequences during renaturation of DNA. During the preannealing step, Cot-1 DNA blocks repetitive sequences in the probe and, thus, prevents hybridisation labelled probe sequences to the target DNA.

18.3 Chromosome Painting Probes

Chromosome painting probes are DNA libraries which recognise specifically the sequences spanning the length of a particular chromosome (Cremer et al. 1988). They can be generated either by flow sorting or by microdissection.

In the case of flow sorting, a dispersed suspension of chromosomes is stained with fluorochromes and passed through a flow cytometer. The sorting is carried out according to the fluorescence intensity of each chromosome, which depends on its DNA content and the AT-to-GC ratio.

Whole chromosomes can be isolated from metaphase spreads by using a technique which is similar to microdissection of chromosomes. This can be either laser-mediated (Monajembashi et al. 1986) or carried out using extended glass needles (Senger et al. 1990; Chudoba and Senger 2002). In both cases, the chromosomal DNA has to be amplified subsequently. A powerful method for representative amplification of small amounts of DNA of which the specific sequences are unknown is the degenerate oligonucleotide-primed polymerase chain reaction (PCR) method (Telenius et al. 1992). Here, the complete DNA is amplified in a PCR with a set of degenerate primers. Consequently, not only single copy sequences are present in the final product, but all kinds of repetitive DNA as well. Therefore, the preannealing with Cot-1 DNA is particularly important when hybridising chromosome painting probes.

Chromosome painting probes can be specific for either a complete chromosome or only parts of it, like arm-specific partial painting probes or even smaller parts such as region-specific or band-specific painting probes. Whole chromosome painting probes hybridise to both chromosome homologues in metaphases. If a translocation is present, the FISH signal of one chromosome is split and distributed among the two translocation partners (Fig. 18.1a). In clinical cytogenetics and tumour cytogenetics, whole chromosome painting probes are mainly used in order to confirm results obtained by other cytogenetic methods.

A more sophisticated use of whole chromosome painting probes was implemented by the methods of multicolour karyotyping: multiplex FISH (mFISH) (Speicher et al. 1996) and spectral karyotyping (Schröck et al. 1996). Both approaches use different image capturing and analysis strategies,

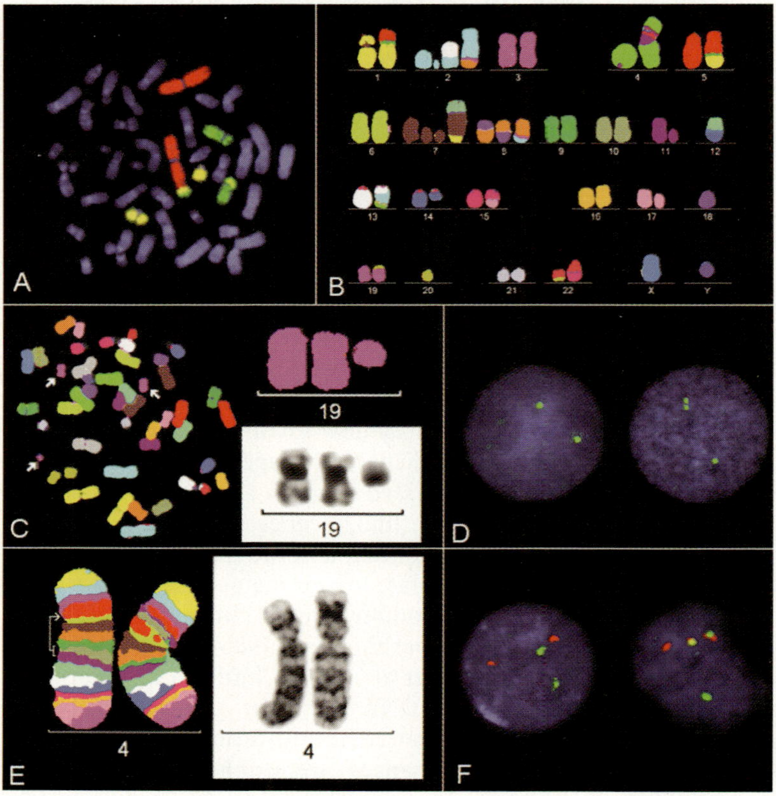

Fig. 18.1 **a** A metaphase after hybridisation with whole chromosome painting probes. Chromosome 1 is labelled in *red*, chromosome 12 in *green* and chromosome 16 in *red* and *green*, resulting in a *yellow* signal. A complex translocation occurred in which fragments of chromosomes 12 and 16 were transferred to chromosome 1 and a fragment from 16 was transferred to chromosome 12. **b** A karyogram in false colour display after multiplex fluorescence in situ hybridisation (FISH). Multiple chromosomal aberrations are observed. **c** Another example for multicolor FISH, where a supernumerary marker chromosome was identified as a derivative of chromosome 19. The metaphase spread and the respective chromosomes 19 in false colour (*upper row*) and inverted 4′,6-diamidino-2-phenylindole display (*lower row*) are shown. **d** Two normal interphase nuclei showing hybridisation signals of a locus-specific probe labelled with fluorescein isothiocyanate (*green spots*). Both nuclei show two spots; however, one of the signals in the nucleus on the *right* is split into two separate spots. **e** A chromosomal aberration of chromosome 4. The original interpretation after GTG-banding (*right*) explained the derivative chromosome 4 as a pericentric inversion. Multicolour banding analysis clearly revealed that a region from the long arm of chromosome 4, comprising the *grey band* and parts of the *purple band* (indicated by the *square brackets* on the chromosome normal 4, *left*), was excised and reinserted into the short arm of the same chromosome, which results in a splitting of the red colour band. **f** Two interphase nuclei after hybridisation with a *BCR/ABL* dual-colour dual-fusion probe. *Left*: A normal signal constellation. *Right*: A fusion between *BCR* and *ABL* is indicated by two fusion signals

but very similar probe kits. They are mixtures of whole chromosome painting probes for all chromosomes and each whole chromosome painting probe is labelled with a combination of different fluorochromes. For a human mFISH kit, five different fluorochromes are necessary to provide a minimum of 24 different combinations (Table 18.2). Multicolour karyotyping has been shown to be a valuable tool in the precise characterisation of complex chromosomal aberrations (Schoch et al. 2002, Fig. 18.1b, c).

Intrachromosomal aberrations cannot be detected using whole chromosome painting probes. For a more specific analysis of, for example, pericentric inversions, hybridisation with arm-specific probes is more suitable. The original

Table 18.2 Example of a labelling scheme for a probe kit for multicolour karyotyping. Here, the scheme for the probe kit 24 XCyte (MetaSystems, Germany) is shown

Chromosome number	DEAC	FITC	Spectrum Orange™	Texas Red®	Cy™5
1					X
2	X				
3				X	
4		X			
5			X		
6		X			X
7	X				X
8				X	X
9			X		X
10	X	X			
11		X		X	
12		X	X		
13	X			X	
14	X		X		
15			X	X	
16	X	X			X
17		X		X	X
18		X	X		X
19	X			X	X
20	X		X		X
21			X	X	X
22	X	X		X	
X	X	X	X		
Y	X		X	X	

DEAC (diethylamino)coumarin, *FITC* fluorescein isothiocyanate

pattern of a normal chromosome with its short arm labelled in red and long arm in green will results in an alternating pattern of red–green–red–green.

Microdissection of region-specific partial chromosome painting probes is the basis for a chromosome-specific multicolour banding, details of which were first published in 1999 (Chudoba et al. 1999). A specific feature of these probe sets is that each partial chromosome painting probe overlaps with its adjacent partial chromosome painting probe. They are labelled with different fluorochromes and are hybridised together. Owing to the overlapping of neighbouring probes, a multitude of different fluorescence intensity ratios are obtained. Fluorescence intensity ratios within defined regions are measured and a false colour is assigned to a certain ratio, which leads to a characteristic and reproducible colour banding pattern (Fig. 18.1e). Multicolour banding analyses are carried out in order to precisely define chromosomal aberrations in clinical (Lemke et al. 2001) and tumour (Lestou et al. 2003) cytogenetics. Furthermore, it was demonstrated that this method provides important information in experimental cytogenetics (Johannes et al. 2004).

18.4 FISH Probes Composed of Repetitive DNA Sequences

There are two different kinds of probes that hybridise to specific chromosomal structures featuring repetitive sequences, namely the centromeres and the telomeres of chromosomes. A preannealing step with Cot-1 DNA should be avoided, because this would suppress the sequences of interest.

18.4.1 Centromere-Specific Probes

Normal human centromeres consist of megabases of alpha or beta satellite DNA; a repeat family based on monomers that are approximately 171-bp long (Willard and Waye 1987). Centromere-specific FISH probes hybridise to these highly repetitive sequences.

Centromere probes are available for most human chromosomes; however, owing to the minimal differences in the centromere DNA sequences of chromosomes 5 and 19, 13 and 21 as well as 14 and 22, there cannot be distinguished by FISH. Usually, alpha/beta satellite centromere probes yield a large, bright signal which makes them useful for chromosome identification in metaphase preparations as well as chromosome enumeration in interphase nuclei.

Examples for hybridisations with centromere-specific probes are the detection of aneuploidies in interphase nuclei, the definition of the mosaic status of sex chromosome aneuploidies, the monitoring of bone marrow transplant in sex mismatch bone marrow transplant and the identification of the origin of marker chromosomes.

18.4.2 Telomere-Specific Probes

Telomeres are the tips of eukaryotic chromosomes. They have important functions in the protection, replication and stabilisation of chromosome ends. In most organisms studied, telomeres are composed of short repeated DNA sequences. These repeats are highly conserved; in fact all vertebrates appear to have the same simple sequence repeat in telomeres: $(TTAGGG)_n$ (Blackburn 1991).

The progressive shortening of telomeres is an important component of senescence and is involved in the control of the cell cycle (Dahse et al. 1997). In cancer cells, the telomere shortening is overcome, resulting in more or less unlimited cell divisions.

For quantification of telomeres, specific probes are available which comprise the aforementioned telomere-specific repeat. Advanced imaging systems are used in order to quantify the fluorescence signals of telomeres.

18.5 Locus-Specific Probes

Unique sequence probes hybridise to single copy DNA sequences in a specific chromosomal region or gene. This kind of probe is derived from cloned sequences. The different cloning vectors differ with respect to the size of the DNA fragment they can harbour, the cloning efficiency and further aspects such as stability and extractability of the DNA (Landegent et al. 1987; Sambrook et al. 2001).

The size and the intensity of fluorescence signals depend on the size of the DNA fragment that was cloned (Table 18.3). FISH probes based on yeast artificial chromosome result in signals that comparable to those of centromere signals in interphase nuclei regarding size and fluorescence intensity. On metaphase chromosomes the signals appear as separated spots on each chromatid. In the interphase state, one spot is usually observed for each locus. If the same locus is present on both chromosome homologues, two spots are observed in interphases. Deletions are indicated by a loss of a spot, and accordingly a gain of the same locus is indicated by an extra spot. Depending

Table 18.3 Types of vectors suitable for cloning

Vector		Maximum size of insert (kb)
YAC	Yeast artificial chromosome	1,000
BAC	Bacterial artificial chromosome	300
PAC	P1 (bacteriophage) derived artificial chromosome	100
Cosmid	A plasmid vector that carries COS sites	45

on the state of the cell cycle, even normal cells occasionally exhibit two separated spots that are derived from a normal locus. This indicates that the cell has undergone the S phase of the cell cycle and the respective region has been duplicated already (Fig. 18.1d).

18.6 Comparative Genomic Hybridisation and Array Comparative Genomic Hybridisation

The DNA probes described already are generally hybridised onto metaphase spreads or interphase nuclei prepared from a patient's blood or other tissues. However, in tumour cytogenetics, especially of solid tumours, the number of metaphases is often limited and the chromosomal morphology quite poor. Here comparative genomic hybridisation (CGH) can provide information about gains and losses of chromosomes and chromosomal regions (Kallioniemi et al. 1992). In contrast to standard cytogenetics techniques, dividing cells for metaphase preparation are not required. The DNA necessary for the experiment can be isolated from fresh or frozen tumour samples or even from archival formalin-embedded tissue samples (Speicher et al. 1993).

The basic prerequisites are the fluorescently labelled DNA from the tumour (test DNA) and a differently labelled normal DNA (control DNA). The two DNA fractions are mixed together in equal proportions and hybridised to metaphase spreads from normal tissue. Test and control DNA compete with each other for hybridisation sites on the target chromosomes. Adapted imaging systems compare the fluorescence intensities of the two different fluorochromes along the individual chromosomes. The resulting ratios are usually displayed as profiles and compared to the ISCN ideogram. Profiles exceeding defined limits in one or the other direction indicate gains and losses for the respective chromosomal regions of the tumour sample.

Even though the original idea of CGH was to gain more insight into tumours that are inaccessible with standard cytogenetic techniques, CGH proved to be a valuable tool as well in the determination of imbalanced constitutional chromosome aberrations (Tönnies et al. 2001). Not only the chromosomal origin of supernumerary marker chromosomes or de novo unbalanced translocations can be defined, but even the band-specific origin can be determined. The major limitation of CGH is the fact that only chromosomal imbalances can be detected and balanced aberrations remain undetected. Furthermore, the presence of normal cells mixed with the tumour tissue or the existence of a supernumerary marker as a mosaic reduce the effect and might even hide the actual presence of a chromosomal imbalance.

In experienced hands the resolution of CGH is rated to be between 3 and 10 Mb (Kirchhoff et al. 1999; Tönnies et al. 2001). Factors influencing the resolution are the quality of the hybridisation (signal background ratio), the quality of the DNA (homogeneity of the signals) and the condensation grade of the target chromosomes.

In order to improve the resolution of CGH recent developments resulted in the establishment of the so-called array-CGH or matrix-CGH (Solinas-Toldo et al. 1997; Pinkel et al. 1998). Metaphase spreads are replaced by specific DNA sequences, spotted in arrays on glass slides. The procedure is quite similar to the already described method of CGH: test DNA and control DNA are hybridised to the array. For genomic analysis bacterial artificial chromosome (BAC) arrays are used in which DNA from human BACs are spotted. The resolution depends on the number of BACs used and the spacing between neighbouring ones. A clone set has been established which comprises more than 2,500 BACs and which results in a resolution of approximately 1 Mb across the whole genome (Fiegler et al. 2003).

18.7 Probe Labelling

In order to make a probe suitable for FISH, the DNA has to be labelled. Labelling can be carried out by use of so-called reporter molecules (indirect labelling) or fluorescent dyes (direct labelling).

18.7.1 Indirect Labelling

In indirect labelling, a probe DNA is marked with reporter molecules. This is achieved by synthesising the complementary DNA strand in the presence of nucleotides conjugated with a reporter molecule. Common reporter molecules are biotin, digoxigenin, oestrogen and dinitrophenyl. The labelled nucleotides are introduced by nick-translation (Langer et al. 1981) or PCR. After hybridisation an indirectly labelled probe has to be detected either with specific antibodies carrying a fluorescent dye or, in the case of biotin, with fluorescently labelled avidin or streptavidin. These are proteins isolated from chicken egg white and have a particular specific affinity to biotin. Both primary and secondary antibodies have been used for the detection of indirectly labelled FISH probes. A primary antibody directly detects the incorporated reporter molecule. If a suitable antibody conjugate is not available, secondary antibodies can be used. In this case, the secondary antibody carries the fluorophore and recognises the unlabelled primary antibody. A major disadvantage of indirect labelling is the time-consuming detection after the hybridisation. However, a clear advantage is the signal amplification, which is especially favourable for relatively small probes (only several kilobases).

18.7.2 Direct Labelling

In the case of direct labelling, fluorescent molecules are directly attached to the DNA. This is the method of choice for most applications, because it is a straightforward and very convenient procedure. After the hybridisation only a stringent wash has to be carried out in order to eliminate the surplus of probe DNA.

18.8 The Hybridisation Procedure

A probe is hybridised to its target DNA under conditions that allow reannealing of specific complementary DNA strands. It is essential that the probe and target DNAs are denatured prior to hybridisation.

The hybridisation temperature is determined by the melting point of DNA. The melting point (T_m) of a given DNA is defined as the temperature where half of the DNA is present in a single-strand form (Sambrook et al. 2001). T_m depends on the GC content of the DNA: the higher the molar ratio of GC, the higher T_m. Sodium ions (monovalent cations) interact electrostatically with nucleic acids, i.e. salt concentrations below 0.4 M decrease the stability of the double-strand form. Divalent cations, on the other hand, strongly stabilise double-strand DNA, and they can be complexed by adding citrate to the hybridisation mixture.

DNA denatures at 90-100°C in presence of 0.1–0.2 M Na+. However, the morphology of target cells or chromosomes is better retained at lower temperatures; therefore, denaturation is usually carried out in the presence of the chaotropic compound formamide. T_m of double-strand DNA is reduced by 0.72°C for each percent of formamide; however, the rate of renaturation decreases in the presence of formamide (Casey and Davidson 1977).

Other factors influencing hybridisation are pH, the probe DNA fragment length and the probe concentration. The rate of hybridisation is relatively independent of the pH in the range between pH 5 and pH 9. However, some of the fluorochromes (e.g. fluorescein isothiocyanate) used for direct labelling are quite sensitive to pH changes and therefore a pH around 7 is recommended. The probe fragment length should be between 150 and 400 bp. The longer the fragments, the more difficult it is for them to penetrate cell plasma structures and the three-dimensional nuclear scaffold to reach the target DNA. Furthermore, long fragments (more than 800 bp) are difficult to dissolve and result in spotlike precipitations of labelled probe on the slide. These precipitates impair the interpretation of results and very often they exhibit high fluorescence intensities which hamper correct image acquisition. For the probe concentration the following rule applies: the higher the probe concentration, the higher the reannealing rate; however, the optimal probe concentration depends largely on the nature of the probe and varies between 2 and 100 ng of DNA per microlitre of hybridisation mix.

According to the factors influencing denaturation and hybridisation, some standard conditions have evolved. Denaturation of the target is usually carried out using 70% formamide in the presence of 2 × saline–sodium citrate (SCC) at pH 7–7.5. SSC is a widely used buffer for hybridisations. It is prepared as a 20 × concentrate of 3 M NaCl and 0.3 M sodium citrate. The pH is adjusted to 7.0. The denaturation temperature varies between 68 and 72°C. It is recommended to prepare the denaturation mixture fresh.

The probe is usually dissolved in the hybridisation mixture. The mixture contains formamide as well in a concentration of 50–70%. SSC is present in a 2 × concentration and a phosphate buffer is added to stabilise the pH. Dextran sulphate is added and acts as a hybridisation enhancer. In aqueous solutions, dextran sulphate is highly hydrated and thus reduces the availability of water for the DNA. This effect results in an apparent increase in probe concentration and consequently higher hybridisation rates.

18.9 Applications

There is a vast diversity of applications for FISH. In clinical cytogenetics, FISH with whole chromosome painting probes as well as locus-specific probes is used. Locus-specific probes are especially of interest in the diagnostic field of microdeletion syndromes (e.g. Smith–Magenis syndrome, DiGeorge syndrome). Here, deleted regions which are below the microscopic resolution of metaphase chromosomes are characterised by locus-specific probes. In general, a reference probe is included which hybridises to another locus or region of the same chromosome and which is labelled in a different colour. The examination is carried out on chromosomes and the reference probe helps to identify the relevant chromosome. Absence of the relevant hybridisation signal on one of the two homologue chromosomes indicates the occurrence of the respective microdeletion syndrome.

FISH on interphase nuclei plays an important role in prenatal diagnosis. Uncultured amniocytes are fixed and spread on microscope slides. FISH is applied using a combination of centromere-specific and locus-specific probes. The probes identify the chromosomes which are most frequently involved in numerical aberrations (13, 18, 21, X and Y).

In tumour cytogenetics, FISH has become an important tool in identifying specific tumour-related aberrations. The benefits of FISH are obvious: on the one hand, the quality of conventional banding preparations in tumour cytogenetics is often suboptimal and, on the other hand, complex chromosomal aberrations obscure important tumour-related alterations. The number of specific aberrations that have been identified has increased tremendously over the last few years and so has the number of specific FISH assays. Furthermore, there is a diversity of mechanisms of aberrations in tumour cytogenetics, such as deletions of tumour-suppressor genes which can be identified in the interphase by loss of a hybridisation signal. Amplification of tumour genes like *Her2/NEU* can be identified in interphases through quantification of the fluorescence signal.

Specific fusion genes like *BCR/ABL* in the Philadelphia chromosome, which is a specific aberrations in chronic myeloid leukaemia (Nowell and Hungerford 1960; Rowley 1973), can be detected by analysing the pattern of differently labelled probes that are specific for the relevant chromosomal loci (Tkachuk et al. 1990). For a translocation two chromosomal breaks must occur. In our example, one occurs in the *BCR* locus and the other in the *ABL* locus. A part of the *ABL* gene is translocated to a part of the *BCR* gene, creating a fusion gene and vice versa. A translocation is identified by the presence of a fusion signal (colocalisation of one of each of the differently labelled spots). However, with increasing knowledge about the underlying complex mechanism accompanying a *BCR/ABL* fusion more sophisticated FISH assays have evolved. The so-called dual colour dual fusion not only identifies the fusion gene by detecting the fusion signal, but also detects deletions eventually occurring at the translocation sites. The selected probes are designed in such a way that they span the breakpoint regions, leading to split signals when a translocation occurs. The two probes are labelled with different colours and a double fusion occurs if a translocation is present. At the same time the total number of spots for each colour increases from two in a normal cell to three in a cell with a simple translocation (Fig. 18.1f). Deletions may occur at either of the derivative chromosomes and can be identified by loss of one of the spots in cells with a translocation. Identification of additional deletions is of utmost importance because deletions on the derivative chromosome 9 characterise a group of patients with significantly higher risk of disease progression (Reid et al. 2003).

In radiation biology, FISH is widely applied for the assessment of chromosomal breaks. Hybridisations are carried out after in vitro exposure to different kinds of damaging radiation (Knehr et al. 1999; Durante et al. 2003; Romm and Stephan 2004). In general, whole chromosome painting probes are used to quantify the DNA damage by counting the occurrence of chromosome breaks. For these applications a representative mix of three whole chromosome painting probes is used and high numbers (several hundreds) of metaphase spreads have to be evaluated in order to establish dose–response curves.

These correlations provide the basis for the interpretation of results obtained from FISH analysis performed on individuals who were exposed to radiation (Edwards et al. 2005). Long-lived, sensitive and specific biomarkers of particular mutagenic agents are much sought after and potentially have broad applications in the fields of cancer biology, epidemiology and prevention (Hande et al. 2005). Therefore, numerous retrospective investigations have been performed on individuals who have been exposed to radiation owing to accidents like the nuclear blast at Chernobyl (Snigiryova et al. 1997; Hande et al. 2003).

18.10 Conclusion

Since its introduction in the 1980s, FISH has undergone a still continuing refinement of its methodology. Thanks to the specificity of the probes and its sensitivity, FISH has developed into a powerful tool for diagnostics and research with capabilities far beyond those of classical cytogenetics. In the early 1980s, cytogenetics could not arouse much excitement anymore and was regarded as an old-fashioned method. But since FISH expanded into standard cytogenetics, the gap between chromosomes and molecular genetics has been bridged.

References

Blackburn EH (1991) Structure and function of telomeres. Nature 350:569–573

Casey J, Davidson N (1977) Rates of formation and thermal stabilities of RNA:DNA and DNA:DNA duplexes at high concentrations of formamide. Nucleic Acids Res 4:1539–1552

Chudoba I, Senger G (2002) Microdissection of chromosomes and reverse FISH. In: Rautenstrauß BW, Liehr T (eds) FISH technology. Springer, Berlin, pp 388–407

Chudoba I, Plesch A, Lörch T, Lemke J, Claussen U, Senger G (1999) High resolution multicolor-banding: a new technique for refined FISH analysis of human chromosomes. Cytogenet Cell Genet 84:156–160

Cremer T, Lichter P, Borden J, Ward DC, Manuelidis L (1988) Detection of chromosome aberrations in metaphase and interphase tumor cells by in situ hybridization using chromosome-specific library probes. Hum Genet 80:235

Cremer T, Popp S, Emmerich P, Lichter P, Cremer C (1990) Rapid metaphase and interphase detection of radiation-induced chromosome aberrations in human lymphocytes by chromosomal suppression in situ hybridization. Cytometry 11:110–118

Dahse R, Fiedler W, Ernst G (1997) Telomeres and telomerase: biological and clinical importance. Clin Chem 43:708–714

Durante M, Snigiryova G, Akaeva E, Bogomazova A, Druzhinin S, Fedorenko B, Greco O, Novitskaya N, Rubanovich A, Shevchenko V, Von Recklinghausen U, Obe G (2003) Chromosome aberration dosimetry in cosmonauts after single or multiple space flights. Cytogenet Genome Res 103:40–46

Edwards AA, Lindholm C, Darroudi F, Stephan G, Romm H, Barquinero J, Barrios L, Caballin MR, Roy L, Whitehouse CA, Tawn EJ, Moquet J, Lloyd DC, Voisin P (2005) Review of translocations detected by FISH for retrospective biological dosimetry applications. Radiat Prot Dosimetry 113:396–402

Fiegler H, Carr P, Douglas EJ, Burford DC, Hunt S, Scott CE, Smith J, Vetrie D, Gorman P, Tomlinson IP, Carter NP (2003) DNA microarrays for comparative genomic hybridization based on DOP-PCR amplification of BAC and PAC clones. Genes Chromosomes Cancer 36:361–374

Hande MP, Azizova TV, Geard CR, Burak LE, Mitchell CR, Khokhryakov VF, Vasilenko EK, Brenner DJ (2003) Past exposure to densely ionizing radiation leaves a unique permanent signature in the genome. Am J Hum Genet 72:1162–1170

Hande MP, Azizova TV, Burak LE, Khokhryakov VF, Geard CR, Brenner DJ (2005) Complex chromosome aberrations persist in individuals many years after occupational exposure to densely ionizing radiation: an mFISH study. Genes Chromosomes Cancer 44:1–9

Johannes C, Horstmann M, Durante M, Chudoba I, Obe G (2004) Chromosome intrachanges and interchanges detected by multicolor banding in lymphocytes: searching for clastogen signatures in the human genome. Radiat Res 161:540–548

Kallioniemi A, Kallioniemi OP, Sudar D, Rutovitz D, Gray JW, Waldman F, Pinkel D (1992) Comparative genomic hybridization for molecular cytogenetic analysis of solid tumors. Science 258:818–821

Kirchhoff M, Gerdes T, Maahr J, Rose H, Bentz M, Dohner H, Lundsteen C (1999) Deletions below 10 megabasepairs are detected in comparative genomic hybridization by standard reference intervals. Genes Chromosomes Cancer 4:410–413

Knehr S, Huber R, Braselmann H, Schraube H, Bauchinger M (1999) Multicolour FISH painting for the analysis of chromosomal aberrations induced by 220 kV X-rays and fission neutrons. Int J Radiat Biol 75:407–418

Landegent JE, Jansen in de Wal N, Dirks RW, Baao F, van der Ploeg M (1987) Use of whole cosmid cloned genomic sequences for chromosomal localization by non-radioactive in situ hybridization. Hum Genet 77:366–370

Langer PR, Waldrop AA, Ward DC (1981) Enzymatic synthesis of biotin-labeled polynucleotides: novel nucleic acid affinity probes. Proc Natl Acad Sci USA 78:6633–6637

Lemke J, Chudoba I, Senger G, Stumm M, Loncarevic IF, Henry C, Zabel B, Claussen U (2001) Improved definition of chromosomal breakpoints using high-resolution multicolour banding. Hum Genet 108:478–483

Lestou VS, Ludkovski O, Connors JM, Gascoyne RD, Lam WL, Horsman DE (2003) Characterization of the recurrent translocation t(1;1)(p36.3;q21.1-2) in non-Hodgkin lymphoma by multicolor banding and fluorescence in situ hybridization analysis. Genes Chromosomes Cancer 36:375–381

Monajembashi S, Cremer C, Cremer T, Wolfrum J, Greulich KO (1986) Microdissection of human chromosomes by a laser microbeam. Exp Cell Res 167:262

Nowell PC, Hungerford DA (1960) A minute chromosome in human granulocyte leukemia. Science 132:1497

Pinkel D, Straume T, Gray JW (1986) Cytogenetic analysis using quantitative, high-sensitivity, fluorescence hybridization. Proc Natl Acad Sci USA 83:2934–2938

Pinkel D, Segraves R, Sudar D, Clark S, Poole I, Kowbel D, Collins C, Kuo WL, Chen C, Zhai Y, Dairkee SH, Ljung BM, Gray JW, Albertson DG (1998) High resolution analysis of DNA copy number variation using comparative genomic hybridization to microarrays. Nat Genet 20:207–211

Reid AG, Huntly BJ, Grace C, Green AR, Nacheva EP (2003) Survival implications of molecular heterogeneity in variant Philadelphia-positive chronic myeloid leukaemia.Br J Haematol 121:419–427

Romm H, Stephan G (2004) Dose dependency of FISH-detected translocations in stable and unstable cells after Cs gamma irradiation of human lymphocytes in vitro. Cytogenet Genome Res 104:162–167

Rowley JD (1973) A new consistent chromosomal abnormality in chronic myelogenous leukaemia identified by quinacrin fluorescence and Giemsa staining. Nature 243:290–293

Sambrook J, Fritsch EF, Maniatis T (2001) Molecular cloning: a laboratory manual. Cold Spring Harbor Laboratories, Cold Spring Harbor

Schoch C, Haferlach T, Bursch S, Gerstner D, Schnittger S, Dugas M, Kern W, Löffler H, Hiddemann W (2002) Loss of genetic material is more common than gain in acute myeloid leukemia with complex aberrant karyotype: a detailed analysis of 125 cases using conventional chromosome analysis and fluorescence in situ hybridization including 24-color FISH. Genes Chromosomes Cancer 35:20–29

Schröck E, du Manoir S, Veldman T, Schoell B, Wienberg J, Ferguson-Smith MA, Ning Y, Ledbetter DH, Bar-Am I, Soenksen D, Garini Y, Ried T (1996) Multicolor spectral karyotyping of human chromosomes. Science 273:494–497

Senger G, Lüdecke H-J, Horsthemke B, Claussen U (1990) Microdissection of banded human chromosomes. Hum Genet 84:507–511

Snigiryova G, Braselmann H, Salassidis K, Shevchenko V, Bauchinger M (1997) Retrospective biodosimetry of Chernobyl clean-up workers using chromosome painting and conventional chromosome analysis. Int J Radiat Biol 71:119–127

Solinas-Toldo S, Lampel S, Stilgenbauer S, Nickolenko J, Benner A, Dohner H, Cremer T, Lichter P (1997) Matrix-based comparative genomic hybridization: biochips to screen for genomic imbalances. Genes Chromosomes Cancer 20:399–407

Speicher MR, du Manoir S, Schrock E, Holtgreve-Grez H, Schoell B, Lengauer C, Cremer T, Ried T (1993) Molecular cytogenetic analysis of formalin-fixed, paraffin-embedded solid tumors by comparative genomic hybridization after universal DNA-amplification. Hum Mol Genet 11:1907–1914

Speicher MR, Gwyn Ballard S, Ward DC (1996) Karyotyping human chromosomes by combinatorial multi-fluor FISH. Nat Genet 12:368–375

Tkachuk DC, Westbrook CA, Andreeff M, Donlon TA, Cleary ML, Suryanarayan K, Homge M, Redner A, Gray J, Pinkel D (1990) Detection of bcr-abl fusion in chronic myelogeneous leukemia by in situ hybridization. Science 250:559–562

Telenius H, Carter NP, Bebb CE, Nordenskjold M, Ponder BA, Tunnacliffe A (1992) Degenerate oligonucleotide-primed PCR: General amplification of target DNA by a single degenerate primer. Genomics 13:718–725

Tönnies H, Stumm M, Wegner RD, Chudoba I, Kalscheuer V, Neitzel H (2001) Comparative genomic hybridization based strategy for the analysis of different chromosome imbalances detected in conventional cytogenetic diagnostics. Cytogenet Cell Genet 93:188–194

Willard HF, Waye JS (1987) Chromosome-specific subsets of human alpha satellite DNA: analysis of sequence divergence within and between chromosomal subsets and evidence for an ancestral pentameric repeat. J Mol Evol 25:207–214

19 The CAB and S&S Systems, and an Approach Towards the Classification of Complex Chromosome Exchanges

John R.K. Savage

Abstract

From the results of fluorescence in situ hybridization (FISH) "painting" of induced *chromosome-type* exchanges, it is now clear that very many of them arise from the interaction between three or more breaks in two or more chromosomes, i.e. they are actually "complex exchanges". This is not obvious from the usual solid-staining methods, so nomenclature, scoring systems and theory have developed on the assumption that two-break exchanges are the norm. Such systems are therefore proving inadequate for recording "painted" exchanges. For general purposes, complex exchanges can be classified using the CAB system. However, every CAB can have many break-end interactions and so each generates a family of different exchange patterns, all of which fall within the same designation. This renders such a method of classification rather crude. If, for a given multicolour FISH painted exchange, the single-paint pattern for each participating chromosome is classified according to the S&S system, the resulting string of pattern codes forms a useful recording/scoring tool (the S&S descriptor) which turns out to be much more specific than the CAB. Moreover, such descriptors usually contain enough information to make a diagrammatic reconstruction of the original overall exchange pattern. This chapter presents a table of all the possible CAB families up to CAB 6/6/6, and drawings of the complete S&S single-paint codes up to CAB 5/5/5. The formation of an S&S descriptor is illustrated, and some of its limitations are discussed.

19.1 Introduction

With the arrival of fluorescence in situ hybridization (FISH) "painting" of individual homologue pairs came the realization that many of the *chromosome-type* exchanges, normally assumed to be simple two-break events, were, in fact, the result of interactions between three or more breaks in two or more chromosomes – i.e. they were "complex exchanges" (Savage 2002). Such complex exchanges are well known for *chromatid-type* aberrations, but were generally believed to be rare for *chromosome-type* ones.

Günter Obe and Vijayalaxmi (Eds.)
Chromosomal Alterations: Methods, Results and Importance in Human Health

Initially, only a few chromosome-specific fluorochromes were available, so most of the observed anomalous patterns came from exchanges with just one participating chromosome painted, and this situation conditioned the early classification attempts.

Now that it is possible to "paint" all the homologue pairs distinctively (multicolour FISH, mFISH, and equivalent techniques; Speicher et al. 1996), the situation is exacerbated because it is clear than many chromosomes and breaks may be involved in a single exchange (Cornforth 2001).

We are now faced with the problem of the classification and recording of these new anomalous patterns, since the established scoring schemes, which are based upon solid-staining and a two-break origin inference (Savage 1976), are no longer valid. Many different combinations of *interchanges* and *intrachanges* can occur within a single complex exchange.

This chapter suggests one possible step towards the solution of this problem.

19.2 Configuration Versus Pattern

A very important principle has emerged from painting that has been largely overlooked in all our previous aberration studies.

Let us define the actual complete structural exchange, with all its related components as the *configuration*.

What we actually see down the microscope is a *pattern*, derived from this configuration. Now, how much of the configuration is revealed in the pattern depends upon the staining protocol we use. With the majority of protocols in regular use, we almost always see only part of the configuration. Since our qualitative and quantitative scoring can only be done from the pattern, our inferences must frequently be wrong. Obviously we have been underestimating the level of chromosome damage produced by aberration-inducing agents, and considerable revision of our past ideas is in order.

Only when every participating chromosome is distinctively coloured are we able to see (in most, but not all cases) the actual configuration and assess the nature of the exchange and the relationships of the participants.

Unfortunately, the commonest protocol, solid-staining, is the least efficient for *chromosome-type* exchanges. Much information is lost, and false identification is frequent. This principle is illustrated in Fig. 19.1.

Nevertheless, within a cell system, quantitative scores of solid-stained *chromosome-type* aberrations have proved to be consistent in relation to radiation dose, and have been found to be very reliable for techniques such as biological dosimetry. It is only when morphology and scores are applied to fundamental studies on the formation and origin of aberrations that the inefficiency assumes importance (Savage 1998). It must always be remembered that any treatment which influences the formation of complex exchanges will inevitably alter the recovered frequencies of solid-stained aberrations.

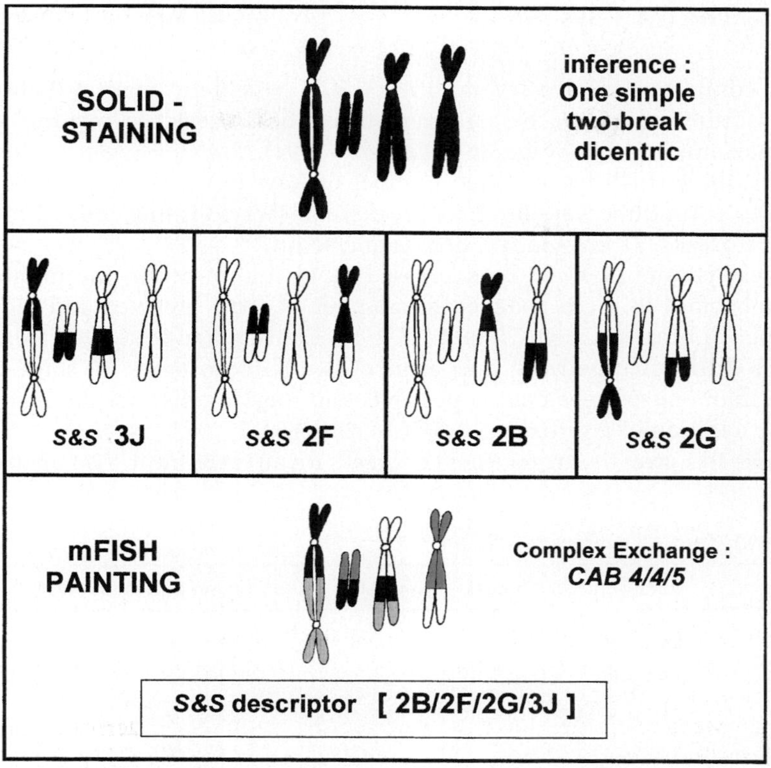

Fig. 19.1 *Top*: A collection of four chromosomes, solid-stained. One asymmetrical exchange (a "dicentric") is present and one infers that the accompanying acentric fragment is related to it, being formed by fusion of the two terminal deletions, i.e. a simple two-break interchange. Other symmetrical changes may be present, but these are not normally seen with solid-staining. *Middle*: Each of the four chromosomes can yield a single-paint S&S pattern, (painted segments shown *solid*), but in an experimental situation, only one will be seen, for each individual cell is unique. *Inference has to be drawn solely from the pattern you happen to get.* Pattern 1 (*3J*) has two parts expected for a simple dicentric, but has an additional insertion in another chromosome; thus, the dicentric cannot be simple but is a complex exchange involving a minimum of four breaks. Patterns 2 and 4 (*2F, 2G*) are anomalous, each requiring at least three breaks, and again confirm that the dicentric is not simple. Pattern 3 (*2B*) is interesting. It suggests a simple reciprocal translocation, but reference to the multicolour fluorescence in situ hybridization (mFISH) pattern shows the unpainted segments derive from thee other chromosomes! This is a classic example of a *pseudosimple* pattern. *Bottom*: mFISH staining shows the configuration generating the patterns of the other staining protocols. It indicates that all four chromosomes are involved in one complex exchange, something impossible to deduce from the other protocols. The $CAB_{observed}$ is 4/4/5. For practical purposes we have to assume that this is also the CAB_{actual} but there may be non-registering hidden changes. Taken together, the panels demonstrate the principle that most staining protocols only reveal part of the actual exchange configuration, and result in an underestimate of the actual damage present. The S&S descriptor is the string of the total S&S codes present in the observed exchange, and is discussed in the text

19.3 The CAB System

Since complexes involve several breaks, it is possible to make a theoretical classification based on the arrangement of these breaks in relation to the chromosome arms (Savage and Simpson 1994). For any given number of breaks, there is always a limited number of ways in which they can be allocated to arms: three ways for three breaks, eight ways for four breaks, 15 ways for five breaks, 31 ways for six breaks and so on.

These possibilities can be expressed by giving, in order, the number of chromosomes in the exchange, the number of arms involved and the total number of breaks (CAB). Thus, CAB 2/4/5 is an exchange derived from five breaks distributed between four arms of two chromosomes. In some cases, more than one arrangement is possible, and then families are distinguished by the addition of a suffix, e.g. 2/3/5(A), 2/3/5(B).

Table 19.1 gives the arrangement of breaks for all CABs from 2/2/3 up to 6/6/6.

Table 19.1 The CAB families classification

$CAB_{initial}$	Arrangement of breaks	$CAB_{initial}$	Arrangement of breaks
2/2/3	0◆2; 0◆1	2/2/6(A)	0◆3; 0◆3
2/3/3	1◆1; 0◆1	2/2/6(B)	0◆4; 0◆2
3/3/3	0◆1; 0◆1; 0◆1	2/2/6(C)	0◆5; 0◆1
		2/3/6(A)	2◆2; 0◆2
2/2/4(A)	0◆2; 0◆2	2/3/6(B)	1◆3; 0◆2
2/2/4(B)	0◆3; 0◆1	2/3/6(C)	2◆3; 0◆1
2/3/4(A)	1◆1; 0◆2	2/3/6(D)	1◆4; 0◆1
2/3/4(B)	1◆2; 0◆1	2/3/6(E)	1◆2; 0◆3
2/4/4	1◆1; 1◆1	2/3/6(F)	1◆1; 0◆4
3/3/4	0◆2; 0◆1; 0◆1	2/4/6(A)	1◆2; 1◆2
3/4/4	1◆1; 0◆1; 0◆1	2/4/6(B)	2◆2; 1◆1
4/4/4	0◆1; 0◆1; 0◆1; 0◆1	2/4/6(C)	1◆3; 1◆1
2/2/5(A)	0◆3; 0◆2	3/3/6(A)	0◆2; 0◆2; 0◆2
2/2/5(B)	0◆4; 0◆1	3/3/6(B)	0◆3; 0◆2; 0◆1
2/3/5(A)	1◆1; 0◆3	3/3/6(C)	0◆4; 0◆1; 0◆1
2/3/5(B)	1◆2; 0◆2	3/4/6(A)	1◆1; 0◆2; 0◆2
2/3/5(C)	2◆2; 0◆1	3/4/6(B)	2◆2; 0◆1; 0◆1
2/3/5(D)	1◆3; 0◆1	3/4/6(C)	1◆2; 0◆2; 0◆1
2/4/5	1◆2; 1◆1	3/4/6(D)	1◆3; 0◆1; 0◆1
		3/4/6(E)	1◆1; 0◆3; 0◆1
		3/5/6(A)	1◆1; 1◆1; 0◆2

(*Continued*)

Table 19.1 The CAB families classification—Cont'd

$CAB_{initial}$	Arrangement of breaks	$CAB_{initial}$	Arrangement of breaks
3/3/5(A)	0◆2; 0◆2; 0◆1	3/5/6(B)	1◆2; 1◆1; 0◆1
3/3/5(B)	0◆3; 0◆1; 0◆1	3/6/6	1◆1; 1◆1; 1◆1
3/4/5(A)	1◆1; 0◆2; 0◆1		
3/4/5(B)	1◆2; 0◆1; 0◆1	4/4/6(A)	0◆2; 0◆2; 0◆1; 0◆1
3/5/5	1◆1; 1◆1; 0◆1	4/4/6(B)	0◆3; 0◆1; 0◆1; 0◆1
		4/5/6(A)	1◆1; 0◆2; 0◆1; 0◆1
4/4/5	0◆2; 0◆1; 0◆1; 0◆1	4/5/6(B)	1◆2; 0◆1; 0◆1; 0◆1
4/5/5	1◆1; 0◆1; 0◆1; 0◆1	4/6/6	1◆1; 1◆1; 0◆1; 0◆1
5/5/5	0◆1; 0◆1; 0◆1; 0◆1; 0◆1	5/5/6	0◆2; 0◆1; 0◆1; 0◆1; 0◆1
		5/6/6	1◆1; 0◆1; 0◆1; 0◆1; 0◆1
		6/6/6	0◆1; 0◆1; 0◆1; 0◆1; 0◆1; 0◆1

◆=centromere

In practice, because of visual limitations, we have to recognize the existence of three distinct CABs:

1. $CAB_{initial}$ is the arrangement of breaks prior to the start of the exchange process. This is used for theoretical calculations, but will always remain hypothetical since some break-end junctions may not register visually.
2. CAB_{actual} is the arrangement of the breaks that do not restitute, but actually take part in illegitimate rejoining.
3. $CAB_{observed}$ is that deduced from the rejoin junctions (usually, but not necessarily bicoloured) observed in the microscopical exchange pattern.

For practical purposes, one has to assume that $CAB_{observed} = CAB_{actual}$, but same-colour hidden junctions (SCHJ) frequently mean that the observed is less than the actual.

The CAB system has proved to be a useful general classification for complex exchange patterns, but remains rather broad.

19.4 The S&S System

Each CAB defines a family of configurations, the number in the family being determined by the number of break-ends available for interaction.

Each interaction removes two break-ends. Thus, for C/A/3 (six break-ends) there are $5 \times 3 \times 1 = 15$ possible configurations; for C/A/4 (eight break-ends) $7 \times 5 \times 3 \times 1 = 105$ possible configurations; for C/A/5 945 possible configurations; for C/A/6 10,395 possible configurations, and so on (in general, $(2b)!/b!2^b$, where b is the number of interactive breaks). These numbers include configurations with break-end restitutions, so not all configurations will be complex exchanges, and some will not reflect the $CAB_{initial}$ (Sect. 19.6.).

It was quickly realized that using a single-chromosome paint, a given configuration can produce a variety of different patterns, some of which may lead to false inferences (*pseudosimple* patterns). This is illustrated in Fig. 19.1. Also it was realized that the same single-paint pattern can arise from totally different configurations within the same or from different CABs (Savage 2002).

Therefore, it became of interest to know just how many unique single-paint patterns existed, and if any were diagnostic for a particular CAB.

From CAB 2/2/3 to CAB 5/5/5 there are 26 families which generate between them 15,060 configurations and 41,895 single-chromosome paintings. All of these configurations and paintings were analysed and only 203 *unique* single-paint patterns emerged. These represent 0.0048 (about 0.5%) of the total, so it is obvious that a large number of patters recur, and none can be regarded as diagnostic for a particular CAB. However, in many cases a minimum number of interacting breaks is required before a particular pattern will be encountered.

Each of these unique single-paint patterns was given an alphanumeric code. Those containing a centric-ring (*interarm intrachange*) have the prefix CR followed by a number. The rest have a numerical prefix corresponding to the actual number of painted elements present, and are distinguished from each other by an alphabetical suffix. These codes comprise the S&S system (Savage and Tucker 1996).

Part of the system has been published (Savage and Simpson 1994), but the complete code series (up to CAB 5/5/5) has hitherto only been available from an unpublished monograph. It is now presented in its entirety in Tables 19.2–19.7.

The patterns illustrated are for structurally complete forms where all painted segments (shown solid in the tables) are potentially visible. For each pattern, the *minimum* number of breaks in the *CAB* required to produce it is shown in the top right-hand corner of each square.

There are additional patterns if "one-way" exchanges are included, which, although known, are not considered here. There may also be some new patterns which will be discovered in higher-order CABs (i.e. above 5/5/5), but the enormous number of configurations which these generate preclude detailed analysis.

19.5 mFISH Painting

It has now become possible to paint all the homologue pairs in the human karyotype distinctively, using techniques like mFISH (Speicher et al. 1996) and its equivalents. So now, for the first time, we can see the complete configuration (or a large part of it) which is underlying and generating the observed exchange painting pattern, and so get a much fuller picture of the relationship of its parts (Fig. 19.1).

Table 19.2 S&S single-paint codes. Painted segments shown *solid*

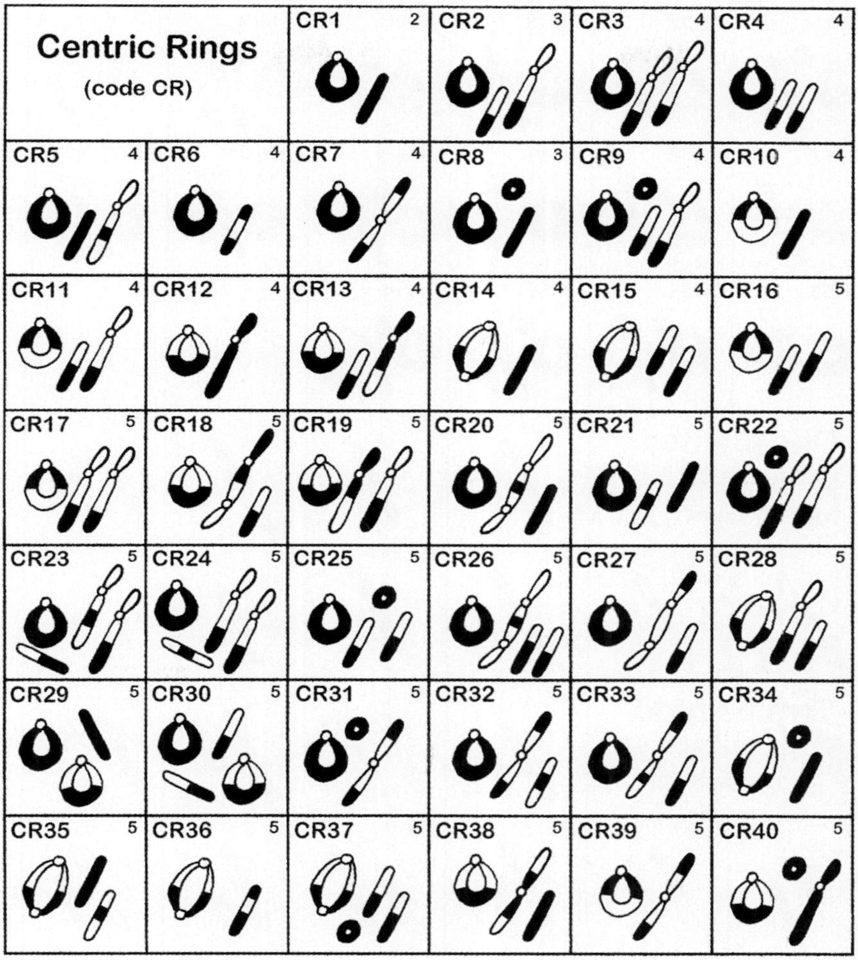

Unfortunately, this increased precision has introduced a whole raft of problems which we have not had to consider before, not least that complex exchanges are more common, and more complicated than we have hitherto realized (Savage 2002).

One pressing problem is how to classify and record the exchange patterns we see. At the moment, for quantitative purposes, counting the total number of bicolour junctions (each of which must represent at least one rejoining event) will give a crude estimate of overall damage, although it must be recognized that valid rejoining events within a given chromosome, or when homologues are involved in the same exchange, will be missed. This approach, however, does not help in recording the form of individual exchanges.

Table 19.3 S&S single-paint codes. Painted segments shown *solid*

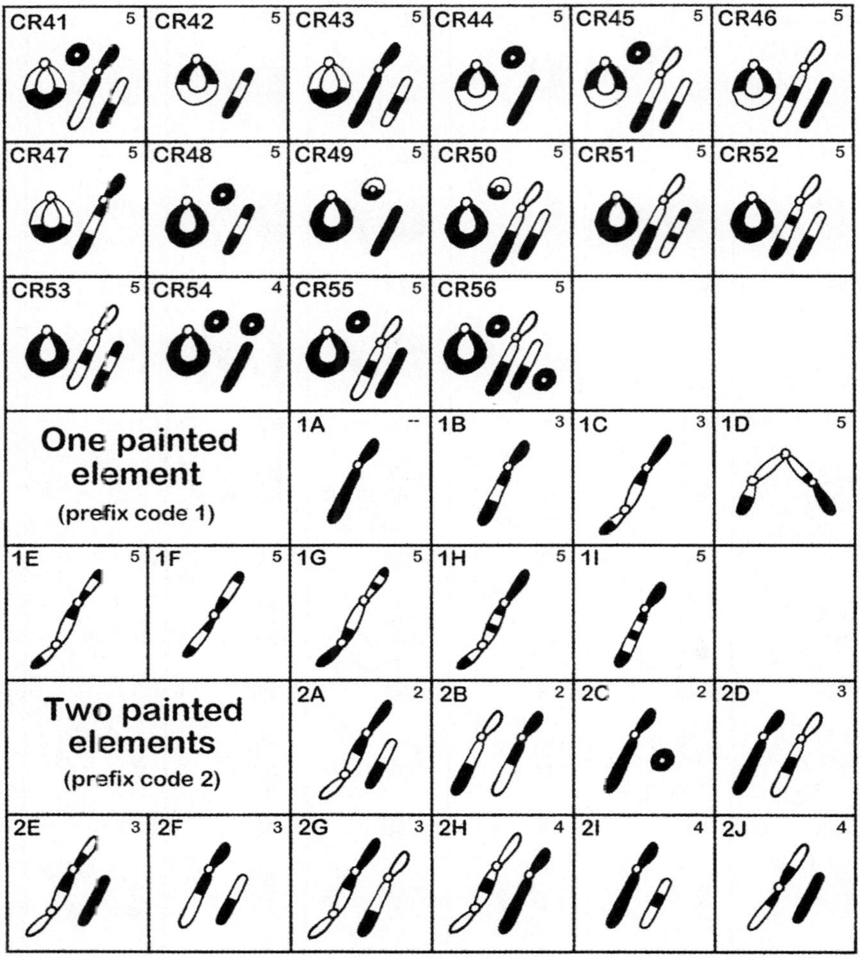

The old scoring schemes, based on solid staining, allowed convenient names for the various types seen, so the scoring criteria like "dicentrics" or "centric-rings" were universally understood, and could be used for valid interlaboratory comparisons (Savage 1976). But now, with mFISH, there is such a plethora of different patterns that simple names for each are out of the question.

We can, of course, classify any exchange pattern using $CAB_{observed}$, but as there are relatively few CAB families, many quite different exchange patterns will be covered by such a designation.

By ensuring that each participating chromosome is identified by a code, a complete description for each painted element in an exchange pattern could

Table 19.4 S&S single-paint codes. Painted segments shown *solid*

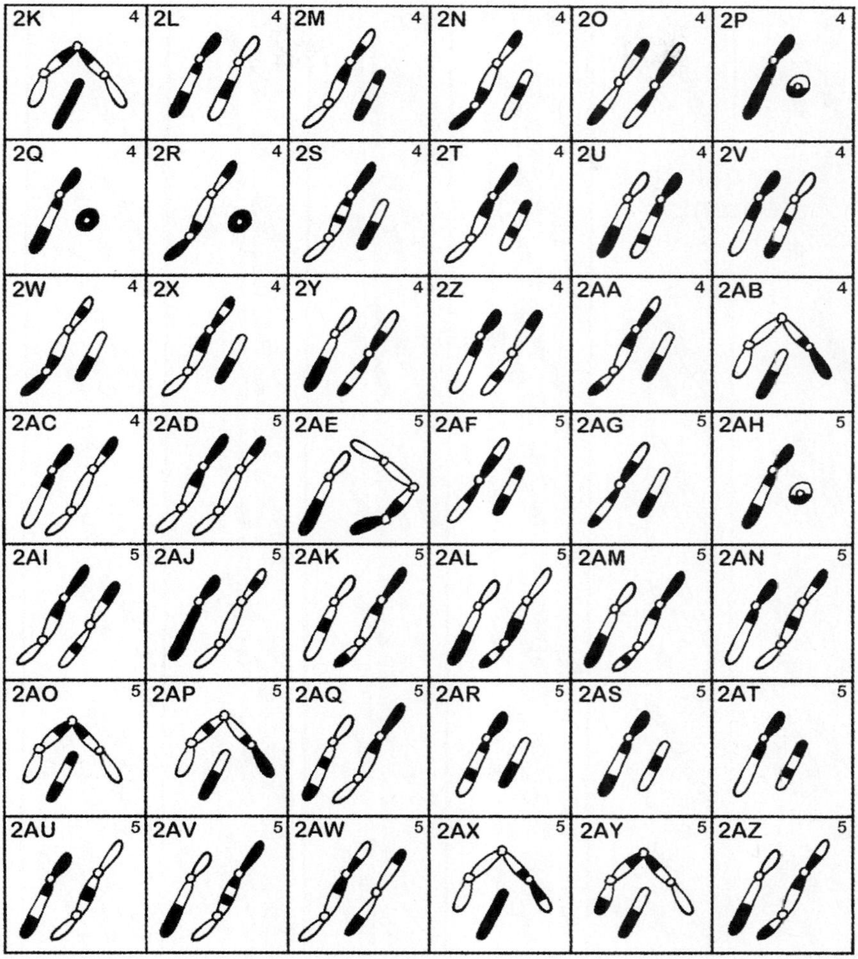

be provided by the PAINT nomenclature (Tucker et al. 1995). The linked elements would then form a descriptor, but this soon becomes very cumbersome when large exchanges are present in a mFISH situation.

19.6 The S&S Descriptor

A more refined approach is to make use of the S&S single-paint codes to construct a descriptor for each observed exchange pattern in its entirety.

Provided that all the chromosomes participating in an exchange are distinctively painted, it is possible to consider the pattern of each one individually, and

Table 19.5 S&S single-paint codes. Painted segments shown *solid*

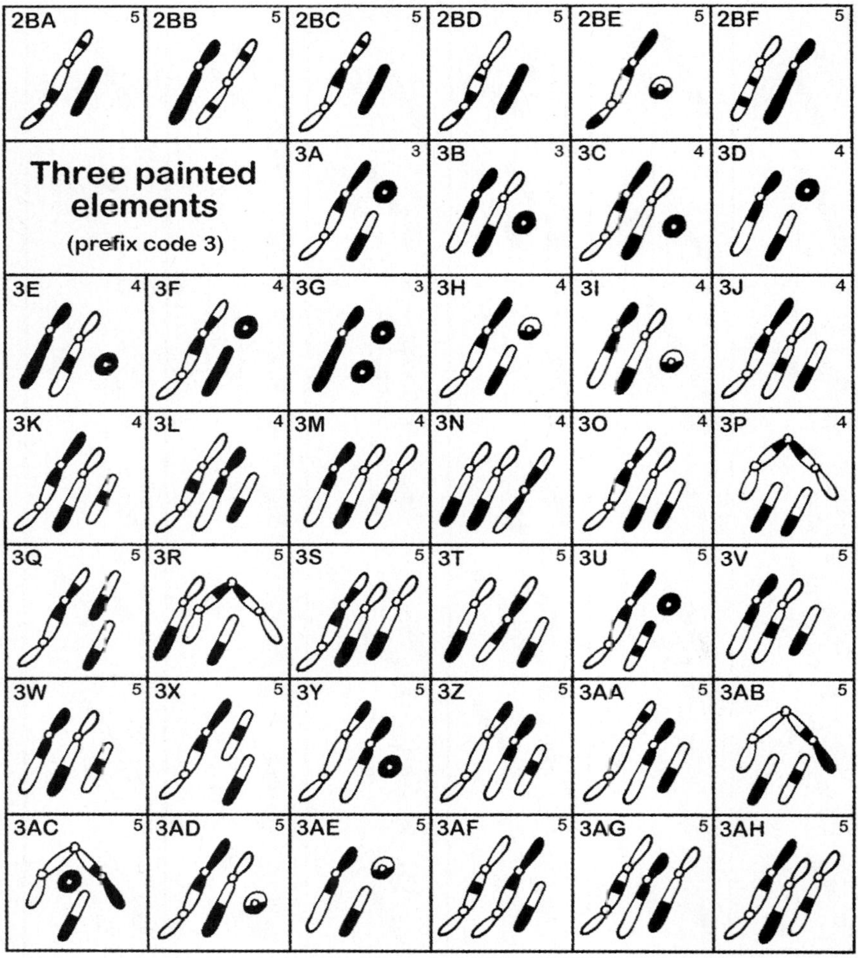

assign to it an S&S single-paint code. The string of codes thus obtained (expressed in the order found in Tables 19.2–19.7) forms a simple and useful descriptor for the whole exchange pattern. An example is given in Fig. 19.1, where the complete exchange from a CAB 4/4/5 becomes a [2B/2F/2G/3J]. The relevant S&S patterns and codes can be found in Tables 19.3 and 19.5.

This descriptor not only facilitates the scoring and classification of complex exchanges, but provides sufficient data for their diagrammatic reconstruction.

It will be noted that for any exchange pattern, the S&S single-paint codes present are not wholly independent. Certain collections are permissible, others impossible. This will be obvious if we consider any two-chromosome exchange

Table 19.6 S&S single-paint codes. Painted segments shown *solid*

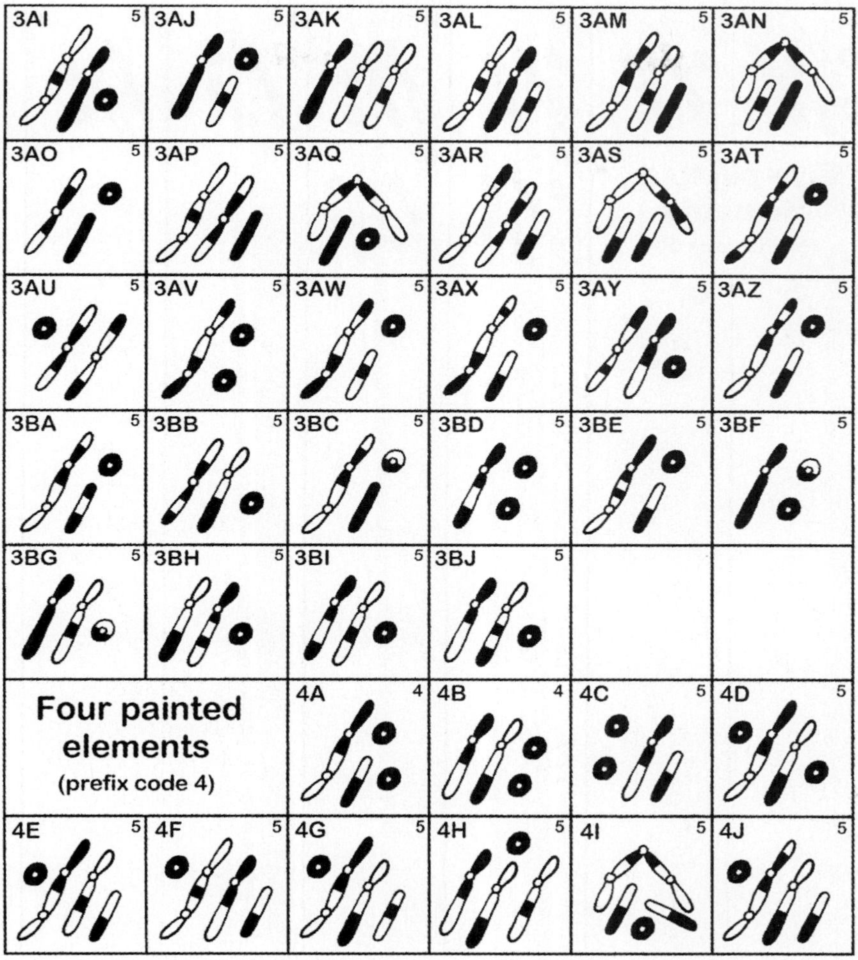

pattern, for here only two S&S codes form a descriptor, and once one is fixed, so also is the other. Consequently, a limited number of recurrent pairs like 2A/2A, 2M/2N and CR2/2F always occur, whilst other combinations like 2A/2F are precluded.

For CABs with several chromosomes, many more code combinations become possible, and the same code may now occur within different descriptors, but the limitation to permissible combinations still holds.

As a consequence of these restrictions, all the theoretical configurations for any CAB fall into a limited number of recurrent descriptors. Some code

Table 19.7 S&S single-paint codes. Painted segments shown *solid*

4K 5 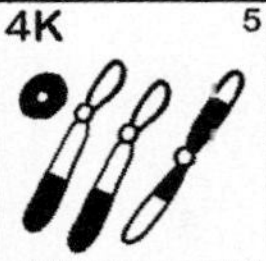	4L 5 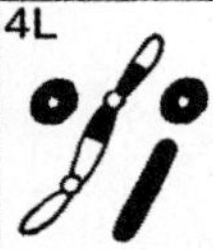	4M 5 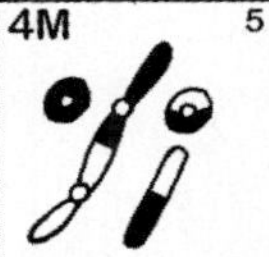	4N 5 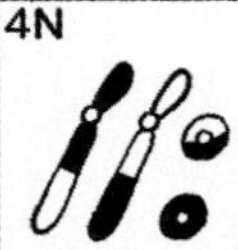	4O 4 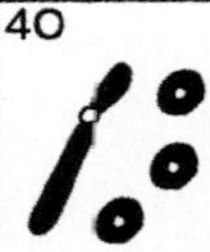	4P 5 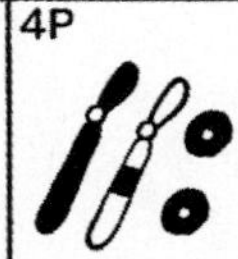
Five painted elements (prefix code 5)		5A 5 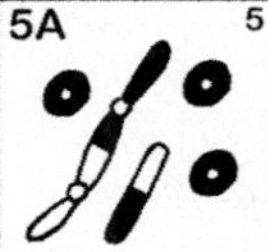	5B 5 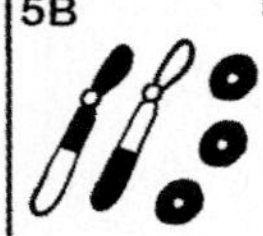		

combinations are very common (e.g. 2A/2F/2G), and are found in several families, whilst a few are confined to one family. Within a family, the actual number of configurations falling within each descriptor varies considerably, depending largely on the number of different break-end interactions possible to satisfy a particular exchange pattern.

The permissible descriptors which classify each configuration can be derived theoretically for each CABinitial family, and it has been shown that these fall into three different categories:

1. *Restitution-reduced configurations.* Any restitution occurring as part of the exchange process effectively lowers the $CAB_{initial}$ by introducing apparently normal chromosomes (code 1A) and/or by producing SCHJ.

As the observer will be unaware of these restitution interactions, the net effect will be to record a lower-order $CAB_{observed}$.

The theoretical number of configurations which involve one or more restitution events can be readily derived form a full expansion of a CAB. This number is solely dependent upon the available number of interacting breaks. Thus, of the 15 configurations possible for CA3 families, seven of them have restitutions. Similarly, CA4 families have 45 in 105, and CA5 families have 401 in 945.

2. *Integrated configurations.* No restitutions; all rejoins are illegitimate, with the majority observable as bicolour junctions. However, even here, internal chromosome interactions, or interactions between any homologues present, can lead to SCHJ, with concomitant underestimation of CAB_{actual}.
3. *Disintegrated configurations.* When the number of breaks equals or exceeds 4, then some complex exchanges can arise from two or more apparently independent lower-order exchanges. These are known as sequential exchange complexes (Savage 2002). Sometimes this can cause physical breakup of a complex exchange, and the lower-order exchanges will be observed and recorded as separate events. For example, in CAB 3/5/5, the theoretical descriptor CR1/1C/2E will actually appear as two separate exchanges, CR1 and 1C/2E. In theory, disintegrated descriptors should be relatively rare, forming only a few percent of the total configurations within a CAB family.

19.7 Conclusions

The S&S descriptor provides an easy method for recording the complete observed pattern of any complex exchange where all the participating chromosomes are distinctively painted. It can be made even more precise by adding the actual chromosome number to each S&S code in the descriptor. The data thus provided enable one to make a fairly accurate diagrammatic reconstruction of the entire pattern.

Whilst the number of legitimate descriptors is limited, it is still very large, but this is inevitable, bearing in mind the enormous random combinatorial possibilities that exist for sets of proximate break-ends.

How useful this approach will prove for numerical studies must await the results of its practical application in experimental situations.

References

Cornforth, MN (2001) Analyzing radiation-induced complex chromosome rearrangements by combinatorial painting. Radiat Rese 155:643–659

Savage JRK (1976) Annotation: classification and relationships of induced chromosomal structural changes. J Med Genet 13:103–122

Savage JRK (1998) A brief survey of aberration origin theories. Mutat Res 404:139–147
Savage JRK (2002) Reflections and meditations upon complex chromosomal exchanges. Mutat Rese 512:93–109
Savage JRK, Simpson P (1994) FISH "painting" patterns resulting from complex exchanges. Mutat Res 312:51–60
Savage JRK, Tucker JD (1996) Nomenclature systems for FISH-painted chromosome aberrations. Mutat Res 366:153–161
Speicher MR, Ballard SG, Ward DC (1996) Karyotyping human chromosomes by combinatorial multi-fluor FISH. Nat Genet 12:368–375
Tucker JD, Morgan WF, Awa AA, Bauchinger M, Blakey D, Cornforth MN, Littlefield GL, Natarajan AT, Shassere C (1995) A proposed system for scoring structural aberrations detected by chromosome painting. Cytogenet Cell Genet 68:211–221

20 Aberration Patterns and Cell Cycle Progression Following Exposure of Lymphocytes to the Alkylating Agent Trenimon

Ursula von Recklinghausen, Christian Johannes, Lothar Riedel, and Günter Obe

Abstract

Chromosomal aberrations are generally analyzed in first metaphases (M1) following exposure to chromosome-breaking agents. During subsequent cell cycles there is strong selection against aberrant cells. Aberrant cells which survive selection exhibit changed aberration patterns. Data on chromosomal aberrations induced with the chemical mutagen Trenimon in the G1 phase of human lymphocytes in culture are discussed. Trenimon leads to various types of DNA damage, but not to DNA double-strand breaks; it is an S phase-dependent chromosome-breaking agent. Aberrations are analyzed in first (M1), second (M2) and third and further (M3+) posttreatment metaphases using preparations block-stained with Giemsa stain or with multicolour fluorescence in situ hybridization. In M1 cells, chromatid-type aberrations, such as chromatid breaks and chromatid translocations occur. In M2 and M3+ cells, the aberration frequencies diminish considerably and in aberrant cells which survived selection, chromosome-type aberrations such as dicentric chromosomes and chromosome translocations appear. Asymmetrical chromatid translocations, where the two centromeres of the participating chromosomes are connected, give rise to dicentric chromosomes and to normal chromosomes. Symmetrical chromatid translocations, in which the centromeres of the participating chromosomes are not connected, give rise to reciprocal translocations and to normal chromosomes or to two types of nonreciprocal translocations. Cells with reciprocal, but not with nonreciprocal translocations are expected to survive because no genetic information is lost. Cells with micronuclei and polyploid cells preferably occur at later divisions and will then be selected.

20.1 Introduction

Chromosomal aberrations should be analyzed in first in vitro metaphases (M1) after exposure to a chromosome-breaking agent. Two major reasons for this are (1) selection of cells with chromosomal aberrations leads to reduced frequencies of chromosomal aberrations in second (M2) or third and later

Günter Obe and Vijayalaxmi (Eds.)
Chromosomal Alterations: Methods, Results and Importance in Human Health

(M3+) metaphases and (2) when cells with chromosomal aberrations divide the aberration pattern changes. Both effects have been repeatedly analyzed in cells exposed to radiation or chemical mutagens (Beek et al. 1980; Manti et al. 2006; Marshall and Obe 1998; Obe and Beek 1982; Riedel 1989; Riedel and Obe 1980). Mutagens which do not directly induce DNA double-strand breaks lead to chromatid-type chromosomal aberrations in M1 when applied before or during DNA synthesis. During subsequent cell cycles, chromatid-type chromosomal aberrations are transformed to chromosome-type chromosomal aberrations which are then visible in M2 or M3+. Transformation of aberration types in connection with selection of most cells with chromosomal aberrations would lead to wrong assumptions on the types of lesions and on the frequencies of chromosomal aberrations induced by the test agent in question.

In this chapter we discuss results of experiments in which lymphocytes in G1 phase of the cell cycle were exposed to the trifunctional alkylating agent Trenimon [2, 3, 5-tris(1-aziridinyl)-1,4-benzoquinone, TRN]. TRN is an S-dependently chromosome breaking agent (Obe and Beek 1979; Rademacher and Obe 1995). Exposure of cells before or during S phase to TRN leads to chromatid-type aberrations such as chromatid interchanges in M1; in M2 and M3+ chromatid-type aberrations are replaced by chromosome-type aberrations, such as dicentric chromosomes (DIC) and translocations (Beek et al. 1980; Marshall and Obe 1998; Obe and Beek 1982; Riedel and Obe 1980). Asymmetrical chromatid interchanges (RB′a) give rise to DIC and acentric fragments; symmetrical chromatid interchanges (RB′s) give rise to reciprocal or nonreciprocal translocations (Marshall and Obe 1998). Many types of human cancers are characterized by specific translocations (Mitelman et al. 1997) and it is interesting that they may occur by transformation of chromatid-type exchanges.

20.2 Methods

The results of two sets of experiments using human lymphocytes are presented. In one study cytogenetic effects were analyzed in Giemsa-stained preparations. In a second study chromosomal aberrations were analyzed in preparations painted with the method of multicolor fluorescence in situ hybridization (mFISH).

20.2.1 Giemsa Staining

Standard lymphocyte cultures of 5 ml were set up with heparinized whole blood (0.4 ml) from a 32-year-old male donor supplemented with Ham's F-10 medium (4 ml), fetal calf serum (0.5 ml), phytohemagglutinin (0.12 ml) and antibiotics (penicillin, streptomycin, 0.02 ml). For each fixation time four cultures were set up, and two of them received 5-bromdeoxyuridine (BrdUrd; final concentration 20 μmol/l) at a culture time of 22 h. Exposure to TRN (0.5 μmol/l) was for 5 h beginning 15 h after the start of culture. After exposure, TRN was washed out by discarding the medium and substitution with medium from untreated cultures set up for this purpose (conditioned medium). All cultures received colcemid (final concentration 0.08 μg/ml) 4 h before fixation. Culture times were 55–119 h in steps of 4 h. Cultures were set up in such a way that fixations and preparations could be performed during the day. Metaphases were prepared using routine procedures.

Preparations obtained from cultures without BrdUrd were block-stained with Giemsa solution; those from cultures with BrdUrd were differentially stained for sister-chromatid differentiation to analyze the frequencies of M1, M2 and M3+ in 200 metaphases (100 per parallel culture per fixation time). Chromosomal aberrations were analyzed in preparations from cultures without BrdUrd. At each fixation time, 500 metaphases (250 per parallel culture) were scored for chromosomal aberrations, 2,000 mononucleated interphases for micronuclei (MN) (1,000 per parallel culture), 1,000 metaphases for polyploid and endoreduplicated cells (500 per parallel culture) and 1,000 metaphases (500 per parallel culture) for premature chromosome condensation (PCC) in MN (MN-PCC).

20.2.2 Multicolor Fluorescence in Situ Hybridization

Lymphocyte cultures were set up using heparinized whole blood of a female donor principally as described in Sect. 20.2.1, but Mc Coy's 5A medium and 20% fetal calf serum were used. At a culture time of 18 h, TRN (final concentration 0.5 μmol/l) was added to the cultures for 2 h. After exposure, cells were washed and incubated in conditioned medium for the remaining culture times of 55, 60, 74, 122, 194 and 242-h. The 192- and 242-h cultures received conditioned medium for a second time at 122 h, which resulted in a better mitotic index. To determine the proportion of M1, M2 and M3+ metaphases after different incubation times, one culture each received BrdUrd (final concentration 20 μmol/l). After adding colcemid (final concentration 0.08 μg/ml) for the last 4 h of incubation, metaphases were prepared using routine procedures. Slides from cultures with BrdUrd were differentially stained in order to analyze the frequencies of M1, M2 and M3+ cells. The other slides were painted with mFISH probes and counterstained with 4′,6-diamidino-2-phenylindole, which allows all chromosomes to be recognized by their specific color spectrum.

Pooled data from two independent experiments are given.

20.2.3 Description of Aberration Types Analyzed

Gaps: Gaps are unstained areas in chromatids which are not longer than the diameter of the chromatids in which they occur. This structural change is usually not considered as a real chromosomal aberration.

Chromatid breaks (B′): Breaks in chromatids with the broken part more or less dislocated from the chromatids from which they descend.

Isochromatid breaks (B″): Both chromatids of a chromosome are broken at the same site and by this produce fragments without centromeres. B″ can also arise from chromosome breaks which occur in unreplicated chromosome regions.

Chromatid translocations (RB′): In most cases RB′ are exchanges between broken chromatids of two chromosomes. In rare cases more than two chromosomes can participate in the formation of an RB′. In RB′ with two chromosomes, two types can be differentiated, namely, symmetrical (RB′s) and asymmetrical (RB′a) ones.

RB′s: Reciprocal exchanges between the broken chromatids with all chromatids linked to a centromere (X-type fusion). The chromatids of an RB′s separate in anaphase without difficulties, but may give rise to different types of translocations in subsequent cell cycles.

RB′a: Exchange between broken chromatids in such a way that both centromeres are connected by the fused chromatids of both chromosomes (U-type fusion). The chromatid fragments without centromeres generally also fuse. In anaphase RB′a may lead to DIC or MN. A rare variety of chromatid exchanges are triradials where a B′ in one chromosome fuses with a B″ in a second chromosome.

DIC: DIC are chromosomes with two centromeres associated with a B″. DIC are generally formed by fusion of two centromere-bearing fragments of two unreplicated chromosomes. Both centromere-less fragments may fuse and by this form a B″. After treatment with TRN, DIC can also be formed when cells with chromatid interchanges divide.

Ring chromosomes (rings): Rings may arise from two breaks within one unreplicated chromosome. They may contain centromeres (centric ring) or not (acentric rings). Centric rings are associated with fragments (B″). Rings are quite rare and may also represent derived aberration types.

Polyploids and endoreduplications: Polyploid metaphases contain more than two sets of chromosomes. One reason for this may be the formation of endoreduplications, in which DNA synthesis is not followed by a segregation of the chromosome set, or by other disturbances of the mitotic mechanism.

MN: When cells divide, chromosome fragments without centromeres are generally not distributed to the cell poles, but enclose themselves with their own nuclear membrane. Whole chromosomes when not properly distributed can also give rise to MN.

MN-PCC: Compared with that of the main nucleus, the cell cycle of MN is slowed. When main nuclei undergo mitosis most MN are in S phase and very few are in G2 or G1 phases. When the main nucleus undergoes mitosis, mitotic factors stimulate MN to enter a prophaselike stage in which the chromosomes are visible. MN-PCC is generally located at the periphery of the metaphases. G1-PCC are long single-stranded structures; G2-PCC are double-stranded structures which are generally Giemsa-banded. S-PCC, by far the most frequent MN-PCC type, are characterized by a pulverized appearance (Obe and Beek 1982).

Translocations: Translocations can only rarely be detected in block-stained preparations. They are easily detectable in FISH-painted preparations. Translocations can be reciprocal (chromosomal material is mutually exchanged between two chromosomes) or nonreciprocal (chromosomal material from one chromosome is transferred to another chromosome, but not vice versa).

20.3 Results

20.3.1 Giemsa Staining

At 55 h, 95% of the metaphases were M1 and only 5% were M2. The maximal frequencies of M2 occurred at fixation times of 75 and 79 h with 48 and 50%, respectively. At these fixation times 32 and 24% of the cells were still in M1, and 28 and 38.5% were already in M3+. High frequencies of M3+ cells occurred at fixation times of 103 h (85%) up to 119 h (94%). B′ including B″ and RB′ (including few triradials) had their maximum at 55 h with 51.0 and 16.0 per 100 metaphases, respectively. The frequencies of B′ as well as RB′ (including triradials) were rare or absent at fixation times of 79–119 h. Gaps are not considered as chromosomal aberrations, but they show a similar pattern to chromatid-type chromosomal aberrations in that their frequencies are high at early fixation times (55 and 59 h: 15.4 and 12.2 per 100 metaphases, respectively) and much lower at later fixation times (63 h: 5.3 per 100 metaphases; 111 h: 0.3 per 100 metaphases). B″ occurred at all fixation times. Their frequencies dropped from 35.2 per 100 metaphases at 59 h to 10.0 or less per 100 metaphases at fixation times of 79 h and later. The decrease of the frequencies of B″ was not as pronounced as that found for B′ and RB′. This indicates that they are not uniform aberration types. B″ can be of chromosome or chromatid origin and in addition they can be derived from RB′a, DIC and centric ring chromosomes. DIC have a maximum frequency of 6.4 per 100 cells at 83 h; at 55 h only 2.0 DIC per 100 cells were found. Surviving cells with DIC can lead to daughter cells with DIC and this may explain why low frequencies of DIC are found at later fixation times. Rings were very rare and had a maximum frequency of 0.8 per 100 metaphases at 111 and 115 h and of 1.0 at 119 h. MN occurred at all fixation times and had their maximum frequency of 70 per 1,000 interphases at 79 h. The maximum frequencies of MN-PCC were 61 per 1,000 metaphases at 83 h and occurred at all fixation times at low frequencies (8–56 per 1000 metaphases). Polyploid metaphases including endoreduplications are very rare at fixation times of 55–63 h (zero to three per 1,000 metaphases) and reach a maximal value of 19 per 1,000 metaphases at 91 h. Endoreduplications do not occur before 83 h of cultivation, but they are very rare at all fixation times (two to five per 1,000 metaphases). The frequency of metaphases with chromosomal aberrations excluding gaps is

39.2% at 55 h and continuously drops to 6.6% at 119 h. The maximum frequencies of chromosomal alterations were found as follows: gaps, B′ including B″, RE′ (including triradials) at 55 h, DIC including rings at 83 h, MN at 79 h, MN-PCC at 83 h, polyploid metaphases and endoreduplications at 91 h. The data clearly show that TRN is an S-dependently acting chromosome breaking agent. It induces mainly chromatid-type aberrations in M1. In M2 and M3+, chromatid aberrations continuously diminish and derived aberrations appear. All data obtained from Giemsa-stained preparations are from Riedel (1989).

20.3.2 Multicolor Fluorescence in Situ Hybridization

The frequencies of M1, M2 and M3+ cells at different fixation times as determined in differentially stained metaphases from cultures with BrdUrd are shown in Fig. 20.1.

At 55 h, 95% M1 and 5% M2 cells were found. The number of M1 cells continuously decreased and they were not detected at 194 h. The first M3+ cells were found at 74 h and reached a value of 96% at 194 h. Preparations

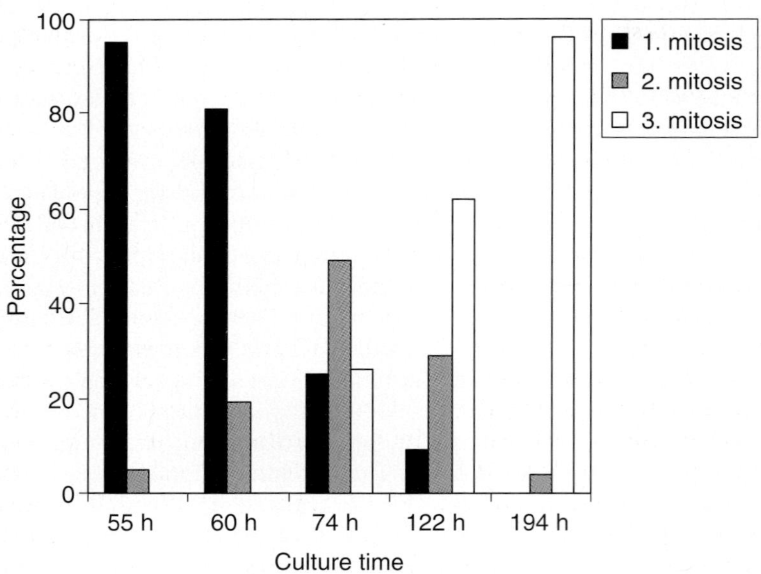

Fig. 20.1 First (M1, uniformly darkly stained chromosomes), second (M2, all chromosomes differentially stained) and third (M3+, different proportion of differentially and uniformly lightly stained chromosomes) metaphases after Trenimon treatment following different culture times. Data are from the multicolor fluorescence in situ hybridization (mFISH) study

Table 20.1 Exchange aberrations per 100 metaphases after Trenimon treatment at different cultivation times following painting with multicolor fluorescence in situ hybridization

Culture time (h)	Number of cells scored	Chromatid interchanges	Dicentrics	Translocations
74	258	10	6	30
122	307	1	17	35
194	309	0	12	33
242	65	0	9	17

painted with mFISH are especially suited for the analysis of exchange-type aberrations. Table 20.1 show the frequencies of exchange-type aberrations at fixation times of 74–242 h. RB′ predominantly occurred at 74 h and only one was found at 122 h. Both DIC and translocations had their maximum at 122 h, but occurred at all fixation times. Translocations were always more frequent than DIC.

Examples of different types of exchange-type aberrations are shown in Fig. 20.2.

In accordance with our data obtained with Giemsa staining (Sect. 20.3.2) the mFISH data show that exposure of human lymphocytes in G1 phase of the cell cycle to TRN leads to chromatid-type aberrations in M1 which give rise to chromosome-type aberrations in M2 and M3+. DIC are derived from RB′a; reciprocal and nonreciprocal translocations are derived from RB′s (Marshall and Obe 1998). Upon mitotic division a cell with an RB′a will give rise to a normal cell and a cell with a DIC plus associated fragment. Assuming a random distribution of chromatids, a dividing cell with an RB′s will lead to undamaged cells, cells with reciprocal translocations or cells with nonreciprocal translocations (Marshall and Obe 1998).

In contrast to nonreciprocal translocations derived from RB′s, translocations directly induced by X-rays are probably all reciprocal, although some of them appear to be nonreciprocal (Obe et al. 2002). Since cells with nonreciprocal translocations are genetically imbalanced, they are prone to be selected during cell divisions. Cells with reciprocal translocations in which no deleterious mutations were induced during their origin and which do not contain other types of chromosomal aberrations are genetically balanced and will survive during ongoing divisions. This seems to be the case in radiation-exposed people in whose lymphocytes the frequencies of reciprocal translocations are used for retrospective dosimetry (Chap. 24 by Edwards et al.). Experimental studies in which DIC and translocations were analyzed in splenocytes of mice up to 120 days after X-irradiation show that DIC disappear much faster than translocations (Natarajan and Obe 1999).

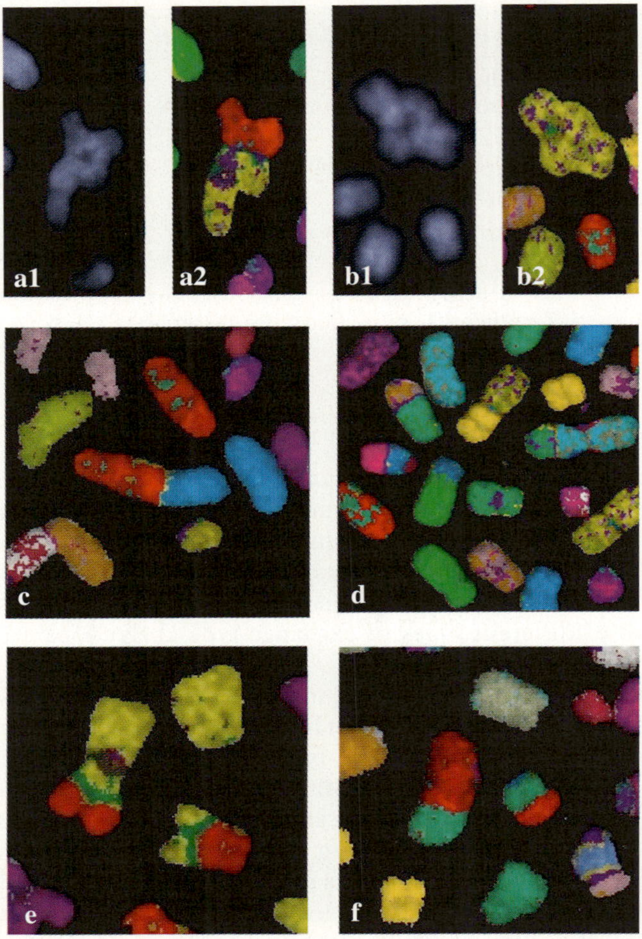

Fig. 20.2 Metaphases of human chromosomes showing Trenimon-induced aberrations stained with mFISH. **a1** Asymmetric chromatid translocation (RB′a) as visible with 4′, 6-diamidino-2-phenylindole (DAPI) counterstain. **a2** Same aberration painted with mFISH shows the participation of chromosomes 1 and 5. **b1** Symmetric chromatid translocation (RB′s), DAPI-stained. **b2** Same aberration painted with mFISH shows the participation of both chromosomes 1. **c** Dicentric chromosome formed by chromosomes 5 and X. **d** Five chromosomes involved in translocations. **e** Reciprocal translocation between chromosomes 1 and 5. **f** Reciprocal translocation between chromosomes 4 and 15

20.3.3 Conclusions

We analyzed chromosomal aberrations induced in G1 lymphocytes by the polyfunctional alkylating agent TRN and found typical changes of the aberrations pattern at different recovery times. A strong selection against chromosomally aberrant cells leads to a considerable decrease of aberration frequencies. Furthermore, chromatid-type exchanges are substituted by chromosome-type exchanges. The appearance of derived reciprocal translocations is of special interest, because they are stable over many cell cycles. Cells with DIC and nonreciprocal translocations are genetically unbalanced and most of them will be selected during ongoing cell divisions.

Our study underlines the utmost importance of analyzing chromosomal aberrations in first posttreatment metaphases (M1) in order to obtain reliable information on aberration quantities and aberration patterns induced by a test agent. Exposure of cells in G1 and analysis of aberration patterns in M1 indicates the types of lesions induced by the test agent. Agents that directly induce DNA double-strand breaks, such as ionizing radiation, lead to chromosome-type aberrations (DIC and translocations). Agents which induce DNA damage other than double-strand breaks lead to chromatid-type aberrations such as chromatid translocations during DNA synthesis. These typical aberration patterns change when M2 or M3+ cells are analyzed. Selective forces wipe out most aberrant cells. Only cells with balanced aberrations such as reciprocal translocations survive.

Acknowledgement: This paper is dedicated to our late colleague and friend Dr. Richard Marshall who left us too early.

References

Beek B, Klein G, Obe G (1980) The fate of chromosomal aberrations in a proliferating cell system. Biol Zentralbl 99:73–84

Manti L, Durante M, Grossi G, Ortenzia O, Pugliese M, Scampoli P, Gialanella (2006) Measurements of metaphase and interphase chromosome aberrations transmitted through early replication rounds in human lymphocytes exposed to low-LET protons and high-LET 12C ions. Mutat Res 596:151–165

Marshall R, Obe G (1998) Application of chromosome painting to clastogenicity testing in vitro. Environ Mol Mutagen 32:212–222

Mitelman, F, Mertens F, Johansson B (1997) A breakpoint map of recurrent chromosomal rearrangements in human neoplasia. Nat Genet 15:417–474

Natarajan AT, Obe G (1999) Biological dosimetry of absorbed radiation dose based on the frequencies of chromosomal aberrations in human lymphocytes. In: Baumstark-Khan C, Kozubek S, Horneck G (eds) Fundamentals for the assessment of risks from environmental radiation. NATO science series. Kluwer, Dordrecht, pp 179–186

Obe G, Beek B (1979) Trenimon: biochemical, physiological and genetic effects on cells and organisms. Mutat Res 65:21–70

Obe G, Beek B (1982) Premature chromosome condensation in micronuclei. In: Rao PN, Johnson RT, Sperling K (eds) Premature chromosome condensation. Application in basic, clinical, and mutation research. Academic, New York, pp 113–130

Obe G, Pfeiffer P, Savage JRK, Johannes C, Goedecke W, Jeppesen P, Natarajan AT, Martinez-Lopez W, Folle GA, Drets ME (2002) Chromosomal aberrations: formation, identification and distribution. Mutat Res 504:17–36

Rademacher P, Obe G (1995) Trenimon: structure and reactivity of a versatile chemical agent. Mutat Res 340:37–49

Riedel L (1989) Chromosomale Aberrationen und Schwesterchromatidenaustausche in menschlichen peripheren Lymphozyten in Abhängigkeit von unterschiedlichen Inkubationszeiten nach Behandlung mit Trenimon. Dissertation, Freie Universität Berlin
Riedel L, Obe G (1980) Trenimon-induced SCEs and structural chromosomal aberrations in early- and late-dividing lymphocytes. Mutat. Res 73:125–131

21 Cytogenetic Analysis and Occupational Health

Radim J. Sram and Pavel Rössner

Abstract

This chapter summarizes the experiences of the Czech Hygiene Service from the mid-1970s until the present time, using cytogenetic analysis as a biomarker of exposure. During this period, the method has proved its ability to detect an occupational hazard. We observed a significant association between the increased frequency of aberrant cells and occupational exposure to carcinogens and mutagens at a group level. Examples of occupational exposures to carcinogens are given from the chemical industry [bis(chloromethyl) ether, epichlorohydrin, vinyl chloride monomer, acrylonitrile, 1,3-butadiene, ethylbenzene], mining (ore mining, soft coal open-cast mining, uranium mining), and health care (cytostatics, halothane, microbiology). The interpretation of the cytogenetic findings in the context of occupational health is discussed. The results of cytogenetic analysis were used to enforce the decrease of carcinogens in the working environment and to establish safer maximum allowable concentrations. The exceptionally broad use of cytogenetic analysis in the Czech Republic was facilitated by the fact that the Hygiene Service accepted the theoretical assumption that the frequency of aberrant lymphocytes is a surrogate indicator for the genetic damage caused by clastogens in target tissues. As a biomarker of exposure, cytogenetic analysis wrote a great chapter in the history of Czech public health and in many cases played a crucial role in improving working conditions in the Czech Republic.

21.1 Introduction

Cytogenetic analysis of peripheral blood lymphocytes (PBLs) has been accepted as the only suitable assay for biological monitoring of the genetic damage induced in somatic cells by excessive exposure to clastogenic agents in the workplace (Buckton et al. 1973; Bochkov et al. 1974; Kilian et al. 1976; Natarajan and Obe 1980). The merits of cytogenetic analysis for industrial hygiene were soon recognized, and Purchase (1978) suggested setting hygienic standards in such a way as to reduce occupational exposure to

Günter Obe and Vijayalaxmi (Eds.)
Chromosomal Alterations: Methods, Results and Importance in Human Health

clastogens to levels that do not increase the frequency of chromosomal aberrations in PBLs over a suitable control level. In the Czech Republic, cytogenetic analysis has proven its utility, particularly in the area of occupational health. To prevent genetic injury in workers and its expected consequences, such as an increase in cancer rates or in the rates of malformations in their children, cytogenetic analysis has been implemented as a component of preventive medical checkups since 1976–1977. Soon it became a particularly useful tool to test the safety of industrial hygiene standards. In many instances, cytogenetic analysis provided the chief argument used to enforce improved working conditions (Sram and Kuleshov 1980; Sram 1981).

The exceptionally broad use of cytogenetic analysis in the Czech Republic was facilitated by the fact that the Hygiene Service accepted the theoretical assumption that the frequency of aberrant lymphocytes is a surrogate indicator for the genetic damage caused by clastogens in target tissues. Finally, the Hygiene Service appreciated the ability of cytogenetic analysis to integrate exposures from different sources mediated by different pathways.

Today, chromosomal aberrations in human PBLs are recognized as a valuable biomarker of effect, probably the only one which has been internationally standardized and validated (Carrano and Natarajan 1988; Sorsa et al. 1992; Albertini et al. 2000). Nowadays it is generally accepted that a high frequency of chromosomal aberrations in PBLs is predictive of an increased risk of cancer (Hagmar et al. 1998, 2004; Bonassi et al. 2000, 2004; Rössner et al. 2005).

21.2 History

The first genetic toxicology laboratory in the Czech Republic was established at the Regional Institute of Hygiene in the city of Usti nad Labem in response to a newly emerged need to check the exposure of workers to epichlorohydrin (ECHH) and bis(chloromethyl) ether (BCME) in the chemical industry in 1974. The results obtained provided important evidence of deleterious changes in human cells induced by overexposure to clastogens. The success of this first laboratory led to the establishment of several other cytogenetic laboratories (Rössner et al. 1995). The necessity for coordinating the activities of an increasing number of laboratories resulted in the establishment of the Reference Laboratory of Genetic Toxicology at the Institute of Hygiene and Epidemiology, Prague, in 1977. The major task of the Reference Laboratory was to support the founding of other genetic toxicology laboratories at the regional level and it was responsible for the development of standard cytogenetic methods used in the Czech Hygiene Service; the first guidelines were released in 1979.

21.3 Standard Method

All the genetic toxicology laboratories of the Hygiene Service of the Czech Republic have been working according to the standard operating procedures published by the Reference Laboratory of Genetic Toxicology (Rössner et al. 1998, 2000). Since 2001, the Reference Laboratory has met the requirements of the CSN EN ISO/IEC 17025 standard and holds the Certificate of Accreditation of the Czech Accreditation Institute.

The modified conventional Hungerford method was used on short-term 48-h cultures with all cells being in the first division. Cultures of PBLs from whole venous blood were established within 24 h after the collection. The cultures were prepared in RPMI 1640 Sevac medium supplemented with 20% calf serum Opavac and 1% phytohemagglutinin (Murex, UK). Two identical cultures were set up from each sample, cultivated at 37°C, and harvested after 48 h of incubation. Two hours before the end of cultivation 0.1 μg/ml colchicine (Fluka) was added. Cells were collected by centrifugation, resuspended in a prewarmed hypotonic solution (0.075 M KCl) for 20 min and fixed in acetic acid/methanol (1:3, v/v) on slides. These were air-dried and stained with 5% Giemsa solution (pH 6.8). Slides from each culture were numbered randomly and scored as "blind" in numerical order. At least 100 well-spread metaphases with 46 ± 1 centromeres were examined per donor.

The following four categories of chromosomal aberrations were evaluated: chromatid and chromosome breaks, and chromatid and chromosome exchanges. Gaps were recorded, but they were not scored as aberrations. Cells bearing breaks or exchanges were marked as aberrant cells.

21.4 Chromosomal Aberrations and Maximum Allowable Concentrations

Cytogenetic analysis of chromosomal aberrations has been suggested as a tool for checking whether a safe maximum allowable concentration (MAC) has really been established (Sram and Kuleshov 1980; Sram 1981). The MAC of a chemical is defined as the maximum concentration of a gas, vapor or aerosol in the working atmosphere that, according to contemporary scientific knowledge, does not adversely affect the health of exposed humans.

The curve in Fig. 21.1 represents a simplified model situation in which the exposure (concentration of the chemical in the working atmosphere integrated over time) is related to a measurable effect such as the frequency of aberrant cells. It is assumed that a clastogenic exposure that does not increase the frequency of adverse effects over the level found in a suitable control

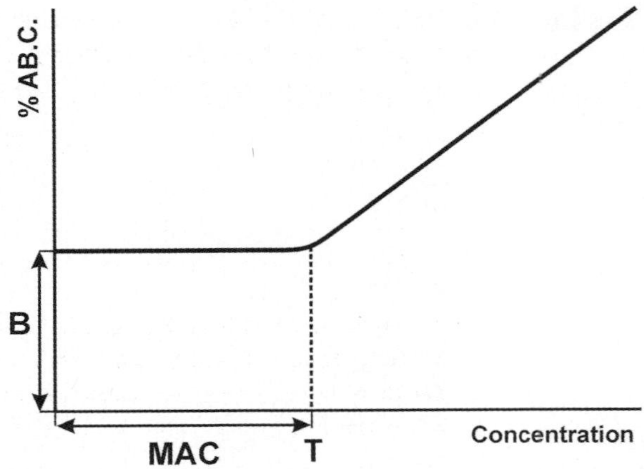

Fig. 21.1 Principle of establishing the maximum allowable concentration (*MAC*). *B* background frequency of aberrant cells determined in controls, *T* threshold limit of clastogenic action, *AB.C* aberrant cells

(unexposed) population is expected to be safe. According to this reasoning, the MAC should be set below the threshold level for clastogenic action.

It is, however, a controversial issue if any threshold can be established for carcinogenic and/or mutagenic effects. We argued in favor of a threshold level in the following way:

> Use of this method for checking MAC safety is based on the assumption that some mutagenic/carcinogenic molecules must be metabolically activated before reaching the target-cell DNA. Moreover, the DNA damage caused by these chemicals may be still eliminated by repair mechanisms. So we have good reasons to believe that also for chemical mutagens/carcinogens a threshold value can be found. This threshold, however, may vary from person to person . . . (Sram 1981).

In spite of the fact that the determination of chromosomal aberrations in PBLs has been conceptualized as a biomarker of early effects, in common practice we have interpreted it solely as a biomarker of exposure objectively indicating a need for reducing occupational exposure.

21.5 Cytogenetic Analysis and Public Health Practice

In the late 1970s several papers demonstrated that occupational exposure to vinyl chloride monomer (VCM) (Ducatman et al. 1975; Funes-Cravioto et al. 1975; Hansteen et al. 1978), chloroprene (Katosova 1973) and styrene (Meretoja et al. 1977) significantly increased the frequency of chromosomal aberrations in PBLs. A decrease in exposure to VCM (Hansteen et al. 1978) and chloroprene (Katosova 1973) was paralleled by a decrease in the frequency of aberrant cells to the level found in unexposed control groups. These findings resulted in the setting of new and safer exposure limits at a level below the observed threshold for clastogenic effects and set a precedent followed by public health authorities in the Czech Republic.

To provide public health authorities with a firm base for the decision-making process, the Reference Laboratory estimated the frequency of chromosomal aberrations in the Czech general population. This estimate, based on data from unexposed controls, was 1.2–2.0% aberrant cells. Svandova and Rössner (1989) analyzed data from 1,201 unexposed controls, collected in the frame of the Czech Hygiene Service in the period 1980–1988. The mean frequency of aberrant cells was 1.75%. More recently, Rössner et al. (1998) summarized the data on the spontaneous frequency of chromosomal aberrations in the general population of subjects in the period 1977–1997 ($N = 5{,}430$). The mean frequency of aberrant cells found in the subset of individuals of age from 20 to 63 years ($N = 2{,}801$) was 1.59 ± 1.34%. Both estimates were based on the analyses of at least 100 well-spread metaphases per subject.

The occupationally exposed groups with a mean frequency of aberrant cells above this reference value were then categorized by the frequencies of chromosomal aberrations (Sram et al. 2004a):

1. A group at increased genotoxic risk bearing on average 2–4% aberrant cells: exposure to acrylonitrile, asbestos, benzene, chromium-welding, cytostatics, dimethylformamide, ECHH, ethylbenzene, formaldehyde, halothane, methyl mercury, polycyclic aromatic hydrocarbons (PAHs): coke oven, glassware, rubber production, soft coal open-cast mining, soft coal pressure gasification, soot production, soft coal underground mining, PCB, propane–butane, radon: ore mining, uranium mining, uranium dressing plant, styrene, synthetic resins, toluene (Fig. 21.2).
2. A group at high genotoxic risk bearing more than 4% aberrant cells: exposure to BCME, cytostatics, mixture of styrene and methyl metacrylate, nickel, nitrous gases, olachindox, PAHs: coal tar production, glassware, VCM: PVC production (Fig. 21.3).

The discovery of occupational groups with elevated levels of chromosomal aberrations (usually initiated by a thorough hygiene inspection of workplaces), coupled with industrial hygiene measurements of the quality of the working environment, allowed estimation of the level of occupational exposure to carcinogens or mutagens. Simultaneously, a database and literature search for the toxicological properties of all agents identified in the working environment was begun. Even if technical and/or organizational measures reduced occupational exposure below the MACs in effect at the time, if repeated cytogenetic analyses still indicated an increase in the frequency of aberrant cells in the same group, then the public health authorities reassessed the safety of the MACs in use, and these were subsequently set below the

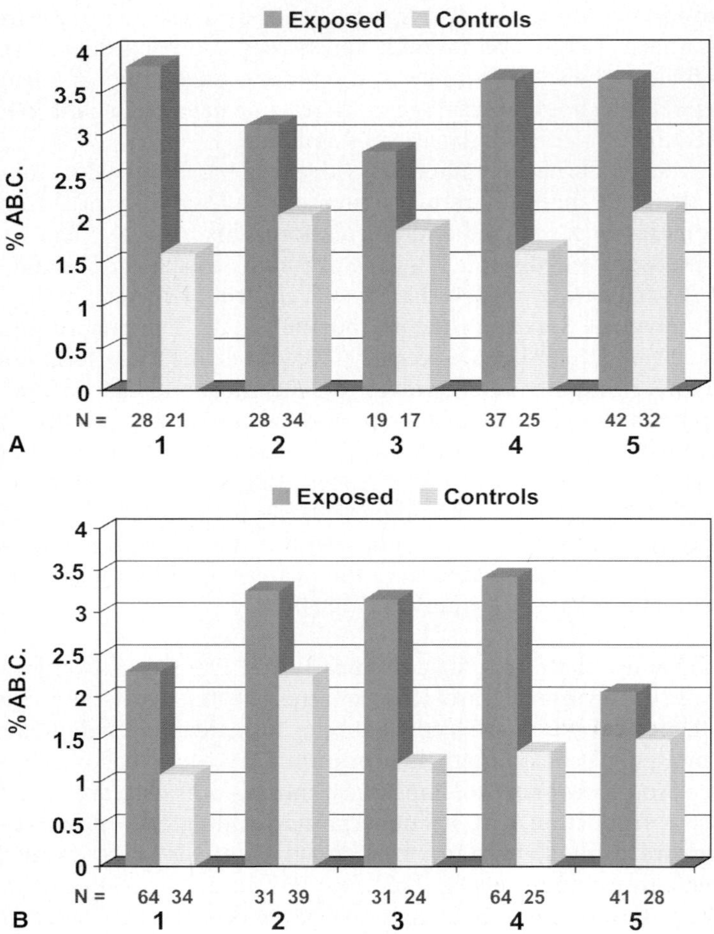

Fig. 21.2 Results of cytogenetic analyses in groups with an increased genotoxic risk. **a** *1* dimethylformamide, *2* epichlorohydrin, *3* styrene, *4* synthetic resins, *5* toluene (printing). **b** Exposure to polycyclic aromatic hydrocarbons: *1* coke oven, *2* rubber production, *3* soft coal open-cast mining, *4* soft coal pressure gasification, *5* soot production

threshold for clastogenic effects. Strict measures were enforced particularly in the case of groups at high genotoxic risk.

The target for public health intervention was not only technological improvements at critical workplaces but also behavioral changes in the exposed workers. This usually promoted positive changes such as more systematic care about the working environment, the voluntary use of personal protective equipment and so on.

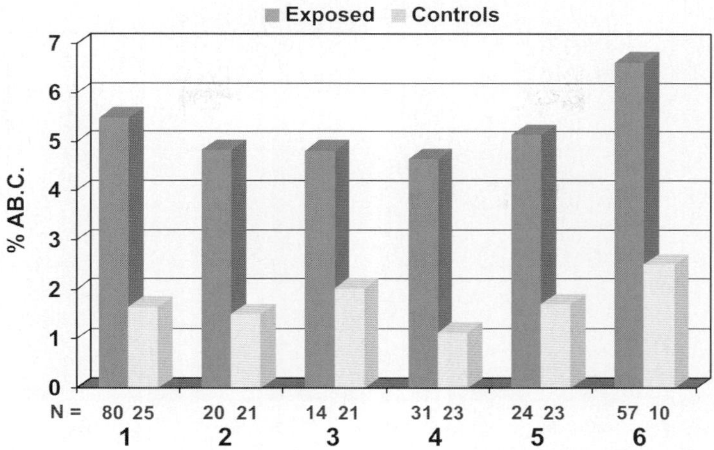

Fig. 21.3 Results of cytogenetic analysis in groups with high genotoxic risk. *1* bis(chloromethyl) ether, *2* coal tar production, *3* cytostatics, *4* glassware production, *5* styrene + methyl methacrylate (button production), *6* PVC production

21.6 Examples of Occupational Exposures to Carcinogens

21.6.1 Chemical Industry

21.6.1.1 Bis(chloromethyl) Ether, Chloromethyl Methyl Ether

BCME and chloromethyl methyl ether (CMME) were used as chlormethylating agents in the production of anionic and cationic ion exchangers. BCME causes lung cancer and is listed among human carcinogens (IARC 1987).

The ion-exchange unit in which workers were exposed to both halogenated ethers was in operation from 1973 to 1987. In 1974 it became the very first workplace in which chromosomal aberrations were used as a biomarker of exposure to evaluate the biological significance of occupational exposure. The results obtained were presented as evidence that ion-exchanger production poses a high genotoxic risk for workers. Therefore, since 1976 cytogenetic analysis has become part of their periodic checkups, and 80–110 workers have been examined each year (Zudova and Landa 1977; Roznickova et al. 1987).

The workers involved in ion-exchanger production were considered to be at high carcinogenic risk. When all available technological and organizational corrective measures were implemented at the workplaces but exposure to BCME still remained, this group became the first target for ascorbic acid supplementation (Sram et al. 1983a; Fig. 21.4). Gebhart et al. (1985) showed a clear-cut ability of ascorbic acid to block the clastogenic action of alkylating

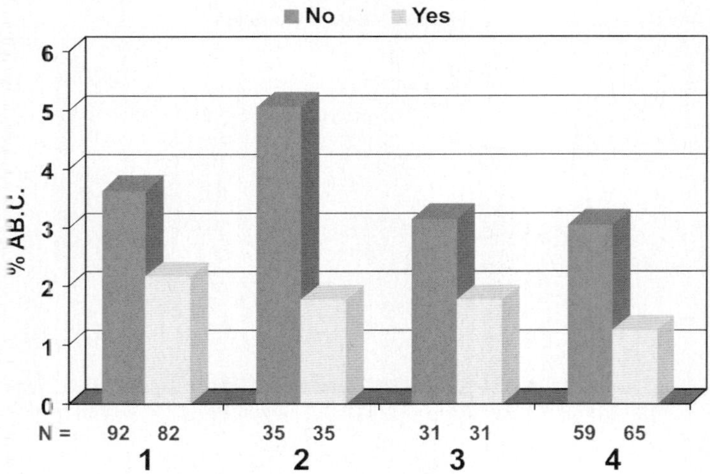

Fig. 21.4 Effect of the ascorbic acid supplementation on the frequency of aberrant cells

agents: therefore, in the early 1980s we believed it appropriate to administer ascorbic acid as a natural protector against the clastogenic (and mutagenic/carcinogenic) effects of environmental chemicals. When supplementation stopped in 1986, the frequency of aberrant cells increased to 4.35 ± 0.42%. The unit was closed in February 1987. After another 6 months, the frequency of aberrant cells decreased in the originally exposed workers (who meanwhile had moved to other production units in the same company) to 2.90 ± 0.15%, and in 1988, after another 12 months, the frequency decreased to 2.07 ± 0.15%.

21.6.1.2 *Epichlorohydrin*

Sram et al. (1980) carried out cytogenetic analysis in three groups of subjects: 28 workers occupationally exposed to ECHH for 4 years in a new production unit, 34 matched controls and 21 external controls (members of the general population). The mean concentration of ECHH of 0.936 mg/m^3 (0.20–1.85) found at workplaces in 1978 was below the MAC of 1 mg/m^3: We found 3.12% aberrant cells in the ECHH-exposed group, 2.06% aberrant cells in matched controls and 1.33% aberrant cells in the external controls. It was concluded that a MAC of 1.0 mg ECHH/m^3 was still too high to avoid the genotoxic risk associated with exposure to ECHH.

This occupational group was followed up for an additional 4 years. During this time ECHH exposure decreased to a mean concentration of 0.384 mg/m^3 (0.05–1.11). At this exposure level, ECHH did not exhibit any significant clastogenic effects. In a group of 33 workers we observed 2.00% aberrant cells vs. 1.68% aberrant cells in a group of 25 controls (Sram et al. 1983b).

21.6.1.3 Vinyl Chloride Monomer

Rössner et al. (1980) studied a group of workers ($N = 31$) in a VCM production unit. At the time of sampling, they had been exposed for 2–3 years to VCM levels below the Czechoslovakian MAC of 10 mg/m^3. In exposed workers they observed 1.60% aberrant cells in 1977 and 1.05% in 1978. In controls ($N = 35$) they found 1.12% aberrant cells.

The authors concluded that if VCM air concentrations in the working environment did not exceed 10 mg/m^3, then no increase in chromosomal aberrations in exposed workers seemed to be detectable.

21.6.1.4 Petrochemical Industry

Acrylonitrile The process of polymerization of India rubber is associated with exposure to acrylonitrile, butadiene and styrene. The most significant chemical regarding human health is considered to be acrylonitrile, because its concentrations in occupational settings often exceed the MAC (0.5 mg/m^3). A study carried out in a chemical plant in 1988 indicated a significant increase in the frequency of aberrant cells in the exposed group ($N = 54$, 2.96% aberrant cells) vs. controls ($N = 25$, 1.63% aberrant cells $P<0.01$). It was concluded that exposure to acrylonitrile posed a genotoxic risk, and this occupational group was examined for chromosomal aberrations every second year (Rössner et al. 1989).

1,3-Butadiene The association between occupational exposure to 1,3-butadiene (BD) and chromosomal damage in PBLs was studied in monomer production and polymerization units (Sorsa et al. 1994).

The clastogenic effects of occupational exposure to BD were studied in 1994 in 19 male workers from the monomer production unit and 19 control subjects from the heat production unit (Sram et al. 1998). The individuals' exposures to BD were measured by passive personal monitors. It was demonstrated that exposure to a median BD concentration of 0.53 mg/m^3 significantly increased the percentage of aberrant cells. The mean frequency of aberrant cells reported in exposed workers was 3.11%, and in controls it was 2.03% ($P<0.01$). The results presented by Sram et al. (1998) showed that although exposure levels in modern plants were substantially reduced, a significant induction of cytogenetic damage could still be demonstrated.

Ethylbenzene Ethylbenzene is produced by benzene alkylation by ethylene. During the production process, workers are exposed to a mixture of several chemicals. The cytogenetic effects in the staff members involved in an ethylbenzene production unit exposed to benzene, ethylbenzene, diethylbenzene, xylene and toluene were studied in 2002 (Sram et al. 2004b). The concentrations of ethylbenzene ranged from 0.2 to 13.1 mg/m^3 and the concentrations of benzene ranged from 0.4 to 15.1 mg/m^3. The mean frequency of aberrant cells reported in exposed workers was 3.28% ($N = 39$), and that in controls was 2.05% ($N = 55$, $P < 0.001$). Using later a new technology (closed system), no significant differences in the frequency of aberrant cells were observed between the exposed and control groups.

21.6.2 Mining

21.6.2.1 Ore Mining

The increased incidence of lung cancer attracted the attention of the Hygiene Service to a tin mine in the early 1970s (Roznickova et al. 1987). In addition to other measures, cytogenetic analysis has been employed as one of the control measures since 1976. The estimated individual exposure did not exceed 15 mSv/year and 8 mSv/3 months.

The differences in the magnitude of chromosomal damage were dependent on radon dose. If a miner's exposure exceeded the yearly exposure limit of 15 mSv/year ($N = 36$) then the mean frequency of aberrant cells was 5.7%. Even higher values were seen in miners who had been excessively exposed for 3 months prior to sampling ($N = 20$, 8.2% aberrant cells). The reported number of aberrant cells in the group of miners below the limit ($N = 199$) was 4.1%.

Thirty-six cases of cancer were reported in this cohort of miners. Cox regression models, which took into account the age at the time of the first cytogenetic assay, radon exposure and smoking, showed significant associations between cancer incidence and the frequency of chromatid breaks and the frequency of aberrant cells. An 1% increase in the frequency of aberrant cells in miners was paralleled by a 62% increase in the relative risk of cancer ($P < 0.001$) (Smerhovsky et al. 2002).

21.6.2.2 Soft Coal Open-Cast Mining

In the 1980s, soft coal open-cast mining operations were characterized by ever-present fumes and fire hazards, resulting in a significant exposure to PAHs. In 1984, two miner groups were examined: group 1 – engaged in stripping operations approximately 20–50 m under the mine surface – and group 2 – engaged in digging operations approximately 50–80 m under the mine surface. The miners were the operators of belt haulage conveyors, giant excavators and overburden dumping machines. Group 1 miners ($N = 27$) showed 2.41% aberrant cells, and group 2 miners showed ($N = 31$) 3.16% aberrant cells. In comparison, the workers in a coal cleaning plant ($N = 23$) had 1.30% aberrant cells and matched controls ($N = 24$) had 1.21% (Sram et al. 1985).

21.6.2.3 *Soft Coal Underground Mining*

In 1985, miners from deep mines using two different blast technologies were examined. In miners using a coyote-hole blast, resulting in a short exposure to nitrous gases, the frequency of aberrant cells in the group studied ($N = 31$) increased to a mean of 2.97%. The miners ($N = 22$) using a wall blast were not exposed to explosive combustion products, and the frequency of aberrant cells found in this group was 1.82%. The clastogenic effect observed in miners using a coyote-hole blast was significantly increased in comparison with a control group ($N = 20$, 1.55% aberrant cells, $P<0.05$) (Roznickova et al. 1987).

21.6.2.4 *Uranium Mining*

A study focused on chromosomal damage in uranium miners was carried out in 1990 in three mines (Hamr, $N = 20$, Tachov, $N = 23$, and Pribram, $N = 23$) located in three different regions of the Czech Republic. The control group was recruited in the town of Pribram ($N = 23$) and the capital, Prague ($N = 33$). The observed frequencies of aberrant cells were as follows: Hamr 3.44%, Tachov 2.69%, Pribram 2.65%. The frequency of aberrant cells in controls selected in Pribram was 2.16%, and that of controls selected in Prague was 1.52%. In miners from Hamr, the frequency of aberrant cells was significantly increased when compared with both control groups ($P<0.05$). Contrary to expectations, no increase in the frequency of dicentric chromosomes was observed (Sram et al. 1993a).

The uranium mines studied were very deep (1,200–1,600 m) with high temperature and humidity, providing suitable conditions for the growth of different molds. In throat swab samples taken from miners ($N = 116$), molds producing mycotoxins were identified. In contrast, mycological examination in controls ($N = 78$) was negative (Sram et al. 1993a). The same molds found in miners were also detected in samples from mine walls and timbers (Sram et al. 1993b). It was hypothesized that mutagenic mycotoxins might be another risk factor responsible for the high cancer risk in uranium miners. All three mines were shut down in 1991; therefore, no further investigations were possible.

21.6.3 Health Care

21.6.3.1 *Cytostatics*

Since 1981, the original cytostatic drug Platidiam (*cis*-diaminodichlor platinum complex, *cis*-DDP) has been produced in the former Czechoslovakia. The health care providers involved in the clinical testing of this drug began complaining about side effects very similar to those seen in treated patients. Because of a lack of other suitable methods to objectify occupational exposure to *cis*-DDP, a decision was taken to use cytogenetic analysis. The first results showed a mean frequency of aberrant cells exceeding 6.0% in the exposed staff. This finding initiated a systematic follow-up of this group and thorough inspection of all wards where antineoplastic drugs were administered. As corrective measures, the Hygiene Service required that infusions be prepared in biohazard hoods and that personal protective equipment such as gloves and masks be used during drug preparation and administration (Rössner et al. 1985; Zudova et al. 1985). Since that time, health care providers administering or preparing antineoplastic drugs have been treated as a group at high carcinogenic risk. Cytogenetic analysis of PBLs has become a component of their periodic checkups. At present, the health care providers involved in the preparation or administration of antineoplastic drugs represent one of the largest groups occupationally exposed to carcinogens (Smerhovsky et al. 2001).

21.6.3.2 Halothane

Srb et al. (1987) used cytogenetic analysis to study the effects of occupational exposure in anesthesiologists in 1981 and 1982. The halothane concentrations in the breathing zone of anesthesiologists measured during pediatric surgery ranged from 76 to 907 mg/m^3, during orthopedic surgery from 31 to 106 mg/m^3, and during neurosurgery from 30 to 49 mg/m^3. The levels of chromosomal damage in this group ($N = 31$) were 1.95% aberrant cells and 2.45% aberrant cells in 1981 and 1982, respectively. In controls, Srb et al. (1987) found 0.93% aberrant cells and 1.61% aberrant cells in 1981 and 1982, respectively ($P<0.01$ and $P<0.05$). These results initiated improvements in the construction of inhalation anesthetic equipment and the installation of air conditioning in operating theaters. Nevertheless, similar frequencies of aberrant cells were observed in anesthesiologists in a hospital in the town of Jihlava many years later (Hanzl 1991, unpublished results given in Sram et al. 2004a).

21.6.3.3 Microbiology

Srb and Kubzova (1985) investigated a group of subjects working in microbiological laboratories, predominantly exposed to various infectious agents. The exposed group ($N = 21$) showed 4.07% aberrant cells, and the control subjects ($N = 65$) showed 1.55% aberrant cells ($P<0.001$). The authors concluded that this type of exposure was associated with a high genotoxic risk.

21.7 Conclusions

We reviewed all scientific papers and reports dealing with cytogenetic analyses published by Czech authors in the period from 1974 to 2005, emphasizing the importance of the cytogenetic method not only for the Czech public health

(Sram et al. 2004a). During this period of 33 years, the method has proved its ability to detect an occupational hazard. The application of cytogenetic analysis was particularly fruitful for checking the clastogenic effects of mixtures of different chemicals and under conditions in which, besides inhalation exposure, other pathways contributed to human exposure. We observed a significant association between the increased frequency of aberrant cells and occupational exposure to carcinogens and mutagens at a group level.

As a biomarker of exposure, cytogenetic analysis wrote a great chapter in the history of Czech public health and in many cases played a crucial role in improving working conditions in the Czech Republic. An excellent example is the case of soft coal pressure gasification (Sram et al. 2004a). The gasification of soft coal was a chemical process used extensively in the Czech Republic because of its large coal reserves. The repeated findings of elevated levels of chromosomal aberrations in workers engaged in this operation became the main impetus to include this technology in the list of proven human carcinogens by the Czech Ministry of Health (1985) (group I according to IARC). Furthermore, Czech genetic toxicologists first reported the genotoxicity of soft coal open-cast mining (Sram et al. 1985), platinum cytostatics in health care providers (Zudova et al. 1985) and asbestos in workers (Srb 1984; Srb et al. 1985).

At the present time we are witnessing the final step in validating conventional cytogenetic analysis as a biomarker of early effects. During the 1990s several European countries accumulated sets of data large enough to test the cancer predictivity of the frequency of chromosomal aberrations (Hagmar et al. 1998, 2004; Bonassi et al. 2000, 2004). The Czech Republic has also contributed to this effort. From the archives of Czech genetic toxicology laboratories, we retrieved data on approximately 20,000 cytogenetic analyses performed on an occupational group of almost 12,000 individuals (Rössner et al. 2005). In this context, we would like to emphasize the fact that contrary to other studies, the incidence of cancer in exposed groups is relatively low, but still significant. We believe this is also due to the activities of the Czech Hygiene Service, which very efficiently used cytogenetic data to put into place effective preventive measures. Perhaps we may speculate that the low cancer incidence observed in our cohort is the best evidence for the importance of cytogenetic analysis for public health.

References

Albertini RJ, Anderson D, Douglas GR, Hagmar L, Hemminki K, Merlo F, Natarajan AT, Norppa H, Shuker DE, Tice R, Waters MD, Aition A (2000) IPCS guidelines for the monitoring of genotoxic effects of carcinogens in humans. Mutat Res 463:111–172

Bochkov NP, Zhurkov VS, Jakovenko KN, Kuleshov NP (1974) Culture of lymphocytes as a material to studying genetic effects in persons exposed to mutagens. Dokl Akad Nauk SSSR 218:463–465 (in Russian)

Bonassi S, Hagmar L, Strömberg U, Montagud AH, Tinnerberg H, Forni A, Heikkilä P, Wanders S, Wilhardt P, Hansteen IL, Knudsen L, Norppa H (2000) Chromosomal aberrations in lymphocytes predict human cancer independently of exposure to carcinogens. Cancer Res 60:1619–1625

Bonassi S, Znaor A, Norppa H, Hagmar J (2004) Chromosomal aberrations and risk of cancer in humans: an epidemiologic perspective. Cytogenet Genome Res 104:376–382
Buckton KE, Evans HJ (1973) Methods for the analysis of human chromosome aberrations. WHO, Geneva
Carrano A, Natarajan AT (1988) Considerations for population monitoring using cytogenetic techniques. Mutat. Res 204:379–406
Czech Ministry of Health (1985) Hygienic guidelines for the work with chemical carcinogens. Decree of the Ministry of Health of the Czech Republic no 64/1985
Ducatman A, Hirschorn K, Selikoff IJ (1975) Vinyl chloride exposure and human chromosome aberrations. Mutat Res 31:163–168
Funes-Cravioto F, Lambert B, Linsten J, Ehrenberg L, Natarajan AT, Osterman-Golkar S (1975) Chromosome aberrations in workers exposed to vinyl chloride. Lancet I 459
Gebhart E, Wagner H, Crziwok K, Behnsen H (1985) The action of anticlastogens in human lymphocyte cultures and their modification by rat-liver S 9 mix. II. Studies with vitamins C and E. Mutat Res 149:83–94
Hagmar L, Bonassi S, Stromberg U, Brogger A, Knudsen L, Norppa H, Reuterwall C (1998) Chromosomal aberrations in lymphocytes predict human cancer: a report from the European Study Group on Cytogenetic Biomarkers and Health (ESCH). Cancer Res 58:4117–4121
Hagmar L, Stromberg U, Bonassi S, Hansteen IL, Knudsen L, Lindholm C, Norppa H (2004) Impact of types of lymphocyte chromosomal aberrations on human cancer risk: results from Nordic and Italian cohorts. Cancer Res 64:2258–2263
Hansteen IL, Hillestad L, Thils-Evensen E, Heldaas SS (1978) Effects of vinyl chloride in man. A cytogenetic follow-up study. Mutat Res 51:271–278
IARC (1987) IARC monographs on the evaluation of carcinogenic risk to humans, overall evaluations of carcinogenicity: an updating of IARC monographs, vols 1–42, supplement 7. IARC, Lyons
Katosova LD (1973) Cytogenetic analysis of peripheral blood of workers engaged in the production of chloroprene. Gig Tr Prof Zabol 17:30–32 (in Russian)
Kilian DJ, Picciano D (1976) Cytogenetic surveillance of industrial populations. In: Hollander A (ed) Chemical mutagens. Principles and methods for their detection, vol 4. Plenum, New York, pp 321–339
Meretoja T, Vainio H, Sorsa M, Härkkonen H (1977) Occupational exposure and chromosomal aberrations. Mutat Res 56:193–198
Natarajan AT, Obe G (1980) Screening of human populations for mutations induced by environmental pollutants: Use of human lymphocyte system. Toxicol Environ Saf 4:468–481
Purchase IFH (1978) Chromosomal analysis of exposed populations: a review of industrial problems. In: Evans HJ, Lloyd DC (eds) Mutagen-induced chromosome damage in man. Edinburgh University Press, Edinburgh, pp 258–267
Rössner P (1985) Genetic impact of chemical compounds present in the environment. Final research report P17-335-457-05. Institute of Hygiene and Epidemiology, Prague (in Czech)
Rössner P (1989) Genetic impact of chemical compounds present in the environment. Research report P12-335-87-31.5. Institute of Hygiene and Epidemiology, Prague (in Czech)
Rössner P, Sram RJ, Novakova J, Lambl V (1980) Cytogenetic analysis in workers occupationally exposed to vinyl chloride. Mutat Res 43:425–427
Rössner P, Cerna M, Bavorova H, Pastorkova A, Ocadlikova D (1995) Monitoring of human exposure to occupational genotoxicants. Cent Eur J Publ Health 3:219–223
Rössner P, Sram RJ, Bavorova H, Ocadlikova D, Cerna M, Svandova E (1998) Spontaneous level of chromosomal aberrations in peripheral blood lymphocytes of control individuals of the Czech Republic population. Toxicol Lett 96,97:137–142
Rössner P, Bavorova H, Ocadlikova D (2000) Methods on biologic monitoring of genotoxic effects of environment – standard methods: cytogenetic analysis of peripheral lymphocytes. AHEM 5:333 (in Czech)
Rössner P, Boffetta P, Ceppi M, Bonassi S, Smerhovsky Z, Landa K, Juzova D, Sram RJ (2005) Chromosomal aberrations in lymphocytes of healthy subjects and risk of cancer. Environ Health Perspect 113:517–520

Roznickova I (1987) Possibilities of the study on genotoxic risk in occupational exposure. Research report. District Institute of Hygiene, Teplice (in Czech)
Smerhovsky Z, Landa K, Rössner P, Brabec M, Zudova, Hola N, Pokorna Z, Mareckova J, Hurychova D (2001) Risk of cancer in an occupationally exposed cohort with increased level of chromosomal aberrations. Environ Health Perspect 109:41–45
Smerhovsky Z, Landa K, Rössner P, Juzova D, Brabec M, Zudova Z, Hola N, Zarska N, Navsimalova E (2002) Increased risk of cancer in radon-exposed miners with elevated frequency of chromosomal aberrations. Mutat Res 514:165–176
Sorsa M, Wilbourn J, Vainio H (1992) Human cytogenetic damage as a predictor of cancer risk. IARC Sci Publ 543–554
Sorsa M, Autio K, Demopolous NA, Järventaus H, Rössner P, Sram RJ, Stephanou G, Vladimiropoulos D (1994) Human cytogenetic biomonitoring of occupational exposure to 1,3-butadiene. Mutat Res 309:321–326
Sram RJ (1981) Cytogenetic analysis of peripheral lymphocytes as a method for monitoring environmental levels of mutagens. In: Gut I, Cikrt M, Plaa GL (eds) Industrial and environmental xenobiotics: metabolism and pharmacokinetics of organic chemicals and methods. Springer, Berlin, pp 187–193
Sram RJ, Kuleshov NP (1980) Monitoring of the occupational exposure to mutagens by the cytogenetic analysis of human peripheral lymphocytes in vivo. Arch Toxicol Suppl 4:11–1
Sram RJ, Zudova Z, Kuleshov NP (1980) Cytogenetic analysis of peripheral lymphocytes in workers occupationally exposed to epichlorohydrin. Mutat Res 70:115–120
Sram RJ, Samkova I, Hola N (1983a) High-dose ascorbic acid prophylaxis in workers occupationally exposed to halogenated ethers. J Hyg Epidemiol Microbiol Immunol 27:305–318
Sram R, Landa K, Samkova I (1983b) Effect of occupational exposure to epichlorohydrin on the frequency of chromosome aberrations in peripheral lymphocytes. Mutat Res 122:59–64
Sram RJ, Hola N, Kotesovec F, Vavra R (1985) Chromosomal abnormalities in soft coal open-cast mining workers. Mutat Res 144:271–275
Sram RJ, Binkova B, Dobias L, Rössner P, Topinka J, Vesela D, Vesely D, Stejskalova J, Bavorova H, Rericha V (1993a) Monitoring genotoxic exposure in uranium miners. Environ Health Perspect 99:303–305
Sram RJ, Dobias L, Rössner P, Vesela D, Vesely D, Rakusova R, Rericha V (1993b) Monitoring genotoxic exposure in uranium mines. Environ Health Perspect 101 (Suppl 3): 155–158
Sram RJ, Rössner P, Peltonen K, Podrazilová K, Mračková G, Demopoulos NA, Stephanou G, Vladimiropoulos D, Darroudi F, Tates AD (1998) Chromosomal aberrations, sister-chromatid exchanges, cells with high frequency of SCE, micronuclei and comet assay parameters in 1,3-butadiene exposed workers. Mutat Res 419:145–154
Sram RJ, Rössner P, Smerhovsky Z (2004a) Cytogenetic analysis and occupational health in the Czech Republic. Mutat Res 566:21–48
Sram RJ, Beskid O, Binkova B, Rössner P, Smerhovsky Z (2004b) Cytogenetic analysis using fluorescence in situ hybridization (FISH) to evaluate occupational exposure to carcinogens. Toxicol Lett 149:335–344
Srb V (1984) Genotoxic acitivity testing in production units with asbestos risk. I. Cytogenetic examination of peripheral lymphocytes. Prac Lek 36:175–178 (in Czech)
Srb V, Kubzova E (1985) Genetic risk at work in microbiological laboratory (In Czech). Cas Lek Ces 124:686–689
Srb V, Kubzova E, Musil M, Havel V, Kubikova K, Hartman M (1985) Testing the genotoxic activity in plant with asbestos risk. II. Cytogenetic monitoring of workers in 1982–1983. Prac Lek 37:76–80 (in Czech)
Srb V, Prochazkova J, Chylkova V, Turkova M et al (1987) Biologic-medical monitoring of members of surgical teams working in total anaesthetic hazards. Collec Sci Commun Charles Univ Hradec Kralove 30:337–412
Svandova E, Rössner P (1989) The contribution to the assessment of spontaneous level of chromosomal aberrations in Czechoslovak population. Prac Lek 41:216–219 (in Czech)

Zudova Z, Koudela K, Cerna M, Kapickova M, Mihula J, Suskavicova L (1985) Occupational exposure of medical personnel in selected health establishments. AHEM Suppl 3:230–235 (in Czech)
Zudova Z, Landa K (1977) Genetic risk of occupational exposure to haloethers. Mutat Res 46:242–243

22 Biological Dosimetry

Günter Stephan, Ursula Oestreicher, and Horst Romm

Abstract

Chromosome analysis is the method of choice in case of biological dosimetry, used for the quantification of exposures to ionising radiation. The advantages and disadvantages of dicentric chromosomes and symmetrical translocations are described. In general, confounding factors on the lower detectable dose limit and scoring criteria for symmetrical translocations are discussed. In the case of acute exposures, scoring of dicentric chromosomes is preferred when the blood sample can be taken shortly after the exposure. The dicentric chromosome is characteristic for ionising radiation and the spontaneous frequency is very low in the healthy general population (about one dicentric per 1,000 cells) and may be influenced by smoking. After whole-body exposure with low linear energy transfer radiation, doses down to about 100 mGy are detectable. In the case of chronic exposure or if the exposure happened years before blood sampling, the analysis of symmetrical translocations in stable cells is favoured. In contrast to dicentric chromosomes, which fade with time, symmetrical translocations can persist, being transmitted from blood-forming tissue to the peripheral blood. The background frequency of symmetrical translocations shows a clear age-dependency, resulting in an about tenfold higher spontaneous frequency and in higher individual variation. Using appropriate calibration curves, one can detect whole-body doses of about 300 mGy for individuals under 40 years of age and about 500 mGy for individuals over 40 years of age.

22.1 Introduction

Biological dosimetry is the quantification of exposures to ionising radiation by means of a biological indicator. Such indicators can provide a means of estimating an individual's exposure to ionising radiation and of predicting possible consequences. Such biological estimates of exposure would enable suitable medical treatment to be planned when information from physical dosimetry cannot be clearly related to radiation exposure or is not available.

Günter Obe and Vijayalaxmi (Eds.)
Chromosomal Alterations: Methods, Results and Importance in Human Health

Over a number of years several biological responses to radiation have been investigated, such as the induction of chromosome aberrations in peripheral blood lymphocytes and the increased level of micronuclei. In addition, other radiation-induced changes have been described using haematological, cytochemical, immunological and biochemical techniques.

Cytogenetic changes in the indicator system 'peripheral blood' have achieved significance for radiological protection purposes. The main breakthrough occurred in 1960 when Nowell (1960) reported a method for stimulating peripheral blood lymphocytes to divide in culture. With the availability of lymphocytes and the ease with which they can be cultured and dropped onto slides, these cells became the material of choice for biological dosimetry purposes. In 1962 it was suggested by Bender and Gooch (1962) that dicentric chromosomes in peripheral blood lymphocytes could well be used for the detection and dose assessment of human radiation exposures and these authors (Bender and Gooch 1966) used this method for the first time in the sense of biological dosimetry on the occasion of the so-called Recuplex criticality accident at Hanford, USA. A further significant development of the method was harlequin staining (Perry and Wolf 1974). With this technique it became possible to distinguish for the first time subsequent mitotic divisions after culture initiation. This discrimination of first and following cell divisions is insofar important as dicentric chromosomes are lost at cell division. Following the development of the fluorescence in situ hybridisation (FISH) technique (Pinkel et al. 1986), symmetrical translocations can easily be detected. This aberration type has the advantage that it can pass cell divisions without being selected. In consequence, dicentric chromosomes are the indicator of choice for recent and acute radiation exposures and symmetrical translocations for past and chronic radiation conditions.

This chapter describes the advantages and disadvantages of dicentric chromosomes and symmetrical translocations for the purpose of biological dosimetry considering confounding factors on the lower detectable dose limit. Furthermore, scoring criteria for symmetrical translocations are discussed.

22.2 Blood Sampling and Culture Conditions

A blood sample should be taken as soon as possible after radiation exposure. The best anticoagulant for lymphocyte cultures is preservative-free lithium heparin. In the case of partial-body or non-uniform exposure, the sample should be taken earliest about 24 h after the exposure in order to ensure a homogeneous mixture of irradiated and unirradiated lymphocytes. For the analysis of dicentric chromosomes, blood should be sampled within 4 weeks after exposure since after this time the frequency begins to decrease (IAEA 2001), but after exposure with lower doses the sampling time can be extended to about 6 months.

T lymphocytes are generally stimulated in vitro to proliferate by the mitogen phytohaemagglutinin during a culture period of about 48 h at 37°C. In order to discriminate between first and following cell divisions, bromodeoxyuridine is added to the culture medium. After the induction of a mitotic block by colcemid for the last 3 h of culture and subsequent fixation, the suspension can be dropped onto slides. The metaphase preparations can be further processed for conventional Giemsa staining (fluorescence plus Giemsa) or for FISH. A detailed protocol for culture setup is given by IAEA (2001).

22.3 Background Frequency of Asymmetrical and Symmetrical Chromosome Exchanges

An important prerequisite for dose reconstruction of exposed persons is the knowledge of the background levels of the corresponding chromosome aberration types determined in blood samples of healthy unexposed persons. Therefore a careful analysis of the baseline frequency of chromosome aberrations is important since this information has a direct bearing on biological dosimetry, especially in cases where individuals are suspected to have been overexposed, but no preexposed background aberration frequency is known.

Numerous studies have been performed to evaluate spontaneous background levels; however, the frequencies vary extensively among different studies for asymmetrical exchanges (dicentrics) (reviewed in Lloyd et al. 1980; Bauchinger 1995). Values for dicentric yields between 0.09 per 1,000 cells (Sevan'kaev et al. 1974) and 2.99 per 1,000 cells (Ganguly 1993) have been reported. Also symmetrical exchanges (translocations) vary among laboratories. It was expected that the use of harlequin staining together with standardisation of culture conditions would eliminate most interlaboratory variations, especially for dicentric chromosomes. It is evident, however, that most of the interlaboratory differences remain unexplained and should be attributed, at least partly, to the scoring criteria of the observer. In consequence, for biological dosimetry purposes each laboratory should have its own control data and its own dose–response curves. If so, in the case of a chromosome analysis of an exposed person, all conditions (culture, preparation, scoring) are identical for the three samples (control, dose–response curve, exposed person). Since the quantification of radiation exposure in the low-dose range is mainly influenced by the background frequency of chromosome aberrations, complete misinterpretation of data obtained for subjects who have received low-level exposure may occur depending to the background levels selected from the literature.

From Fig. 22.1 it becomes clear that most of the 53 individuals investigated show zero and one dicentric per 1,000 cells. Individuals with three and four dicentrics per 1,000 cells are smokers with a daily consumption of more than 30 cigarettes. The background level of this group (2.5 ± 0.6 dicentrics per 1,000 cells)

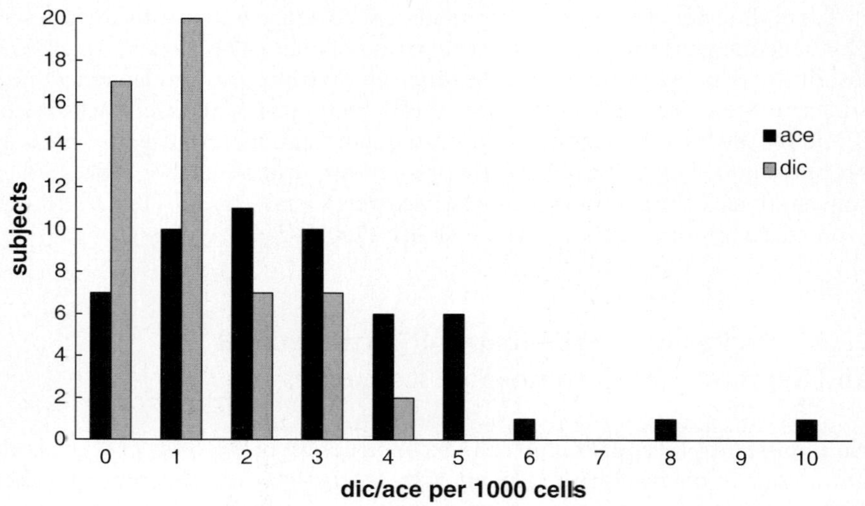

Fig. 22.1 Variation of the background frequency of dicentric chromosomes (*dic*) and acentric fragments (*ace*) between individuals (Stephan and Pressl 1999)

is significantly higher than that for non-smokers (0.9 ± 0.2 dicentrics per 1,000 cells). In the control group, with an age range from 20 to 73 years, no relation to age was observed; furthermore, there was no difference between men and women in the frequency of dicentric chromosomes (Stephan and Pressl 1999). In children, however, the background frequency of dicentrics was found to be lower (0.4 ± 0.2 dicentrics per 1,000 cells) than in adults (unpublished data).

The background frequency of symmetrical translocations is about 10 times higher than the dicentric frequency (Pressl et al. 1999). Whitehouse et al. (2005) have summarised background levels of eight working groups. It became evident that gender has no influence, smoking has only a marginal influence on the background frequency of symmetrical translocations, but age shows a very obvious effect. There is a significant difference in the background level between people younger than 40 years and those older than 40 years. As can be seen from Fig. 22.2, with increasing age the variability between individuals also increases, which, of course, influences the lowest detectable dose limit.

Sometimes the question is posed whether an increased level of chromosome aberrations has some health consequences. Since symmetrical translocations can be transmitted to descendent cells they may bear a potential risk and it is well documented that cancer is generally recognised to involve chromosomal changes. It is, however, not possible to predict some risk associated with an increased level of radiation-induced chromosome aberrations on the individual level. It was, however, shown that in populations individuals with an

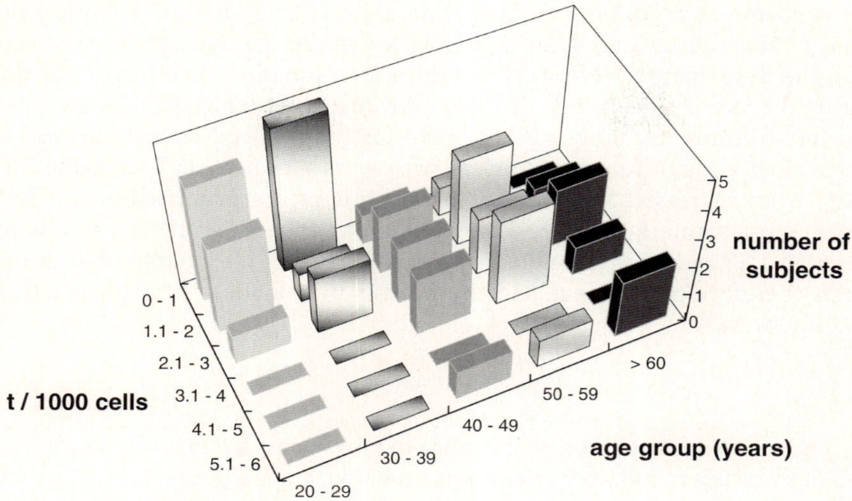

Fig. 22.2 Variation of the background frequency of symmetrical translocations (*t*) between individuals and dependency on age (Pressl et al. 1999)

elevated background frequency of aberrations have a higher risk of developing a carcinoma than individuals with a lower frequency (Hagmar et al. 1998).

22.4 Dose–Response Relationships

Dose–response curves for the yield of exchange aberrations are generated by irradiation of unstimulated lymphocytes in vitro. The relation between acute doses of low linear energy transfer (LET) radiation and yield of exchange aberrations can be described by the following equation:

$$y = c + \alpha D + \beta D^2,$$

where y is the yield of aberrations per cell, c is the background frequency, α is the coefficient for exchange aberrations produced by single electron tracks, β is the corresponding coefficient for aberrations produced as a consequence of two electron tracks and D is the absorbed dose in the cell. The quotient α/β is equal to the dose at which the linear and quadratic components contribute equally to the formation of exchange aberrations. Below this dose, the majority of aberrations will be produced by single tracks. As the interacting lesions must be produced almost simultaneously when the ionising particle traverses the nucleus, aberrations associated with the linear term are independent of the effects of dose rate or fractionation. The effect of fractionating the exposure or modifying the dose rate of low-LET radiation will affect the yield of aberrations mainly at higher doses where the damage is associated with the dose-squared term (βD^2) of the above equation. The reason for this is that the two breaks in two different chromosomes which are the supposition for forming exchange aberrations must be induced temporally and in space close enough for a random pairwise interaction. With increasing LET there is an increased likelihood that two breaks are produced by a single ionising track and the relative contribution of αD becomes more prominent. For high-LET radiations, the dose–response relation for exchange aberrations tends therefore to be linear and the yield is independent of exposure time and fractionation:

$$y = c + \alpha D.$$

It is of course of practical importance to know what are the relative efficiencies of the types of radiation of various qualities that are in use, or are likely to be in use, in medicine, or are in everyday use in industry. The relative biological effectiveness (RBE)–dose relations for the production of dicentric chromosomes are shown in Fig. 22.3 for various radiation qualities. It is obvious that the indicated X-ray quality possesses a higher effectivity for the induction of dicentric chromosomes in comparison with Cs γ-rays. The biological efficiency of X-rays is influenced by filtration since different filters/filter combinations produce various photon energy spectra and the biological effectivity of photon radiation is dependent on photon energy (Krumrey et al. 2004). It is well established that neutrons have a higher RBE than photon radiation, but also the RBE of neutrons is influenced by their energy (Nolte et al. 2005). In consequence, for biological dosimetry purposes an appropriate calibration curve must be used for dose estimation.

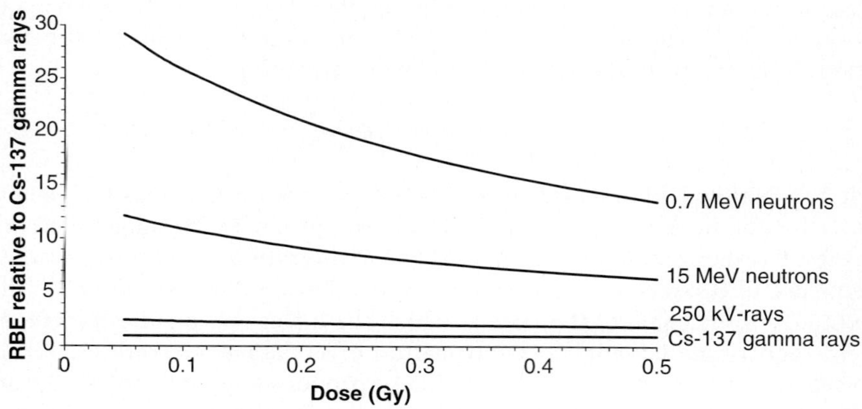

Fig. 22.3 Dose—relative biological effectiveness (*RBE*) relation for the induction of dicentric chromosomes by different radiation qualities

22.5 In Vitro and in Vivo Radiation-Induced Aberration Frequencies

The advantages of measuring radiation-induced chromosome aberrations produced in human circulating lymphocytes are obvious. The cells are in the same phase of the cell cycle at the time the blood sample is taken; therefore, only chromosome-type aberrations are produced. This fact makes quantification of the data reasonably simple. The circulating lymphocytes offer the opportunity to perform direct quantitative studies of the effect of ionising radiation on chromosomes. For the conversion of an observed frequency of aberrations into a dose, calibration curves are used which are established by in vitro exposures of whole blood. In order to extrapolate to humans from this system with confidence, results from controlled comparisons of in vivo and in vitro exposures must be known. There are some data which show that there is no significant difference in the aberration frequencies after whole-body irradiation of patients and their blood in vitro with X-rays (Buckton et al. 1969; Leonard et al. 1995). In comparative in vivo and in vitro investigations after ^{60}Co γ-irradiation, again, comparable results were obtained for the two different exposure conditions (Schmid et al. 1974). In a more recent paper (Stephan et al. 2005), results obtained in ankylosing spondylitis patients who had been treated with ^{224}Ra were in good agreement with published dose–response curves established by in vitro exposure to α-particles (Edwards et al. 1980; Schmid et al. 1996). These data demonstrate that calibration curves established by in vitro exposure can be used for the conversion of an observed frequency of exchange aberrations in an exposed individual into a dose.

22.6 Dicentric Chromosomes or Symmetrical Translocations for Dose Estimation

Biological dosimetry using the measurement of the yield of dicentric chromosomes in human lymphocytes is an established technique. It works well for acute radiation exposures when the blood sample is taken within a few weeks after exposure. This technique is capable of estimating whole-body doses of low-LET radiation down to about 100 mGy. It is less sensitive, however, for chronic exposures or if the exposure occurred years prior to the blood sampling. This is because the yield of dicentric chromosomes decreases with time after exposure depending on the cell turnover characteristics of the exposed person. A half-time of about 3 years can be applied to estimate the initial dicentric yield for dose estimation. But those estimates may be affected by uncertainties influenced by individual factors like those stimulating the cell turnover. Cells containing symmetrical translocations are not substantially eliminated from circulating blood with increasing postexposure time, since translocations do not impose a direct mechanistic hindrance to cell division and therefore they can be transmitted from blood-forming tissue to peripheral lymphocytes. For this reason, it would appear that the most reliable retrospective biological dosimetry can be expected from determining frequencies of this aberration type. Translocations can be recognised by FISH with composite whole-chromosome-specific DNA probes (Pinkel et al. 1986). There are a number of studies which have shown results suggesting that symmetrical translocations do not persist with time (Tucker et al 1997; Matsumoto et al. 1998; Natarajan et al. 1998), whereas others confirm persistence (Salassidis et al. 1995; Lucas et al. 1996; Stephan and Pressl 1997; Lloyd et al. 1998). An explanation for the conflicting results may be that in all the studies quoted only the painted chromosomes were scored. This means that translocations occurring in cells which carry also dicentric chromosomes, acentric fragments and centric rings (unstable cells) in the counterstained material are lost at cell division. Lindholm and Edwards (2004) showed, however, that when translocations in only stable cells were scored, the frequency remained at the same level for years. These results imply that retrospective dosimetry and calibration should be performed using stable cells only. In contrast to dicentric chromosomes, symmetrical translocations show an about tenfold higher background frequency and also a high interindividual variation (Fig. 22.2) and, furthermore, they show a distinct dependency on age. In consequence, the lower detectable dose limit is about 300 mGy in individuals under 40 years and about 500 mGy for individuals older than 40 years.

22.7 Summary

Scoring of chromosome aberrations in metaphase preparations of human T lymphocytes is an established method for estimating individual whole-body doses of radiation exposures. For acute exposures where the blood sample can be taken shortly after the exposure, scoring of dicentric chromosomes is the method of choice. For retrospective biological dosimetry, scoring of dicentric chromosomes is less precise since this aberration type is eliminated from peripheral blood with increasing postexposure time. Symmetrical translocations can pass cell division; therefore, they can be transmitted from blood-forming tissue to the peripheral blood and the frequency remains at the same level for years. In contrast to dicentric chromosomes, which show a low background frequency (about one dicentric per 1,000 cells), symmetrical translocations have a about tenfold higher spontaneous frequency and a higher individual variation. Furthermore, symmetrical translocations show a strong dependency on age and have an influence on the lowest detectable dose limit. Using appropriate calibration curves, we can quantify whole-body doses of about 100 mGy of low-LET radiation by dicentric chromosomes. Symmetrical translocations can detect doses to about 300 mGy for individuals under 40 years of age and about 500 mGy for individuals over 40 years of age.

References

Bauchinger M (1995) Quantification of low-level radiation exposure by conventional chromosome aberration analysis. Mutat Res 339:177–189

Bender MA, Gooch PC (1962) Types and rates of x-ray-induced chromosome aberrations in human blood irradiated in vitro. Proc Natl Acad Sci USA 48:522–532

Bender MA, Gooch PC (1966) Somatic chromosome aberrations induced by human whole-body irradiation: the "Recuplex" criticality accident. Radiat Res 29:568–582

Buckton KE, Langlands AO, Smith PG, Looby PC, Woodcock GE, McLelland J (1969) Chromosome aberrations induced in human peripheral blood by 2 MeV x irradiation to the whole body and in vitro. In: IAEA symposium on radiation induced cancer. Proceedings of the symposium in Athens. International Atomic Energy Agency, pp 135–146

Edwards AA, Purrott RJ, Prosser JS, Lloyd DC (1980) The induction of chromosome aberrations in human lymphocytes by alpha-radiation. Int J Radiat Biol 38:83–91

Ganguly BB (1993) Cell division, chromosomal damage and micronucleus formation in peripheral lymphocytes of healthy donors: related to donor's age. Mutat Res 295:135–148

Hagmar L, Bonassi S, Stromberg U, Brogger A, Knudsen LE, Norppa H, Reuterwall C (1998) Chromosomal aberrations in lymphocytes predict human cancer: a report from the European Study Group on Cytogenetic Biomarkers and Health. Cancer Res 58:4117–4121

IAEA (2001) Cytogenetic analysis for radiation dose assessment. Technical report series no 405. International Atomic Energy Agency, Vienna

Krumrey M, Ulm G, Schmid E (2004) Dicentric chromosomes in monolayers of human lymphocytes produced by monochromatized synchroton radiation with photon energies from 1.83 keV to 17.4 keV. Radiat Environ Biophys 43:1–6

Leonard A, Baltus I, Leonard ED, Gerber GB, Richard F, Wambersie A (1995) Dose-effect relationship for in vivo and in vitro induction of dicentric aberrations in blood lymphocytes of children. Radiat Res 141:95–98

Lindholm C, Edwards A (2004) Long-term persistence of translocations in stable lymphocytes from victims of a radiological accident. Int J Radiat Biol 80:559–566

Lloyd DC, Purrot RJ, Reeder EJ (1980) The incidence of unstable chromosome aberrations in peripheral blood lymphocytes from unirradiated and occupationally exposed people. Mutat Res 72:523–532

Lloyd DC, Moquet JE, Oram S, Edwards AA, Lucas JN (1998) Accidental intake of tritiated water: a cytogenetic follow-up case on translocation stability and dose reconstruction. Int J Radiat Biol 73:543–547

Lucas JN, Hill FS, Burk CE, Cox AB, Straume T (1996) Stability of the translocation frequency following whole-body irradiation measured in rhesus monkeys. Int J Radiat Biol 70:309–318

Matsumoto K, Ramsey MJ, Nelson DO, Tucker JD (1998) Persistence of radiation-induced translocations in human peripheral blood determined by chromosome painting Radiat Res 149:602–613

Natarajan AT, Santos SJ, Darroudi F, Hadjidikova V, Vermeulen S, Chatterjee S, Berg M, Grigorova M, Sakamoto-Hojo ET, Granath F, Ramalho AT, Curado MP (1998) 137 Cesium-induced chromosome aberrations analysed by fluorescence in situ hybridization: eight years follow up of the Goiania radiation accident victims. Mutat Res 400:299–312

Nolte R, Mühlbradt KH, Meulders JP, Stephan G, Haney M, Schmid E (2005) RBE of quasi-monoenergetic 60 MeV neutron radiation for induction of dicentric chromosomes in human lymphocytes. Radiat Environ Biophys 44:210–209

Nowell PC (1960) Phytohemagglutinin: an initiator of mitosis in cultures of normal human leukocytes. Cancer Res 20:462–466

Perry P, Wolff S (1974) New Giemsa method for differential staining of sister chromatids. Nature 251:156–158

Pinkel D, Straume T, Gray JW (1986) Cytogenetic analysis using quantitative high-sensitivity fluorescence hybridisation. Proc Nat Acad Sci USA 83:2934–2938

Pressl S, Edwards A, Stephan G (1999) The influence of age, sex and smoking habits on the background level of fish-detected translocations. Mutat Res 442:89–95

Salassidis K, Georgiadou-Schumacher V, Braselmann H, Müller P, Peter RU, Bauchinger M (1995) Chromosome painting in highly irradiated Chernobyl victims: a follow-up study to evaluate the stability of symmetrical translocations and the influence of clonal aberrations for retrospective dose estimation. Int J Radiat Biol 68:257–262

Schmid E, Bauchinger M, Bunde E, Ferbert HF, v Lieven H (1974) Comparison of the chromosome damage and its dose response after medical whole-body exposure to 60Co-gamma-rays and irradiation of blood in vitro. Int J Radiat Biol Relat Stud Phys Chem Med 26:31–37

Schmid E, Hieber L, Heinzmann U, Roos H, Kellerer AM (1996) Analysis of chromosome aberrations in human peripheral lymphocytes induced by in vitro alpha-particle irradiation. Radiat Environ Biophys 35:179-184

Sevan'kaev AV, Kozlov VM, Guseev GG, Izmailova NN (1974) Frequency of spontaneous chromosome aberrations in the culture of human lymphocytes. Genetics (USSR) 10:114–120

Stephan G, Pressl S (1997) Chromosome aberrations in human lymphocytes analysed by fluorescence in situ hybridization after in vitro irradiation, and in radiation workers, 11 years after an accidental radiation exposure. Int J Radiat Biol 71:293–299

Stephan G, Pressl S (1999) Chromosomal aberrations in peripheral lymphocytes from healthy subjects as detected in first cell division. Mutat Res 446:231–237

Stephan G, Kampen WU, Nosske D, Roos H (2005) Chromosomal aberrations in peripheral lymphocytes of patients treated with radium-224 for ankylosing spondylitis. Radiat Environ Biophys 44:23–28

Tucker JD, Breneman JW, Briner JF, Eveleth GG, Langlois RG, Moore DH 2nd (1997) Persistence of radiation-induced translocations in rat peripheral blood determined by chromosome painting. Environ Mol Mutagen 30:264–272

Whitehouse CA, Edwards AA, Tawn EJ, Stephan G, Oestreicher U, Moquet JE, Lloyd DC, Roy L, Voisin P, Lindholm C, Barquinero J, Barrios L, Caballin MR, Darroudi F, Fomina J (2005) Translocation yields in peripheral blood lymphocytes from control populations. Int J Radiat Biol 81:139–145

23 Statistical Methods for Biological Dosimetry

Marta Szłuińska, Alan A. Edwards, and David C. Lloyd

Abstract

Statistical methods most commonly employed in biological dosimetry by analysis of chromosomal aberrations in blood lymphocytes are reviewed and discussed. They are presented, together with worked examples, to illustrate the principles underlying the approaches to assessing data from differing types of radiation overexposures. These techniques have been developed from experience gained over many years from dealing with many radiological incidents and with in vitro experiments that explore possible irradiation scenarios. Together, they serve to illustrate practical aspects of the techniques, ranging from fitting curves to dose–response calibration data through to specifying uncertainties on individual dose estimates. A novel method of computation of dose and the corresponding 95% limits, taking into account both the Poisson-derived uncertainties in induced aberration yield as well as uncertainties associated with the calibration curve, is presented. A very good agreement of this stringent Bayesian-like calculation approach with a more simplified one, which merely takes into consideration the Poisson uncertainties in yield, validates the approximations used over many years for dose assessment.

23.1 Introduction

Human T lymphocytes are long-lived cells that can be easily obtained from a blood sample. They comprise a population of G_0-phase cells, which however can be stimulated in vitro by the mitogen phytohaemagglutinin (Nowell 1960) and can be stopped at their first metaphase with colcemid after about 45 h of culture at 37°C. Slides are then prepared and stained with Giemsa or fluorescence in situ hybridisation (FISH) probes and scored for aberrations. A check for contamination with second-division metaphases is recommended and is carried out by incorporating bromodeoxyuridine in conjunction with fluorescence plus Giemsa staining. The full details of these techniques can be found in an IAEA (2001) manual.

With conventional Giemsa staining, the scored aberrations comprise mainly dicentrics, centric rings and acentric fragments. For biological dosimetry,

Günter Obe and Vijayalaxmi (Eds.)
Chromosomal Alterations: Methods, Results and Importance in Human Health

dicentrics are the preferred aberration owing to their low background incidence, approximately 0.5–1 dicentric per 1,000 cells, and the reliability of detection. Centric rings are approximately 5% as frequent as dicentrics (Lloyd and Purrott 1981). The main disadvantage of the dicentric assay is that the damage is unstable as damaged cells are removed from the pool of peripheral blood lymphocytes at the rate with which cells are replaced. It is presently believed that for persons with normal haematological parameters, the disappearance rate is exponential with a half-life of approximately 3 years (Lloyd et al. 1980), although significant individual variation is observed. In this respect, the introduction of the FISH technique (Pinkel et al. 1986) permits the more persistent aberrations, such as stable translocations, to be scored, providing a reliable retrospective and long-term exposure dosimetry system (Lindholm and Edwards 2004; Darroudi 2000). The progress in biological dosimetry using FISH cytogenetics has recently been reviewed by Edwards et al. (2005) and Whitehouse et al. (2005).

An additional technique occasionally employed in biodosimetry is the cytokinesis blocked micronucleus assay (IAEA 2001). This identifies fragments, or sometimes whole chromosomes, that do not segregate properly at mitosis to daughter nuclei. They are excluded into the cytoplasm as discrete chromatin positive bodies seen alongside the two daughter nuclei. The most important advantage of this technique stems from the speed with which micronuclei can be scored, which, it has been suggested, is particularly convenient in the case of initial triage of large numbers of persons involved in a major radiation incident (Kóteles 1996). However, the value of the technique is limited mainly by the larger and more variable background frequency of micronuclei as compared with dicentrics.

The aim of this chapter is to review and discuss the statistical methods employed at the Health Protection Agency (HPA) (formerly the National Radiological Protection Board) for the analysis of chromosome aberrations in blood lymphocytes. Worked examples are included, which serve to illustrate practical aspects of the techniques. A novel method of computation of dose and the corresponding 95% limits, taking into account both the Poisson-derived uncertainties in induced aberration yield as well as uncertainties associated with the calibration curve, is also presented. A very good agreement of the latter approach with a more simplified one, which merely takes into consideration the Poisson uncertainties in yield, further validates the approximations used over many years for dose assessment.

23.2 Statistical Approach

The treatment of cytogenetics data invariably relies on tests obtained from standard textbooks of statistics. Thus, methods such as χ^2 and t tests are used when testing for homogeneity or establishing differences or trends between data points or patients. Uncertainties may be expressed as standard deviations, standard errors of the mean or confidence limits based on Poisson, binomial or normal distributions. These may involve conventional or Bayesian techniques. As always, the key to data analysis is to know which is the appropriate test for the purpose and how to interpret the result. This most certainly applies to the use of cytogenetics in radiation biodosimetry and this chapter attempts to summarise experience gained over several decades.

23.2.1 Dose–Response Calibration Curves

In the majority of cases, accidental overexposures are to X-radiation or γ-radiation sources; therefore, the determination of in vitro dose–response calibration curves for these types of radiation are the priority for laboratories involved in biological dosimetry and ought to be supported by reliable and accurate physical dosimetry. Important considerations, discussed in detail in IAEA (2001), to bear in mind are (1) the uniform irradiation of the blood sample, which can be ensured by adjusting the distance between the source and the sample to be sufficiently long; (2) the production of charged particle equilibrium, achieved by surrounding the sample with materials of similar composition to blood; (3) the dose rate for the in vitro simulation of an acute exposure, which should be chosen such that the required dose is delivered in less than 15 min; and (4) the geometry and materials of and surrounding the ionisation chamber should be selected such that the corrections for distance, absorption, scattering and the mismatch at the blood–holder interface are minimised.

From a statistical point of view, in order to carry out a satisfactory calibration curve fitting procedure, a sufficient number of degrees of freedom (DF) is necessary. This translates to around ten doses in the range 0.1–5.0 Gy for low linear energy transfer (LET) radiation such as X-rays or γ-rays. Doses higher than 5 Gy show progressive deviation from a curve of simple form, eventually producing saturation of the aberration yield (see the discussion at the end of this section; Lloyd and Edwards 1983). It is still a matter of some debate as to how to treat the background level of aberrations in the fitting procedure. The problem arises owing to the very low incidence of background dicentrics that may give a zero yield in a scored data point. The inclusion of the point of zero in the curve fitting may lead to a fitted negative control level, which is not biologically meaningful. The data point in question could be omitted from the fitting procedure and the curve forced to pass through the origin; however, this too is unrealistic because there is a large body of control data showing that, while low, the background is positive. Therefore, at HPA, the solution that has been chosen is to fit the dose–response curve with a small but positive yield with a large uncertainty assigned to the zero-dose data point.

The yield of chromosome aberrations (Y) is related to dose (D) by the following equation:

$$Y = c + \alpha D + \beta D^2, \tag{23.1}$$

Table 23.1 Summary of values of α and β and their standard errors (*SE*) for different types of radiation

Reference	Radiation source	$\alpha \pm$ SE (Gy^{-1})	$\beta \pm$ SE (Gy^{-2})
Lloyd et al. (1986)	250-kVp X-rays	0.046 ± 0.005	0.065 ± 0.003
Lloyd et al. (1986)	^{60}Co γ-rays	0.014 ± 0.004	0.076 ± 0.003
Prosser et al. (1983)	^{3}H β-particles	0.054 ± 0.005	0.067 ± 0.004
Lloyd et al. (1984)	14.9-MeV neutrons	0.195 ± 0.018	0.119 ± 0.020
Lloyd et al. (1976)	Cyclotron-generated neutrons ($\bar{E}$ = 7.6 MeV)	0.482 ± 0.040	0.064 ± 0.022
Lloyd et al. (1978)	^{252}Cf fission neutrons ($\bar{E}$ = 2.13 MeV)	0.600 ± 0.019	–
Lloyd et al. (1976)	Fission neutrons ($\bar{E}$ = 0.9 MeV)	0.733 ± 0.024	–
Lloyd et al. (1976)	Fission neutrons ($\bar{E}$ = 0.7 MeV)	0.832 ± 0.032	–
Lloyd et al. (1988)	24-keV neutrons	0.821 ± 0.031	–
Edwards et al. (1980)	^{242}Cm α-particles	0.286 ± 0.015	–
Purrott et al. (1980)	^{239}Pu α-particles	0.375 ± 0.024	–
Edwards et al. (1986)	8.7-MeV protons (L = 5.1 keV μm⁻1)	0.044 ± 0.075	0.058 ± 0.057
Edwards et al. (1986)	23.5-MeV ^{3}He ions (L = 22 keV μm^{-1})	0.394 ± 0.018	–

$\bar{E}$ mean energy, L linear energy transfer

where c, α and β are the fitted coefficients. The control value c for dicentrics is approximately 0.0005–0.001. Some values of α and β measured at HPA are summarised in Table 23.1; however in practice, a synthesis of various experimental sets of data, judged to be the best to date, is used to produce the coefficients routinely used for the purposes of biological dosimetry.

Chromosome aberrations are a good example of a biological effect caused by ionising radiation where it is possible to discriminate differences between radiation qualities that are generally termed low LET. These are qualities for which in radiological protection a deliberate decision has been taken to ascribe a weighting factor (w_R) of 1; thus, no attempt is made to distinguish between their biological effectiveness (ICRP 1990). Nevertheless, with dicentrics, differences can be seen and are reflected in the linear yield coefficient α and are thus most evident at low doses. In practice, laboratories tend to calibrate with orthovoltage X-rays (200–250 kVp) and with a convenient γ-ray source, usually cobalt-60. However, when confronted with the need to estimate a dose in a case that involves exposure to other low-LET sources, the practice at HPA is to assume an α coefficient by extrapolating or interpolating between established calibration data. For example, when dealing with lower-energy diagnostic X-rays, such as 30 kVp, an α coefficient of about 0.05 is assumed. For cobalt-60 γ-ray sources ($\bar{E}$ = 1.2 MeV) an α value of 0.02 is used, but for lower-energy sources, also commonly used in industrial radiography, such as caesium-137 ($\bar{E} = 660$ keV) or iridium-192 ($\bar{E} = 317$ keV), α values of 0.025 and 0.03 are used, respectively.

Values of α and β from Table 23.1 have been used to produce an illustrative comparison of calibration curves for some chosen types of radiation shown in Fig. 23.1. It is worth noting that while the dose–response curves depend very strongly on the energy in the case of neutrons, they are rather similar in the case of photon radiation, but, as discussed earlier, with γ-rays being less effective at producing aberrations at low doses than X-rays. The reason for this stems from differences in the actual physical interaction of ionising radiation of different qualities with tissue. In the case of high doses of low-LET, or sparsely ionisation radiation, the ionisations are randomly distributed among cells and a large number of tracks are present; hence, assuming an equal probability that any DNA damage can interact with a nearby similar break and be converted into an aberration, aberrations should be randomly distributed among cells. Indeed, it has been shown that aberrations for low-LET radiation conform closely to the Poisson distribution; however, this is not the case for high-LET radiation (e.g. neutrons, α-particles), where more cells with multiple and no aberrations are observed than expected from the Poisson distribution. This is because ionisations are non-randomly distributed across the cell, as high-LET energy is deposited in discrete tracks. From this it follows that while aberrations produced by a single track have a frequency that is proportional to a linear function of dose (β in Eq. 23.1 equals zero; e.g. for lower-energy neutrons), those induced by two independent tracks have a frequency proportional to the square of dose. The ratio of α/β of

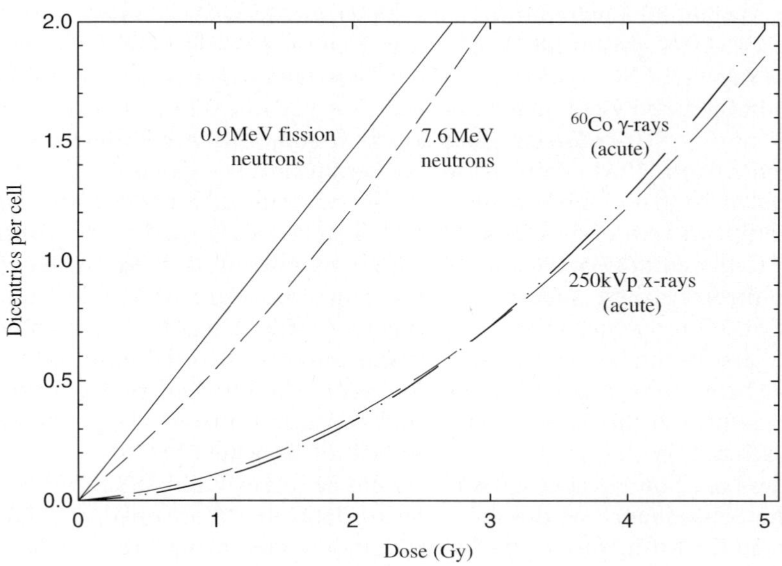

Fig. 23.1 Dose–response curves for four selected radiation qualities

the coefficients in Eq. 23.1 is equivalent to the dose at which linear and quadratic components contribute equally to the formation of aberrations.

In practice, prior to fitting a polynomial defined by Eq. 23.1 to the data points given at each dose, a test on the distribution of dicentrics is carried out in order to determine whether the yields at each data point conform to the Poisson distribution. Following the method of Savage (1970), the so-called *u* test of the goodness of fit is employed (CtamPoiss computer program available from HPA). This relies on the value of the dispersion index, which provides a measure of how well the sample's variance corresponds to that predicted by the Poisson distribution. Therefore, any deviation of the variance-to-mean ratio from unity (i.e. underdispersion or overdispersion compared with the Poisson distribution) can subsequently be taken into account in the fitting procedure by adjusting the weights of the data points appropriately (by multiplying the Poisson-derived variance of the data points by the ratio of the variance and the mean). Although some laboratories tend to use the individual ratios of the variance and the mean in their curve fitting, at HPA it is common practice to check for any trends in the values of variance-to-mean ratios with increasing dose. This is preferable because the ratios themselves have large statistical uncertainties. If no trends are observed, then a mean ratio is calculated and applied to all the data points.

The importance of weighting in curve fitting cannot be overstressed, as more accurate points have to be considered more seriously than the inaccurate ones and, indeed, many commercially available curve-fitting programs simply ascribe equal weights to each point. This has led to erroneous dicentric response curves being published. The most common approach is to minimise the sum of squared differences (SSD) between the observed and fitted values, with weights of the data points given by the inverse of their variance. It was suggested by Papworth (1975) to base the estimates of variance on the value of the fitted yield rather than the observed one. This can be done iteratively until a self-consistent solution is found (MlPol computer program available from HPA). This method of iteratively reweighted least squares is equivalent to that of the maximum likelihood (Papworth 1975). The goodness of the fit can be tested assuming that SSD follows a χ^2 distribution with DF equal to the number of data points less the number of fitted parameters (normally three: c, α, β). A close correspondence between SSD and DF implies a good fit. Whilst values of SSD significantly greater than DF suggest an uncertainty due to the lack of fit, SSD considerably less than DF hint to the data points being too close, within their errors, to the fitted curve. Generally, the uncertainties in the parameters should be based on the Poisson estimate of the variance; however, in the case of lack of fit, where SSD>DF, the uncertainties based on Poisson statistics should be increased by $(\mathrm{SSD/DF})^{1/2}$.

The consequence of the inclusion of data points obtained at 5 Gy and above in the fitting procedure can be demonstrated using data from an early classic experiment of Norman and Sasaki (1966), who irradiated human

Table 23.2 The effect of sequentially removing the higher dose levels on the results of fitting the data of Norman and Sasaki (1966)

Highest dose (Gy)	$\alpha \pm \text{SE} \times 10^{-3}\ \text{Gy}^{-1}$	$\beta \pm \text{SE} \times 10^{-2}\ \text{Gy}^{-2}$	χ^2	Degrees of freedom
30	180.0 ± 37.0	0.95 ± 0.24	205	10
25	149.0 ± 34.0	1.35 ± 0.28	142	9
20	120.0 ± 31.0	1.79 ± 0.32	99	8
12	49.0 ± 21.0	3.20 ± 0.35	30	7
8	16.0 ± 15.0	4.18 ± 0.36	11	6
6	3.4 ± 13.1	4.67 ± 0.37	5.9	5
5	−5.9 ± 4.4	5.16 ± 0.44	2.5	4

lymphocytes with X-rays to doses up to 30 Gy. As mentioned before, approximately from 5 Gy onwards, a saturation of dose–response curves is observed and, therefore, special care must be taken in the selection of the data points used in the fit. As shown in Table 23.2, the inclusion of higher-dose data points in the fit to the quadratic model (Eq. 23.1) using the maximum-likelihood method leads to a decrease in the values of the β coefficient and a simultaneous increase in the α coefficient. This is due to the fact that α and β are correlated and, hence, lower values of one must be compensated by enlarged values of the other. From Table 23.2, it can be deduced that the best fit in this case is obtained when doses up to 6 Gy are included in the fit.

23.2.2 Dose Assessment in Whole-Body Exposure

Deriving the dose from a given yield of aberrations does not present any difficulty as it follows directly from Eq. 23.1 that

$$D=\left(-\alpha+\sqrt{\left[\alpha^2+4\beta\left(Y-c\right)\right]}\right)\Big/2\beta. \tag{23.2}$$

However, there are a number of approaches that permit calculation of uncertainties. These are customarily expressed as confidence limits, with 95% being the limit chosen most often. The 95% confidence limits define an interval that embraces the true dose on 95% of occasions. The difficulty in determining the confidence limits arises from the fact that there are two components to the uncertainty: one originates from the Poisson nature of aberrations seen in blood samples from irradiated persons, and the other is associated with uncertainties in the calibration curve. This problem was addressed by Savage and Papworth (2000) and resulted in two fairly complex methods of estimating the confidence limits. In the first approach, the confidence limits can be obtained from differentiation of Eq. 23.2 and expressing the variance in dose (var$D$) in terms of variances and covariances of the fitted parameters α, β and c. Then, 95% confidence limits are given by $D \pm 1.96\mathrm{SE}(D)$, where $\mathrm{SE}(D) = \sqrt{\mathrm{var}\, D}$. The second approach is similar, but relies on uncertainties in yield rather than dose. Full details of the two methods can be found in IAEA (2001).

In this chapter, it is intended to emphasise the two methods of the confidence limit estimation that are routinely employed at HPA. It has recently been shown that the simplified approach, described in detail below, produces results that are in very good agreement with a more sophisticated explicit calculation. The simplified approach, illustrated in Fig. 23.2, is a modification of the method proposed by Merkle (1983) and involves the following steps:

- Determining the yields corresponding to the lower and upper 95% confidence limit on the observed yield (Y_L and Y_U), assuming the Poisson distribution.
- Determining the dose at which Y_L crosses the dose–response curve. This corresponds to the lower confidence limit (D_L).
- Determining the dose at which Y_U crosses the dose–response curve. This corresponds to the upper confidence limit (D_U).

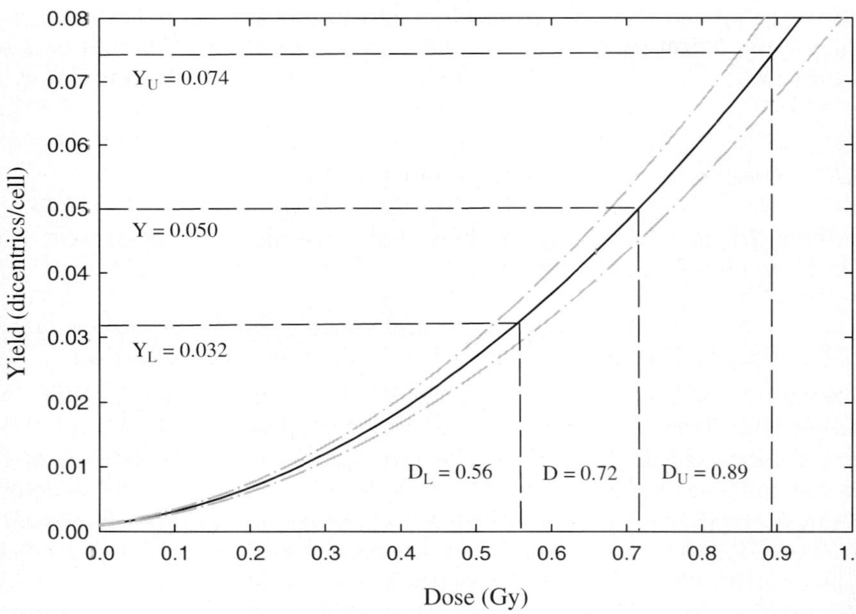

Fig. 23.2 A dose–response calibration curve used to estimate uncertainties for the simplified approach. The *dot-dashed lines* represent 95% confidence limits of the calibration curve. *Y* yield, *D* dose, *U* upper 95% confidence limits, *L* lower 95% confidence limit

Example: Twenty-five dicentrics are observed in 500 cells from a person overexposed to γ-rays. The parameters of the γ-ray calibration curve, shown in Fig. 23.2, are $c = 0.001$, $\alpha = 0.014$ and $\beta = 0.076$.

- From standard statistical tables or using a computer spreadsheet, the 95% lower and upper confidence limits on 25 observations are 16.18 and 36.91, respectively; hence, $Y_L = 16.18/500 = 0.032$ and $Y_U = 36.91/500 = 0.074$ dicentrics per cell.
- Y_L intersects the calibration curve at $D_L = 0.56$ Gy.
- Y_U intersects the calibration curve at $D_U = 0.89$ Gy.

It is important to recognise that the simplified approach does not take account of uncertainties associated with the dose–response curve. An initial approach to take into consideration both the uncertainties in the calibration curve as well as the Poisson nature of the yield was made by Merkle (1983) by reading off the value of dose for which $Y_L(Y_U)$ intersects the upper (lower) 95% confidence limit of the dose–response calibration curve. This approach overestimated the effect of uncertainties in the calibration curve.

The second approach discussed here, developed specifically to test the departure of the results of the simplified method from an explicit calculation accounting for all uncertainties, relies on a systematic convolution of the possible values of yield with the possible values of the calibration curve. This is a Bayesian type of approach as uncertainties in the observed yield and in the calibration curve are expressed as continuous probability distributions. More specifically, as shown schematically in Fig. 23.3, the values of possible yields are described by the Poisson likelihood function for the particular number of observed dicentrics (dark grey curve in the Fig. 23.3, in this case 25 dicentrics seen in 500 cells). This Poisson likelihood function is then convoluted with a distribution of possible values of dose given, for each value of yield (Y), by dose likelihood functions (light grey curves in Fig. 23.3). These are assumed to be normal and are determined using the knowledge of the standard errors on each data point of the calibration curve (obtained in turn from the curve-fitting procedure, described earlier in Sect. 23.2.1). The standard deviation of the normal distribution can be approximated by $\sigma = (D_{UL} - D_{LL})/2$, where D_{LL} and D_{UL} are the lower and upper limits of dose computed from the standard error in Y (Fig. 23.3). In the case of γ-rays, the standard errors on the yields at individual data points incorporated in the calculation varied from very large at very low doses to approximately 2–3% on doses from 0.8 Gy onwards. The summation of all the dose likelihood functions, weighted according to the likelihood given in dark grey, produces the final distribution of dose (thick black curve), from which a mean, a most likely or a median value, as well as 95% limits, can be calculated. The development of the computer program required checks of convergence of the dose results depending on the increments in yield and dose used for the calculation. Moreover, in order to simulate a biologically

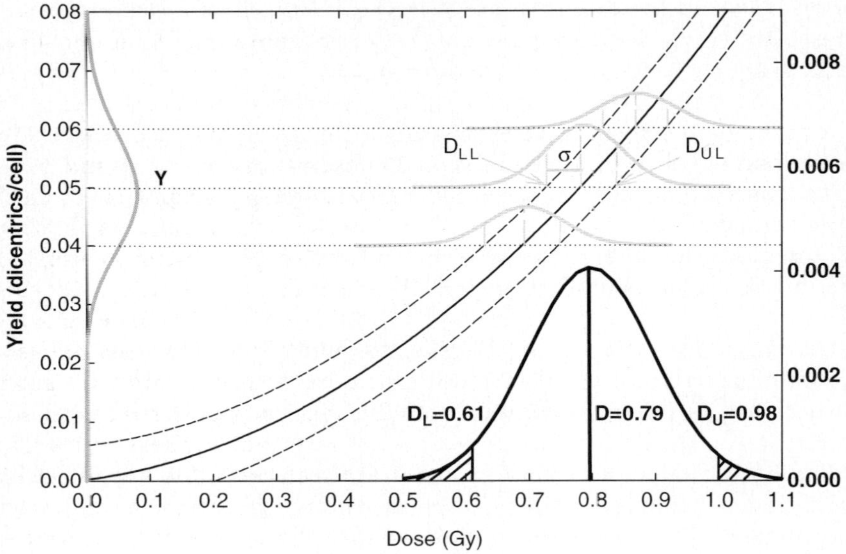

Fig. 23.3 Illustration of the explicit dose calculation which takes into account both the Poisson nature of the yield distribution and the uncertainties in the calibration curves: *solid black curve* γ-ray calibration curve $Y = 0.0005(\pm 0.0004) + 0.0128(\pm 0.0031)\ D + 0.064(\pm 0.0021)\ D^2$; *dashed black curves* -standard error on the calibration curve (exaggerated for clarity); *dark grey curve* the yield likelihood function for 25 dicentrics scored in 500 cells; *light grey curves* dose likelihood functions; *thick black curve* dose distribution. For more details see the text

realistic situation, it has been assumed that all negative dose values correspond to zero dose. In Table 23.3, the estimated doses obtained employing the simplified method are compared with those from the explicit calculation. Strictly, the 95% confidence limits in the simplified method and the 95% limits in the calculation are not equivalent; nevertheless, a very good agreement between the two set of data for the commonly encountered numbers of aberrations entirely justifies the use of the simplified method for the purposes of biological dosimetry. However, for very high and very low yields of aberrations, notable departures from the results of the simplified approach may be expected. Whilst at low doses the difference is due to negative doses not being permitted in the Bayesian calculation, at high doses it is the errors in the curve that become increasingly significant. It was concluded by Lloyd et al. (2007) that by the traditional method of scoring 1,000 metaphases, the lowest measurable γ-ray dose is around 100 mGy. The present investigation has also confirmed the general feeling that the uncertainties on a *good* calibration curve contribute little to the total convoluted uncertainty of the estimated dose, in contrast to the uncertainties associated with the Poisson nature of the yield distribution. Indeed in the case of γ-rays for the yields considered in Table 23.3, it has been verified that σ

Table 23.3 A comparison between the dose and the 95% lower (*LCL*; *LL*) and upper (*UCL*; *UL*) limits obtained using the simplified approach and the explicit calculation for various numbers of dicentrics scored in 500 cells in the case of γ-ray and fission neutron irradiation

	Simplified method			Explicit calculation		
Dicentrics	95% LCL	Mean dose (Gy)	95% UCL	95% LL	Mean dose (Gy)	95% UL
γ-rays: $Y = 0.0005(\pm 0.0004) + 0.0128(\pm 0.0031)D + 0.064\ (\pm 0.0021)D^2$						
0	0	0	0.242	0	0.070	0.249
5	0.129	0.297	0.505	0.164	0.326	0.506
20	0.519	0.692	0.883	0.538	0.704	0.882
50	0.978	1.151	1.336	0.990	1.154	1.334
Neutrons: $Y = 0.0005(\pm 0.0005) + 0.832(\pm 0.009)D$						
0	0	0	0.008	0	0.002	0.008
5	0.003	0.011	0.028	0.005	0.013	0.025
20	0.029	0.048	0.074	0.028	0.046	0.068
50	0.089	0.120	0.158	0.083	0.112	0.148

would have to be increased by a factor larger than 3 in order to observe considerable differences in the 95% limits of dose.

The procedures discussed above for the dicentric assay, such as principles of curve fitting, distribution testing and derivation of uncertainties, apply equally to the FISH assay as well as the micronucleus assay. In the latter case, it is worth noting that micronuclei are inherently overdispersed. For FISH, statistical analysis using the frequency of aberrations detected with chromosome painting probes depends on, amongst others, the number of cells analysed, the cocktail of chromosomes used and the exposure time. The full genome translocation frequency is given by $F_G = F_p/2.05f_p\,(1 - f_p)$, where F_p is the aberration frequency detected by FISH and f_p is the fraction of genome hybridised (Lucas 1992) taking into account the gender of the subject. Values of f_p can be obtained from Morton (1991). It is important to note that the control level of aberrations in the case of FISH analysis is much higher than in the case of the dicentric assay and is variable, with age being the main confounding factor. This is primarily because FISH is a retrospective technique and thus reflects a cumulative lifetime dose to the bone marrow. In Fig. 23.4, the results of the most comprehensive study to date of the dependence of the background levels of stable translocations on age is shown, where the age group of 80 years or more produced approximately 15 translocations per 1,000 genome-equivalent cells. To summarise, the calibration curves used in the case of FISH for the dose assessment differ from those for the dicentric assay in that the *c* coefficient in Eq. 23.2 is age-dependent and the quadratic term is equal to zero for all qualities of radiation as the exposure is protracted, i.e. the probability of a translocation being produced by two or more different tracks is negligible.

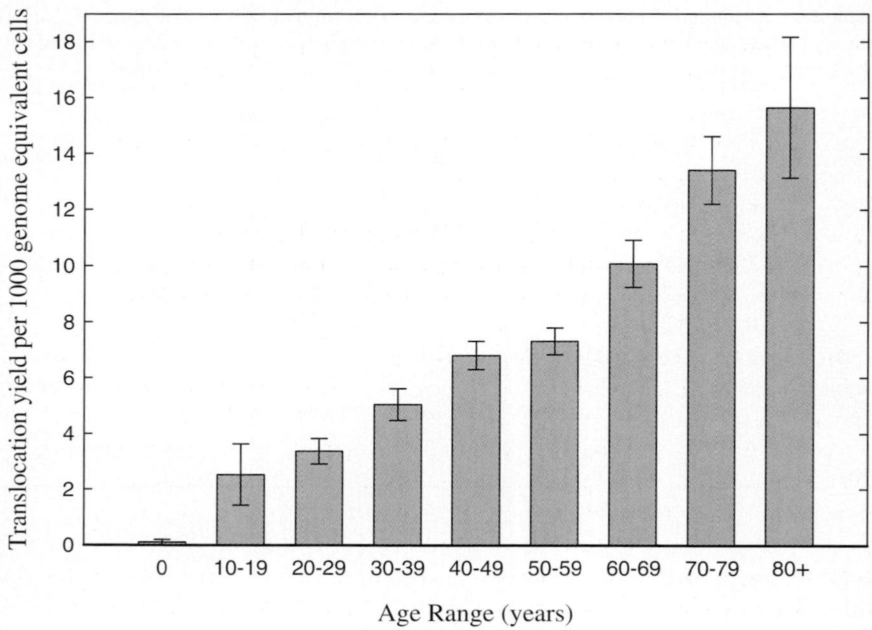

Fig. 23.4 Dependence of the background level of stable translocations on age measured with fluorescence in situ hybridisation (Whitehouse et al. 2005)

23.3 Examples of Dose Estimation

23.3.1 Criticality

In a criticality accident the body is irradiated by both neutrons and γ-rays. Since the ratio of doses due to neutrons and γ-rays is often known from physical measurement, the separate doses due to the two types of radiation can be obtained iteratively as follows:

1. Initially, all aberrations are assumed to be due to neutrons, and from the measured yield of dicentrics, a dose is estimated.
2. The ratio of neutrons to γ-rays is used to estimate the dose due to γ-rays.
3. From the estimate of γ-ray dose, a yield of dicentrics due to γ-rays is obtained.
4. The neutron yield is obtained by subtracting the yield of dicentrics in stage 3 from the measured yield.
5. A new estimate of neutron dose is made, and stages 2–5 are repeated until a self-consistent solution is found.

Example: Following a criticality accident 100 cells are scored and 100 dicentrics are observed. The neutron to γ-ray ratio available from physical measurements is 2:3. Cytogenetic dose estimates are to be made with 0.7-MeV fission neutron ($Y = 0.0005 + 0.83D$) and ^{60}Co γ-ray ($Y = 0.0005 + 0.014D + 0.076D^2$) calibration curves:

1. 100 dicentrics per 100 cells = 1.0 dicentric per cell, which is equivalent to 1.20 Gy neutron dose.
2. 1.20 × 3/2 = 1.81 Gy γ-rays.
3. 1.81 Gy γ-rays is equivalent to 0.275 dicentrics per cell.
4. 1.0–0.275 = 0.725, which is the dicentric yield attributable to neutrons.
5. 0.725 dicentrics per cell is equivalent to 0.87 Gy neutron dose.

In Table 23.4, the sequence of steps 1–5 is laid out. After a few iterations a solution is found with doses of approximately 1.0 and 1.5 Gy due to neutrons and γ-rays, respectively.

An alternative approach developed for criticality incidents involves an elaborate Bayesian type of calculation (Brame and Groer 2003), where uncertainties in all parameters can be rigorously taken into account by expressing them in terms of probability distributions. This technique, although time-consuming, is particularly useful in situations where the γ-ray to neutron ratio is uncertain (Voisin et al. 2004). However, the two methods give results in good agreement.

23.3.2 Low Dose Overexposure

Example: A case of a worker who returned a thermoluminescence dosimeter badge which recorded a dose of 66 mSv is considered. This was reported to be due to γ-rays of energy around 100 keV and it was not possible to discount it as a false reading owing to various circumstances. Subsequently, the dose was allocated to the worker's dose record with a consequence of the worker being removed from radiological work. This case was referred for a cytogenetic test, which resulted in an observation of one dicentric in 1,000 metaphases. From the calibration curve $Y = 0.001 + 0.04D + 0.06D^2$ a dose estimate of zero with

Table 23.4 Sequence of steps for estimation of doses for mixed neutron and γ-ray irradiation

Steps 1 and 5: neutron dose (Gy)	Step 2: γ-ray dose (Gy)	Step 3: γ-ray yield (dicentrics/cell)	Step 4: neutron yield (dicentrics/cell)
1.20	1.81	0.275	0.725
0.87	1.31	0.149	0.851
1.02	1.54	0.202	0.798
0.96	1.44	0.178	0.822
0.99	1.48	0.188	0.812

95% confidence limits ranging from 0 to 100 mGy was obtained and, thus, a dose of 66 mSv could not be discounted. However, there is a different way in which results can be presented. If only two possibilities are taken into consideration, i.e. zero dose or 66 mGy, then the odds in favour of zero dose come out as 4.5:1. This is obtained from the relative probabilities of observing one dicentric from the two doses. If the dose was zero, then the expected number of dicentrics to be observed is 1 in 1,000 (the presumed background). For a dose of 66 mGy, 3.9 dicentrics are expected in 1,000 cells by reference to the calibration curve. Therefore, the chances of observing one dicentric when one and 3.9 are expected, derived from a Poisson distribution, are $e^{-1} = 0.368$ and $3.9e^{-3.9} = 0.08$, respectively. The ratio of the two values is approximately 4.5.

In this particular case, the latter approach of presenting the results might be more appropriate because on the balance of probabilities this was a dose received by the badge and not the person. Whilst long-term health effects at such low doses are unlikely, there are legal and social consequences. In Table 23.5, 95% confidence limits for various low yields of dicentrics in 1,000 cells are shown together with odds ratios representing the probability of 0 or 100 mGy dose.

Experience has shown that when results of biodosimetry investigations are reported to physicians or even directly to patients, it is often difficult for them to comprehend the concept of confidence limits. The odds method, when appropriate, is much more easily understood. However, for situations where the odds method cannot be used, a new approach expressing a normalised distribution in dose (Fig. 23.5) has been developed in an effort to visually aid the presentation of the results (Lloyd et al. 2006). The method for deriving this distribution is to set up the probability distribution in yield which could have produced the observed number of dicentrics, which is

Table 23.5 Ninety-five percent upper and lower confidence limits (*UCL*; *LCL*) on various low yields of dicentrics in 1,000 cells and the odds ratios showing the likelihood (p) of the observation given that the dose is 0 or 100 mGy. The assumed calibration is $Y = 0.001 + 0.04D + 0.06D^2$

Observations (dicentrics)	Dose (Gy) LCL	Mean	UCL	Odds ratio $p(0\ \text{Gy})/p(0.1\ \text{Gy})$
0	–	0	0.06	180:1
1	0	0	0.10	25:1
2	0	0.02	0.13	4:1
3	0	0.05	0.16	1:1
4	0	0.07	0.18	1:9
5	0.02	0.10	0.21	1:50
6	0.03	0.11	0.23	1:310
7	0.04	0.13	0.25	1:1,900

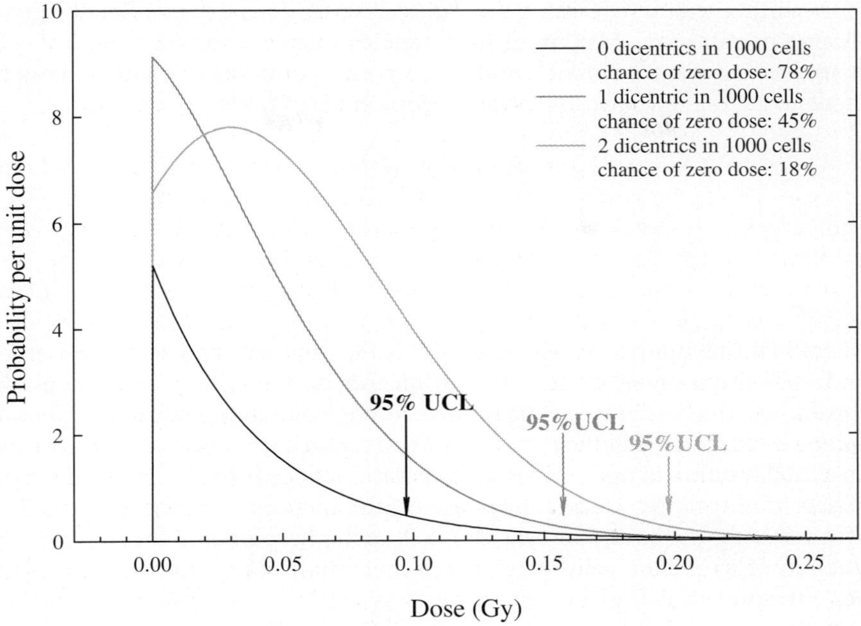

Fig. 23.5 A probability distribution in dose derived from the observed number of dicentrics in 1,000 cells, calculated for high-energy γ-rays (e.g. cobalt-60). *UCL* upper confidence limit

numerically identical to the likelihood function. Then, for each value of yield, a value of dose can be assigned, leading to the distribution in dose. Arbitrarily, all situations where the yield is less than the control level have been assigned a dose of zero and, thus, the proportion of the full bell-shaped curve that represents doses below zero has not been drawn.

23.3.3 Partial Body Exposure

In the case of a partial body exposure a more realistic estimate of dose may be given than simply stating the mean dose to the whole body. For partial body exposures, where a high dose is received inhomogeneously over the body, the distribution of dicentrics amongst all scored cells is overdispersed and this is used to estimate the extent of partial body irradiation. The observed distribution of cells is assumed to comprise (1) a Poisson distribution, which represents the irradiated fraction of the body, and (2) the remaining unexposed part. It should be stressed that it is a simplified way of approximating a real situation where a gradient of doses is expected. Undamaged cells include two subpopulations, those from the unexposed fraction and irradiated cells representing the first term of the Poisson series (e^{-Y}, i.e. irradiated but no dicentrics observed). Maximum-likelihood estimates of the fraction, f, of cells scored which were irradiated and the corresponding mean yield, Y, may be made using the following equations (Lloyd et al. 1991):

$$\frac{Y}{1-e^{-Y}} = \frac{X}{N-n_0} \tag{23.3}$$

and

$$f = \frac{X}{NY}, \tag{23.4}$$

where N is the number of cells scored, X is the number of dicentrics observed and n_0 is the number of cells free of dicentrics. Equation 23.3 can only be solved iteratively. Then, the mean dose to the irradiated fraction of the body can be estimated using the standard dose–response curves. In order to derive an actual fraction of body exposed to radiation, f needs to be corrected for the effects of mitotic delay and interphase death (apoptosis), as these are the factors reducing the ability of cells to reach metaphase in a 48-h culture. If the fraction of irradiated cells which reach metaphase was p, the fraction of the body irradiated, F, is given by

$$F = \frac{f/p}{1-f+f/p}. \tag{23.5}$$

At HPA, the computer program CtamPoiss (available on request) is used to solve the contaminated Poisson problems. For more details on the limitations of the aforementioned approach, see IAEA (2001).

An alternative approach to estimate a dose delivered in a partial body exposure was proposed by Sasaki and Miyata (1968) and is known as the Qdr method. In this case, the yield of dicentrics and rings in just the damaged cells is considered, assuming that these cells were present at the time of the irradiation. Then, Qdr is the expected yield of dicentrics and rings (X) amongst the damaged cells (N_d) and is given by

$$\mathrm{Qdr} = X/N_\mathrm{d} = Y_1/\left[1-\exp\left(-Y_1-Y_2\right)\right], \tag{23.6}$$

where Y_1 and Y_2 are yields of dicentrics plus rings and excess acentrics, respectively. Despite the limitations of the original Qdr method, which include the lack of any information on the size of the fraction irradiated as well as incorrect assumptions that excess acentric aberrations conform to a Poisson distribution and that the background frequency of dicentrics plus rings is zero, the agreement between the two methods described is good.

Additionally, in an attempt to reduce the oversimplification of regarding the exposure geometry as two fractions, an unirradiated and a uniformly irradiated component, Sasaki (2003) suggested an approach whereby the chromosome aberration distribution is unfolded into a mixed Poisson distribution and, from that, into a dose-distribution profile.

Example: An in vitro experiment was performed to simulate partial body exposure, whereby 50% of irradiated cells (5 Gy, 250 kVp X-ray) were mixed with unirradiated ones from the same donor (Lloyd et al. 1991). The number of dicentrics scored was 45 in 247 cells, with the following distribution: 225 cells with zero, four with one, 13 with two and five with three dicentrics. This produced the ratio of variance to mean of 2; therefore, an overdispersion expected from partial body exposure was revealed. The application of the contaminated Poisson method using the dose–effect curve of $Y = 0.0005 + 0.04D + 0.06D^2$ and the value of $D_0 = 3$ Gy ($p = e^{-D/D_0}$), which represents the relative selection against irradiated cells reaching metaphase, resulted in an estimate of the mean dose of (4.9 ± 0.5) Gy delivered to an estimated (40 ± 6)% of the body. Only (0.11 ± 0.02)% of the scored cells appeared to have been irradiated, which demonstrates the need to compensate for the reduction due to interphase death and mitotic delay. In order to carry out the Qdr analysis, the number of dicentrics plus rings, 48 in this case, as well as the number of damaged cells, 31, and the following dose–response calibration curves are required: $Y_{\text{dicentrics+rings}} = 0.0005 + 0.043D + 0.063D^2$ and $Y_{\text{acentrics}} = 0.0035 + 0.051D + 0.11D^2$. By substitution into Eq. 23.6,

$$\text{Qdr} = 48/31 = 1.55 = \frac{0.0005 + 0.043D + 0.063D^2}{1 - \exp\left(-0.0035 - 0.051D - 0.11D^2\right)}. \tag{23.7}$$

Equation 23.7 can be solved for D iteratively and gives an estimate of 4.4 Gy. Sasaki and Miyata (1968) did not state how uncertainties on dose could be calculated and it is not at all obvious how this should be done. The problem lies in the fact that both the number of dicentrics and rings as well as the number of damaged cells carry uncertainties, which are correlated. It might be possible to use Monte Carlo simulation techniques to solve the problem but this has not, to our knowledge, been done. This problem becomes even more complex when the errors in the calibration curves are considered. Nevertheless, the contaminated Poisson approach and the Qdr method yield consistent results.

23.3.4 Protracted and Fractionated Exposure

Protraction or fractionation of exposure can result in reduced numbers of observed chromosome aberrations. Whilst it is not the case for high-LET radiation, because the dose–response curve is linear, in the case of low-LET radiation the β coefficient in Eq. 23.1 is reduced. This is due to repair mechanisms that have time to operate during the course of the protracted exposure. It has been shown that the decrease in the frequencies of aberrations appears to follow an exponential function with a mean time of approximately 2 h.

Lea and Catcheside (1942) considered a time-dependant factor, known as the G function, to facilitate the modification of the dose-squared coefficient as follows:

$$Y = \alpha D + \beta G(x)D^2, \tag{23.8}$$

where $G(x) = 2[x - 1 + \exp(-x)]/x^2$ and $x = t/t_0$, where t is the time over which the radiation occurred and t_0 is the mean lifetime of the breaks (about 2 h). Applying these ideas to protracted and fractionated cases requires a certain amount of information about the incident, such as the time over which exposure happened and the constancy of the dose rate. The analysis is only worthwhile in instances where large doses and long exposure times (up to few days) are experienced. For a dose being delivered over a very long period of time, $G(x)$ reaches zero since Eq. 23.1 reduces to $Y = c + \alpha D$. For interfraction times shorter than about 6 h, the G function can be written as exp $(-t_1/t_0)$, where t_1 is the time between fractions.

Example: In 1998 a serious accident occurred in Turkey when a ^{60}Co teletherapy source was sold as scrap metal. Ten persons were exposed, mainly during 1 day, with exposure times ranging from 2 to 7 h. Here, a case of one of the subjects exposed during 6 h is considered. The pooled results from three laboratories yielded 474 dicentrics scored in 1,363 cells. From the acute dose–response curve $Y = 0.001 + 0.02D + 0.06D^2$, the acute dose (± standard error) of 2.2 ± 0.1 Gy is obtained. The application of the G function with $x = t/t_0 = 6/2 = 3$ gives $G(x) = 0.46$; hence, the dose–response curve becomes $Y = 0.001 + 0.02D + 0.028D^2$ and the dicentric frequency corresponds to a 6 h exposure of 3.2 ± 0.1 Gy.

23.4 Summary

Statistical methods in biological dosimetry are well developed as over the years a wealth of information has been gleaned through the experience of dealing with a number of different radiological incidents as well as in vitro experiments designed to explore possible irradiation scenarios. In this chapter, appropriate statistical techniques have been reviewed and discussed in order to highlight the scope of their application. These were accompanied by worked examples, often extracted from actual incidents, which further elucidate the principles underlying each of the approaches. It is worth emphasising that different radiological scenarios require different sets of information in order to carry out meaningful analyses. It is also pleasing to confirm that approximations used routinely in the calculation of 95% confidence limits on dose are in very good agreement with a stringent Bayesian-like calculation which takes into account both the Poisson-derived uncertainties in yield as well as uncertainties in the dose–response calibration curve.

References

Brame RS, Groer PG (2003) Bayesian methods for chromosome dosimetry following a criticality accident. Radiat Prot Dosim 104:61–63

Darroudi F (2000) The use of FISH translocations analyses for retrospective biological dosimetry: How stable are stable chromosome aberrations? Radiat Prot Dosim 88:101-109

Edwards AA, Purrott RJ, Prosser JS, Lloyd DC (1980) The induction of chromosome aberrations in human lymphocytes by alpha-radiation. Int J Radiat Biol 38:83–91

Edwards AA, Lloyd DC, Prosser JS, Finnon P, Moquet JE (1986) Chromosome aberrations induced in human lymphocytes by 8.7 MeV protons and 23.5 MeV helium-3 ions. Int J Radiat Biol 50:137–145

Edwards AA, Lindholm C, Darroudi F, Stephan G, Romm H, Barquinero J, Barrios L, Caballin MR, Roy L, Whitehouse CA, Tawn EJ, Moquet J, Lloyd DC, Voisin P (2005) Review of translocations detected by FISH for retrospective biological dosimetry applications. Radiat Prot Dosim 113:396–402

IAEA (2001) Cytogenetic analysis for radiation dose assessment. Technical report series no. 405. International Atomic Energy Agency, Vienna

ICRP (1990) Recommendations of the international commission on radiological protection. (Publication 60). Ann ICRP 21

Kóteles GJ (1996) The human lymphocytes micronucleus assay. A review on its applications in occupational and environmental medicine. Cent Eur J Occup Environ Med 2:12–30

Lea DE, Catcheside DG (1942) The mechanism of the induction by radiation of chromosome aberrations in Tradescantia. J Genet 44:216–245

Lindholm C, Edwards AA (2004) Long-term persistence of translocations in stable lymphocytes from victims of a radiological accident. Int J Radiat Biol 80:559–566

Lloyd DC, Edwards AA (1983) Chromosome aberrations in human lymphocytes: effect of radiation quality, dose and dose rate. In: Ishihara T, Sasaki MS (eds) Radiation-induced chromosome damage in man. Liss, New York, pp 23–49

Lloyd DC, Purrott RJ (1981) Chromosome aberration analysis in radiological protection dosimetry. Radiat Prot Dosim 1:19–28

Lloyd DC, Purrott RJ, Dolphin GW, Edwards AA (1976) Chromosome aberrations induced in human lymphocytes by neutron irradiation. Int J Radiat Biol 29:169–182

Lloyd DC, Purrott RJ, Reeder EJ, Edwards AA, Dolphin GW (1978) Chromosome aberrations induced in human lymphocytes by radiation from 252Cf. Int J Radiat Biol 34:177–186

Lloyd DC, Purrott RJ, Reeder EJ (1980) The incidence of unstable chromosome aberrations in peripheral blood lymphocytes from unirradiated and occupationally exposed people. Mutat Res 72:523–532

Lloyd DC, Edwards AA, Prosser JS, Bolton D, Sherwin AG (1984) Chromosome aberrations induced in human lymphocytes by D-T neutrons. Radiat Res 98:561–573

Lloyd DC, Edwards AA, Prosser JS (1986) Chromosome aberrations induced in human lymphocytes by in vitro acute X and gamma radiation. Radiat Prot Dosim 15:83–88

Lloyd DC, Edwards AA, Prosser JS, Finnon P, Moquet JE (1988) In vitro induction of chromosomal aberrations in human lymphocytes, with and without boron 10, by radiations concerned in boron neutron capture therapy. Br J Radiol 61:1136–1141

Lloyd DC, Edwards AA, Natarajan AT, Darroudi F (1991) Biological dosimetry applied to in vitro simulated partial body irradiation. In: CEC report. Treatment and biological dosimetry of exposed persons. European Commission, Luxembourg

Lloyd DC, Edwards AA, Moquet JE, Hone PA, Szłuińska M (2006) Doses in radiation accidents investigated by chromosomal aberration analysis XXIV: review of cases investigated, 2003–2005. HPA-RPD-012. Health Protection Agency, Chilton

Lloyd DC, Edwards AA, Szłuińska M (2007) The minimum detectable dose by biodosimetry in a radiation accident. Mutat Res (in press)

Lucas JN (1992) Rapid translocation frequency analysis in humans decades after exposure to ionising radiation. Int J Radiat Biol 62:53–63

Merkle W (1983) Statistical methods in regression and calibration analysis of chromosome aberration data. Radiat Environ Biophys 21:217–233

Morton NE (1991) Parameters of the human genome. Proc Natl Acad Sci USA 88:7474–7476

Norman A, Sasaki MS (1966) Chromosome-exchange aberrations in human lymphocytes. Int J Radiat Biol 11:321–328

Nowell PC (1960). Photohemagglutinin – an initiator of mitosis in cultures of normal human leukocytes. Cancer Res 20:462–466

Papworth D (1975) Curve fitting by maximum likelihood. Radiat Bot 15:127–131

Pinkel D, Straume T, Gray JW (1986) Cytogenetic analysis using quantitative high-sensitivity fluorescent hybridisation. Proc Nat Acad Sci USA 83:2934–2938

Prosser JS, Lloyd DC, Edwards AA, Stather JW (1983) The induction of chromosome aberrations in human lymphocytes by exposure to tritiated water in vitro. Radiat Prot Dosim 4:21–26

Purrott RJ, Edwards AA, Lloyd DC, Stather JW (1980) The induction of chromosome aberrations in human lymphocytes by in vitro irradiation with α-particles from plutonium-239. Int J Radiat Biol 38:277–284

Sasaki MS (2003) Chromosomal biodosimetry by unfolding a mixed Poisson distribution: a generalized model. Int J Radiat Biol 79:83–97

Sasaki MS, Miyata H (1968) Biological dosimetry in atom bomb survivors. Nature 220:1189–1193

Savage JRK (1970) Sites of radiation induced chromosome exchanges. In: Ebert M, Howard A (eds) Current topics in radiation research vol 6. North-Holland, Amsterdam, pp 129–194

Savage JRK, Papworth DG (2000) Constructing a 2B calibration curve for retrospective dose reconstruction. Radiat Prot Dosim 88:69–76

Voisin P, Roy L, Hone PA, Edwards AA, Lloyd DC, Stephan G, Romm H, Groer PG, Brame R (2004) Criticality accident dosimetry by chromosomal analysis. Radiat Prot Dosim 110:443–447

Whitehouse CA, Edwards AA, Tawn EJ, Stephan G, Oestreicher U, Moquet JE, Lloyd DC, Roy L, Voisin P, Lindholm C, Barquinero J, Barrios L, Caballin MR, Darroudi F, Fomina J (2005) Translocations yields in peripheral blood lymphocytes from control populations. Int J Radiat Biol 81:139–145

24 Retrospective Biological Dosimetry by FISH

Alan A. Edwards, David C. Lloyd, and Marta Szłuińska

Abstract

This chapter reviews the present status of applying fluorescence in situ hybridisation (FISH) for retrospective biological dosimetry by chromosomal analysis. The technique helps to overcome a major drawback, the limited persistence of signal, of the more traditional method of dicentric analysis. FISH enables the more persistent so-called stable translocations to be visualised. Criteria developed by a consensus of European laboratories are described. They attempt to define which cells and which types of translocations to score for both in vitro calibration and case investigation. Also reviewed are the control levels of translocations which are highly age dependent and constitute the major determinant of the low-dose sensitivity of the method. Evidence for the stability of the technique with time is discussed. This needs further verification by following up new accident cases for many years.

24.1 Introduction

For many years, ionising radiation has been known to cause chromosome aberrations in cells. Their application for biological dosimetry has centred on the dicentric aberration in human lymphocytes and this has been used successfully for estimating doses following radiation accidents. The one disadvantage of the dicentric is that it is unstable and does not pass successfully though cell division. This means that the dicentric yield will decrease with time following the accident, making it unsuitable for exposures received either many years before blood sampling or continuously over decades.

The translocation does not suffer the same fate and the advent of fluorescence in situ hybridisation (FISH) has made the scoring of translocations much easier than with the earlier banding techniques. It is now easier to investigate the ability of cells containing translocations to pass through cell division. However, more complex rearrangements have also been revealed, which has created additional problems in deciding how to use FISH techniques for retrospective biological dosimetry. This decision requires more information on persistence of translocations and international agreement

Günter Obe and Vijayalaxmi (Eds.)
Chromosomal Alterations: Methods, Results and Importance in Human Health

concerning the various judgements involved. A special issue of a journal edited by Edwards (2000) gave preliminary views of a number of laboratories attempting to solve these problems. Updated information has led to consensus views on how to use FISH for biological dosimetry (Edwards et al. 2005; Whitehouse et al. 2005). A follow-up of an accident which occurred in Estonia has proved particularly helpful (Lindholm et al. 1998; Lindholm and Salomaa 2000; Lindholm and Edwards 2004). This chapter recalls the consensus views and the reasons for those views.

24.2 Method

A blood specimen is taken from a person for whom a dose estimate is required and the cells are cultured using techniques that enable the chromosomes to be observed. These have been published repeatedly and described in detail (IAEA 2001). In brief, phytohaemagglutinin is added to stimulate the small lymphocytes to divide and after 45 h of culture at 37°C, colcemid is added to stop the cells at their first metaphase. At about 48 h, metaphase spreads are harvested and slides prepared according to standard techniques. Chromosomes are then stained by FISH techniques using whole-chromosome and pancentromeric probes (IAEA 2001).

24.3 Which Chromosomes To Paint?

To date, most laboratories agree that the centromeres of each chromosome should be clearly visible to distinguish dicentrics from translocations. Early work (Nakano et al. 1993) showed that some dicentrics could be misclassified as translocations when viewed with chromosome paints alone. Some laboratories can identify the centromere because the two chromatids of each chromosome and the single indentation or "waist" are clearly visible. Some consider the extra brightness of the centromere when using 4′,6-diamidino-2-phenylindole counterstain is sufficient for its location and yet others prefer to specifically FISH-paint the centromeres. Apart from that, most laboratories usually paint three of the larger chromosome pairs in the groups A-C i.e., numbers 1–12, either in a single or in multiple colours, and counterstain the remainder. When about 20% of the genome is painted in a single colour, this permits 30–35% of the total translocations present to be detected. One laboratory painted chromosome pairs 1–6, three in one colour and three in another, leading to an improved translocation detection efficiency of about 56% (Tucker 2000). There were early indications that certain chromosomes did not take part in exchanges in proportion to that expected from their DNA content (Knehr et al. 1996; Boei et al. 1997). Later, more comprehensive experiments indicated better agreement with the expected values (Barquinero et al. 1998). Thus, the formulae of Lucas and Deng (2000), which rely upon the probability that a chromosome will take part in an exchange being proportional to its DNA content, have been shown to give acceptable conversions to full genome yields. The full genome translocation frequency is given by $F_G = F_p/2.05f_p(1-f_p)$, where F_p is the aberration frequency detected by FISH and f_p is the fraction of genome hybridised in a single colour taking into account the gender of the subject. Values of f_p can be obtained from Morton (1991). This enables direct comparisons to be made between laboratories that routinely paint different chromosome combinations and the equation can be expanded to account for painting in two or more colours.

24.4 What To Score

Differences in the scoring criteria used strongly influence interlaboratory variability in reported results. The scoring of translocations requires two judgements. The first involves whether or not to score the cell and the second concerns which particular aberrations relate to a retrospective dose.

Experience has shown that in a suspected case of past exposure, at least 1,000 genome-equivalent cells would need to be scored, corresponding to about 3,000 cells when three of the larger chromosomes are painted. It would be quite inefficient to select cells and score them using the same criteria as for dicentrics, i.e. only cells with exactly 46 chromosomes (IAEA 2001). A more rapid and therefore less rigorous approach has evolved whereby, when searching for suitable metaphases at low microscope magnification, most scorers ensure at a glance that there are about 46 chromosomes in each metaphase. Then at high magnification most insist that the chromosomes are well spread, that the painted chromosomes (usually six) are clear and that any that are shorter than they should be are accompanied by the deleted fragment, either as a separate object or attached to another chromosome. In other words the total length of each painted chromosome is present somewhere in the cell. Some laboratories have allowed the scoring of cells containing shortened painted chromosomes provided the corresponding centromere is present. It has now been demonstrated that including these cells that are deficient in some of the painted material can lead to abnormally high translocation yields, particularly when considering low induced frequencies near control levels (Lloyd et al. 2001; Edwards et al. 2002).

Most laboratories record two-way and one-way translocations separately, because of an early report that two-way translocation yields are more stable with time than the one-way variety (Lindholm et al. 1998). The terminology of "one-way" and "two-way" was invented as a purely descriptive way of naming aberrations without implying any underlying mechanism. Later research with improved resolution showed that many of the apparently one-way translocations were in fact also reciprocal (Fomina et al. 2000). Many laboratories record only those aberrations that are seen to involve the painted chromosomes. Some (Finnon et al. 1995; Rodriguez et al. 2004), however, have adopted a routine of recording all the aberrations seen, including those involved in the counterstained material. This enables them to distinguish stable from unstable cells using full genome information. Interestingly this concept has reinvented the designations of C_u and C_s introduced many years ago (Buckton et al. 1967) and used, for example, in the classic follow-up studies of the irradiated ankylosing spondylitis patients. With the realisation that lymphocytes sampled many years after irradiation were derived from stem cells when irradiated, it is strictly only the stable daughter cells from the irradiated population that have survived through the subsequent divisions. So strictly it is translocations in stable cells which should be measured, and when this is done translocation frequencies for an exposed individual show better constancy with time (see later.) To save scoring time when yields in the region of control levels are measured, only cells that contain aberrations involving the painted chromosomes are scanned for other unstable aberrations among the remaining chromosomes. The inclusion of a small number of unstable cells with cells that have no visible translocations has only a marginal effect on the overall yield and so pragmatically any bias can be ignored.

24.5 Control Levels

Control levels of translocations have been measured by several laboratories and the data have been reviewed and analysed (Sorokine-Durm et al. 2000). This database has been recently updated, with stricter inclusion criteria (Whitehouse et al. 2005), and subjected to a more detailed analysis. It might be expected that dicentric yields in unirradiated persons should not vary substantially with time because an approximate equilibrium would be established. The rate of production of dicentrics is balanced by the rate of loss. By contrast, translocation yields should show a continuous increase with time, and therefore age, because translocations are not removed. This has indeed been found and applies to both one-way and two-way translocations (Lucas and Deng 2000; Tucker 2000; Sorokine-Durm et al. 2000). The same applies to translocations in cells containing only simple aberrations (Whitehouse et al. 2005). The increase is from about three translocations per 1,000 genome-equivalent cells at age 20 to around 15 at age 80 (Whitehouse et al. 2005); thus, if translocation frequencies are to be used for retrospective biological dosimetry, then an age-dependent control level must be taken into account. Even after correcting for age, there still exists extra variation in translocation yield between unexposed individuals, and the causes of this require further investigation (Sorokine-Durm et al. 2000; Whitehouse et al. 2005). The effect that smoking habits have on control translocation yields is equivocal; One survey involving seven laboratories reported no observable effect (Whitehouse et al. 2005), whereas another laboratory reported a small effect in the region of 30% (Jones et al. 2002). Overall, smoking is not a serious confounding factor that significantly elevates the background frequency of translocations. Lymphocytes from control persons generally do not contain many unstable aberrations, so it makes little difference whether translocations are measured in stable cells or in all cells and this greatly simplifies the microscope analysis. However, the inclusion of cells with incomplete painted material can cause a serious overestimate of control translocation yields (Lloyd et al. 2001; Edwards et al. 2002).

24.6 Persistence of Translocations

For retrospective biological dosimetry by FISH, failure for translocation yield to persist in vivo has two possible causes. The first is that on passage through a cell division aberrations that are unstable are likely to disappear because the daughter cells are not viable; therefore, if translocations are more likely to occur in unstable than in stable cells, this will cause a decrease in yield. The second is that the sensitivity for the production of aberrations in stem cells in bone marrow is not the same as in mature lymphocytes. Initially one-way and two-way translocations were distinguished because the two-way type appeared to be more stable. Later, both types of translocations appeared more constant with time provided scoring was limited to stable cells (Lindholm and Edwards 2004). A review of data on persistence (Darroudi 2000) indicated that experiments which followed cells through several cycles showed that translocations are very stable, particularly the two-way variety. There is some evidence that the one-way types decrease either because they are associated with complex rearrangements, most of which are unstable, or because they are incomplete exchanges with associated fragments and so the cell as a whole is unstable. Experiments in vivo with laboratory animals or the small amount of human experience have led to conflicting impressions. A follow-up of the victims of the Goiania accident shows a large reduction (about a factor of 3) in translocation yield compared with dicentric yield measured at the time of the accident. This applied to those persons with doses of 1 Gy and above, but below 0.8 Gy, dicentric and translocation yields were much closer. However, no attempt was made to distinguish one-way and two-way translocations or stable and unstable cells. A follow-up of reactor crew from the Chernobyl accident showed reasonable consistency between translocation yield 10–13 years after the accident and dicentric measurements made at the time for doses up to 3 Gy. Above that, the dose estimate from translocations was low, implying some reduction in translocation yield. Again no attempt to distinguish one-way and two-way translocations or stable and unstable cells was made (Sevan'kaev et al. 2005).

Three persons from the Estonian accident, which occurred in 1994, could be followed using FISH techniques from the beginning. The follow-up shows that both one-way and two-way translocations decrease with time, but that one-way translocations decrease faster (Lindholm et al. 1998; Lindholm and Salomaa 2000). The Estonian cases have now been followed for 7 years and the translocation yield is seen to level off after about 2 years following the initial reduction (Lindholm and Edwards 2004). However, a further observation was that if translocations in stable cells are scored, then the yield does not fall with time after exposure. The yield in stable cells up to 2 years following exposure is the same as the observations in all cells in samples taken several years after exposure (Lindholm and Edwards 2004). This is understandable because at later times cell division will have removed most unstable cells. Follow-up of a tritiated water intake by a worker has shown no decrease of translocation yield with respect to the original dicentric yield for either one-way or two-way translocations over a period of 11 years (Lloyd et al. 1998). An overall message is emerging that translocation yields are stable with time provided that the translocations are measured in stable cells. There is in vitro evidence that, following simulated part-body exposures, both types of translocation yield reduce because translocations are confined to the exposed group of cells, which are more likely to be unstable by virtue of also containing dicentrics and fragments (Guerrero-Carbajal et al. 1998). This might explain, at least in part, why some translocation yields in the Goiania subjects reduced so much with time. The message that comes from this limited set of human data, accumulated with differing scoring criteria, is that future accidents need to be followed up using FISH techniques, noting on the score sheets whether the damaged cells are stable or not. It is essential to obtain further evidence for this stable cell hypothesis.

A few experiments with mice and monkeys have shown a decrease in translocation yield with time, but some of these experiments involved partial body exposure and most doses were high. Acute whole-body doses up to 2 Gy show good persistence of two-way translocations (Darroudi 2000). Clearly further work on persistence of translocations is necessary, concentrating on translocations in stable cells.

24.7 Calibration

As has been pointed out previously, there is no guarantee that the induced yield of translocations per unit dose in stem cells is the same as in mature lymphocytes. However, as experiments are now suggesting that translocations do persist well in vivo following whole-body exposure, particularly when scored in stable cells, this implies near equality of sensitivity between stem cells and mature lymphocytes. This is because at early times the lymphocytes scored are those that were irradiated, while at later times they are derived from irradiated stem cells which have passed through many cell divisions. With this assumption it is therefore possible to calibrate a system of retrospective biological dosimetry using translocations induced in vitro in lymphocytes. The literature now contains quite a lot of dose–response curves, albeit not all confined to stable cells. A frequent problem, however, is to generate enough data at sufficiently low doses to define the linear term with a reasonable statistical uncertainty. It is of course the retrospective discrimination of relatively low doses or high doses received at low dose rates which will be the greatest application for FISH biodosimetry. High acute doses tend to be appreciated early because of tissue injury signs well within the time frame for successful use of the simpler dicentric assay. Calibrating for low doses requires a lot of cells to be scored and a lowest dose of 0.5 Gy is not sufficient.

Despite early reports that immediately after irradiation translocations considerably outnumber dicentrics (Schmid et al. 1992), there is now a consensus view (e.g. Nakano et al. 1993; Bauchinger et al. 1993) that reciprocal translocations and dicentrics are induced with a more-or-less equal frequency. Moreover, one laboratory has scored translocations and all other aberrations involving painted and unpainted chromosomes so that full genome correction formulae and ratios of translocations to dicentrics may be checked (Finnon et al. 1995). These results are encouraging because the correction for the full genome has been shown to apply also to dicentric yields, at least for the chosen chromosome combination. It has also been shown that the ratio of induced two-way and one-way translocations to all dicentrics is constant, independent of dose. This being the case, the linear term for translocations may be derived from the relatively better defined linear term for dicentrics without the need for a large amount of scoring at low doses. In the longer term one would wish to see FISH calibrations taken down to lower doses, but certainly as an interim measure a laboratory could use the analogy with dicentrics with reasonable confidence.

24.8 Sensitivity

Based on knowledge to date and the assumptions about calibration, an estimate of the lowest dose that can be measured by translocations may be made. Converted to genome equivalents, the total (two-way plus one-way) translocation yield produced by 1 Gy chronic irradiation from cobalt-60 is about 15 per 1,000 cells. For a person aged 60 years, the control level is about nine translocations per 1,000 cells; thus, a dose of about 0.5 Gy would nearly double the control level and it could be argued that it is not reasonable to claim to have measured a dose unless the control level is doubled. The control level itself is probably known to an accuracy of about 20% at any age, although some extra variation from person to person still exists. Assuming that Poisson statistics can be used, it is necessary to score about 1,000 genome-equivalent cells (about 3,000 actual cells with six painted chromosomes) from a 60-year-old person exposed to 0.5 Gy to have a translocation yield which approaches being significantly higher than that in a control. A similar calculation for a young person, under 20 years, where the control yield is around two translocations per 1,000 genome-equivalent cells, shows that a dose of about 0.3 Gy can be measured by scoring around 1,000 genome-equivalent cells.

24.9 Conclusion

There are decades of experience to show that the measurement of dicentrics in human lymphocytes produces a reasonable estimate of recent absorbed dose to a person. For exposures which occur some years prior to blood sampling and which may additionally be delivered over a long period of time, the dicentric is not so useful. The persistence of translocation yields in stable cells is indicated, but more data from humans, particularly from the follow-up of accidents, are required. More work on the causes of extra variation in control subjects, apart from age, is also required. The removal of cells which do not have the full complement of the painted chromosomes has helped improve consistency in control groups. The protocol for the scoring of translocations for retrospective biological dosimetry has reached the consensus stage, at least within laboratories in Europe.

References

Barquinero JF, Knehr S, Braselmann H, Figel M, Bauchinger M (1998) DNA-proportional distribution of radiation-induced chromosome aberrations analysed by fluorescence *in situ* hybridization painting of all chromosomes of a human female karyotype. Int J Radiat Biol 74:315–323

Bauchinger M, Schmid E, Zitzelberger H, Braselmann H, Nahrstedt U (1993) Radiation-induced chromosome aberrations analysed by two colour fluorescence in situ hybridisation with composite whole chromosome-specific DNA probes and a pancentromeric DNA probe. Int J Radiat Biol 64:179–184

Boei JWA, Vermeulen S, Natarajan AT (1997) Different involvement of chromosomes 1 and 4 in the formation of chromosomal aberrations in human lymphocytes after X-irradiation. Int J Radiat Biol 72:139–145

Buckton K, Smith PG, Court Brown WM (1967) The estimation of lymphocyte lifespan from studies on males treated with X-rays for ankylosing spondylitis. In Evans HJ, Court Brown WM, McLean AS (eds) Human radiation cytogenetics. North Holland, Amsterdam, pp 106–114

Darroudi F (2000) The use of FISH translocations analyses for retrospective biological dosimetry: How stable are stable chromosome aberrations? Radiat Prot Dosim 88:101–109

Edwards AA (ed) (2000) Fluorescence *in situ* hybridisation (FISH) biological dosimetry. Radiat Prot Dosim 88:1–112

Edwards AA, Lindholm C, Darroudi F, Stephan G, Romm H, Barquinero J, Barrios L, Caballin M-R, Roy L, Whitehouse CA, Tawn EJ, Moquet JE, Lloyd DC, Voisin P (2005) Review of translocations detected by FISH for retrospective biological dosimetry applications. Radiat Prot Dosim 113:396–402

Edwards A, Maznik N, Moquet J, Hone P, Vinnikov V, Lloyd D, Cox R (2002) Choosing metaphases for biological dosimetry by fluorescence in situ hybridization (FISH). Radiat Res 157:467–471

Finnon P, Moquet JE, Edwards AA, Lloyd DC (1995) The ^{60}Co gamma ray dose-response for chromosomal aberrations in human lymphocytes analysed by FISH; applicability to biological dosimetry. Int J Radiat Biol 75:1215–1222

Fomina J, Darroudi F, Boie JWA, Natarajan AT (2000) Discrimination between complete and incomplete chromosome exchanges in X-irradiated human lymphocytes using FISH with pan-centromeric and chromosome specific DNA probes in combination with telomeric DNA probe. Int J Radiat Biol 76:807–813

Guerrero-Carbajal YC, Moquet JE, Edwards AA, Lloyd DC (1998) The persistence of FISH translocations for retrospective biological dosimetry after simulated whole or partial body irradiation. Radiat Prot Dosim 76:159–168

IAEA (2001) Cytogenetic analysis for radiation dose assessment. A manual. Technical reports series no 405. IAEA, Vienna

Jones IM, Galick H, Kato P, Langlois RG, Mendelsohn MI, Murphy G, Pleshnov P, Ramsey MJ, Thomas CB, Tucker JD, Tureva L, Vorobsova I, Nelson DG (2002) Three somatic genetic biomarkers and covariates in radiation-exposed Russian cleanup workers of the Chernobyl nuclear reactor 6-13 years after exposure. Radiat Res 158:424–442

Knehr S, Zitzelberger H, Braselmann H, Nahrstedt U, Bauchinger M (1996) Chromosome analysis by fluorescence *in situ* hybidization: further indications for a non-DNA proportional involvement of single chromosomes in radiation-induced structural aberrations. Int. J Radiat Biol 70:385–392

Lindholm C, Edwards AA (2004) Long-term persistence of translocations in stable lymphocytes from victims of a radiological accident. Int J Radiat Biol 80:559–566

Lindholm C, Salomaa S (2000) Dose assessment of past accidental or chronic exposure using FISH chromosome painting. Radiat Prot Dosim 88:21–25

Lindholm C, Tekkel M, Veidebaum T, Ilus T, Salomaa S (1998) Persistence of translocations after accidental exposure to ionizing radiation. Int J Radiat Biol 74:565–571

Lloyd DC, Moquet JE, Oram S, Edwards AA, Lucas JN (1998) Accidental intake of tritiated water: a cytogenetic follow-up case on translocation stability and dose reconstruction. Int J Radiat Biol 73:543–547

Lloyd DC, Lucas JN, Edwards AA, Deng W, Valente E, Hone PA, Moquet JE (2001) A study to verify a reported excess of chromosomal aberrations in blood lymphocytes of Namibian uranium miners. Radiat Res 155:809–817

Lucas JN, Deng W (2000) Views on issues in radiation biodosimetry based on chromosome translocations measured by FISH. Radiat Prot Dosim 88:77–86

Morton NE (1991) Parameters of the human genome. Proc Natl Acad Sci 88:7474–7476

Nakano M, Nakashima E, Pawel DJ, Kodama Y, Awa A (1993) Frequency of reciprocal translocations and dicentrics induced in human blood lymphocytes by X-irradiation as determined by fluorescence in situ hybridization. Int J Radiat Biol 64:565–569

Rodriguez P, Montoro A, Barquinero JF, Caballín MR, Villaescusa I, Barrios L (2004) Analysis of translocations in stable cells and their implication in retrospective biological dosimetry. Radiat Res 162:31–38

Schmid E, Zitzelberger H, Braselmann H, Gray JW, Bauchinger M (1992) Radiation-induced chromosome aberrations analysed by fluorescence in situ hybridisation with a triple combination of composite whole chromosome-specific DNA probes. Int J Radiat Biol 62:673–678

Sevan'kaev AV, Lloyd DC, Edwards AA, Khvostunov IK, Mikhailova GF, Golub EV, Shepel NN, Nadejina NM, Galstian IA, Nugis VY, Barrios L, Caballin MR, Barquinero JF (2005) A cytogenetic follow-up of some highly irradiated victims of the Chernobyl accident. Radiat Prot Dosim 113:152–161

Sorokine-Durm I, Whitehouse C, Edwards A (2000) The variability of translocation yields amongst control populations. Radiat Prot Dosim 88:93–99

Tucker JD (2000) Evaluation of chromosome translocations by FISH for radiation biodosimetry: a view from one laboratory. Radiat Prot Dosim 88:87–92

Whitehouse CA, Edwards AA, Tawn EJ, Stephan G, Oestreicher U, Moquet JE, Lloyd, DC, Roy L, Voisin P, Lindholm C, Barquinero J, Barrios L, Caballin MR Darroudi F, Fomina J (2005) Translocation yields in peripheral blood lymphocytes from control populations. Int J Radiat Biol 81:139–145

25 Chromosomal Aberrations in Astronauts

Kerry George and Marco Durante

Abstract

During spaceflight, astronauts are exposed to very penetrating radiation consisting primarily of high-energy charged particles, mostly protons, and secondary radiation created by interaction of nuclei with shielding material. Significant uncertainty still exists regarding the biological effects of exposure to space radiation. Cytogenetic biodosimetry analyses provide direct measurements of space radiation induced biological damage that take into account individual radiosensitivity and the influence of stress conditions, including microgravity. A review of current data on in vivo induced chromosome damage in the blood lymphocytes of astronauts strongly indicates that space missions of a few months or more induce measurable increases in the yield of chromosome damage that can be used to estimate equivalent dose and radiation risk. Recent studies indicate that biodosimetry estimates from single spaceflights lie within the range expected from physical dosimetry and biophysical models, but single individual measurements have large uncertainties and the total sample population remains low. Retrospective doses are more difficult to estimate because of the fairly rapid time-dependent loss of "stable" aberrations in blood lymphocytes. Also, biodosimetry estimates from individuals who participate in repeated missions, or very long missions, may be complicated by an adaptive response to space radiation and/or changes in lymphocyte survival and repopulation. No cytogenetic biomarkers of space radiation exposure have been uncovered so far, owing in part to the statistical uncertainties inherent in these techniques. A discussion of published data is presented and specific issues related to space radiation biodosimetry protocols are discussed.

25.1 Introduction

The duration of manned spaceflight missions has steadily increased over the years. From brief forays during the Gemini and Apollo era that lasted up to only a few days, missions have extended to weeks on board the Space Shuttle, and months aboard Mir and the International Space Station (ISS). In addition,

Günter Obe and Vijayalaxmi (Eds.)
Chromosomal Alterations: Methods, Results and Importance in Human Health

some proposed future Mars exploration missions could extend to as long as 3 years. This increased time in space means that crewmembers are exposed to higher radiation doses, and concerns have been raised about the potential health effects from these exposures. Radiation exposure limits are based on the lifetime excess cancer risk, because cancer is the main health concern related to space radiation exposure (Cucinotta and Durante 2006). Currently accepted radiation exposure limits restrict excess lifetime cancer mortality risk to no more than 3% (NCRP 2000).

The space radiation environment is a complex mixture of different types of radiation that is very different from the terrestrial radiation environment. The primary sources of radiation in outer space derive from galactic cosmic rays, protons and electrons trapped in the Earth's magnetic field, and infrequent solar particle events. In addition, secondary radiation, such as neutrons and light charged particles, are created by interactions of high-charge (*Z*) and high-energy (*E*) nuclei (HZE particles) with spacecraft shielding material and with the human body itself (NCRP 2000). Although HZE particles contribute only a few percent to the particle flux in space, the significantly higher ionizing energy and greater penetration power of this type of radiation could contribute significantly to biological effects.

Health risks to crewmembers are currently assessed from physical measurements of radiation dose measured using passive thermoluminescent dosimeters and plastic nuclear track detectors. These measurements provide accumulated absorbed skin doses that are combined with computerized anatomical models and radiation transport codes that estimate the effects of shielding. These data are used to calculate radiation doses to specific organs and to assess the corresponding health risk. However, substantial limitations arise with this process that lead to significant uncertainty in estimating the biologically important dose (Cucinotta and Durante 2006). Quality factors for space radiation risk coefficients (mostly derived from the A-bomb survivors database), and the dose and dose-rate effectiveness factor are all quantities affected by large uncertainty. In addition, the location of crew dosimeters cannot always be assured, individual radiosensitivity is totally neglected in the risk models, and synergism between microgravity and radiation is still poorly understood. Recently, biodosimetry programs have been introduced that derive equivalent dose or risk from direct measurements of cytogenetic damage in the peripheral blood lymphocytes of crewmembers (Durante 2005). These data have the added advantage of taking into account individual radiosensitivity in addition to the influence of microgravity effects and other stress conditions, and can be very useful to benchmark the current models used to estimate the risk from physical dosimetry measurements (NCRP 2002).

Cytogenetic biodosimetry methods have been used extensively for assessing radiation exposures, and remain the most sensitive in vivo indicator of dose available to date (IAEA 2001). Dicentric chromosomes have been used for many years to evaluate accidental radiation exposures because they can be easily identified in Giemsa-stained chromosome preparations and preexposure background levels are very low. However, the astronaut population is unique in that astronauts consent to considerably higher occupational radiation exposures than terrestrial radiation workers and, unlike the situation with accident victims, preexposure levels of chromosome damage can be accurately assessed in astronauts, allowing for greater sensitivity in biological dose estimations.

25.2 Measured Yields of Exchanges in the Peripheral Blood Lymphocytes of Astronauts

The first in vivo studies of chromosome damage in the peripheral blood lymphocytes of crewmembers were conducted in the 1960s during the Gemini program. Bender et al. (1967, 1968) used solid Giemsa staining techniques to assess dicentrics, rings, fragments, and chromatid aberrations in blood samples obtained from astronauts who participated in the Gemini-III and Gemini-XI missions, which lasted about 5 h and 3 days, respectively. No significant increase in damage was detected when samples collected before and after flight were compared. More recently these techniques were tested on crewmembers participating in longer-duration flights on board the Mir space station (Testard et al. 1996; Obe et al. 1997; Federenko et al. 2001). Testard et al. (1996) studied seven crewmembers, two of whom participated in flights lasting 3 weeks or less and showed no significant elevation of chromosome damage after flight, and five cosmonauts of whom participated in a 6-month mission and had elevated chromosome damage after flight. Biological dose estimates for the longer-duration flights were calculated using established in vitro calibration curves for X-rays, and varied among individuals, with values ranging from 130 to 455 mGy. Obe et al. (1997) studied crewmembers who participated in spaceflights lasting from 120 to 198 days and reported significant elevation in chromosome damage after flight for six of the seven individuals studied. Pooled data for the six individuals who had increased chromosome damage showed a postflight dicentric yield that was roughly 3.5 times preflight values. Individual biological doses were estimated using published in vitro X-ray-induced dose–response curves, and values ranged from 50 to 290 mGy depending on the individual. Limited physical dose measurements were reported for the two studies described above, but data indicated that the frequencies of aberrations roughly correlated with absorbed radiation dose. A study by Federenko et al. (2001) of 22 individuals who participated in space missions lasting an average of 4–6 months also reported that the mean yields of dicentrics were about threefold greater after flight than before flight; however, large interindividual variations were reported for this group, with some individuals having no appreciable increase in dicentric yield after flight. The biological dose estimates for the individual crewmembers with increased dicentric yields varied from 20 to 280 mGy. No comparisons with physical dose estimates were given for this study.

Evidence from these cytogenetic studies strongly indicates that the radiation dose accumulated during a long-duration flight can induce measurable increases in the yield of chromosome damage, whereas the doses encountered during shorter missions of a few weeks or less are below the threshold for detection using this technique; however, the type of chromosome aberration evaluated is an important factor when considering these data. All the studies described above focus on the detection of dicentric chromosomes, and it has been well documented that radiation-induced dicentrics undergo time-dependent decay in blood lymphocytes. Dicentrics have been estimated to have an average half-life of about 3 years in human peripheral blood lymphocytes, but recent studies indicate large variability in decay rate among individuals (Durante 2005). It would follow that the use of dicentric aberrations for biodosimetry analysis after missions lasting several months or more may not be reliable owing to the loss of aberrant cells during the mission. In addition, any significant individual variability in the rate of loss of cells with dicentrics would indicate that yields could not be accurately corrected for elimination of this type of exchange from the blood.

The introduction of the fluorescence in situ hybridization (FISH) chromosome painting technique radically improved the ease and accuracy of assessing yields of monocentric aberrations such as translocations (Lucas 1997). These types of so-called stable chromosome aberrations have a higher probability of survival after cell division than unstable aberrations, and are therefore commonly believed to persist in peripheral blood cells for many years. This indicates that translocations would be a more appropriate tool than dicentric analysis for assessing the effect of chronic radiation exposures such as those encountered during long-duration missions, as well as for assessing retrospective radiation doses. Background levels of translocations are higher than those of dicentrics; they are known to increase with age and show considerable variation between individuals (Tucker 2001). Although this can be a challenge for situations where preexposure levels of translocations are not available, such as after accidental radiation exposure, it is not a problem for spaceflight biodosimetry because samples can be obtained before flight.

George et al. (2001a) used chromosome painting to assess the yield of chromosome exchanges in the peripheral lymphocytes of six crewmembers before and after they participated in space missions lasting 3 months or more, and in two crewmembers who participated in a 10-day Space Shuttle mission. Postflight increases in chromosome damage were detected in all the crewmembers of the longer-duration missions, whereas no significant change was detected in crewmembers of the shorter Space Shuttle mission. Doses were determined using preflight in vitro dose–response curves for γ-rays that were generated using the individual crewmembers' blood lymphocytes.

Preflight levels of translocations were shown to vary considerably between individuals. Doses estimated from total chromosome damage ranged from 31 to 166 mGy, depending on the individual, and these values roughly correlated with the physical measurements of dose. Average relative biological effectiveness (RBE) was estimated by comparing biological dose estimates with measured physical doses. Average RBE values were in excellent agreement with the average quality factor of 2.5 that was calculated from physical readings corrected for the HZE component of the exposure and adjusted to representative doses at organ locations. These data confirm the results of an earlier study by Yang et al. (1997) in which FISH chromosome painting was used to assess aberrations in three crewmembers who participated in a space mission lasting 115 days.

25.3 Biologically Based Risk Estimates

As we noted in Sect. 25.1, astronauts are exposed to radiation in space under conditions substantially different from those experienced on Earth (Fig. 25.1). The radiation field is totally different, and the crewmembers are simultaneously exposed to several other stressors, such as microgravity. Given these differences, energy deposition (physical dose) may not be representative of the true space radiation risk, and this is why the use of biomarkers of risk is so important for astronauts (Durante 2005). A direct comparison of risk estimates from physical dose measurements and application of models (NCRP 2000) with biological measurements is possible because chromosome aberrations in peripheral blood lymphocytes have been validated as biomarkers of cancer risk (see Chap. 31 by Bonassi).

Risk projections for long-term missions in low Earth orbit (LEO) can be obtained from biodosimetry data using the quantitative approach derived from the European Study Group in Cytogenetic Biomarkers and Health (ESCH) database (Bonassi et al. 2005). The postflight relative increase in aberration frequency can then be used directly to estimate the cancer risk, and this biologically based estimate can be compared with conventional risk estimates based on physical dosimetry and models (NCRP 2000). Through analysis of the cytogenetic data described above, it has been determined that the average excess relative risk (ERR) of cancer induction after long-term missions in LEO is 0.09, with large interindividual variations (ranging from –0.08 to 0.42) (Durante et al. 2001; Durante 2005). Individual risk estimates in this same cohort using conventional assessment leads to an average ERR of 0.05, with a range from 0.02 to 0.12. Despite the large spread in the "biological" values, it is remarkable that the two estimates lie in the same range. In comparison with equivalent doses calculations in the NASA study (George et al. 2001a, c), these results suggest that current average risk estimates for astronauts in LEO are accurate to within a factor of 3 or less (Durante 2005).

Fig. 25.1 An astronaut involved in extravehicular activities during STS-121 mission. The astronauts are exposed to radiation plus many other stressors (including microgravity) during spaceflights. (Photo courtesy of NASA)

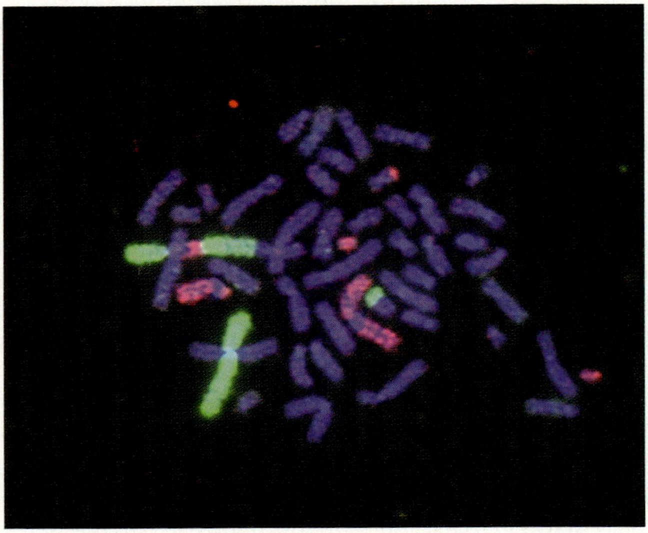

Fig. 25.2 Complex-type exchange involving chromosomes 1 and 2, detected by fluorescence in situ hybridization (FISH) painting in a blood lymphocyte of an astronaut after a long-term space mission

ERRs derived from chromosome aberrations present a wider scatter than those calculated from dose. Certainly, statistical error involving aberration measurements accounts for most of the widened distribution, but interindividual variability in risk may also contribute to this variance.

25.4 Cytogenetic Signatures of Space Radiation Exposure

In addition to the detection of simple reciprocal-type chromosome exchanges, chromosome painting can be used to detect complex-type exchanges (Fig. 25.2), which are defined as exchanges involving three or more breaks in two or more chromosomes. Many studies have demonstrated that no significant increase in complex exchanges is detected after low linear energy transfer (LET) radiation such as γ-rays and X-rays at exposures below 2 Gy, whereas low doses of high-LET radiation, such as the HZE ions in space, are capable of producing extremely complex patterns of chromosome damage (George et al. 2003a; Durante et al. 2002). This evidence suggests that complex-type damage can be induced by a single particle's traversal of a cell and indicates that this type of damage can be a biomarker of exposure to low doses of high-LET radiation (Anderson et al. 2002). It follows, then, that any increase in complex-type chromosome damage detected in blood lymphocytes of crewmembers after spaceflight would be due entirely to the high-LET component of the exposure.

However, one confounding factor in the assessment of high-LET radiation-induced chromosome damage is the influence of delays in progression of the cell cycle from G2 to metaphase. Recent ground-based studies have indicated that high-LET-induced chromosome damage undergoes a complicated pattern of mitotic delay (George et al. 2001b), and it is assumed that this is due to the less homogenous nature of this type of exposure compared with X-ray or γ-ray irradiation. When chromosomes in the G2 phase of the cell cycle are used for the analysis of high-LET-induced cytogenetic damage, much higher yields of damage are obtained than with standard metaphase analysis. This is due to extensive G2–metaphase cell-cycle delays. It is therefore possible that postflight expression of complex chromosome damage in crewmembers' lymphocytes could be influenced by a complicated pattern of mitotic delay after exposure to the mixed radiation field encountered in space and consequently it could be underestimated in metaphase samples collected at one time point. To ensure that chromosome damage is not underestimated in astronauts' blood lymphocytes, it may be more appropriate to use the chemically induced premature chromosome condensation (PCC) technique for analysis, in which chromosomes in the G2 phase as well as metaphase are assessed.

Cytogenetic analysis using FISH painting of lymphocytes from ten astronauts after long-term space missions of 3 months or more revealed that the frequency of complex damage was very low and in most cases only one or two complex exchanges were detected in several thousand PCC samples analyzed for each individual (George et al. 2003b). A low background level of complex damage was also detected in preflight samples, hampering statistical significance. Similar results were obtained in a cohort of Russian cosmonauts (Durante et al. 2004). In a different study, Horstmann et al. (2005) used a multiplex FISH technique (MFISH) to assess all chromosomes simultaneously in each cell and detected no complex-type exchanges in a total of 3,590 lymphocyte cells pooled from astronauts involved in short- and long-duration space missions. However, this study was conducted using standard metaphase analysis and therefore does not discount the possibility that complex damage was not expressed owing to extensive cell-cycle delay. Testard et al. (1996) and Obe et al. (1997) also reported a very low frequency of highly damaged metaphase cells after long-duration missions; however, these studies were limited to so-called rogue cells containing multiple dicentric and acentric fragments because analysis was conducted using solid staining techniques, and again cells were analyzed using standard metaphase techniques. The nature of the "rogue" cells is still unclear, but they are generally considered to be unrelated to radiation exposure (Mustonen et al. 1998).

Intrachromosomal exchanges (inversions) have also been suggested as a biomarker of exposure to densely ionizing radiation (Hande et al. 2003). Only one group so far (Horstmann et al. 2005) has attempted to assess the frequency of intrachromosomal exchanges in blood lymphocytes after spaceflight. For this study, a high-resolution multicolor fluorescence banding (mBAND) probe for chromosome 5 was used, and no inversions were detected in 2,800 cells pooled from 11 astronauts involved in long- and short-duration missions.

25.5 Persistence of Space Radiation Induced Cytogenetic Damage

It was commonly believed that chromosome translocations would persist in the blood lymphocytes for many years; however, as more data become available, evidence suggests that this may not always be the case (reviewed in Tucker 2001; IAEA 2001; Durante 2005). Some limited data are available on the persistence of chromosome aberrations in astronaut peripheral blood lymphocytes. Durante et al. (2003) clearly demonstrated that dicentric yields in astronauts decay with time after flight. The authors could not measure the complete time dependence of translocations using FISH painting, but their results show a lack of correlation between time in space and translocation yields (Fig. 25.3). No significant correlation was observed between yields of dicentrics or translocations and duration of flight or absorbed dose. The yield of aberrations declined rapidly during the time interval between flights, and after repeated flights translocation yields were similar to background levels before the first mission (Fig. 25.4).

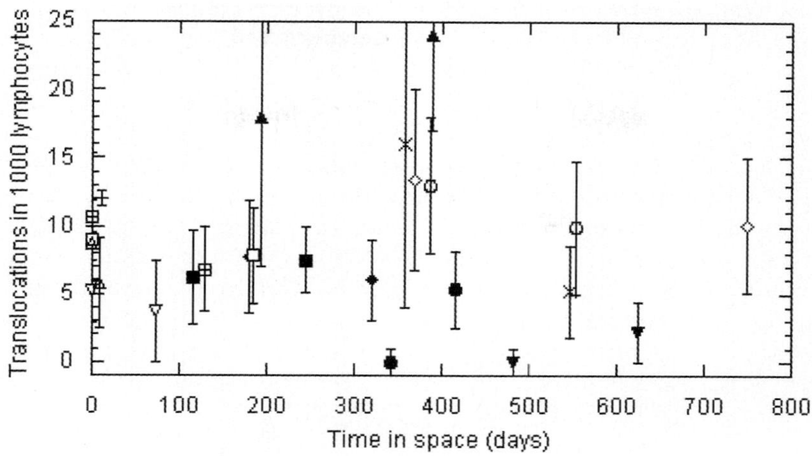

Fig. 25.3 Relationship between translocation frequency (measured by FISH and converted to whole-genome equivalents) and total duration of space sojourns for cosmonauts involved in multiple spaceflights. Each *symbol* represents a different cosmonaut. (Plots from Durante et al. 2003, reproduced with permission of S. Karger AG, Basel)

George et al. (2005) analyzed blood samples collected at various times, beginning directly after return from space to more than 5 years after flight. For five of the six astronauts studied, the analysis of individual time courses for translocations revealed a temporal decline of yields with various half-lives ranging from 10 to 58 months. Since the level of stable aberrations depends on the interplay between natural loss of circulating T lymphocytes and replenishment from the stem or progenitor cells, the differences in the rates of decay could be explained by interindividual variation in lymphocyte turnover. The results indicate possible complications with the use of stable aberrations for retrospective dose reconstruction or risk estimates from biological data. It should be stressed that space is a unique environment and these results may not be indicative of terrestrial radiation exposures. It is possible that confounding factors from exposure to the space environment, such as stress and microgravity, could induce cell instability. It has been shown that spaceflights induce drastic changes in the immune system, and the ability to respond to challenges such as infection is reduced in the space environment.

The complex time dependence of the chromosome aberration yield is apparently a severe limitation on the use of cytogenetic end points to estimate equivalent dose or risk. In fact, the result of the biological test will be strongly dependent on the time after sampling. However, it can be argued that the "time factor" does not necessarily reflect a technical limitation of the cytogenetic biomarker, but rather provides an *individual* risk assessment rather than the *average* risk estimate based solely on absorbed dose (Durante 2005). As in the case of individuals who quit smoking at a certain age after

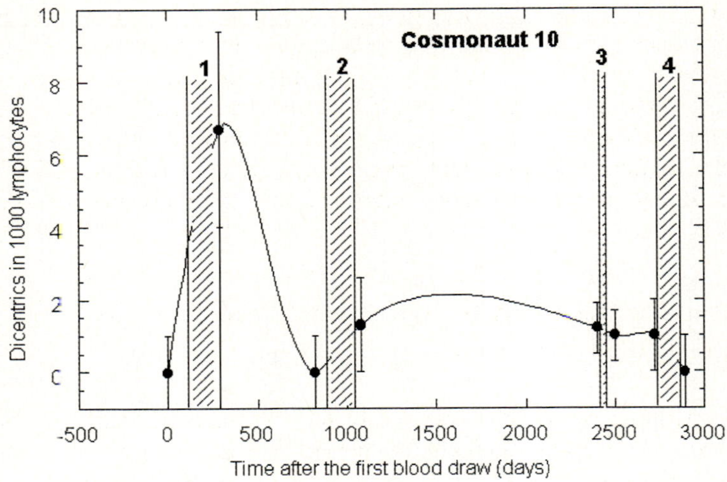

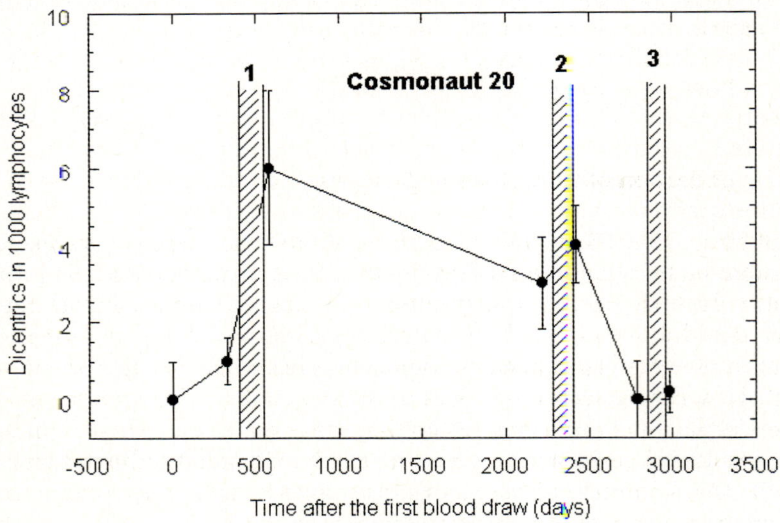

Fig. 25.4 Time course of dicentrics in two Russian cosmonauts involved in multiple spaceflights. The dicentric frequency, measured in metaphases of lymphocytes after Giemsa staining, is plotted vs. the time from the blood draw used for the first cytogenetic test. *Shaded areas* indicate the space mission time, and the *numbers between the bars* the flight number for the cosmonaut. In a period of about 10 years, cosmonaut 10 made four spaceflights and cosmonaut 20 three flights. The *line connecting the datapoints* is a guide for the eye. (Plots from Durante et al. 2003, reproduced with permission of S. Karger AG, Basel)

years of chronic exposure to carcinogens in the cigarette smoke, radiation relative risk decreases with time after exposure (as shown in the recent analysis of A-bomb survivors; Preston et al. 2003), and the decrease is likely to show a large interindividual variability, caused by physiological, environmental, and genetic factors. Since these factors are not taken into account in biophysical models of radiation action, a reliable picture of the relative risk for an individual at a given time after exposure may be derived only by biomarkers of risk.

25.6 Methodology Considerations and Suggestions for Standardized Protocols

After all available data have been considered, it appears that cytogenetic biodosimetry can be used reliably to reconstruct radiation dose after protracted exposure to space radiation, and these data can potentially be used to assess individual health risk. However, it is becoming clear that space radiation biodosimetry has some unique challenges, compared with terrestrial-based methods, which must be addressed before a standard procedure can be established. It is apparent that several major areas should be considered, as addressed in the following sections.

25.6.1 Collection of Chromosome Samples

Lymphocytes must be examined in blood samples that were collected before flight to establish accurate preflight yields of damage. This not only provides an individual baseline for translocations, which are highly variable between individuals, but also increases the dose sensitivity of the assay. Considering that chromosome damage may undergo a fairly rapid time-dependent decay, postflight samples must be collected within 1 or 2 weeks after the mission. Some studies have used samples collected the day after return from space; however, it has been noted that response of lymphocytes to phytohemagglutinin (PHA), a compound that is used to induce lymphocytes to undergo cell division, is severely suppressed on the day of return from space. The mitotic index was drastically lower in samples collected from three crewmembers on the day of return from a 3-month space mission than it was in preflight samples collected from the same individuals (Yang et al. 1997). The number of cycling cells was much lower, and progression through the cell cycle was much slower. This phenomenon, which may be related to changes observed in the immune response and/or other stress and microgravity effects, seems to last only a few days. The growth kinetics of PHA-stimulated lymphocytes collected 9 days after flight were almost the same as those in preflight samples. Given the challenges of collecting chromosome samples within a few days of return from flight, a later sampling time (7–14 days after flight) is preferable.

The analysis of chemically induced PCC samples, which contain G2-phase as well as metaphase cells, is preferable to standard metaphase analysis. PCC analysis will ensure that any cells undergoing extensive G2–metaphase cell-cycle delay will still be included in the analysis. As mentioned above, this method has been shown to help avoid population bias after high-LET radiation in vitro and may ensure correct yields of complex damage induced in vivo, although this has yet to be proven. In addition, PCC samples generally have a higher index of analyzable cells, so using them makes it easier to obtain sufficient cells from samples that are consistently difficult to culture.

25.6.2 Assessment of Chromosome Damage

Since it has been well documented that radiation-induced dicentrics undergo fairly fast time-dependent decays and will certainly be lost from the blood during a longer spaceflight mission, stable monocentric aberrations will provide more accurate assessments of dose. A combination of three FISH painting probes for the larger chromosome pairs that comprise about 20% of the genome is recommended for analysis. This technique allows damage to be scored quickly, and if each chromosome is painted with a different color, the detection of complex damage will be maximized, while exchanges within the painted chromosomes can also be detected. Data for analysis of individual combinations of chromosomes can be extrapolated to whole-genome-equivalent values using the appropriate calculations (George et al. 2001a). Use of MFISH techniques is not recommended for routine biodosimetry analysis because analysis of individual cells takes a considerable amount of time.

The analysis of complex-type exchanges could provide valuable information about the high-LET component of radiation exposure, as any significant increase will likely be due to heavy ions. However, it appears this type of exchange is quite rare and it is unlikely that valuable data can be obtained using the average number of cells analyzed for these tests.

One other important consideration when scoring chromosome damage for dose estimations is that clonal exchanges must be identified and excluded. Several recent studies have indicated that some individuals' blood lymphocytes possess cells with balanced chromosome abnormalities that have clonally expanded in vivo, resulting in a population of cells with identical rearrangements. Including these clonal exchanges can result in an overestimation of radiation-induced damage and, therefore, influence the accuracy of dose calculations. Clonal exchanges may be more prevalent than previously thought. In fact, in one study clonal aberrations were identified in three of 12 astronauts (George et al. 2004). These clonal cells were present in samples collected both before and after spaceflight, and yields were higher than those previously reported for healthy individuals in this age range (40–52 years of age). The individuals with clonal aberrations were followed up over several months and the yields of all clones decreased during this period. It is possible that the inclusion of clonal exchanges could explain some of the interindividual biological differences in dose estimates for astronauts. Since clonal aberrations may be associated with increased risk of tumorigenesis, it is important to accurately identify cells containing clonal rearrangements for risk assessment as well as biodosimetry.

25.6.3 Dose Limitations

The radiation dose incurred during shorter spaceflight missions is below the threshold for currently available cytogenetic biodosimetry techniques. The lowest dose that can be assessed seems to be around 10 cGy. This generally restricts the use of biodosimetry to missions lasting at least 3 months. Using the FISH technique, several thousand cells should be analyzed in both preflight and postflight samples to obtain statistically significant data, although exact numbers can vary depending on the individual's background frequencies of damage.

25.6.4 Individual Sensitivity

Recent data indicate that in vitro dose response for chromosome damage may vary for some individuals (Fig. 25.5). If this is the case, then one standard calibration curve cannot be used for all individuals and it is necessary to generate a preflight in vitro calibration curve for blood lymphocytes from each crewmember before their participation in their respective space mission. This will indicate if the particular individual has a greater predisposed sensitivity to radiation and should be monitored more closely. In addition, if the same combination of chromosomes is assessed before and after flight and individual dose–response curves are used, then the results can be compared directly without the need for subtracting background yield of chromosome damage. This results in a substantially decreased error in dose estimations.

Recent estimates have put the overall percentage of individuals heterozygous for the ataxia telangiectasia (A-T) gene at around 3% of the human population. These individuals have defective DNA repair capabilities that result in a higher sensitivity to radiation and they are clearly at greater risk of health effects due to radiation exposure. If these percentages are correct, it is very likely that the astronaut population includes some A-T heterozygous individuals. Clearly the issue of individual sensitivity to space radiation is very important and data are limited. More study is required in this area.

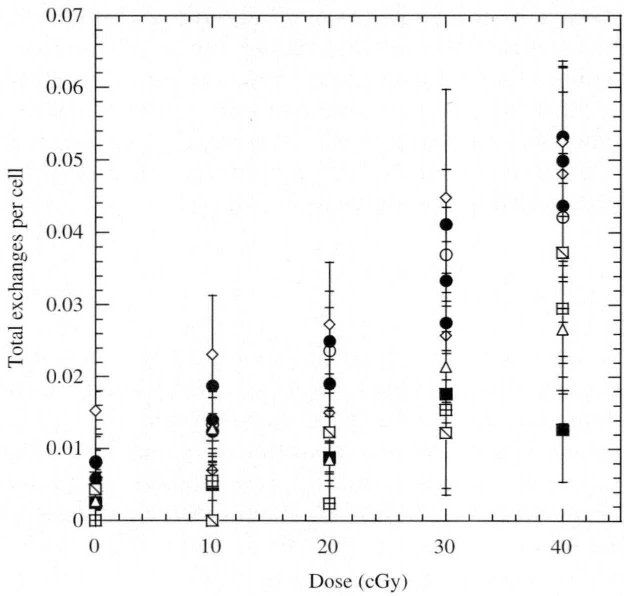

Fig. 25.5 Dose–response curves for chromosome damage in astronauts' peripheral blood lymphocytes induced after in vitro irradiation with γ-rays. Each *symbol* represents a different crewmember. *Bars* present standard errors of the mean values

25.6.6 Dose Estimations

As mentioned already, if the same combination of chromosomes is assessed for each sample, then postflight increases can be obtained by subtracting preflight background yields from the postflight yield and biologically determined dose can be calculated directly using the curve fit for the preflight calibration curve. At least two doses below 10 cGy should be used for the in vitro calibration to increase the accuracy of the curve fitting for doses within the range encountered during spaceflight. At least five different doses are required for the calibration curve, and using the probes that comprise 20% of the whole genome, several thousand cells must be analyzed to generate sufficient data for curve fitting.

25.6.7 Influence of Repeat Flights

The recent study of crewmembers after repeated flights (Durante et al. 2003) indicates that aberrations in individuals who have been involved in several missions may not be additive (Fig. 25.4), perhaps because of adaptive response to space radiation exposure or loss of radiosensitive cell subpopulations.

Given that these individuals may have less of an increase in chromosome damage than expected after repeated flights, it may not be possible to accurately assess their dose using these methods. It is important, therefore, to research an individual's previous radiation exposure from earlier spaceflights and possibly medical procedures, and to take this into consideration when evaluating biodosimetry data. This study of adaptive response is very interesting and it is important that the phenomenon be investigated further.

25.7 Conclusions

It now appears that if blood is collected within 1 or 2 weeks of return from space, cytogenetic biodosimetry can be used reliably to estimate equivalent dose and radiation risk after protracted exposure to space radiation for a few months or more. Results indicate that biodosimetry estimates from single spaceflights lie within the range expected from physical dosimetry and biophysical models. However, very large uncertainties are associated with single individual measurements and the total sample population remains low.

Retrospective doses may be more difficult to estimate because of the fairly rapid time-dependent loss of "stable" aberrations in blood lymphocytes. Also, biodosimetry estimates from individuals who participate in repeated missions, or very long (interplanetary) missions, may be complicated by an adaptive response to space radiation and/or changes in lymphocyte survival and repopulation. Suggested protocols for space radiation biodosimetry and issues that need to be addressed were discussed.

References

Anderson RM, Stevens DL, Goodhead DT (2002) M-FISH analysis shows that complex chromosome aberrations induced by alpha -particle tracks are cumulative products of localized rearrangements. Proc Natl Acad Sci USA 99:12167–12172

Bender MA, Gooch PC, Kondo S (1967) The Gemini-3 S-4 spaceflight-radiation interaction experiment. Radiat Res 31:91–111

Bender MA, Gooch PC, Kondo S (1968) The Gemini XI S-4 spaceflight-radiation interaction experiment: the human blood experiment. Radiat Res 34:228–238

Bonassi S, Ugolini D, Kirsh-Volders M, Stromberg U, Vermeulen R, Tucker JD (2005) Human population studies with cytogenetic biomarkers: review of the literature and future prospectives. Environ Mol Mutagen 45:258–270

Cucinotta, FA, Durante M (2006) Cancer risk from exposure to galactic cosmic rays: implications for space exploration by human beings. Lancet Oncol 7:431–435

Durante M, Bonassi S, George K, Cucinotta FA (2001) Risk estimation based on chromosomal aberrations induced by radiation. Radiat Res 156:662–667

Durante M, George K, Wu H, Cucinotta FA (2002) Karyotypes of human lymphocytes exposed to high-energy iron ions. Radiat Res 158:581–590

Durante M, Snigiryova G, Akaeva E, Bogomazova S, Druzhinin S, Ferenfo B. Greco O, Novitskaya N, Rubanovich A, Shevchenko V, von Recklinghausen U, Obe G (2003) Chromsoome aberration dosimetry in cosmonauts after single or multiple flights. Cytogenet Genome Res 103:40–46

Durante M, Ando K, Furusawa Y, Obe G, George K, Cucinotta FA (2004) Complex chromosomal rearrangements induced in vivo by heavy ions. Cytogenet Genome Res 104:240–244

Durante M (2005) Biomarkers of space radiation risk. Radiat Res 164:467–473

Federenko B, Druzhinin S, Yudaeva L, Petrov V, Akatov Y, Snigiryova G, Novitskaya N, Shevchenko V, Rubanovich (2001) Cytogenetic studies of Blood lymphocytes from cosmonauts after long-term space flights on Mir station. Adv Space Res 27:355-359

George K, Durante M, Wu H, Willingham V, Badhwar G, Cucinotta, FA (2001a) Chromosome aberrations in astronauts' blood lymphocytes after space flight. Radiat Res 156:731–738

George K, Wu H, Willingham V, Furusawa Y, Kawata T, Cucinotta FA (2001b) High- and low-LET induced chromsome damage in human lymphocytes: a time course of aberrations in metaphase and interphase. Int J Radiat Biol 77:175–183

George K, Wu H, Willingham V, Cucinotta FA (2001c) The effect of space radiation on the induction of chromosome damage. Phys Med 17:224–228

George K, Durante M, Willingham V, Wu, Yang TC, Cucinotta FA (2003a) Biological effectiveness of accelerted particles for the induction of chomosome damage measured in metphase and interphase human lymphoyctes. Radiat Res 160:425–435

George K, Durante M, Wu H, Willingham V, Cucinotta FA (2003b) In vivo and in vitro measurements of complex-type chromosomal exchanges induced by heavy ions. Adv Space Res 31:1525–1535

George K, Durante M, Willingham V, Cucinotta FA (2004) Chromosome aberrations of clonal origin are present in astronauts' blood lymphocytes. Cytogenet Genome Res 104:245–251

George K, Willingham V, Cucinotta FA (2005) Stability of chromosome aberrations in the blood lymphocytes of astronauts after space flight by FISH chromosome painting. Radiat Res 164:474–480

Hande, MP, Azizova TV, Geard CR, Burak LE, Mitchell CR, Khokhryakov VF, Vasilenko EK, Brenner DJ (2003) Past exposure to densely ionizing radiation leaves a unique permanent signature in the genome. Am J Hum Genet 72:1162–1170

Hortsmann M, Duante M, Johannes C, Pieper R, Obe G (2005) Space radiation does not induce a significant increase of intrachromosomal exchanges in astronauts' lymphocytes. Radiat Environ Biophysics 44:219–224

IAEA (2001) Cytogenetic analysis for radiation dose assessment. Technical report no 405. International Atomic Energy Agency, Vienna

Lucas JN (1997) Dose reconstruction for individuals exposed to ionizing radiation using chromosome painting. Radiat Res 148:33–38

Mustonen R, Lindholm C, Tawn EJ, Sabatier L, Salomaa S (1998) The incidence of cytogenetically abnormal rogue cells in peripheral blood. Int J Radiat Biol 74:781–785

NCRP (2000) Radiation protection guidance for activities in low-Earth orbit. NCRP report no. 132. NCRP, Bethesda

NCRP (2002) Operational radiation safety program for astronauts in low-Earth orbit: a basic framework. NCRP report no 142. NCRP, Bethesda

Obe G, Johannes I, Johannes C, Hallman K, Reitz G, Facius R (1997) Chromosomal aberrations in blood lymphocytes of astronauts after long-term space flights. Int J Radiat Biol 72:727–734

Preston DL, Pierce DA, Shimizu Y, Ron E, Mabuchi K (2003) Dose response and temporal patterns of radiation-associated solid cancer risks. Health Phys 85:43–46

Testard L, Ricoul M, Hoffschir F, Flury-Herard A, Dutrillaux B, Federenko B, Gerasimenko V, Sabatier L (1996) Radiation-induced chromosome damage in astronauts' lymphocytes. Int J Radiat Biol 70:403–411

Tucker JD (2001) FISH cytogenetics and the future of radiation biodosimetry. Radiat Prot Dosim 9755–60

Yang TC, George K, Johnson AS, Durante M, Fedorenko BS (1997) Biodosimetry results from space flight Mir-18. Radiat Res 48:17–23

26 Internet Information System and Literature Database on Biomedical Effects of Electromagnetic Fields: Cancer and Cytogenetic

Jiri Silny, Roman Wienert, Frank Klubertz, and Sarah Driessen

Abstract

In order to assess potential health effects due to electromagnetic field exposure the outcome of in vitro studies must be compared with the results from existing epidemiological investigations, animal and in vivo experiments. For this purpose, sometimes several hundred publications related to specific exposure conditions must be researched and evaluated in a time-consuming process. The knowledge-based scientific database "*EMF-Portal*" presented here facilitates this process by providing standardized descriptors of exposure characteristics, material, method and results which have been extracted from individual biological/medical as well as epidemiological publications. In this chapter, an evaluation process concerning the endpoint cancer/genotoxicity, particularly in connection with mobile phone exposure, is presented as an example. Some predominantly epidemiological and a few in vitro investigations have reported to have found effects. According to our assessment, however, there is no clear consistency within the categories (in vitro, in vivo and epidemiology). The *EMF-Portal* currently contains more than 10,000 publications mainly from peer-reviewed scientific journals; on average, more than 600 new articles are added per year. A comprehensive glossary and a number of additional tools provide the user with valuable medical and technical assistance. A query structure, alternatively general or more detailed, allows the scope of the literature search to be narrowed down to individual publications regarding one particular subject. The *EMF-Portal* is already available in German and English; other languages can be added. All information collected is available via Internet access and free of charge at http://www.emf-portal.org.

26.1 Introduction

In the general populations of industrialized countries there are growing concerns and even fears of adverse health effects due to technical electromagnetic fields (EMF) occurring in everyday life. With the introduction of new techniques and technologies, the exposure of humans to technical EMF is becoming more and more diversified and the total intensity is increasing gradually.

Günter Obe and Vijayalaxmi (Eds.)
Chromosomal Alterations: Methods, Results and Importance in Human Health

Everyday fields are sometimes associated with subjective feelings of discomfort such as headache, sleeplessness and dizziness but also with serious illnesses like multiple sclerosis, Parkinson's disease or even cancer. This association is maintained and increasingly circulated in public appearances by influential social groups and commercial companies and institutes, without a critical discussion based on actual current scientific knowledge taking place. Business with public fears of EMF is a booming market; already enormous amounts of money are spent on unnecessary measurements conducted by so-called building biologists and on pointless countermeasures such as "protective" foils and chips.

There is a critical debate of these alleged health hazards by scientists and the general public but also by other groups, such as physicians, lawyers and politicians, who are increasingly confronted with these problems and must deal with them.

The most reliable basis for the assessment of current knowledge is the worldwide scientific literature which so far already amounts to more than 10,000 publications, even when counting only those published in peer-reviewed journals.

Investigations on chromosomal changes due to electromagnetic exposure are an important cornerstone in the demonstration of health effects. Their results must be compared with the results from epidemiological studies and animal experiments, particularly in case of the frequently discussed endpoint "cancer". Taken together, there are several hundred publications which must be traced and obtained prior to scrutiny and evaluation.

Among other information sources, the Internet-based *EMF-Portal* (accessible free of charge at http://www.emf-portal.org) provides valuable assistance in the assessment of the current status of knowledge in the research area of bioelectromagnetic interaction with about 10,000 included publications so far. In a first step, these publications must be divided according to physical criteria and medical/biological endpoints end as well as according to the investigation methods applied into categories and classes before a comprehensive evaluation of the results in individual publications and categories can be carried out. However, a majority of these are of a more technical and theoretical nature and deal predominantly with questions of dosimetry as well as with hypotheses on mechanisms of action.

Following the explanation of the necessary medical and physical differentiation of the publications, it is subsequently illustrated how the *EMF-Portal* can be advantageously used for literature research by laymen as well as by experts.

26.2 Classification of Publications

26.2.1 Physical Properties

Electromagnetic waves represent a major group of physical phenomena which follow comparable physical rules but must be differentiated with regard to their potential effects on the organism in several ways. Following the established effects of electromagnetic waves in the organism, one must distinguish primarily between non-ionizing and ionizing electromagnetic waves. The elementary energy of electromagnetic waves increases with frequency; in case of X-rays or γ-rays it reaches a point where the binding energy of biological molecules and atoms is exceeded. These ionizing waves damage the bindings of biomolecules and initiate irreversible biochemical reactions. Specific medical problems are associated with ionizing electromagnetic waves; they are not treated here further.

The subject matters of this chapter are non-ionizing electromagnetic waves with frequencies lower than 300 GHz. They are, according to their frequencies and their established effects in the organism, divided into low-frequency electric and magnetic fields (0–10 MHz) as well as radiofrequency fields and microwaves (10 MHz–300 GHz), as shown in Fig. 26.1.

Strong low-frequency electromagnetic waves have been shown to stimulate nerves, muscles and sensory receptors, whereas strong radiofrequency fields lead to heating of the tissue. Indeed, the effects of low-frequency and radiofrequency EMF are very different; both, however, can initiate a substantial health disturbance or even a life-threatening situation when the applied intensity is very high. The thresholds for effects depend on a number of features of the applied EMF, for example frequency, modulation, strength, spatial extension and duration of the exposure; their precise definition is a prerequisite for an equivalent comparison of publications.

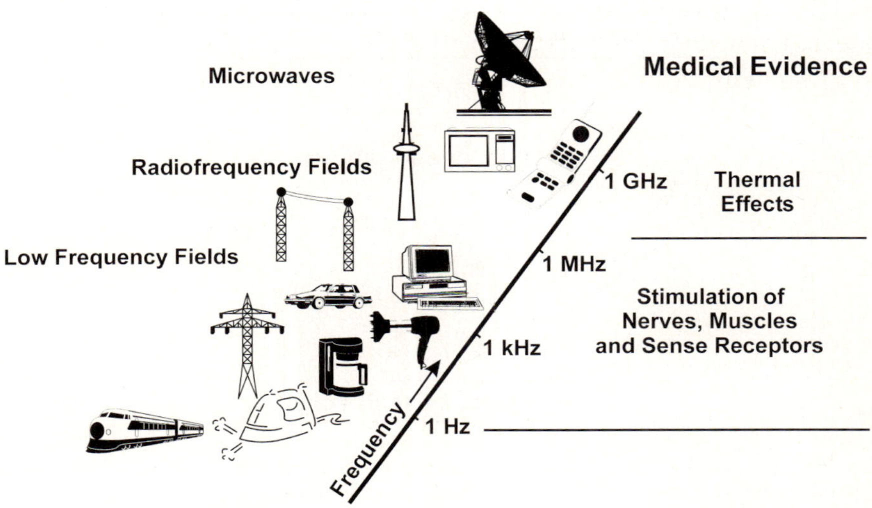

Fig. 26.1 Differentiation of non-ionizing electromagnetic fields according to their possible effects in humans

26.2.2 Medical/Biological Discrimination

For the medical/biological studies a working hypothesis is phrased on the basis of postulated working mechanisms or of scientific results but also on the basis of preliminary reports. The working hypothesis must contain an assumption that defined EMF can promote or initiate an effect or illness (Fig. 26.2).

A confirmation or non-confirmation of the working hypothesis in a single study alone is not sufficient for its acceptance or rejection. Biological variations or even systematic or random failures could be a reason for false results; therefore, only concurrent findings regarding predefined fields, a medical endpoint

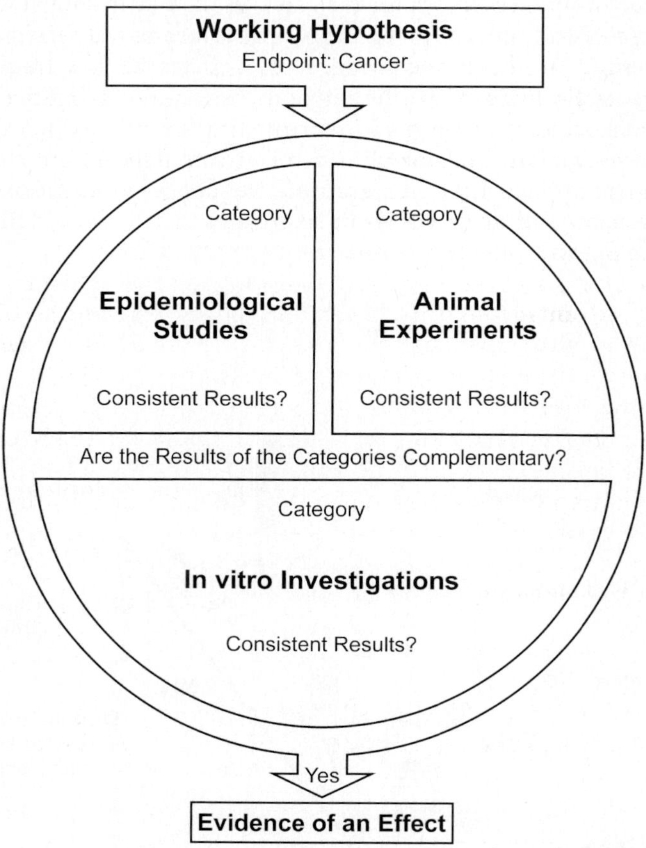

Fig. 26.2 The contribution of publications, originating from the use of different investigation methods (categories) and with the same medical endpoint, to current knowledge. The literature shows clear evidence of an effect only if the results are consistent in each category and complementary between the categories

and literature categories elaborated by a number of independent research groups can be accepted as sustainable results. For this reason the first step of the review is to collect all publications pertaining to one subject, consulting medical databases, for example PubMed, or problem-oriented databases such as IVI's (Information Ventures' *EMF Database*) *EMF Database*, the WHO EMF Research Database or our femu's (Research Center of Bioelectromagnetic Interaction) *EMF-Portal*, which are available at different conditions.

According to the existing literature categories, a broad division is carried out between epidemiological investigations, provocation studies, animal experiments and in vitro investigations. All these investigation classes have certain advantages and disadvantages which must be taken into account in the final evaluation. For example, the advantage of epidemiological and provocation studies is their focus on humans. Their weakness generally lies in their shortcomings regarding the retrospective data collection, the statistical evaluation in small groups, as well as in the missing causality of an observed effect. Therefore, animal experiments or in vitro investigations must always be carried out additionally in order to corroborate the causality of an effect that was found in an epidemiological study.

Animal experiments allow studies with big populations of homogeneous groups in well-standardized attempts, thus yielding more reliable results. A disadvantage of animal studies is the problem of transferability of the results from animal to human.

The cause of an effect and the medical/biological mechanisms can be investigated advantageously in in vitro investigations. On the other hand, results from in vitro investigations must be reviewed with regard to their transferability to animals and, ultimately, humans.

Each of the previously stated classes of biological investigations utilizes different procedures to which in each case a greater or lesser explanatory power individually or in combination with one another can be attributed.

In the following sections the evaluation of the category "in vitro investigations" with the subject "chromosomal changes" (or "genotoxicity" in general, respectively) within the scope of the endpoint "cancer" is demonstrated exemplarily utilizing our *EMF-Portal*.

26.3 The *EMF-Portal*: Internet Information System on the Effects of Electromagnetic Fields

Within the framework of the Internet Information Portal on the effects of EMF on humans and the environment (which is accessible free of charge at http://www.emf-portal.org), scientific research results are summarized comprehensibly (*EMF-Portal* 2006). An "introduction to the fundamentals of the effects of EMF", a "database of exposure sources occurring in everyday life" and a "glossary" are available as supporting tools.

Scientific investigations are an important basis for the assessment of current knowledge on the consequences of electromagnetic exposure for the health of persons or the environment. But with scientific studies one often meets the following problems:

- They are not readily available to everyone
- They are not easy to understand
- Individual studies frequently contradict each other
- Some publications contain an abundance of expert terminology which tells the scientist from other research areas only little and the layman nothing at all
- They occur often in a greater context of high complexity which requires extensive background knowledge.

Moreover, it is difficult and time-consuming to find suitable publications in a collection of more than 10,000 studies and the constantly growing body of literature is too specific regarding the subjects currently discussed and the variety of questions posed.

Currently, only the *EMF-Portal* introduced here offers concise, comparable and comprehensible summaries of scientific research results (in English and German) covering studies published in the last 25 years.

26.3.1 Literature Database – the Principal Item

The principal item of the *EMF-Portal* is the literature database. Within the scope of the *EMF-Portal*, articles in the area of bioelectromagnetic interaction of low-frequency and radiofrequency fields with the organism are researched in more than 1,300 scientific journals on a regular basis, and more than 600 studies are added per year. Moreover, included in the database are mainly published articles from peer-reviewed journals or detailed working reports and research reports if they describe the methodology applied and the experiments conducted comprehensively.

In the literature database these publications are sorted according to study types, categorized according to thematic profiles and represented in different levels of detail.

Individual articles can be found via the menu item "publication query" (Sect. 26.3.2).

26.3.2 How to Find a Publication?

There are two different ways to find publications. In the "standard query", for example, one can enter terms like "mobile phone" and "chromosome aberration". The search engine looks for related concepts and adds them to the query (e.g. "cellular phone", "handheld phone", "GSM", "TETRA", or "aberration", "chromosome break", "inversion", "translocation" and so on, respectively). In addition, the frequency utilized by the device (mobile phone from 890 MHz to 2.1 GHz) is included in the query automatically by the search engine.

The "detailed query" makes it possible to narrow down the contents of the query precisely. In a first step, a restriction of the query can be made according to bibliographical information (author, title, abstract, journal, year and keywords). By selecting the items "exposure parameters" and/or "general data" additional choices become available and the query can be limited further.

For bibliographical data, articles are searched in all collected publications (currently about 10,000 articles). In contrast, by inclusion of the "exposure parameters" and/or "general data", the search is restricted to the publications which have already been summarized. The query within the "exposure parameters" is possible within about 3,000 publications, while the query within the "general data" is possible within about 1,400 publications.

Found publications are returned in the form of a "hit list". Figure 26.3 shows the first three publications found by the query on the subject of "mobile phone" (including the search terms "mobile phone", "cellular phone", "GSM" and "UMTS") and "genotoxicity". For a better overview, the list contains not only the bibliographical data but also some important summarized details of the respective study (aim, endpoint, frequency range).

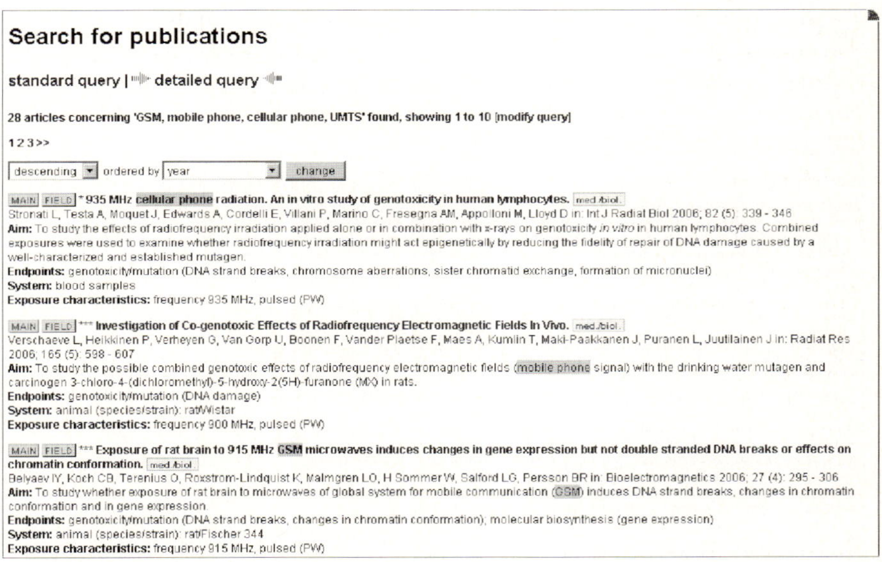

Fig. 26.3 Generated result list (of the first three results) following a detailed query for "mobile phone" (with the search terms "mobile phone", "cellular phone", "GSM" and "UMTS") and "genotoxicity"

The associated summaries can be viewed by selecting (clicking) an article from the "hit list". Figure 26.4 shows the representation of the first article from the result list exemplarily.

Using the German or English summaries of individual publications, the user gets a first impression concerning the essential contents of the processed publications without the need for access to the original full text.

Medical/biological Study (experimental study)

Exposure of rat brain to 915 MHz GSM microwaves induces changes in gene expression but not double stranded DNA breaks or effects on chromatin conformation.

by Belyaev IY, Koch CB, Terenius O, Roxstrom-Lindquist K, Malmgren LO, H Sommer W, Salford LG, Persson BR
published in: Bioelectromagnetics 2006; 27 (4): 295 - 306 (**PubMed Entry**)

Aim of study (according to author)
To study whether exposure of rat brain to microwaves of global system for mobile communication (GSM) induces DNA strand breaks, changes in chromatin conformation and in gene expression.

Endpoints

- genotoxicity/mutation: DNA strand breaks, changes in chromatin conformation
- molecular biosynthesis: gene expression

Exposure

Field characteristics	Parameters
915 MHz pulsed (PW) exposure duration: 2 h	SAR: 0.4 W/g power: 2 W max value (max. output power of the GSM900 test mobile phone)

FIELD **View further expo parameters**

Exposed system:
animal (species/strain): rat/Fischer 344
whole body exposure

Methods
Measurement parameters:

- genotoxicity/mutation: DNA double-strand breaks (pulsed-field gel electrophoresis), changes in chromatin conformation (as indicator for stress response and genotoxicity; AVTD)
- molecular biosynthesis: gene expression (Affymetrix U34 GeneChips (DNA microarray)); protein expression of stress response gene HSP70 (Western Blot)

investigated material: DNA/RNA (in vitro), cell suspensions; cell lysates
investigated organ system: brain/CNS, spleen, thymus

time of investigation: after exposure

Main outcome of study (according to author)
Effects of microwave exposure were found on neither conformation of chromatin nor DNA double-strand breaks.
In cerebellum from all exposed animals, 11 genes were upregulated in a range of 1.34-2.74 fold and one gene was downregulated 0.48-fold. The induced genes encode proteins with diverse functions including neurotransmitter regulation, blood-brain barrier, and melatonin production.
The results show that GSM microwaves at 915 MHz did not induce DNA double-strand breaks or changes in chromatin conformation, but affected gene expression in rat brain cells.
Because the nonthermal effects of microwaves appear to be dependent of many physical and biological variables, the data of this study should be compared with care with other studies where these variables are different.

(Study character: medical/biological study, experimental study, full/main study, pilot/exploratory/preliminary study)

Related articles

- **Markova E et al. (2005)**: Microwaves from GSM mobile telephones affect 53BP1 and gamma-H2AX foci in human...
- **Belyaev I (2005)**: Non-thermal Biological Effects of Microwaves.
- **Crumpton MJ et al. (2004)**: Are environmental electromagnetic fields genotoxic?
- **Sarimov R et al. (2004)**: Nonthermal GSM Microwaves Affect Chromatin Conformation in Human Lymphocytes...
- **Hossmann KA et al. (2003)**: Effects of electromagnetic radiation of mobile phones on the central nervous...
- **Pacini S et al. (2002)**: Exposure to global system for mobile communication (GSM) cellular phone...
- **Goswami PC et al. (1999)**: Proto-oncogene mRNA levels and activities of multiple transcription factors in...
- **Lai H et al. (1997)**: Melatonin and a spin-trap compound block radiofrequency electromagnetic...

Glossary: animals, AVTD, biosynthesis, blood-brain barrier, brain, cells, cell suspensions, cerebellum, chromatin, CNS, conformation, DNA, DNA microarray, double-strand breaks, downregulated, exposed, exposure, full/main study, gel electrophoresis, gene, gene expression, genotoxicity, GSM, HSP70, indicator, induced, in vitro, lysates, melatonin, MHz, microwave, mobile phone, molecular, mutation, neurotransmitter, nonthermal, pilot/exploratory/preliminary study, power, Proteins, pulsed, PW, RNA, SAR, species, spleen, strand breaks, stress, thymus, upregulated, Western blot

Fig. 26.4 Example of the summary of an experimental study

So far, epidemiological and experimental studies from the medical/biological area have been evaluated. The structure of the representation of the epidemiological studies or the experimental investigations, respectively, is always the same. This serves the purpose of a quick orientation as well as an optimum comparability of the studies with one another.

In the case of reviews (survey studies), only the main subject focus and frequency range are extracted so that they can be found more easily and allow the user to decide quickly whether the respective review is of interest.

26.3.3 An Example on the Subject of "Genotoxicity/Cancer"

Currently, there are 204 experimental studies in the database from different frequency ranges concerning the topic "genotoxicity" (Fig. 26.5).

If one wants to put the subject "genotoxicity" in a broader context in order to be able to evaluate the possible effects described in the studies more globally, regarding, for example, "cancer", it is advisable to differentiate between in vitro investigations, animal experiments and epidemiological studies (Sect. 26.2).

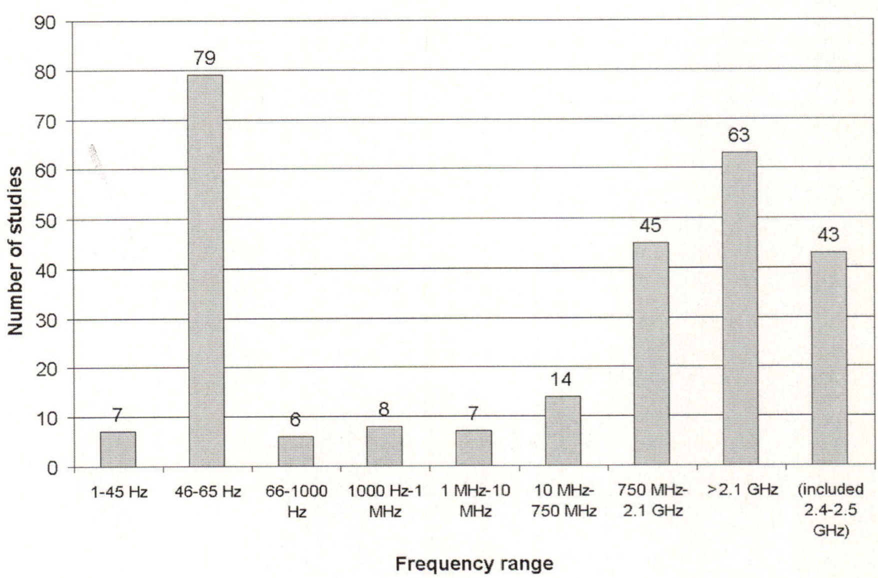

Fig. 26.5 Studies with the main topic "genotoxicity", distributed according to the frequency applied. If several frequencies or broader frequency bands were used in individual studies, the studies of each frequency range are added separately; hence, the sum of all studies listed in the chart is more than 204 (actual sum of existing studies)

With the help of the detailed query (Sect. 26.3.2) one can search for different exposure types and frequency ranges, for different exposed materials (in vitro systems, animals, plants or humans) and for different endpoints.

Figure 26.6 shows, for example, the results of different detailed queries:

- For mobile phone (search words "GSM", "cellular phone", "mobile phone" or "UMTS")
- In each case divided into categories according to the exposed in vitro system or animal
- For the biological endpoints "cancer" or "genotoxicity".

In total there are currently 52 studies on the effects of EMF in the mobile phone frequency range regarding the endpoints "genotoxicity" or "cancer": 18 in vitro investigations and seven animal studies deal with the endpoint "genotoxicity", 17 experimental animal studies and 35 epidemiological studies address the endpoint "cancer".

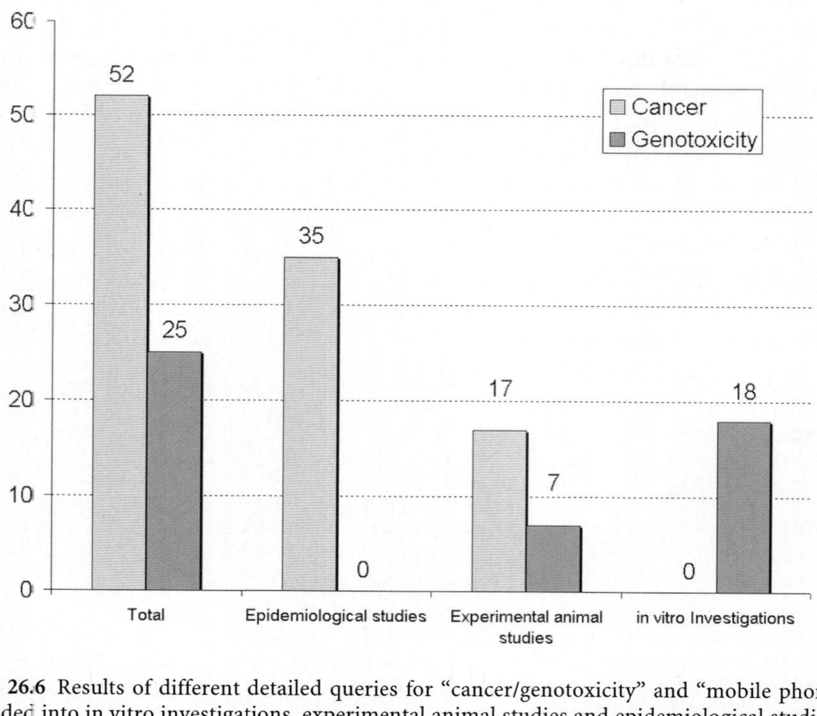

Fig. 26.6 Results of different detailed queries for "cancer/genotoxicity" and "mobile phone", divided into in vitro investigations, experimental animal studies and epidemiological studies

- There are no consistent data from the in vitro investigations regarding the endpoint "genotoxicity". There are about the same number of publications that report effects of exposure to EMF in the frequency range related to mobile phones and that do not. The studies differ greatly with respect to their frequency ranges (835–1,900 MHz), the different types of modulation (TDMA, CDMA, FDMA, etc.) and specific absorption rates (SAR; from 0.005 to 10 W/kg) as well as the different cell systems applied and measurement parameters investigated (chromosome alteration, sister-chromatid exchange, micronuclei formation, DNA stability, DNA strand breaks). An evaluation and a comparison of these inconsistent data are very difficult and nearly impossible for non-experts. Standardized investigation methods should be required for future studies.
- Additionally, there are seven experimental animal studies regarding the endpoint "genotoxicity". These studies also differ greatly with respect to their frequency ranges (900–1,747 MHz), the applied SAR values (0.07–33 W/kg, whole body), the mobile phone systems and types of modulation (GSM, PDC, IRIDIUM, FDMA, TDMA) that are taken into account, the different animals investigated (a variety of mouse and rat strains) as well as their utilized measurement parameters (DNA strand breaks, chromatin structure, formation of micronuclei, DNA sequence analysis, chromosomal alterations). Moreover, the exposure duration varies between 2 h/day for 1 day and for 2 years. The results of the animal studies are equally inconsistent. Three out of seven studies report observed effects; four studies do not. Only experts in this field are competent enough to compare and interpret these data adequately (Sect. 26.4).
- In contrast, the 17 experimental animal studies addressing the endpoint "cancer" provide more consistent data. Although the frequency ranges (835–1,439 MHz) investigated as well as the modulation and mobile phone systems (GSM, PDC, CDMA, FDMA, TDMA, NMT, UMTS) differ in the various studies, the applied SAR values (whole body) were in the range 0.4–1 W/kg in the majority of investigations. Even though different mouse and rat strains as well as different types of cancer (lymphoma, skin cancer, mammary carcinoma, liver cancer) were investigated, all publications (with the exception of one) conclude that EMF exposure in the frequency range related to mobile phones neither initiates nor promotes or copromotes cancer.
- Of the 35 epidemiological studies, 15 have found a weak but significantly increased risk of malignant brain tumours, acoustic neuroma, leukaemia and uveal melanoma in the users of mobile phones. However, 14 of these studies were published by the same Swedish group, and they predominantly address a since-outdated analogue mobile phone system. The remaining 20 studies from different research groups have not shown any association between the short- or medium-term exposure by microwaves emitted by mobile phones and brain tumours, acoustic neuroma, cancer of the salivary gland, leukaemia or malignant melanomas of the eye.

26.4 Discussion

The body of scientific literature regarding chromosomal alterations due to EMF provides important and detailed insights into possible effects and influence mechanisms. But for the evaluation of potential health effects a prudent comparison of the findings of epidemiological studies, animal experiments and in vitro investigations resulting from comparable EMF exposure is necessary.

The review of health effects due to EMF is a demanding and time-consuming process even if the physical and medical expert knowledge is available. In the course of this process all publications dealing with a defined EMF range and a particular medical endpoint are collected and evaluated individually. Today, literature research is generally carried out utilizing Internet databases which – beyond the mere bibliographical data – provide a variety of additional information and are available under specific terms and conditions. For example, through the journal literature search system *PubMed* (2006), the by far largest medical database, virtually anyone can access more than 15,000,000 biomedical journal citations of publications since 1966. Currently, it contains about 9,000 articles related to EMF.

As an alternative, specific databases pertaining to EMF can be used. IVI's *EMF Database* (2006) provides about the same scope of information as *PubMed*, but additionally collects records of poster sessions and meeting abstracts. However, the use of its Internet interface is limited to a number of basic operations and the viewing of a handful of sample records; the use of the complete CD version containing third-party abstracts is subject to a fee.

In contrast, the Internet-based WHO *EMF Project Database* (2006) of the WHO International EMF Project and the IEEE *EMF Citation List* (2006), are free of charge and provide additional descriptors such as the type of the study and the endpoints pursued; unfortunately, however, as yet they have only a small stock of records. The Swiss database *ELMAR* (2006), also contains additional descriptors regarding endpoints, materials, methods and results, but is small and is available only in German. Currently, only femu's *EMF-Portal* contains comprehensive information beyond bibliographical standard data as it includes detailed descriptions of the biological or medical studies conducted as well as their results and exposure parameters. Moreover, the Internet access to the information covering practically the whole body of literature pertaining to bioelectromagnetic interaction since 1980 is completely free of charge and access restrictions.

Individual studies can be evaluated objectively under the condition that only explicitly cited data from published works are taken into account. But the next step, the validation of the consistency of a postulated effect in the profiles of literature resulting from comparable investigations is - to a certain extent – a subjective process in which a high level of expert knowledge and the ability to interpret scientific results are essential.

The same problems and requirements arise when trying to assess the complementarities of the results reported between the categories; therefore a comprehensive evaluation like that should preferably be conducted independently by more than one expert, if possible.

Owing to the rapid growth of the body of literature in this area of research with an annual increment of more than 600 publications, this evaluation process should be repeated at least every 4 or 5 years.

The question whether – or if so to what extent – EMF can promote or even initiate cancer is of crucial interest for highly industrialized societies. Here, scientific publications answering the questions concerning chromosomal alterations in EMF are of paramount importance. Several hundred epidemiological studies, animal studies as well as in vitro investigations have been carried out. Some predominantly epidemiological and a few in vitro investigations report to have actually found effects. But there is no clear consistency within the categories and there is no complementary support between the groups. This conclusion can be derived for the low-frequency as well as the radiofrequency range of EMF (Gezondheidsraad Jaarbericht 2005; Health Protection Agency 2006; NRPB 2004). So far, the scientific literature does not provide any sustainable evidence for an association between EMF occurring in everyday life and the various forms of cancer.

References

ELMAR (2006) Dokumentationsstelle *ELMAR* (ELectroMAgnetic Radiation), Basel, Switzerland. http://www.elmar.unibas.ch/elmar_suche.html. Cited 19 July 2006

IVI's *EMF Database* (2006) Information Ventures, Philadelphia, USA. http://infoventures.com/emf/database/. Cited 19 July 2006

EMF-Portal (2006) Research Center for Bioelectromagnetic Interaction – femu, Aachen. http://www.emf-portal.org/_index.php?l=e. Cited 19 July 2006

WHO *EMF Project Database* (2006) World Health Organization, Geneva, Switzerland. http://www.who.int/peh-emf/research/database/emfstudies. Cited 19 July 2006

Gezondheidsraad Health Council of the Netherlands (2005) Electromagnetic fields: annual update 2005. Jaarbericht van de Gezondheidsraad no 2005/14. Gezondheidsraad Health Council of the Netherlands, The Hague

Health Protection Agency (2006) Power frequency electromagnetic fields, melatonin and the risk of breast cancer. Report of an independent advisory group on non-ionising radiation. Documents of the Health Protection Agency. Series B: radiation, chemical and environmental hazards. HPA series B:RCE(1). Health Protection Agency, Chilton, pp 1–180

IEEE EMF Citation List (2006) http://www.who.int/peh-emf/research/database/IEEEdatabase. Cited 19 July 2006

NRPB (2004) Mobile phones and health 2004. Report by the board of NRPB. Doc NRPB 15(5):1–114

PubMed (2006) US National Library of Medicine, Bethesda. http://www.ncbi.nlm.nih.gov/entrez/query.fcgi?DB=pubmed. Cited 19 July 2006

27 Interpreting Studies on Magnetic Fields and the Risk of Childhood Leukaemia: an Ongoing Challenge

JOACHIM SCHÜZ AND TRACY LIGHTFOOT

Abstract

Childhood cancer under the age of 15 years is a rare event accounting for less than 1% of all malignancies diagnosed each year in developed countries. Overall the risk of developing acute leukaemia before the age of 15 years is around 1:2,000. Acute leukaemia is a heterogeneous disease and the advent of improved molecular diagnostic techniques has provided greater insight into its underlying molecular biology. The presence of chromosomal abnormalities in neonatal blood spots from children with leukaemia supports the hypothesis for a prenatal origin of the disease. Furthermore, identification of these abnormalities at a rate higher than the proportion of children who will develop the associated leukaemia subtype has provided evidence that childhood leukaemia is at least a two-step process with a need for additional genetic events and/or exposures. Improved knowledge of chromosomal alterations and methods for their analysis offers new opportunities for understanding the relevance of environmental exposures and the importance of gene–environment interactions in the development of the disease. A prominent example is the ongoing challenge of interpreting findings from epidemiological studies on exposure to magnetic fields and childhood leukaemia risk, where a doubling of leukaemia risk was consistently seen across most studies at average magnetic field exposures of 0.3–0.4 μT. The interpretation of the magnetic-field-related findings is particularly delicate because possible explanations range from methodological artefacts to a causal link. To date, neither of these two contradictory explanations can be completely ruled out.

27.1 Introduction

In 2002, within the framework of their Monography Program on the evaluation of carcinogenic risks to humans, the International Agency for Research on Cancer (IARC) classified extremely low-frequency magnetic fields as a possible carcinogen to humans (IARC 2002). This classification was based on epidemiological studies that investigated the relationship between exposure to residential magnetic fields and the risk of childhood leukaemia. In these

Günter Obe and Vijayalaxmi (Eds.)
Chromosomal Alterations: Methods, Results and Importance in Human Health

studies, about a doubling of the leukaemia risk was found with average exposure above 0.3–0.4 μT (Greenland et al. 2000; Ahlbom et al. 2000). These findings were considered as being credible, but chance, bias and confounding could not be ruled out with reasonable confidence. No support came from either experimental animal or cell studies and no plausible explanation for the epidemiological findings has been put forward (IARC 2002). Thus, the overall evidence was considered to be too weak to provide sufficient evidence of carcinogenicity; however, the epidemiological findings were not dismissed and merit clarification. The disagreement between results from the observational and experimental studies could be due to the failure of experimental studies to provide an appropriate model for the complex origin of childhood leukaemia; this will be discussed in Sect. 27.2. Alternatively, it could be due to methodological flaws of the observational studies or other mechanisms unrelated to magnetic fields that nevertheless create a correlation between exposure and disease risk; this will be discussed in Sect. 27.3 following an account of the epidemiological evidence. Finally, overall conclusions will be drawn and the role of the identification of chromosomal alterations in clarifying this issue will be presented (Sect. 27.4).

27.2 Leukaemia in Children

27.2.1 Description of Epidemiology and Biological Heterogeneity

Cancer in children under the age of 15 years is a rare event accounting for less than 1% of all malignancies diagnosed each year in developed countries. Of those cancers seen in children, leukaemias, in particular acute leukaemias, are the most common, accounting for a third of all diagnoses. The incidence rate of acute leukaemias ranges from 30–50 per million children per year (Steliarova-Foucher et al. 2004). Overall the risk of developing acute leukaemia before the age of 15 years is around 1:2,000 (Greaves 2002), with the disease being more prevalent in boys than in girls (UKCCS 2000). The distinctive age distribution characterised by a rapid increase in incidence after birth, peaking between 2–5 years of age before declining to a steady rate seen for childhood leukaemia, has resulted in many aetiological hypotheses (UKCCS 2000), most notably those involving exposure to infections (Greaves 2002; Kinlen 1995). The most recent data on incidence trends indicate a small annual increase of 0.7% in European countries in the last three decades (Steliarova-Foucher et al. 2004); however, this increase is likely to be attributable to improvements in the classification and registration of childhood cancers.

The biological heterogeneity of childhood leukaemia is well documented (Biondi et al. 2000) and it is therefore unlikely that there is a single causal exposure or mechanism. The major subtypes diagnosed in children are acute lymphoblastic leukaemia (ALL) and acute myeloid leukaemia (AML). They account for 80% and around 15–17% of cases in white populations, respectively, with a slightly higher proportion of AML cases reported in Asia and black populations of North America (Parkin et al. 1988). By contrast, chronic forms of leukaemia in children are extremely rare in all populations and seldom exceed 4% of all leukaemia diagnoses.

Historically, one of the problems faced when trying to unravel the potential determinants of childhood leukaemia was the inability to classify different subtypes of leukaemia such that all cases were pooled together, which may have masked any potential associations. However, over the last 25 years our knowledge of immunology has increased, leading to an improvement in diagnostic techniques. Furthermore, these advances in classifying the different biological subtypes combined with the recognition of prognostic variables and improved chemotherapeutic agents has resulted in more than 80% of children diagnosed with acute leukaemia surviving into adulthood (Pui et al. 2004); however, some children remain unresponsive to treatment (Biondi et al. 2000), and the long-term health consequences of survivors is the subject of much ongoing research.

Subtypes of leukaemia are characterised by genetic alterations including point mutations, deletions and inversions resulting in gross chromosome changes such as translocations or hyperdiploidy or hypodiploidy (Look 1997). As with other cancers, chromosomal translocations are initiated by DNA double-strand breaks. In contrast to epithelial carcinomas, however, the translocations are generally balanced or reciprocal and often result in chimeric or fusion genes which generate hybrid proteins with altered properties such as activated kinase activity or novel transcriptional regulation (Look 1997). Although over 200 genes have been associated with the occurrence of chromosomal abnormalities, some are more frequently observed than others and are related to particular morphologic or phenotypic subtypes (Greaves and Wiemels 2003; Table 27.1). Those most commonly seen include *MLL*, *TEL* and *AML1* – all of which have been shown to be essential for normal haemopoiesis in mice knockout studies (Okudo et al. 1996) [the acronyms display gene symbols that represent the positional mapping of a gene, like *MLL* being located on 11q23, which is band 2 subband 3 on the long arm (q) of chromosome 11].

In infants under 12 months old, a distinct type of ALL with a pro-B phenotype has been identified such that over 80% of cases have 11q23 chromosomal translocations. These translocations involve the breakage and recombination of the *MLL* gene, which is involved in regulation of gene expression and maintenance of chromatin stability (Yu et al. 1998), with one of several potential partner genes (Heerema et al. 1994). The marked peak in incidence seen between 2 and 5 years of age is associated with common B-cell precursor ALL where the most common abnormalities reported are hyperdiploidy and the *TEL-AML1* gene fusion which account for 35% and 25% of cases, respectively (Greaves 2005). The question arises as to how and when these translocations occur. However, there is now substantive evidence that some, if not all, of the translocations seen in childhood leukaemia originate in utero.

Table 27.1 The major biological subtypes of childhood leukaemia and associated chromosomal changes. (Adapted from Greaves and Wiemels 2003)

Subtype	Cell type involved	Chromosome abnormality	Molecular lesion	Frequency (%)	Functional product
ALL	B-cell progenitor-monocytic[a] (infants)	11q23 translocations	*MLL-AF4*	~85 (of infant ALL)	Modified transcription factor[b]
			MLL-ENL and other fusions	~5 (of total ALL)	
	B-cell precursor	Hyperdiploidy	Increased gene dose	~35 (of B-cell precursor ALL)	Unknown
		t(12;21)(p13;q22)	*TEL-AML1* fusion	~20 (of B-cell precursor ALL)	Chimeric transcription factor[c]
		t(1;19)(q23;p13)	*E2A-PBX1* fusion	~5 (of B-cell precursor ALL)	Chimeric transcription factor
		t(9;22)(q34;q11)	*BCR-ABL* fusion	~5 (of B-cell precursor ALL)	Activated kinase
	T-cell precursor	1q deletion; t(1;14)(p32;q11)	*SIL-SCL* fusion	~25 (of T-cell precursor ALL)	Dysregulated transcription factor *(SCL/TAL1)*
AML	In infants	11q23 translocations	*MLL-AF6*, *MLL-AF9*, *MLL-AF10* or other fusions	~50 (of infant AML)	Modified transcription factor[b]
		t(8;21)(q22;q22)	*AML-ETO* fusion	~15 (of total AML)	Chimeric transcription factor[c]

ALL acute lymphoblastic leukaemia, *AML* acute myeloid leukaemia

[a]This subtype has phenotypic features of both early B-lineage progenitors and monocytes.

[b]*MLL* may modify chromatin structure and control the expression of key developmental genes (such as *HOX*).

[c]*AML-1*, which is an important positive transcriptional factor, may be switched to a transcriptional repressor in these fusions.

27.2.2 Origins of Childhood Leukaemia

Leukaemia is a clonal disease arising from the neoplastic transformation of a single cell which is likely to be either a pluripotent stem cell or a more mature progenitor cell (Greaves 2005). The single cell, via repeated cell division, generates an expanding clone of neoplastic cells which is highly malignant. As well as the nature of the leukaemic clone, the stage of disease progression at the time the pathological symptoms are recognised, the diagnosis made and treatment regimen implemented are also important for the clinical outcome of the disease. Any delay in diagnosis increases the probability that the clone will have progressed to the point where additional mutations will have accrued, including those involved with drug resistance, which can make treatment more difficult.

Identification of the origin of chromosomal translocations has been facilitated by the design of analytical tests for specific fusion genes which can detect as few as one to 20 leukaemic cells in a blood spot. The presence of the same fusion gene sequence in neonatal blood spots or Guthrie cards and the patient's leukaemic cells at diagnosis provides compelling evidence that chromosomal translocations can originate in utero and that leukaemia can be initiated prenatally. Indeed at least six different types of paediatric leukaemia have been identified where the chromosomal rearrangements, including hyperdiploidy, have been shown to originate in utero (Greaves 2005; McHale and Smith 2004). Data from studies of identical twins and children with concordant ALL support this hypothesis (Greaves 2006).

Screening of newborn cord blood samples suggests that around 1% have a functional *TEL-AML1* fusion gene, which is 100 times the proportion of children that will develop ALL with a *TEL-AML1* gene fusion before the age of 15 years (Mori et al. 2002). This implies that the frequency of conversion of the preleukaemic clone to overt disease is low, and that development of ALL is a multistep process requiring at least one prenatal initiating event in combination with other prenatal and/or postnatal events (Fig. 27.1). These events may involve additional genetic changes arising either before or after the formation of the chromosomal translocation, as well as exposure to biological, chemical

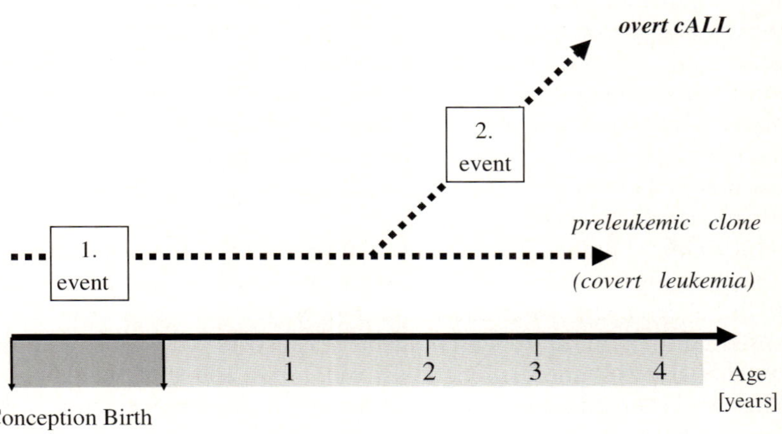

Fig. 27.1 A two-step model for the natural history of the most common subtype of childhood leukaemia, common acute lymphoblastic leukaemia (*cALL*). (Slightly modified from Greaves 1999)

or physical agents (Greaves 2006; Roman et al. 2006, 2007). Whilst the initiating in utero event is believed to be common, it is thought that the subsequent events act as the rate-determining steps in disease development. Assuming that the "two-hit" model for childhood leukaemia is correct, Greaves (2002) suggests that for every child diagnosed with ALL, there are at least 20 "healthy" children who also have a chromosomal translocation generated in utero. There is clearly a need to identify the possible exposures or molecular events that cause these chromosome breaks which are inefficiently or incorrectly repaired. Owing to the biological diversity of childhood leukaemia it is, however, unlikely that there is a single cause for the disease.

27.2.3 Aetiology of Childhood Leukaemia

In adults, carcinogens can accumulate over a long period of time, whereas in children the potential exposure period is much shorter. Accordingly, exposures acting before birth and early in life have long been thought to be important determinants of leukaemia, and the list of suspected chemical, physical and biological agents continues to increase. There are several mechanisms by which exposure to both exogenous and endogenous agents may impact on the risk of childhood leukaemia.

With respect to parental exposures prior to conception, exposure to possible carcinogenic substances early in the mother's life may be important since oocyte maturation begins during gestation. Accordingly, germ-cell mutations arising in the offspring could be attributable to exposures acting on the maternal grandmother while she was pregnant with the index child's mother. Conversely, paternal exposures may play a more direct role since spermatogenesis continues from puberty to old age. Thus, there is a greater period of time during which the accumulation of mutations may occur.

During pregnancy, exposures to agents such as ionising radiation may act directly, whilst others may act indirectly via placental transfer. However, it is not only the timing of exposure that is important but also the mother's and the embryo's ability to metabolise and detoxify the agents to which they are exposed. Furthermore, maternal preconceptional exposure to agents that have a slow metabolic clearance could result in the mother retaining embryotoxic doses into the early stages of pregnancy.

Whilst animal models have shown that preconceptional exposure of germ cells to carcinogens and mutagens results in excess tumours in offspring, no such links to cancer in humans have been confirmed. Indeed the only accepted human in utero carcinogenic exposures are ionising radiation (Stewart et al. 1956) and diethylstilbestrol (Herbst and Ulfelder 1971).

After birth, the infant can still be affected by parental exposures, which in the case of breast-fed infants may have occurred many years in the past, as well as being influenced by direct environmental exposures such as magnetic fields, which will be dealt with separately in Sect. 27.3.

To date the major focus has been with respect to infection. Indeed, the role of infection in the development of childhood ALL is the focus of much current research. The "Kinlen" hypothesis speculates that acute leukaemia may arise as a consequence of exposure to a specific infection, particularly evident at times of unusual population movement and mixing (Kinlen 1995). However, whilst data continue to accrue in this area, as yet no virus-dependent mechanism has been positively identified. By contrast, the "Greaves" hypothesis proposes that a deficit of exposure to infectious agents in infancy and subsequent "delayed" infectious challenge may be causal in the development of B-cell precursor common ALL (c-ALL) (Greaves 2006). However, recent findings suggest that increased exposure to infections early in life may, in some children, also play a critical role (Roman et al. 2006). Much has also been documented regarding the effects of breast-feeding on childhood leukemia risk since immunological benefits are provided to the offspring that may contribute to their response to infection. Whilst many studies have reported weak protective associations between breast-feeding and childhood ALL, data from the UK Childhood Cancer Study and the Oxford Survey of Childhood Cancers failed to support the hypothesis (Lightfoot and Roman 2004). The interpretation of these findings is not straightforward since breast-feeding may also allow transfer of genotoxins along with viruses from the mother to the child.

Many parental exposures have also been suggested as determinants of childhood leukaemia, including general lifestyle factors such as smoking and alcohol intake and exposure to DNA topoisomerase inhibitors and folic acid (Lightfoot and Roman 2004). The adverse health effects of cigarette smoking and excessive alcohol consumption are well documented - both with respect to the health of the individual and with respect to the health of the offspring; however, whilst many studies have reported on this topic there is no consistent evidence to confirm links with childhood leukaemia (Lightfoot and Roman 2004). *MLL* rearrangements are frequently seen in adult cases of leukaemia following chemotherapy treatment with topoisomerase II inhibitors. It is therefore possible that in utero exposure to topoisomerase inhibitors (including pesticides, benzene metabolites, bioflavanoids, some laxatives and antibiotics) could also be important in infant leukaemia; however, the data are sparse and rely entirely on self-report (Lightfoot and Roman 2004).

Folic acid levels and polymorphisms in folate-dependent enzymes have been linked to various types of cancer, including ALL in both adults and children. Folic acid deficiency and polymorphisms which mimic the conditions of folic acid deficiency alter patterns of DNA methylation and impair DNA synthesis and repair and may therefore influence leukaemogenesis. Research has focused on the polymorphic enzyme methylenetetrahydrofolate reductase which is key in regulating the balance between DNA methylation and DNA synthesis. The number of studies to date that have investigated methylenetetrahydrofolate reductase polymorphisms in relation to childhood ALL are limited and based on small numbers (Lightfoot and Roman 2004).

Unfortunately, the evidence regarding the majority of suggested exposures is limited and often contradictory and the search for the candidate exposure – assuming it exists – continues.

27.3 Exposure to Magnetic Fields and the Risk of Childhood Leukaemia

27.3.1 Summary of the Epidemiological Findings

Magnetic fields have been studied as a risk factor for childhood leukaemia since the late 1970s (Wertheimer and Leeper 1979). Since then more than 20 epidemiological studies have investigated this topic, with significant improvements in methods of exposure assessment over time. These include (1) simple distance estimations between power lines and the residence, (2) classifications of different types of power lines and their nearest distance to the residence (wire code), (3) magnetic field calculations based on power load data of power lines and (4) various measurement protocols, ranging from outdoor spot measurements to 24–48-h measurements in the child's bedroom and the use of personal dosimeters. Recently, these studies have been pooled in two separate meta-analyses. Greenland et al. (2000) pooled all available studies in this area and used the original data to estimate a pooled relative risk estimate; however, they had to combine various exposure indices into one single metric; hence, the microtesla estimate for the child's exposure in this pooling study is based on the somewhat different exposure assessment methods used across studies. Greenland et al. (2000) reported a combined relative risk estimate for leukaemia of 1.7 (95% confidence interval, CI, 1.2–2.3) in children exposed to average magnetic fields above 0.3 μT compared with those exposed below 0.1 μT. Ahlbom et al. (2000) only pooled studies that fulfilled certain inclusion criteria such as a defined population base for case ascertainment and control recruitment and usage of long-term measurements or historical magnetic field calculations for exposure assessment. Ahlbom et al. (2000) reported a relative risk estimate of 2.0 (95% CI 1.3–3.1) for exposures of 0.4 μT and above compared with exposures below 0.1 μT. The studies demonstrate a high degree of consistency; however, a major contribution to the overall relative risk estimate comes from a US study (Linet et al. 1997) and no association was seen in a large study in the UK (UKCCS 1999) (Table 27.2). A shortcoming was that the possibilities for dose–response analysis were limited, as only very few children were exposed to average magnetic fields of 0.4–1 μT or higher; hence, the data were too sparse to reliably predict the course of a dose–response curve for magnetic fields higher than 0.4 μT and, based on available data, this course is more or less compatible with trends ranging from a further increase in risk, to a constant risk or a downward gradient. Findings from these two pooled analyses

Table 27.2 Results from nine studies included in a pooled analysis of studies on residential magnetic fields and the risk of childhood leukaemia. (Adapted from Ahlbom et al. 2000)

	Odds ratio (95% CI) ≥ 0.4 μT vs. <0.1 μT	Leukaemia cases	
		Observed ≥ 0.4 μT	Expected[a] ≥ 0.4 μT
Canada	1.55 (0.65–3.68)	13	10
USA	3.44 (1.24–9.54)	17	5
UK	1.00 (0.30–3.37)	4	4
Norway	0 cases, 10 controls	0	3
Germany[b]	3.53 (1.01–12.3)	7	2
Sweden	3.74 (1.23–11.4)	5	2
Finland	6.21 (0.68–56.9)	1	0
Denmark	2 cases, 0 controls	2	0
New Zealand	0 cases, 0 controls	0	0
Total[b]	2.04 (1.33–3.15)	49	26

CI confidence interval
[a]Rounded to whole numbers
[b]Update of Ahlbom et al. (2000) with a second German study (Schüz et al. 2001)

were those referred to in recent reviews of claims of consistency in the epidemiological evidence (IARC 2002). In particular, the approach by Ahlbom et al. (2000) seemed to strengthen the evidence, since by introducing more uniformity into the analytical strategies, there was more consistency in the meta-analysis than in the originally published results. However, if studies suffer from the same systematic bias, this bias will be amplified in a pooled analysis rather than eliminated.

27.3.2 Interpreting the Empirical Association

27.3.2.1 Methodological Artefacts

In an IARC (2002) review it was stated that chance, bias and confounding cannot be ruled out with reasonable confidence as an explanation for the epidemiological association between exposure to magnetic fields and the risk of childhood leukaemia. While chance alone seems unlikely, given the results of the two meta-analyses (Ahlbom et al. 2000; Greenland et al. 2000), further investigations are needed as to whether confounding, exposure misclassification and selection bias have a strong impact.

With respect to confounding, which is where the effect of the exposure of interest itself is affected by another factor, many candidates have been proposed, including traffic density, use of pesticides in the vicinity of power lines, environmental tobacco smoke, to name only a few. However, none have been identified as true confounders to date (Schüz 2007). In statistical simulations it has been shown that a plausible confounder must itself be quite a strong risk factor for childhood leukaemia and, at the same time, must be highly correlated with magnetic field strengths from all sources of elevated magnetic fields (Langholz 2001). No such a factor has currently been identified (Sect. 27.2.3). While confounding by an unknown factor is always a possibility taking the currently available knowledge into consideration, this seems to be an unlikely explanation.

With regard to exposure assessment methods there is huge potential for exposure misclassification, for example classifying children as unexposed when they were in reality exposed or vice versa. Even long-term measurements in case-control studies only extend over at most 2 days and averages of such measurements are used to categorise past exposure of children. However, it is unlikely that the magnitude of misclassification from most of the exposure assessment methods was associated with disease status, i.e. the probability of misclassification being different for cases and controls. In such situations of nondifferential exposure misclassification, the expected bias in the relative risk estimate is towards an underestimation of an association; hence, it is very unlikely that the exposure misclassification apparent in the epidemiological studies was producing a spurious positive association. Conversely, should the association between magnetic fields and childhood leukaemia be causal, it will be a concern that exposure misclassification led to an underestimation of the strength of the association.

With regard to selection bias, the low participation rates in many studies were of concern, especially in studies using measurements for exposure assessment (Mezei and Kheifets 2006; Schüz 2007). Data suggested that families with a lower social status were particularly underrepresented among controls. As the likelihood of measuring higher fields increased with the number of apartments per building, and buildings with a higher number of apartments were more often inhabited by families of lower social status, the association may have been overestimated. In a simulation analysis using a German case-control data set (Schüz et al. 2001), selection bias explained about 66% of the risk increase observed in the study. A considerable impact of selection bias has also been demonstrated for a large US study (Hatch et al. 2000). These findings leave no doubt that selection bias plays a role in studies using measurement devices. The studies using calculated magnetic fields were not affected by selection bias, as no contact with the families was necessary and participation was virtually complete. These studies also confirm an association between magnetic fields and the risk of childhood leukaemia, however based on much smaller numbers of exposed children (see Table 3 in Ahlbom et al. 2000).

Taking all these considerations into account, the probability is that selection bias alone is not sufficient to explain the entire association. However, since the small increased risk is based on relatively small numbers of exposed children, a combination of selection bias, confounding and chance cannot be ruled out as a plausible explanation for the observed association.

27.3.2.2 Alternative Explanations

The absence of a reproducible mechanism for the interaction of magnetic fields below 100 μT with biological systems has led researchers to suggest an indirect involvement of magnetic fields in leukaemogenesis; however, hypotheses proposed during the last few decades have only been speculative. One recent hypothesis is based on the assumption that exposure to magnetic fields suppresses the nocturnal production and release of the hormone melatonin, which is assumed to have oncostatic capabilities (Henshaw and Reiter 2005); however, studies regarding a magnetic-field-related suppression of melatonin are controversial (IARC 2002) and a recent pooled analysis on nighttime exposure to magnetic fields and the risk of childhood leukaemia found no support for the melatonin hypothesis (Schüz et al. 2007). A second recent hypothesis claims that contact currents occur with greater probability in residences with higher magnetic fields and that contact currents via contact with metallic water fixtures during bathing of the child lead to higher bone marrow doses of induced currents rather than magnetic field exposures alone (Kavet 2005); however, so far there have been no studies to support the role of contact currents in leukaemogenesis. Whilst this hypothesis fits very well with the characteristics of the US power distribution system, it is unclear how well it corresponds to the situation in Europe, where the same association between magnetic fields and leukaemia risk has been observed. This indirect mechanism must also be extremely strong to be detected with only a weakly related level of residential magnetic field exposure.

27.3.2.3 A Causal Link

As mentioned in Sect. 27.2.2, the development of childhood leukaemia is a unique process and it is not certain whether the animal models used in genotoxicity or cell proliferation studies are appropriate. The long-term animal carcinogenicity studies provide little evidence that exposure to magnetic fields alone could induce any type of cancer, including haemopoietic cancers (IARC 2002). Results from studies examining the growth of transplanted tumour cells and coexposure studies with known carcinogens have also tended to be negative. However, recent findings on the stimulation of cell proliferation and inhibition of apoptosis have been of interest since environmental agents can also promote the development of cancer through non-genotoxic mechanisms. Unfortunately, these results are not fully understood and as yet have not been reproduced (Fedrowitz et al. 2002; Kumlin et al. 2002). Furthermore, a recent review of in vitro studies provided no further clarification (Vijayalaxmi and Obe 2005). Data from a series of in vitro studies carried out by a European consortium (Reflex project), showed the occurrence of chromosomal damage in some, but not all cell types (including lymphocytes), and with intermittent rather than continuous exposures. A recent meta-analysis of in vitro studies and short-term animal studies that combined magnetic fields with known carcinogens or other toxic physical or chemical agents found some positive results. However, since most of the studies used magnetic fields of 100 μT or higher, the findings do not provide an adequate explanation for the epidemiological findings as the associations seen were based on much lower exposures (Juutilainen et al. 2006). Overall, evidence from animal studies or in vitro studies neither supports nor contradicts the epidemiological findings.

27.4 Conclusions

Childhood leukaemia is a biologically heterogeneous disease. The advent of more powerful molecular diagnostic techniques has provided greater insight into the underlying molecular biology of the disease, which has been critical for improving the survival of children with acute leukaemia. Furthermore, the knowledge that different subtypes of ALL tend to segregate by age has perhaps provided some rationale as to why there are marked prognostic differences between infants, children and adults.

Whilst over 200 genes have been found to be involved in chromosomal abnormalities, many are rare and only certain ones have been associated with childhood leukaemia. However, the identification of specific chromosomal translocations has been of great importance since the chimeric fusion genes derived act as unique markers for each patient's leukaemic clone, and do not only function as prognostic markers but also enable the origins of the disease to be tracked.

The retrospective analyses of Guthrie cards from children diagnosed with leukaemia for the presence of chromosomal abnormalities has provided evidence for a prenatal origin of the disease. Furthermore, identification of chromosomal translocations in Guthrie cards at a rate much higher than the proportion of children who will develop the associated subtype of leukaemia demonstrated that childhood leukaemia is at least a two-step process with a need for additional genetic events and/or exposures – the search for which has continued over many years.

Although much has been learnt from the identification of specific chromosomal translocations associated with subtypes of acute leukaemia for tailoring therapies and improving outcome, progress in understanding the determinants of childhood leukaemia has not been as profitable. Since many of the translocations are rare, some studies have not been sufficiently large enough to enable analyses of these subgroups. Furthermore, there is widespread geographical variation in routine analysis of biological samples; thus, collection of these parameters has not been a feature of all epidemiological studies. This has made it difficult, in fact in many instances impossible, to pool data meaningfully in order to look at exposures such as magnetic fields. Hence, many of the findings with respect to the cause of childhood leukaemia are based only on ALL and AML and do not take into account specific features associated with chromosomal translocations. Nevertheless improved knowledge of chromosomal alterations offers new opportunities for understanding the relevance of environmental exposures and the importance of gene–environment interactions in determining the development of the disease.

A prominent example is the ongoing challenge of interpreting findings from epidemiological studies on exposure to magnetic fields and the risk of childhood leukaemia. Whilst the finding of a doubling of the leukaemia risk at average magnetic field exposures of 0.3–0.4 μT is relatively consistent across studies, it is not clear whether there is a dose–response relationship and if one were to exist what would it be. Sparse data prevent any firm conclusions about the magnitude of risk at field levels higher than 0.4 μT. Furthermore, for exposures in the range from 0.1 to 0.4 μT, the data are ambiguous, which is compatible with the suggestion of a dose–response trend or a threshold effect at 0.3–0.4 μT. The interpretation of the magnetic-field-related findings is particularly delicate because possible explanations range from methodological artefacts to a causal link. To date neither of these two contradictory explanations can be completely ruled out; hence, it becomes imperative to combine knowledge of the biology of the disease with evaluations of the accuracy of the observational studies findings.

In Fig. 27.2 an attempt is made to summarise the current state of knowledge by clarifying the likelihood of the various possible explanations. With regard to methodological flaws, it is clear that the association between magnetic fields

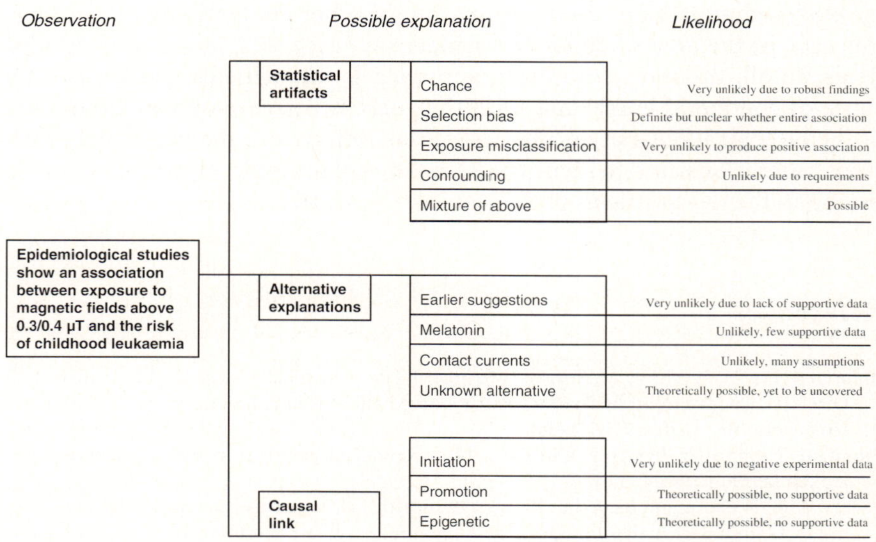

Fig. 27.2 Summary of possible mechanisms to explain the epidemiological association between exposure to magnetic fields and the risk of childhood leukaemia

and leukaemia risk is overestimated owing to selection bias, but it is uncertain whether this bias accounts for the association in its entirety. Furthermore, chance, bias from exposure misclassification and confounding are unlikely on their own to be responsible for producing a spurious positive association; however, a combination of all factors is possible. With regard to indirect mechanisms, many are speculative and the most recent hypotheses on contact currents and the disruption of the melatonin production process are mainly theoretical considerations and are not well supported by data. Although experimental studies to date have failed to provide supportive evidence, they do not contradict the epidemiological results, and it is somewhat disconcerting that no additional evidence for or against a causal link has emerged from the various studies. Whilst there is some evidence against an initiation of leukaemia, either promotion of the disease or an epigenetic mechanism is theoretically possible; however, there are few supportive data for this assumption. Conversely, it must be noted that it is more or less accepted that low-level magnetic fields do not transfer enough energy to cells to directly cause DNA damage. Furthermore, at field levels of 0.3–0.4 μT no other interaction with biological systems apart from the epidemiological childhood leukaemia findings has been identified. It is also important to note that the average exposure levels to the Earth's natural magnetic field are much higher than average exposures to man-made magnetic fields (but only the latter one is alternating, of course).

Be that as it may, there are important lessons to learn, with respect to both the cause of childhood leukaemia and the need for improving epidemiological methods to identify presumably weak associations. Thus, the identification of a severe bias producing such a robust statistical association would have a huge impact on the interpretation of other epidemiological findings, not only in the area of childhood leukaemia (Sect. 27.2.3) but also in other fields. In all, the most promising approach is intensified research into the cause of childhood leukaemia, with the goals to increase our understanding of the biological diversity of the disease, as reflected in the various chromosomal alterations, and to identify interactions of multiple exposures and their timing in the development of this complex disease.

References

Ahlbom A, Day N, Feychting M, Roman E, Skinner J, Dockerty J, Linet M, McBride M, Michaelis J, Olsen JH, Tynes T, Verkasalo PK (2000) A pooled analysis of magnetic fields and childhood leukaemia. Br J Cancer 83:692–698

Biondi A, Cimino G, Pieters R, Pui CH (2000) Biological and therapeutic aspects of infant leukemia. Blood 96:24–33

Fedrowitz M, Westermann J, Löscher W (2002) Magnetic field exposure increases cell proliferation but does not affect melatonin levels in the mammary gland of female Sprague Dawley rats. Cancer Res 62:1356–1363

Greaves M (1999) Molecular genetics, natural history and the demise of childhood leukaemia. Eur J Cancer 35:1941–1953

Greaves M (2002) Childhood leukaemia. BMJ 324:283–287

Greaves M (2005) In utero origins of childhood leukaemia. Early Hum Dev 81:123–129

Greaves M (2006) Infection, immune responses and the aetiology of childhood leukaemia. Nat Rev Cancer 6:193–203

Greaves MF, Wiemels J (2003) Origins of chromosome translocations in childhood leukaemia. Nat Rev Cancer 3:1–11

Greenland S, Sheppard AR, Kaune WT, Poole C, Kelsh MA (2000) A pooled analysis of magnetic fields, wire codes, and childhood leukemia. Childhood Leukemia-EMF Study Group. Epidemiology 11:624–634

Hatch EE, Kleinerman RA, Linet MS, Tarone RE, Kaune WT, Auvinen A, Baris D, Robison LL, Wacholder S (2000) Do confounding or selection factors of residential wiring codes and magnetic fields distort findings of electromagnetic fields studies? Epidemiology 11:189–198

Heerema NA, Arthur DC, Sather H, Albo V, Feusner J, Lange BJ, Steinherz PG, Zeltzer P, Hammond D, Reaman GH (1994) Cytogenetic features of infants less than 12 months of age at diagnosis of acute lymphoblastic leukemia: impact of the 11q23 breakpoint on outcome: a report of the Childrens Cancer Group. Blood 83:2274–2284

Henshaw DL, Reiter RJ (2005) Do magnetic fields cause increased risk of childhood leukemia via melatonin disruption? Bioelectromagnetics 7:S86–97

Herbst AL, Ulfelder H (1971) Adenocarcinoma of the vagina: association of maternal stilbestrol therapy with tumour appearance in young women. N Engl J Med 284:446–450

IARC (2002) Monographs on the evaluation of carcinogenic risks to humans, vol 80: non-ionizing radiation, part 1: static and extremely low-frequency (ELF) electric and magnetic fields. IARC Press, Lyon

Juutilainen J, Kumlin T, Naarala J (2006) Do extremely low frequency magnetic fields enhance the effects of environmental carcinogens? A meta-analysis of experimental studies. Int J Radiat Biol 82:1–12

Kavet R (2005) Contact current hypothesis: summary of results to date. Bioelectromagnetics 7:S75–85

Kinlen LJ (1995) Epidemiological evidence for an infective basis in childhood leukaemia. Br J Cancer 71:1–5

Kumlin T, Heikkinen P, Kosma VM, Alhonen L, Janne J, Juutilainen J (2002) p53-independent apoptosis in UV-irradiated mouse skin: possible inhibition by 50 Hz magnetic fields. Radiat Environ Biophys 41:155–158

Langholz B (2001) Factors that explain the power line configuration wiring code-childhood leukemia association: what would they look like? Bioelectromagnetics 5:S19–31

Lightfoot TJ, Roman E (2004) Causes of childhood leukaemia and lymphoma. Toxicol Appl Pharmacol 199:104–17

Linet MS, Hatch EE, Kleinerman RA, Robison LL, Kaune WT, Friedman DR, Severson RK, Haines CM, Hartsock CT, Niwa S, Wacholder S, Tarone RE (1997) Residential exposure to magnetic fields and acute lymphoblastic leukemia in children. N Engl J Med 337:1–7

Look AT (1997) Oncogenic transcription factors in the human acute leukaemias. Science 278:1059–1064

McHale CM, Smith MT (2004) Prenatal origin of chromosomal translocations in acute childhood leukaemia: implications and future directions. Am J Hematol 75:254–257

Mezei G, Kheifets L (2006) Selection bias and its implications for case-control studies: a case study of magnetic field exposure and childhood leukaemia. Int J Epidemiol 35:397–406

Mori H, Colman SM, Xiao Z, Ford AM, Healy LE, Donaldson C, Hows JM, Navarrete C, Greaves M (2002) Chromosome translocations and covert leukemic clones are generated during normal fetal development. Proc Natl Acad Sci USA 99:8242–8247

Okuda T, van Deursen J, Hiebert SW, Grosveld G, Downing JR (1996) AML1, the target of multiple chromosomal translocations in human leukaemia, is essential for normal fetal liver hematopoiesis. Cell 84:321–330

Parkin DM, Stiller CA, Draper GJ, Bieber CA, Terracini B, Young JL (1988) International incidence of childhood cancer. IARC scientific publications 87. IARC Press, Lyon

Pui CH, Relling MV, Downing JR (2004) Acute lymphoblastic leukaemia. N Engl J Med 350:1535–1548

Roman E, Doyle P, Lightfoot T, Ansell P, Simpson J, Allan JM, Eden TOB (2006) Molar pregnancy, childhood cancer and genomic imprinting – is there a link? Hum Fertil 9:171–174

Roman E, Simpson J, Ansell P, Kinsey S, Mitchell CD, McKinney PA, Birch JM, Greaves M, Eden TOB (2007) Childhood acute lymphoblastic leukaemia and infections in the first year of life: a report from the UKCCS. Am J Epidemiol 165:496–504

Schüz J (2007) Implications on protection guidelines from epidemiologic studies on magnetic fields and the risk of childhood leukemia. Health Phys (in press)

Schüz J, Grigat JP, Brinkmann K, Michaelis J (2001) Residential magnetic fields as a risk factor for childhood acute leukaemia: results from a German population-based case-control study. Int J Cancer 91:728–735

Schüz J, Svendsen AL, Linet MS, McBride ML, Roman E, Feychting M, Kheifets L, Lightfoot T, Mezei G, Simpson J, Ahlbom A (2007) Night-time exposure to electromagnetic fields and childhood leukemia: an extended pooled analysis. Am J Epidemiol (in press)

Steliarova-Foucher E, Stiller C, Kaatsch P, Berrino F, Coebergh JW, Lacour B, Parkin M (2004) Geographical patterns and time trends of cancer incidence and survival among children and adolescents in Europe since the 1970s (the ACCIS project): an epidemiological study. Lancet 364:2097–105

Stewart A, Webb J, Giles D, Hewitt D (1956) Malignant disease in childhood and diagnostic radiation in utero. Lancet 2:447

UKCCS (1999) Exposure to power-frequency magnetic fields and the risk of childhood cancer. Lancet 354:1925–1931

UKCCS (2000) The United Kingdom Childhood Cancer Study: objectives, materials and methods. Br J Cancer 82:1073–1102

Vijayalaxmi, Obe G (2005) Controversial cytogenetic observations in mammalian somatic cells exposed to extremely low frequency electromagnetic radiation: a review and future research recommendations. Bioelectromagnetics 26:412–430

Wertheimer N, Leeper E (1979) Electrical wiring configurations and childhood cancer. Am J Epidemiol 109:273–284

Yu BD, Hanson RD, Hess JL, Horning SE, Korsmeyer SJ (1998) MLL, a mammalian trithorax-group gene, functions as a transcriptional maintenance factor in morphogenesis Proc Natl Acad Sci USA 95:10632–10636

28 Cytogenetic and Carcinogenic Effects of Exposure to Radiofrequency Radiation

James P. McNamee and Pascale V. Bellier

Abstract

Radiofrequency radiation (RFR) is a portion of the electromagnetic spectrum with frequencies of 3 kHz–300 GHz. RFR is produced by many man-made sources, including mobile phones and base stations, television and radio broadcasting facilities, radar, medical equipment, microwave ovens, radiofrequency heaters as well as a diverse variety of other electronic devices within our living and working environments. Owing to ongoing public concern and the increasing prevalence of RFR-emitting devices, a great deal of research has been conducted over the past 50 years to evaluate the biological and/or health effects of thermalizing and non-thermalizing RFR exposures. In the absence of decisive epidemiological evidence to support or refute an association between RFR exposure and cancer risk, laboratory studies of possible mechanisms of carcinogenesis by RFR are important. The scientific literature on this subject is full of conflicting results and the question of whether RFR exposure can contribute to cancer risk remains unresolved. This chapter contains a literature review of the evidence for RFR-induced cytogenetic effects, but also a critique of the literature, highlighting deficiencies in the design of some studies that should be taken into account when assessing the health risk of RFR.

28.1 Introduction

Electromagnetic radiation is emitted by many natural and man-made sources and is a fundamental aspect of our lives. We are warmed by electromagnetic radiation emitted from the sun and our eyes can detect the visible light portion of the electromagnetic spectrum. The source of all electromagnetic radiation is accelerating charges and electromagnetic radiation manifests itself as oscillating electric and magnetic fields. The frequency of this oscillation (measured in hertz; 1 Hz is one oscillation per second) determines the physical properties of these fields. Radiofrequency radiation (RFR) is a portion of the electromagnetic spectrum with frequencies of 3 kHz–300 GHz, which is below that of visible light and above that of extremely low frequency fields (Fig. 28.1). RFR is

Günter Obe and Vijayalaxmi (Eds.)
Chromosomal Alterations: Methods, Results and Importance in Human Health

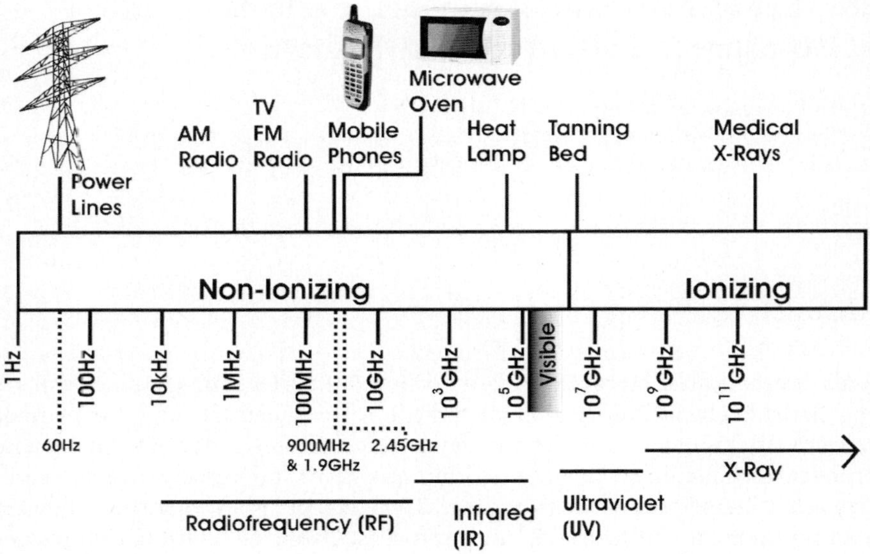

Fig. 28.1 In the electromagnetic spectrum, radiofrequency radiation (RFR) encompasses frequencies ranging from 3 kHz to 300 GHz. Typical man-made sources of RFR include broadcast AM/FM/TV, mobile phones and base stations, microwave ovens and radar

produced by many man-made sources, including mobile phones and base stations, television and radio broadcasting facilities, radar, medical equipment, microwave ovens, radiofrequency heaters as well as a diverse variety of other electronic devices within our living and working environments.

Public anxiety about exposure to RFR has existed for over 80 years. Initially, concern focused on RFR emissions from broadcast AM (530 kHz–1.6 MHz) and FM (88–108 MHz) RFR. This was followed by concerns over RFR emissions from radar installations (1–12.5 GHz), broadcast TV (54–890 MHz), microwave ovens (2.45 GHz) and video-display terminals (3–30 kHz). During the past decade, the increased use of mobile phones (850 MHz–1.9 GHz) and the emergence of new wireless technologies have reinvigorated public apprehension about the safety of human exposure to RFR. Owing to ongoing public concern and the prevalence of RFR-emitting devices, a great deal of research has been conducted over the past 50 years to evaluate the biological and/or health effects of thermalizing and non-thermalizing RFR exposures.

Unlike ionizing radiation, RFR does not have sufficient energy to directly break chemical bonds (Meltz 2003). However, sufficiently intense RFR can cause heating of materials with finite conductivity, including biological tissue. A number of well-established biological effects and adverse health effects from acute exposure to intense RFR have been documented. For the most part, these effects relate to localized heating of sensitive organs/tissues from intense RFR exposure; however, non-thermal effects such as nerve and muscle stimulation are also known to occur. The specific biological responses to RFR are generally related to the rate of RFR energy absorbed. The rate and distribution of RFR energy absorbed depends strongly on the frequency, intensity and orientation of the incident fields as well as the body size and its constitutive properties (dielectric constant and conductivity). At frequencies above 1 MHz, RFR absorption is commonly described in terms of the specific absorption rate (SAR), which is a measure of the rate of energy deposition per unit mass of body tissue and is usually expressed in units of watts per kilogram. Based on a large amount of historical knowledge, national and international exposure limits have been established to protect the general public against adverse effects associated with acute RFR exposures. However, the issue of safety of exposure to long-term, low-level RFR remains controversial and the risk of development of cancer remains the primary public health concern.

At present, there is a relative paucity of epidemiological evidence related to cancer risks associated with long-term, low-level RFR exposure. In the absence of decisive epidemiological evidence to support or refute an association between RFR exposure and cancer risk, laboratory studies of possible mechanisms of carcinogenesis by RFR are important. The scientific literature on this subject is full of conflicting results and the question of whether RFR exposure can contribute to cancer risk remains unresolved. This chapter will review the current state of the literature with respect to cytogenetic and carcinogenic effects of RFR. Since the thermal effects of RFR are well understood and cytogenetic alterations are known to be induced by heat, this chapter will focus on the evidence for cytogenetic effects from low-level, non-thermalizing RFR.

28.2 Challenges When Conducting RFR Research

There are many inherent challenges that must be overcome when conducting RFR bioeffects research which relate to the physical properties of this radiation. One of the most important interactions between RFR and matter is the rotational friction caused when small polar molecules (such as those in water and biological tissues) reorient themselves in the presence of the oscillating field, thereby producing heat. The rate at which heat energy is applied to the sample must not exceed the rate at which it is removed, otherwise the temperature within the sample will rise and thermal confounding may occur. If RFR energy is applied to biological samples at a low rate (e.g., below 1 W/kg) or for a short duration, then passive cooling and/or sweating may permit the tissue/body/sample temperature to remain within a normal physiological range. However, if higher SARs (above 1 W/kg) are employed and/or prolonged, then some form of active cooling mechanism will likely be required to ensure that excessive sample heating does not occur. When performing RFR experiments, it is important to include appropriate temperature controls within the experimental setup to alleviate thermal confounding in the study results. To this end, the use of non-perturbing temperature probes and active cooling with either forced air or water is particularly beneficial.

Another significant challenge that must be addressed when studying the biological effects of RFR relates to the homogeneity of energy absorption within the sample (SAR distribution) and the maintenance of a relatively homogeneous, non-perturbing temperature throughout the biological sample. These are two separate, yet related, phenomena. While controlling volume-averaged temperature within the biological sample is important, it is also important to ensure that RFR energy is deposited in a homogeneous manner or "hot spots" within the biological sample can occur. This phenomenon is akin to focusing light energy on an object, whereby the localized temperature at the focal point is considerably elevated, while that of the surrounding area is not. In such a situation with RFR, the local temperature for some part of the sample may be greatly elevated, while the volume-averaged temperature remains relatively unaffected owing to heat convection and diffusion. With some in vitro exposure systems, heterogeneous RFR absorption can also cause convective flow within aqueous biological samples owing to the presence of temperature gradients within the sample. Such convection currents in a RFR-exposed sample, relative to a control, may be a source of confounding owing to a resultant stirring of the RFR-exposed samples that does not occur within the matched sham control.

Since RFR behaves in a similar fashion to light, with respect to its ability to reverberate/reflect off metallic surfaces, care must be taken to ensure that samples are exposed in a reverberant-free environment or at least that all sources of reflection are taken into account in the dosimetry analysis. This is particularly problematic for in vitro studies, where cells must be maintained under appropriate conditions (37°C, 50–60% humidity, 5% CO_2/95% air) in a tissue culture incubator. Most incubators are built with a stainless-steel-coated interior which acts as an ideal reflector for RFR. As a result, when conducting in vitro RFR research, great care must be taken to either contain RFR within a closed exposure system inside a standard tissue culture incubator or design alternative non-RFR perturbing tissue culture environments. In either case, maintenance of sample temperature, humidity, osmolarity and pH is a challenge and must be carefully controlled.

Owing to the above considerations, it is critical that RFR bioeffects studies encompass certain experimental controls to ensure that confounding variables do not result in artefactual biological observations. For in vitro studies, monitoring the temperature, pH and cell viability within the sample is paramount. For both in vivo and in vitro studies, empirical and/or numerical analysis of SAR distribution patterns within the sample is fundamentally important. The inclusion of unexposed (negative) controls, sham (non-RFR exposed) controls and positive controls is also imperative to ensure that the assay methodology is responding appropriately and experimental conditions are identical. Inclusion of these controls is particularly important when evaluating studies where RFR-induced effects were observed. On the other hand, analysis of statistical power in studies where no evidence of RFR-induced effects were observed is important to determine whether a sufficient number of samples (independent experiments) were employed to allow the detection of significant differences between groups. For all studies, the investigators should be blinded with respect to the exposure status of the samples until all laboratory experiments, data acquisition and statistical analysis are complete.

In summary, when evaluating the literature related to RFR-induced cytogenetic and carcinogenic effects, all of the above considerations should be taken into account. This chapter contains not only a literature review but also a critique of the literature, highlighting certain deficiencies in the design of some of the studies that should be taken into account when evaluating the significance of certain RFR research findings.

28.3 Cytogenetic Effects of RFR

28.3.1 In Vitro Studies

Changes involving DNA and/or chromosome structure in somatic cells are of great relevance owing to their association with cancer and other diseases. Despite the general consensus that RFR is not sufficiently energetic to directly damage DNA, a large number of in vitro studies have been conducted to assess the genotoxic potential of RFR at a variety of frequencies. It is generally agreed that when the in vitro RFR cytogenetic studies published over the past 30 years are taken together and subjected to a weight-of-evidence evaluation, there is no compelling information to suggest that RFR exposure has any direct mutagenic, genotoxic or clastogenic effects (Brusick et al. 1998; Vijayalaxmi and Obe 2004; Moulder et al. 2005; Verschaeve 2005). In this section, the scientific literature relating to the evidence of in vitro cytogenetic effects of RFR will be reviewed and evaluated with particular focus on recent papers and/or pivotal studies (Table 28.1).

The majority of studies examining the cytogenetic effects of RFR during the last 5 years have been conducted in human blood cultures. Of these studies, only a few have observed a "positive" effect resulting from RFR exposures using the cytokinesis block micronucleus (CBMN) assay. Studies by d'Ambrosio et al. (2002), Tice et al. (2002) and Zotti-Martelli et al. (2000, 2005) have all reported increased micronuclei (MN) in RFR-exposed human blood cultures using the CBMN assay. In the study by Tice et al. (2002), significant increases in MN frequency were observed following exposure of unstimulated blood cultures exposed to 837 MHz and 1.9 GHz RFR (with various paradigms) at SARs of 5 and 10 W/kg at the longest RFR exposure time of 24 h. However, the authors identified thermal confounding as a possible explanation for these results as higher localized temperatures may have occurred within the sample region containing the blood cells. It is very

Table 28.1 Summary of in vitro studies that have investigated the possible cytogenetic effects of radiofrequency radiation

Authors	RFR conditions	SAR; duration	Tissue /cell line	End point	Results	Comments
Zotti-Martelli et al. (2000, 2005)	1.8, 2.45 or 7.7 GHz; CW	No SAR given; 15-, 30-, 60-, 120- or 180-min exposures	Human whole blood	MN	Increased MN	Insufficient details on exposure design and dosimetry; no temperature monitoring in actual samples; insufficient details on control conditions
Maes et al. (2001)	900 MHz; various modulations	SARs ranging from 0.4 to 10 W/kg; 2-h exposure	Human whole blood	CA and SCE	No effect	Low sample number for most exposures ($N = 2$); no sham or positive controls
Vijayalaxmi et al. (2001a, b)	835.62 or 874.74 MHz; CW	SARs ranging from 4.4 to 5.5 W/kg; 24-h exposure	Diluted human blood	MN and CA	No effect	Sample number ($N = 2$ for each SAR tested)
d'Ambrosio et al. (2002)	1.748 GHz; CW and phase-modulated	Average SAR of 2.25 W/kg; 15-min exposure	Mitogen-stimulated diluted human blood	MN	Increased MN following exposure to phase-modulated RFR	No sham or positive controls
Tice et al. (2002)	837 MHz, 1.9 GHz; analogue, PCS, CDMA and TDMA	SARs ranging from 1 to 10 W/kg; 3- and 24-h exposures	Diluted human blood	MN	Increased MN at highest SARs after 24-h exposure period	Low sample number ($N = 1$ for each condition); possible thermal confounding due to localized hot spots

Bisht et al. (2002)	847 MHz CDMA-modulated and 835 MHz FDMA-modulated	SARs ranging from 3.2 to 5.1 W/kg; 3-, 8-, 16- or 24-h exposures	C3H 10T1/2 cells	MN	No effect	
McNamee et al. (2002a, b, 2003)	1.9 GHz; CW and pulse-modulated	SARs ranging from 0.1 to 10 W/kg; 2- and 24-h exposures	Diluted human blood	MN	No effect	No comment
Zeni et al. (2003, 2005)	900 or 925 MHz; CWand GSM-modulated	SARs of 0.2, 1 or 1.6 W/kg; various exposure periods	Whole or diluted human blood and mitogen-stimulated lymphocytes	MN, CA and SCE	No effect	No sham or positive controls in the 2003 study
Mashevich et al. (2003)	830 MHz; CW	SARs ranging from 1.6 to 8.8 W/kg; 72-h exposure	Mitogen-stimulated diluted human blood	Aneuploidy	Increase in chromosome 17 aneuploidy	Non-homogenous exposure; sample flasks were thermally insulated from the incubator and no active cooling was employed, resulting in a possible thermal effect
Koyama et al. (2003)	2.45 GHz	SARs ranging from 13 to 100 W/kg; 18-h exposure	CHO-K1 cells, following incubation with cytochalasin B for 24 h	MN	Increased MN at SARs of 78 and 100 W/kg; no effect below 50 W/kg	High SAR levels with little relevance towards risk assessment; did not report the number of repeat experiments

(*Continued*)

Table 28.1 Summary of in vitro studies that have investigated the possible cytogenetic effects of radiofrequency radiation—Cont'd

Authors	RFR conditions	SAR; duration	Tissue /cell line	End point	Results	Comments
Komatsubara et al. (2005)	2.45 GHz CW and pulse-modulated	SARs ranging from 5 to 100 W/kg; 2-h exposure	Mouse m5S cells	CA	No effect	High SAR levels with little relevance towards risk assessment; apparently a single sample for each RFR exposure condition
Stronati et al. (2006)	935 MHz; GSM-modulated	SARs of 1 and 2 W/kg; 24-h exposure	Mitogen-stimulated human whole blood	MN, CA and SCE	No effect	

CA chromosome aberrations, *CDMA* code-division multiple access, *CHO* Chinese hamster ovary, *CW* continuous wave, *FDMA* frequency-division multiple access, *GSM* global system for mobile communication, *MN* micronuclei, *PCS* personal communication services, *RFR* radiofrequency radiation, *SAR* specific absorption rate, *SCE* sister-chromatid exchange, *TDMA* time-division multiple access

important to note that blood samples from only on to two donors were used for each experimental condition; therefore, only limited conclusions can be drawn from this study. Zotti-Martelli et al. (2000, 2005) also observed increased MN in the CBMN assay following exposure of unstimulated whole-blood cultures to 1.8, 2.45 and 7.7 GHz continuous-wave (CW) RFR for up to 3 h at power densities ranging from 5 to 30 mW/cm^2. However, very limited information regarding the RFR exposure system and dosimetry was provided for these studies. Furthermore, while the authors measured the temperature variation within "pseudosamples" during RFR exposure, water and not whole blood was used in these experiments. This is problematic since water has very different dielectric and conductivity properties from whole blood, and much more energy would be absorbed within whole-blood cultures than in water. As such, at the RFR exposure levels described by the authors, it is possible that the observed increases in MN frequency may have been due to hyperthermia. In a well-controlled study by d'Ambrosio et al. (2002), increased MN frequency was reported in phytohemagglutinin (PHA) stimulated human lymphocytes exposed to phase-modulated 1.748 GHz RFR, but not to 1.748 GHz CW RFR. This result suggested that the cellular response to RFR may depend upon the modulation characteristics of the applied RFR field.

On the other hand, many other research groups have failed to find any evidence of RFR-induced cytogenetic effects in human blood cell cultures regardless of the modulation paradigm. Vijayalaxmi et al. (2001a, b) exposed unstimulated whole-blood cultures to 835.62 MHz frequency division multiple access (FDMA) modulated and 847.74 MHz code division multiple access (CDMA) modulated RFR for 24 h at mean SARs of either 4.4 or 5.0 W/kg and 4.9 or 5.5 W/kg, respectively. The authors found no significant changes in either cytogenetic end point between the sham and RFR-exposed samples for either of these RFR exposure conditions. Similarly, McNamee et al. (2002a, b, 2003) exposed unstimulated human whole-blood cultures to 1.9 GHz CW and pulse-modulated RFR for 2–24 h at SARs ranging from 0.1 to 10 W/kg. The formation of MN was assessed using the CBMN assay and no significant changes were found in the binucleated cell frequency, the frequency of binucleated cells with MN or the total MN incidence at any SAR level tested when compared with the sham cells. Unlike many previous studies, the studies by McNamee et al. incorporated exposure of human blood samples from five separate donors to sham and five different RFR exposure levels, allowing the possibility of a dose–response relationship to be assessed. Furthermore, real-time temperature measurements were performed every 60 s within each of the RFR-exposed and sham cultures, thereby excluding thermal effects as a possible confounding variable.

Maes et al. (2001) and Zeni et al. (2003, 2005) examined the effect of 900 MHz global system for mobile communication (GSM) modulated RFR exposure at SARs ranging from 0.2 to 10 W/kg in either PHA-stimulated or unstimulated blood cultures and found similar results. However, while the studies by Maes et al. (2001) and Zeni et al. (2003) found no evidence that

RFR exposure caused any genotoxic effects; these studies did not include the necessary controls. Since both studies lacked positive controls, the relative sensitivity of the cytogenetic assays in these laboratories to detect an effect is difficult to interpret and therefore the significance of these negative results is diminished. Furthermore, these studies failed to include sham controls and relied only upon unexposed controls. Although the importance of a sham control is obvious when a positive effect is detected, it is important to point out that the appropriate comparison for all RFR-exposed samples should be the sham control, rather than the unexposed control, since the sham and RFR-exposed samples should have been treated identically except for the RFR exposure. This comparison ensures that the exposure system and exposure environment are not the cause of possible observed effects, but rather that the RFR exposure itself is responsible. In a follow-up study, Zeni et al. (2005) included all the necessary controls when exposing unstimulated whole-blood cultures to 900 MHz GSM-modulated RFR for 2 h at 1 W/kg. No evidence of chromosomal damage, as measured by analysis of chromosome aberrations and sister-chromatid exchange (SCE), was observed. Very recently, Stronati et al. (2006) exposed PHA-stimulated whole blood to 935 MHz GSM-modulated RFR at SARs of 1 and 2 W/kg for 24 h and observed no increase in chromosome aberrations, MN frequency or SCE in RFR-exposed samples when compared with sham controls.

Several in vitro studies have also assessed the effect of RFR exposure on cytogenetic end points using a variety of animal-derived cell lines. Bisht et al. (2002) exposed a mouse embryo fibroblast derived cell line (C3H 10T1/2) to 835.62 MHz FDMA-modulated or 847.74 MHz CDMA-modulated RFR for up to 24 h at SARs of 3.2–5.1 W/kg, in both the plateau phase and the exponential growth phase of these cells. The authors found no evidence that RFR exposure had any effect on MN frequency at any SAR, modulation paradigm or exposure duration. Koyama et al. (2003) exposed Chinese hamster ovary (CHO) K1 cells to 2.45 GHz CW RFR for 18 h at SARs ranging from 13 to 100 W/kg. While the authors reported no evidence of increased MN frequency in RFR samples exposed at SARs up to 50 W/kg, an increased incidence in MN frequency was reported in cells exposed at SARs of 78 and 100 W/kg. The authors measured the average temperature of the highest RFR exposed group (100 W/kg) to be approximately 39°C. When positive control samples (exposed to a temperature of 39°C for 18 h) were assessed, a similar increased incidence of MN was reported. Thus, the increased MN frequency at the higher SARs assessed in this study appears to be due to thermal effects. Another recent study by this group (Komatsubara et al. 2005) assessed the incidence of chromosome aberrations in murine (embryonal skin derived) m5S cells following a 2-h exposure to 2.45 GHz CW RFR at SARs of 5–100 W/kg or 2.45 GHz pulse-modulated RFR at SARs of 50 and 100 W/kg. The authors reported no evidence of increased chromatid-type or chromosome-type aberrations in RFR-exposed samples at any SAR tested when compared with sham cells. It is interesting to note that the shorter exposure time (2 h) to an elevated temperature of

39–41°C had no apparent clastogenic effect in this study, while the prolonged (18-h) exposure to 39°C in the study by Koyama et al. (2003) appeared to stimulate heat-induced MN formation. In any event, the high SAR levels employed in these studies have little relevance when assessing the relative risk of RFR towards the general population since such intense exposures are not typically encountered in our working or living environment.

In an interesting study by Mashevich et al. (2003), diluted human whole blood was stimulated with 3% PHA and then exposed to 830 MHz CW RFR for 72 h at SARs ranging from 1.6 to 8.8 W/kg. Following the culture period, the cells were harvested for cytogenetic analysis and the authors labeled the α-satellite DNA of chromosome 17 using fluorescence in situ hybridization. A linear increase in chromosome 17 aneuploidy as a function of increasing SAR was reported. The authors suggested that this effect was non-thermal in nature on the basis of their assumption that the temperature within the RFR-exposed samples did not exceed 38°C, while separate control experiments demonstrated that prolonged culturing at temperatures below 38.5°C did not affect chromosome 17 aneuploidy. However, Chou and Swicord (2003) argued the assumption that RFR sample temperatures remained below 38°C since the RFR-exposed flasks were enclosed in Plexiglas and were thereby thermally insulated from the incubator temperature. Chou and Swicord (2003) noted that such a situation would create non-uniform heating within the samples owing to inefficient heat exchange and the localized temperatures within the samples may have exceeded 40°C. As a result, the observed changes in chromosome 17 aneuploidy could have been related to thermal effects. Since no active cooling was employed in this study to ensure the sample temperature was maintained within an acceptable range and no temperature measurements were recorded during the actual exposure period, thermal confounding is a likely explanation for these observations.

28.3.2 In Vivo Studies

Despite an exhaustive number of in vitro RFR studies, there have been relatively few studies examining the cytogenetic effects in animals following in vivo RFR exposure (Table 28.2). Many early studies on RFR cytogenetic effects focused on the ability of 2.45 GHz microwave radiation to cause alterations in meiotic chromosomes in spermatogonia, as microwave hyperthermia was considered as a possible means of male contraception (Fahim et al. 1975). However, inadequate control of thermal effects has been identified as a possible confounding variable in many of the early studies that demonstrated RFR-induced genotoxic effects (Brusick et al. 1998; Royal Society of Canada 1999; Vijayalaxmi and Obe 2004; Moulder et al. 2005; Verschaeve 2005).

Among the early studies where thermal effects were unlikely, Manikowski-Czerska et al. (1985) evaluated the effect of 0.5–20 W/kg 2.45 GHz RFR exposure on mouse spermatogonia. Adult CBA/CAY mice were exposed for

Table 28.2 Summary of in vivo studies that have investigated the possible cytogenetic effects of radiofrequency radiation

Authors	RFR conditions	SAR; duration	Tissue/cell line	End point	Results	Comments
Manikowski-Czerska et al. (1985)	2.45 GHz; CW	Whole body average SAR ranging from 0.05 to 20 W/kg; 30 min/day, 6 days/week, 2 weeks	CBA/CAY mice	CA in sperm	Dose-response increase in translocations and univalent chromosomes	Restrained animals; no positive control
Beechey et al. (1986)	2.45 GHz	Estimated SARs of 0.05–20 W/kg; 30 min/day, 6 days/week, 2 weeks	Hybrid F1(C3H/HeH × 101/H) male mice	CA in sperm	No effect	Restrained animals; no positive control
Saunders et al. (1988)	2.45 GHz	SAR of 5 W/kg; 6 h/day, over 8 weeks, for total exposure of 120 h	Male C3H mice	CA in sperm	No effect	Animals partially restrained, no vertical movement; no positive control
Vijayalaxmi et al. (1997, 1998)	2.45 GHz; CW	Whole body average SAR of 1 W/kg; 20 h/day, 7 days/week, 18 months	C3H/HeJ mammary cancer prone female mice	MN-PCE in blood and bone marrow	Increase in MN-PCE in bone marrow and blood	Increase observed was not deemed to be biologically significant since it was small and within the normal spontaneous range of MN-PCE in mice
Vijayalaxmi et al. (2001c)	2.45 GHz; CW	Whole body average SAR of 12 W/kg; 24-h exposure	Male Sprague-Dawley rats	MN-PCE in blood and bone marrow	No effect	
Sykes et al. (2001)	900 MHz; pulse-modulated	Whole body average SAR	Spleen cells of pKZ1 transgenic mice	Somatic intra-chromosomal	Decrease in rate of spontaneous	Restrained animals; no positive control;

		of 4 W/kg; 30 min/day for 1, 5 or 25 days		recombination	recombination after 25 days of exposure	decrease observed at 25 days after exposure probably due to pooling of all non-exposed control data, thus comparisons between sham and exposed mice were not properly matched
Trosic et al. (2002, 2004)	2.45 GHz; CW	Whole body average SAR of 1–2 W/kg; 2 h/day, 7 days/week for up to 30 days	Male Wistar rats	MN-PCE in blood or bone marrow	Increase in MN-PCE after 8- and 15-day exposures	No positive control; MN-PCE levels in RFR-exposed and sham groups were within the normal range for rat peripheral blood
Vijayalaxmi et al. (2003)	1.6 GHz; CW	Brain-average SAR of 0.16 or 1.6 W/kg; 2 h/day, 2 years	Fisher 344 rats in utero and until 2 years after birth	MN-PCE in bone marrow	No effect	Restrained animals
Görlitz et al. (2005)	902 MHz GSM-modulated, and 1.747 GHz DCS-modulated	Whole body average SAR of 2.8–24.9 W/kg; 2 h/day, 5 days/week for 1 or 6 weeks	Male and female B6C3F1 mice	MN-PCE in blood (6 weeks only) and bone marrow; MN in keratinocytes and spleen lymphocytes	No effect	Restrained animals

DCS digital cordless standard, *MN-PCE* micronuclei in polychromatic erythrocytes

30 min/day, 6 days/week for two consecutive weeks and then chromosome number and structure were evaluated in isolated spermatogonia. At the end of the exposure period, a dose-responsive increase in the number of translocations and univalent chromosomes at meiotic metaphase were reported in the RFR-exposed mice, while no translocations were observed in sham mice. In an attempt to replicate these findings, Beechey et al. (1986) employed a similar irradiation protocol using hybrid male mice but observed no evidence of increased chromosome aberrations in spermatogonia from RFR-exposed mice, where increases in rectal temperatures did not exceed 1°C. Saunders et al. (1988) examined this effect further by exposing C3H mice to 2.45 GHz RFR for 6 h/day over an 8-week period at a SAR of approximately 5 W/kg. No evidence of increased reciprocal translocations or increased numbers of cells with autosome or sex chromosome univalents was observed.

The majority of in vivo cytogenetics studies on RFR have focused on evaluating the incidence of MN in rodent peripheral blood or bone marrow. Only a limited number of chronic and/or lifetime RFR exposure studies have been conducted in experimental animals. Of these, Vijayalaxmi et al. (1997) exposed mammary cancer-prone C3H/HeJ female mice to 2.45 GHz CW RFR at a whole-body SAR of 1 W/kg for 20 h/day, 7 days/week for 18 months. Peripheral blood and bone marrow smears were assessed for the incidence of MN in polychromatic erythrocytes (PCE) and normochromatic erythrocytes (NCE). At this time-point, a chronic level of MN-PCE and/or MN-NCE would be anticipated in both the peripheral blood and the bone marrow of RFR-exposed mice, if RFR exposure induced a clastogenic effect. The authors reported a small, yet statistically significant increase in the number of MN-PCE in both bone marrow and peripheral blood of RFR-exposed mice; however, the authors cautioned that this difference (an increase of 0.05%) was not biologically significant since the MN-PCE frequencies for both the RFR and the control groups were within the normal range of the spontaneous incidence of MN-PCE in mice (Vijayalaxmi et al. 1998). Vijayalaxmi et al. (2003) also evaluated the incidence of MN-PCE in peripheral blood and bone marrow smears of Fischer 344 rats exposed (since birth) for 2 h/day for 2 years to a 1.6 GHz Iridium wireless communication signal at brain-averaged SARs of 0.16 and 1.6 W/kg. In this study, no significant differences were observed between RFR-exposed male and female rats relative to the sham controls. More recently, Görlitz et al. (2005) exposed B6C3F1 mice to 902 MHz GSM-modulated or 1,747 MHz digital cordless standard (DCS) modulated RFR for either 1 week at 2 h/day at whole-body SARs of 3.7–33.2 W/kg or for 6 weeks at 2 h/day for 5 days/week at SARs of 2.8–24.9 W/kg. After either a 1-week or a 6-week exposure protocol, MN in bone marrow erythrocytes, keratinocytes and spleen lymphocytes were evaluated in both male and female mice. No significant differences in the incidence of MN were observed in any of the RFR-treated groups relative to the matched sham controls.

A series of studies have also evaluated the effect of acute and/or subacute in vivo RFR exposure on cytogenetic end points in experimental animals.

Vijayalaxmi et al. (2001c) exposed male Sprague-Dawley rats continuously for 24 h to 2.45 GHz CW RFR at a whole body average SAR of 12 W/kg. At 24 h after the end of RFR exposure, the frequency of MN-PCE in peripheral blood and bone marrow smears was evaluated. RFR-exposed rats displayed elevated, yet statistically insignificant, MN-PCE in both peripheral blood and bone marrow compared with sham animals, but once again the MN-PCE frequencies in both groups were within the normal expected range for rats. In a similar study, Trosic et al. (2002, 2004) exposed adult male Wistar rats for 2 h/day, 7 days/week for up to 30 days to 2.45 MHz CW RFR at an approximate SAR of 1–2 W/kg. The authors examined MN-PCE in peripheral blood and bone marrow at various times of exposure and observed a significant increase in MN-PCE between RFR-exposed animals relative to sham controls at 8 and 15 days of exposure, but no statistically significant differences were observed after 2 or 30 days' exposure. While the authors indicated that RFR exposure may have induced a genotoxic response at days 8–15 which was subsequently masked by an adaptive response, there are no experimental data to support this hypothesis. Since the relative MN-PCE levels in both the RFR-exposed and the sham groups were within the normal range for rat peripheral blood and there was no supporting evidence of elevated MN-PCE at other time-points, these data must be considered equivocal.

In an interesting study, Sykes et al. (2001) used pKZ1 transgenic mice, which possessed a β-galactosidase reporter gene in reverse orientation, to assess somatic intrachromosomal recombination inversion events. The authors exposed the mice to 900 MHz pulse-modulated RFR for 30 min/day for 1, 5 or 25 days at a whole body average SAR of 4 W/kg. While no effect of RFR exposure was observed on intrachromosomal recombination in spleen cells of pKZ1 mice after 1 or 5 days' exposure, a significant decrease in the rate of spontaneous recombination was observed after a 25-day RFR exposure period. However, this result appears to have arisen from the statistical approach employed in this study. The authors found no significant difference in the number of recombination events between the non-exposed control groups at days 1, 5 and 25 and therefore chose to combine control data from all time-points into a single pooled control group for comparison with the RFR-exposed groups at each time-point. As a result, the RFR-exposed samples for each time-point were not directly compared with their matched non-exposed time-point controls. Comparison of the RFR-exposed samples with the time-point matched controls would not likely have yielded significant differences. In any event, it is unclear what the biological significance of a reduced rate of spontaneous intrachromosomal recombination may be.

28.3.3 Human Studies

Only a limited number of studies have evaluated cytogenetic aberrations in human blood lymphocytes following occupational exposure to RFR (Table 28.3).

Table 28.3 Summary of human studies that have investigated the possible cytogenetic effects of radiofrequency radiation

Authors	RFR conditions	SAR; duration	Tissue/cell line	End point	Results	Comments
Garson et al. (1991)	0.4 MHz–20 GHz	SAR unknown, occupational exposure	Radio linemen and age-matched controls	CA	No effect	
Fucic et al. (1992)	1.25–1.35 GHz	SAR unknown, occupational exposure	Non-smoking men	MN	Increased MN frequency	
Garaj-Vrhovac and Fucic (1993)	1.25–1.35 GHz; pulsed	SAR unknown, occupational exposure	Male air traffic control workers	CA	Increased CA in all subjects	
Maes et al. (1995)	450–900 MHz	SAR unknown, occupational exposure	Male transmission antennae maintenance workers	CA	No effect	Pilot study, only 6 exposed and 6 control individuals
Lalic et al. (2001)	Frequencies reaching 8 GHz	SAR unknown, occupational exposure	Subjects working in radio-relay stations	CA	Increased CA relative to background, based on literature references	Authors did not assess the background aberration frequencies in their laboratory to ensure it was similar to literature values; insufficient information on exposure conditions

LaGroye et al. (2004a)	2.45 GHz; CW	Average SAR of 1.9 W/kg; 2-h exposure	Mouse C3H 10T1/2 fibroblasts	DNA strand breaks	No effect	
Hook et al. (2004)	813.6–847.7 MHz; various modulations	SARs ranging from 0.002 to 3.2 W/kg; various exposure times up to 24 h	Molt-4 T-lymphoblastoid cells	DNA strand breaks	No effect	Exponentially growing cells
Diem et al. (2005)	1.8 GHz; intermittent or continuous CW and GSM	SARs of 1.2 and 2 W/kg; 4-, 16- or 24-h exposure	Human diploid fibroblasts and rat granulosa cells	DNA strand breaks	Increased single-strand and double-strand breaks in both cell types	Classification of comets into 5 categories followed by transformation by arbitrary factors; no control for differences in S/G2-phase or apoptotic cells which may have resulted in their classification in an elevated “damage” category
Zeni et al. (2005)	900 MHz; GSM-modulated	SARs of 0.3 and 1 W/kg; 2-h exposure	Human peripheral blood lymphocytes	DNA strand breaks	No effect	
Sakuma et al. (2006)	2.14 GHz; CW and W-CDMA-modulated	SARs of 0.08, 0.25 and 0.8 W/kg; 2- and 24-h exposures	Human glioblastoma A172 cells and human IMR-90 fibroblasts	DNA strand breaks	No effect	
Stronati et al. (2006)	935 MHZ; GSM-modulated	SARs of 1 and 2 W/kg; 24-h exposure	Human whole blood	DNA strand breaks	No effect	

iDEN Integrated Digital Enhanced Network, *W-CDMA* wideband CDMA

Studies by Maes et al. (1997), Vijayalaxmi et al. (2000), Tice et al. (2002), McNamee et al. (2002a, b, 2003), Zeni et al. (2005) and Stronati et al. (2006) have also assessed the ability of non-thermal RFR, at a variety of frequencies and modulations, to induce DNA damage in human peripheral lymphocytes using the alkaline comet assay. The study by Tice et al. (2002) possessed an inadequate number of independent experiments ($N = 1$–2 for each exposure paradigm) upon which to accurately assess a possible RFR-induced genotoxic effect. However, the remaining studies have provided strong evidence that non-thermal RFR exposure has no effect on inducing DNA damage in cultured human lymphocytes. In contrast, Diem et al. (2005) reported increased DNA damage (DNA single-strand and double-strand breaks) in human fibroblasts and rat granulosa cells following in vitro exposure of these cells to 1.8 GHz RFR at SARs of 1.2–2 W/kg for 4–24 h, under a variety of exposure conditions. In this study, the authors classified DNA "comets" into one of five categories based upon visible inspection, then applied a transformation factor to achieve a "normalized tail factor" for each sample. Vijayalaxmi et al. (2006) raised several concerns for this approach since such transformation of each classification group by arbitrary transformation factors did not control for differences in S/G2-phase or apoptotic cells. These cells would be classified into the "higher" damage categories owing to either the presence of DNA replication forks or orderly DNA fragmentation, respectively. As such, small differences in the rate of cell growth/death between control and RFR-exposed groups may have been responsible for the observed differences between the sham and RFR-exposed samples. In any event, additional studies are required to validate these observations.

In summary, there is no consistent evidence in the literature for non-thermal, RFR-induced DNA damage in either mammalian-derived cell lines or primary human lymphocytes. This information is supported by a large database of studies using traditional bacterial/mammalian cell lines and recent transgenic in vivo mutagenesis assays, where there is virtually no evidence of any mutagenic effects from RFR exposure.

28.4.2 In Vivo Studies

Analysis of RFR-induced DNA strand breaks (using the alkaline comet assay) in animal tissue has been much more controversial (Table 28.5). An early study by Sarker et al. (1994) reported RFR-induced alterations in *Hinf*I digested genomic DNA, relative to control DNA, in both brain and testis from mice exposed to 2.45 GHz RFR at a SAR of 1.18 W/kg for 2 h/day for 120–200 days. This was followed by a series of highly publicized studies by Lai and Singh (1995, 1996, 1997), who reported increased levels of primary DNA damage (which may have included DNA single-strand and double-strand breaks, alkali-labile sites and DNA cross-links) in rat brain cells at 0–4 h after a 2-h in vivo exposure to 2.45 GHz CW or pulse-modulated RFR at SARs

Table 28.5 Summary of in vivo studies that have investigated the possible DNA-strand breaks effects of radiofrequency radiation

Authors	RFR conditions	SAR; duration	Tissue/cell line	End point	Results	Comments
Sarker et al. (1994)	2.45 GHz; CW	Average SAR of 1.18 W/kg; 2 h/day, up to 200 days	Inbred male Swiss albino mice; isolated cells from brain and testis	DNA transformations	Distinct DNA alterations in exposed samples	Restrained animals; sample number $N = 2$ per time-point
Lai and Singh (1995, 1996, 1997)	2.45 GHz; PW or CW	Whole body average SAR 0.6 and 1.2 W/kg; 2-h exposure	Male Sprague-Dawley rats; isolated brain cells	DNA strand breaks immediately and 4 h after exposure	Increased DNA damage 4 h after CW and PW exposure and 0 h after CW exposure	Non-restrained animals; no positive control; injection of free-radical scavengers blocked RFR DNA damaging effect; used proteinase K in lysis buffer
Malyapa et al. (1998)	2.45 GHz; CW	Whole body average SAR of 1.2 W/kg; 2-h exposure	Male Sprague-Dawley rats; isolated brain cells	DNA strand breaks immediately and 4 h after exposure	No effect	Non-restrained animals; observed an increased background level of DNA damage in rats euthanized with CO_2 as done in the Lai and Singh studies; observed that the time from euthanasia to dissection needs to be minimized, where increased time results in increased damage; did not use proteinase K in lysis buffer

(*Continued*)

Table 28.5 Summary of in vivo studies that have investigated the possible DNA-strand breaks effects of radiofrequency radiation—Cont'd

Authors	RFR conditions	SAR; duration	Tissue/cell line	End point	Results	Comments
LaGroye et al. (2004b)	2.45 GHz; PW	Whole body average SAR of 1.2 W/kg; 2-h exposure	Male Sprague-Dawley rats; isolated brain cells	DNA strand breaks 4 h after exposure	No effect	Non-restrained animals; lysis buffer with and without proteinase K
Aitken et al. (2005)	900 MHz	Whole body average SAR of 0.09 W/kg; 12 h/day, 7 days	Male CD1 Swiss mice; isolated spermatozoa	DNA strand Breaks and DNA integrity	No effect on strand breaks, but observed DNA lesions in a segment of the β-globin gene and in the mitochondrial genome	Non-restrained animals; no positive control
Paulraj and Behari (2006)	2.45 and 16.5 GHz; amplitude modulated	Whole body average SAR of 1 or 2 W/kg; 2 h/day, 5 days/week, 7 weeks	Male Wistar rats; isolated brain cells	DNA strand breaks	Increased DNA damage in exposed rats	Restrained animals; no positive control
Belyaev et al. (2006)	915 MHz; GSM	Whole body average SAR of 0.4 W/kg; 2-h exposure	Male Fisher 344 rats; isolated cells from spleen, thymus and brain	DNA double-strand breaks and chromatin conformation	No effect	Non-restrained animals; no positive control

PW pulsed wave

of 0.6–1.2 W/kg. Of particular interest was the observation that injection of melatonin or the spin-trap compound *N-tert*-butyl-α-phenylnitrone (PBN), either immediately before or after RFR exposure, could prevent the occurrence of RFR-induced DNA damage, indicating a role for free radicals in this effect. These studies generated a flurry of controversy, alarming media reports and research to confirm these findings and to identify a mechanism by which RFR exposure may have influenced DNA damage formation and/or repair.

In an attempt to confirm these observations, Malyapa et al. (1998) exposed rats to 2.45 GHz CW RFR for 2 h at a SAR of 1.2 W/kg, but found no evidence of immediate or delayed RFR-induced DNA damage in the rodent brain. However, the authors did note that the background level of DNA damage increased in brain cells from CO_2-asphyxiated animals relative to that in guillotined animals and that the time from CO_2 asphyxiation until dissection was an important parameter in the extent of background DNA damage. This is considered an important observation since the studies by Lai and Singh (1995, 1996, 1997) employed CO_2 asphyxiation and included numerous wash steps to remove red blood cells, resulting in an apparent large time-lag between CO_2 asphyxiation and processing of samples via the alkaline and neutral comet assays. As a result, there is considerable potential that the observations by Lai and Singh were not related to RFR exposure, but rather were the result of post-euthanasia confounding by the time-dependent generation of free radicals in the tissue extracts, despite the inclusion of PBN in the mincing buffer. On the other hand, unlike Lai and Singh, Malyapa et al. (1998) did not use proteinase K in their comet assay protocol. Since proteinase K digestion removes nuclear proteins, DNA from nuclear-protein associations and DNA–protein cross-links would be free to migrate and the studies by Lai and Singh may have detected DNA damage that Malyapa's group could not observe. To address this possibility, LaGroye et al. (2004a) used both versions of the alkaline comet assay to evaluate the ability of in vivo 2.45 GHz pulsed-wave RFR exposure at a SAR of 1.2 W/kg to induce DNA damage in the rat brain at 4 h after an acute 2-h exposure. In this study, animals were killed one at a time by guillotine and no DNA damage was detected following RFR exposure, indicating that the comet assay methodology did not appear to be a significant factor in explaining the conflicting results between these laboratories.

More recently, Paulraj and Behari (2006) reported increased DNA damage in brain cells of Wistar rats following exposure to 2.45 or 16.5 GHz RFR for 35 days at 2 h/day at SARs of approximately 1.0 and 2.0 W/kg, respectively. In contrast, Belyaev et al. (2006) found no evidence of DNA double-strand breaks or changes in chromatin conformation in rat brain following a 2-h acute exposure to 915 MHz GSM-modulated RFR at 0.4 W/kg. Similarly, Aitken et al. (2005) observed no evidence of single-strand or double-strand DNA breaks in spermatozoa of mice exposed to 900 MHz RFR for 12 h/day for 7 days; however, Aitken et al. (2005) noted altered PCR amplification of a

9-kb fragment of the β-globin gene and a 10-kb fragment of the mitochondrial genome in RFR-exposed mouse spermatozoa, relative to control animals. The authors argued that these changes were indirect evidence of RFR-induced DNA lesions, which may have been undetectable using the conventional genotoxicity assays. Obviously, additional in vivo studies are required to delineate the inconsistencies in the literature.

28.5 Epigenetic Effects of RFR

Despite the apparent inability of non-thermal RFR to induce direct (genotoxic) DNA and/or cytogenetic damage, it is possible that RFR may interact with biological systems in an indirect (non-genotoxic or epigenetic) fashion to modulate the response of cells or tissues to known genotoxins. Proposed mechanisms for RFR-induced epigenetic effects include changes in the fidelity and/or efficiency of DNA repair processes, changes in cell-cycle progression and/or cell proliferation, changes in the expression of certain genes (such as heat-shock proteins and proto-oncogenes) and changes in cell survival. The evidence for these RFR-induced effects is contradictory and outside the focus on this review. This section will focus solely on the cytogenetic evidence of RFR-induced epigenetic effects.

Relatively few studies have examined the consequence of combined effects of exposure to RFR and known physical and/or chemical mutagens on cytogenetic alterations. Early studies by Balcer-Kubiczek and Harrison (1985, 1989, 1991) reported that 2.45 GHz RFR exposure at SARs of 0.1–4.4 W/kg for 24 h, could increase the rate of neoplastic transformation of C3H 10T1/2 cells when coexposed to either benzo[*a*]pyrene or X-rays in the presence of the tumor promoter 12-*O*-tetradecanoylphorbol-13-acetate (TPA). Subsequent studies by Cain et al. (1997) and Roti Roti et al. (2001) failed to observe similar effects, although these studies employed lower SARs and different RFR frequencies. More recently, Wang et al. (2005a) exposed C3H 10T1/2 cells to 2.45 GHz RFR at SARs of 5–200 W/kg for 2 h, either with RFR alone or in the presence of RFR and 3-methylcholanthrene. No evidence of RFR-induced neoplastic transformation was observed at any SAR level when cells were exposed to RFR in the presence or absence of TPA; however, increased transformation was observed in cells co-exposed to 3-methylcholanthrene and RFR at SAR levels greater than 100 W/kg. While the temperature in the 100–200 W/kg groups reached 40 and 44°C respectively, the authors argued that the RFR-induced increase in neoplastic transformation was not due to thermal effects since control samples were exposed to similar temperatures but no increase in neoplastic transformation was observed in these samples. It is important to point out that while the volume-averaged temperature was reported not to have exceeded 40–44°C, the presence of hot spots within the sample cannot be ruled out since insufficient information on SAR uniformity was reported.

Initial studies by Maes et al. (1996, 1997) reported a strong synergistic increase in the frequency of SCE in human whole-blood cultures following exposure to 900- or 954 MHz RFR at SARs of 0.3–1.5 W/kg and mitomycin C. Similar epigenetic effects were reported by Zhang et al. (2002) and Wang et al. (2005b), where increased levels of MN and primary DNA damage were observed in human whole-blood cultures following co-exposure to RFR and a variety of mutagenic compounds. Unfortunately, these studies suffer from a number serious deficiencies, including inadequate sample size, inappropriate statistics and insufficient detail relating to the exposure system design, dosimetry and SAR distribution to exclude the possibility of thermal effects. On the other hand, an increased frequency of MN was also reported in CHO-K1 cells following coexposure to bleomycin and 2.45 GHz RFR at high SARs (78–100 W/kg), in a well-controlled study by Koyama et al. (2003). It is interesting to note from this study that high RFR SAR exposures seem to induce non-thermal epigenetic cytogenetic effects; however, the relevance of these findings to health risk assessment is questionable owing to the magnitude of these exposures.

In contrast, several recent studies where dosimetry was well defined and possible thermal influences were eliminated failed to detect evidence of RFR-induced epigenetic effects on cytogenetic end points. In a follow-up study to their original work, Maes et al. (2001) exposed human blood cultures to mitomycin C or X-rays and 900 MHz RFR with a variety of modulation paradigms for 2–24 h at SARs of 0.4–10 W/kg, but failed to find any evidence of excess chromosome aberrations or SCE in RFR-exposed samples. Similarly, LaGroye et al. (2004b) reported no evidence of increased primary DNA damage (as measured by the alkaline comet assay) in C3H 10T1/2 cells co-exposed for 2 h to 2.45 GHz RFR at a SAR of 1.9 W/kg and either ^{137}Cs γ-rays or cisplatin. More recently, Stronati et al. (2006) assessed the ability of 935 MHz GSM-modulated RFR exposure at a SAR of 1–2 W/kg for 24 h to alter the magnitude of the cytogenetic and primary DNA damage induced by 1 Gy X-irradiation in human blood cultures. No evidence of increased unstable chromosome aberrations, SCE, MN or primary DNA damage was observed in the RFR plus X-ray co-exposed groups, relative to the X-ray only controls.

Recently, Verschaeve et al. (2006) were the first group to evaluate the ability of in vivo RFR exposure to modulate the cytogenetic effects induced by a known mutagen, namely, 3-chloro-4-(dichloromethyl)-5-hydroxy-2(5*H*)-furanone (MX). In this study, female Wistar rats were supplied with drinking water containing 19 μg/ml MX and were either sham or exposed to 900 MHz RFR for 2 years at 2 h/day, 5 days/week at SARs of 0.3 and 0.9 W/kg. Blood samples were taken at 3, 6 and 24 months of exposure and brain and liver samples were acquired at the end of the study. No evidence of either increased RFR-induced primary DNA damage or MN formation was observed when compared with MX-only controls. However, it is important to point out that MX treatment also had no significant effect on primary DNA damage levels in the blood or liver of exposed animals relative to controls and was only found to increase DNA damage in the rat brain samples at 24 months. Similarly, there was no evidence of an increased frequency of MN in PCEs in the MX-treated rats at any time-point relative to control animals, suggesting that the dose of MX applied to the drinking water may not have been optimal for evaluation of the epigenetic effects of RFR in vivo. Thus, while this study is novel in design and scope, the importance of this study is limited for the evaluation of the human health risks of RFR exposure.

28.6 Cancer Studies in Animals

Although the analysis of human populations has been a powerful tool for detecting previously unidentified human carcinogens, current epidemiological studies of RFR have been hindered by the relatively recent introduction (within the last 10 years) of a wide variety of RFR-emitting devices and the long latency phase (more than 20 years) for the development of most humans cancers. Furthermore, since new wireless technologies and applications are emerging at a tremendous rate, it is difficult for any epidemiology study to capture the effect of long-term human exposure to specific RFR conditions. As such, cancer studies in experimental animals are particularly important in evaluating whether an association between RFR exposure and cancer exists as laboratory animal studies provide an integrated system in which RFR experimental variables can be controlled, specific hypotheses can be investigated and exposures can be precisely measured. In this section, the evidence for a causal association between long-term, low-level RFR exposure and cancer in rodents will be reviewed (Tables 28.6–28.8).

Prior to 1997, only a limited number of studies assessed the effect of long-term, low-level RFR exposure in animals and most of these studies suffered from inadequate dosimetry, confinement or thermal stress and high rates of inherent disease in the animals (Moulder et al. 2005); however, two studies merit consideration. Chou et al. (1992) reported that following 25 months' exposure of Sprague-Dawley rats for 21.5 h/day to 2.45 GHz pulse-modulated RFR at a whole body averaged SAR of 0.15–0.4 W/kg, a significant increase in overall malignant tumor incidence was observed in the RFR-exposed rats when compared with the sham group. However, the authors cautioned that the biological significance of these results was limited as there were no accompanying differences in lifespan, total number of tumors or type of tumor. Wu et al. (1994) exposed Balb/c mice to 2.45 GHz RFR for 3 h/day, 6 days/week for 5 months at SARs of 10–12 W/kg in the presence of the known colon cancer inducing compound dimethylhydrazine. The authors did not find any evidence that RFR exposure had an effect on the incidence or size of chemically induced tumors, or on the incidence of protuberant or infiltrative tumors in tumor-bearing mice when compared with control groups.

Table 28.6 Summary of cancer studies in animals that have investigated the effects of radiofrequency radiation – 1992–2000

Authors	RFR conditions	SAR; duration	Animal model	End point	Results	Comments
Chou et al. (1992)	2.45 GHz; pulse-modulated	Whole body average SAR between 0.15 and 0.4 W/kg; 21.5 h/day for 25 months	Male Sprague-Dawley rats; 100 rats per group	Tumor incidence in all major organs	Increased incidence of primary malignant tumors in RFR-exposed rats	No differences in lifespan, total number of tumors or tumor type; unrestrained animals; all animals were histopathologically assessed; no positive control
Wu et al. (1994)	2.45 GHz; CW	SARs of 10–12 W/kg; 3 h/day, 6 days/week for 5 months	Dimethylhydrazine-induced colon cancer in Balb/c mice; 26–32 mice per group	Incidence of colon cancer	No effect	Animal movement partially restricted during exposure; all animals were histopathologically assessed; low number of animals per group; no positive control
Repacholi et al. (1997)	900 MHz; pulse-modulated	Average SARs ranged from 0.13 to 1.4 W/kg; 2×30 min/day for 18 months	Eμ-*Pim1* (lymphoma-prone) mice; 100 mice per group	Incidence of lymphoma; tumor incidence in all major organs	Increased incidence of lymphoma in RFR-exposed mice	Unrestrained animals; not all animals were histopathologically assessed; non-homogeneous SAR between and within animals; no positive control

(Continued)

Table 28.6 Summary of cancer studies in animals that have investigated the effects of radiofrequency radiation – 1992–2000—Cont'd

Authors	RFR conditions	SAR; duration	Animal model	End point	Results	Comments
Toler et al. (1997)	435 MHz; pulse-modulated	Whole body average SAR of 0.32 W/kg; 22 h/day, 7 days/week for 21 months	Female C3H/HeJ (mammary tumor prone) mice; 200 mice per group	Tumor incidence in mammary glands and all major organs	No effect	Unrestrained animals; all animals were histopathologically assessed; no positive control
Frei et al. (1998a, b)	2.45 GHz; CW	Whole body average SAR of 0.3 and 1.0 W/kg; 20 h/day, 7 days/week for 18 months	Female C3H/HeJ (mammary tumor prone) mice; 100 mice per group	Tumor incidence in mammary glands and all major organs	No effect	Unrestrained animals; all animals were histopathologically assessed; no positive control
Adey et al. (1999, 2000)	836 MHz; CW or pulse-modulated	Brain-average SAR of 1.0–1.6 W/kg; in utero and postweaning exposure for 2 years	ENU-treated and untreated Fischer 344 rats; 60–90 rats per group	Incidence of primary CNS tumors	No effect	Restrained animals; histopathological assessment performed on CNS tissue from all animals

CNS central nervous system, *ENU* ethylnitrosourea

Table 28.7 Summary of cancer studies in animals that have investigated the effects of radiofrequency radiation – 2001–2002

Authors	RFR conditions	SAR; duration	Animal model	End point	Results	Comments
Zook and Simmens (2001, 2006)	860 MHz; CW and pulse-modulated	Brain average SAR of 1.0 W/kg; 6 h/day, 5 days/week for 2 years	ENU-treated and untreated Sprague-Dawley rats; 60 rats per group	Incidence of tumors in CNS and in all major organs	No effect	Restrained animals; all animals were histopathologically assessed
Heikkinen et al. (2001)	902 MHz; CW or GSM-modulated	Whole body average SARs of 0.35 and 1.5 W/kg; 1.5 h/day, 5 days/week for 78 weeks	Female CBA/S mice initiated with ionizing radiation; 50 mice per group	Incidence of neoplastic lesions in all major organs	No effect	Restrained animals; all animals were histopathologically assessed
Imaida et al. (2001)	1.5 GHz; pulse-modulated	Whole body average SAR of 0.08 W/kg; 90 min/day, 5 days/week for 19 weeks	DMBA-induced skin tumors in female ICR mice; 30–48 mice per group	Incidence of tumors on skin and in several major organs	No effect	Restrained animals; all animals were histopathologically assessed
Utteridge et al. (2002)	898.4 MHz; GSM-modulated	SARs of 0.25, 1.0, 2.0 and 4.0 W/kg; 1 h/day, 5 days/week for 104 weeks	Eμ-*Pim1* (lymphoma-prone) mice; 120 mice per group	Incidence of lymphoma; tumor incidence in all major organs	No effect	Restrained animals; all animals were histopathologically assessed; ENU-positive control included
Bartsch et al. (2002)	900 MHz; GSM-modulated	Whole body average SAR ranging from 0.017 to 0.07 W/kg; continuous exposure for 1 year	DMBA-induced mammary tumors in female Sprague-Dawley rats; 60 rats per group	Incidence of mammary tumors	No effect	Unrestrained animals; histopathological assessment only performed when mammary tumors became 1–2 cm in diameter

DMBA 7,12-dimethybenz[*a*]anthracene

Table 28.8 Summary of cancer studies in animals that have investigated the effects of radiofrequency radiation – 2003–2006

Authors	RFR conditions	SAR; duration	Animal model	End point	Results	Comments
Anane et al. (2003)	900 MHz; GSM-modulated	Whole body average SARs of 0.1–3.5 W/kg; 2 h/day, 5 days/week for 9 weeks	DMBA-induced mammary tumors in female Sprague-Dawley rats; 16 rats per group	Incidence of mammary tumors	Equivocal results	Animal movement partially restricted during exposure; histopathological assessment only performed on mammary glands with visible (macroscopic) tumors; insufficient number of animals per group; inconsistency across replicate experiments
La Regina et al. (2003)	835 MHz FDMA- and 847 MHz CDMA-modulated	Brain-average SAR of 1.3 W/kg; 4 h/day, 5 days/week for 2 years	Fischer 344 rats; 80 rats per group	Incidence of tumors in all major organs	No effect	Restrained animals; no positive control; all animals were histopathologically assessed
Heikkinen et al. (2003)	849 or 902 MHz; pulse-modulated	Whole body average SAR of 0.5 W/kg; 1.5 h/day, 5 days/week for 52 weeks	Female transgenic and normal K2 mice coexposed to UV radiation; 8–27 mice per group	Incidence of skin tumors	No effect	Restrained animals; only macroscopic skin lesions were histopathologically assessed; low animal numbers (8–27 per group)
Sommer et al. (2004)	900 MHz; GSM-modulated	Whole body average SAR of 0.4 W/kg; continuous exposure for 46 weeks	Female AKR/J (lymphoma-prone) mice; 160 mice per group	Incidence of lymphoma	No effect	Unrestrained animals; all animals were histopathologically assessed; no positive control

Anderson et al. (2004)	1.6 GHz; Iridium signal	Brain-average SAR ranging from 0.16 to 1.6 W/kg; in utero and postweaning exposure (2 h/day) for 2 years	Fischer 344 rats; 90 rats per group	Incidence of tumors in CNS and all major organs	No effect	Restrained animals; all animals were histopathologically assessed; no positive control
Shirai et al. (2005)	1.6 GHz; pulse-modulated	Brain-average SAR of 0.67 and 2.0 W/kg; 90 min/day, 5 days/week for 104 weeks	ENU-treated and untreated Fischer 344 rats; 100 rats per group	Incidence of tumors in CNS and all major organs	No effect	Restrained animals; all animals were histopathologically assessed
Huang et al. (2005)	849 MHz or 1.763 GHz; CDMA-modulated	Whole body average SAR of 0.4 W/kg; 2×45 min/day, 5 days/week for 19 weeks	DMBA-induced skin tumors in male ICR mice; 20 mice per group	Incidence of skin tumors	No effect	Insufficient information on animal exposure conditions; low animal numbers (20 per group)
Yu et al. (2006)	900 MHz; GSM-modulated	Whole body average SARs of 0.44, 1.33 and 4.0 W/kg; 4 h/day, 5 days/week for 26 weeks	DMBA-induced mammary tumors in female Sprague-Dawley rats; 100 rats per group	Incidence of mammary tumors	No effect	Restrained animals; histopathological assessment performed on mammary glands from all animals

UV ultraviolet

Since 1997, a relatively large number of studies have assessed the effect of long-term, low-level RFR exposure on the induction or promotion of tumors in both normal and tumor-prone rodents. Repacholi et al. (1997) exposed unrestrained mice genetically predisposed to develop lymphomas (Eμ-*Pim1*) to 900 MHz pulse-modulated RFR for 2 × 30 min/day for up to 18 months at SARs ranging from 0.008–4.2 W/kg. In this highly publicized study, the authors reported that while lymphoblastic lymphoma was not significantly affected, an increased incidence of non-lymphoblastic lymphoma was observed in RFR-exposed mice when compared with sham controls. This study was criticized for its poor RFR exposure homogeneity (SAR range 0.008–4.2 W/kg) among the experimental animals and inadequate histopathological assessment (e.g., only ill or dead animals were examined pathologically), which may have inadvertently introduced bias into the analysis. In an effort to address these shortcomings, Utteridge et al. (2002) exposed restrained Eμ-*Pim1* transgenic mice and wild-type mice to 898.4 MHz GSM-modulated RFR for 1 h/day, 5 days/week for 104 weeks at SARs of 0.25, 1.0, 2.0 and 4.0 W/kg. Under tightly controlled exposure conditions and following pathological examination of all animals, the authors reported that long-term RFR exposure had no significant effect on lymphoma induction in normal or lymphoma-prone mice when compared with sham controls. While this study was also criticized since it was not a direct replication of the study by Repacholi et al. (1997) and therefore could not discount the previous findings by this group, the findings from the current study deserve merit in their own regard. Recently, Sommer et al. (2004) continuously exposed unrestrained lymphoma-prone AKR/J mice to 900 MHz GSM-modulated RFR for 46 weeks at a whole body average SAR of 0.4 W/kg. The AKR/J mouse strain was chosen since it is genetically predisposed to develop thymic lymphoblastic lymphoma as its genome carries the AK virus. The authors found no evidence of decreased survival rates or increased incidence of lymphoma in RFR-exposed mice, relative to sham mice, at any SAR level tested. While these results are consistent with those of Utteridge et al. (2002), it must be noted that Repacholi et al. (1997) reported an increased incidence of non-lymphoblastic lymphoma which was not evaluated in this study.

Six recent studies have assessed the ability of long-term RFR exposure to influence the development of mammary tumors in rodent models. Frei et al. (1998a, b) exposed C3H/HeJ mammary tumor-prone mice to 2.45 GHz CW RFR for 18 months at 20 h/day, 7 days/week at SARs of 0.3 and 1 W/kg. In both studies, there were no significant differences between RFR-exposed and sham mice with respect to tumor incidence, latency to tumor onset, growth rate of mammary tumors or animal longevity. In a related study by the same group, Toler et al. (1997) found no evidence that latency to tumor onset, growth rate of mammary tumors or animal longevity was affected by a 21-month exposure (22 h/day, 7 days/week) of C3H/HeJ mammary tumor-prone mice to 435 MHz pulse-modulated RFR at a SAR of 0.32 W/kg. In an attempt to investigate the possible tumor-promoting capability of RFR, several groups have used the well-characterized 7,12-dimethylbenz[*a*]anthracene (DMBA) induced mammary tumor model to study this issue in female Sprague-Dawley rats. Bartsch et al. (2002) reported the results of three independent experiments, initiated in three consecutive years, where unrestrained rats were continuously exposed to 900 MHz GSM-modulated RFR at whole body averaged SARs ranging from 0.017 to 0.07 W/kg. Histopathological evaluation of all tumors was performed as the tumors reached 1–2 cm in diameter during the course of the study and the milk line was verified for the presence of any additional tumors in all animals. The authors found a significantly delayed malignant tumor median latency in RFR-exposed rats in their first experiment that was absent in their other two studies. However, as a whole, the authors concluded that RFR exposure had no effect on either DMBA-induced mammary tumor latency or the cumulative tumor incidence since this prolonged latency to tumor onset was not replicated in their subsequent studies. In a similar study, Anane et al. (2003) performed two independent studies where female Sprague-Dawley rats with DMBA-induced mammary tumors were exposed to 900 MHz GSM-modulated RFR for 2 h/day, 5 days/week for 9 weeks at SARs of 0.1–3.5 W/kg. A total of 16 animals were included in each exposure group. All macroscopic mammary tumors were histopathologically evaluated; however, mammary glands without visible tumors were not evaluated and microscopic tumors may have been missed. The authors reported no differences in the latency to tumor onset between the sham and RFR-exposed groups in each study, nor between either of the two studies. A significant increase in tumor incidence was observed in rats exposed to RFR at SARs of 1.4 and 2.2 W/kg in the first experiment, but the opposite effect was observed in their second experiment, where a decreased tumor incidence was reported in rats exposed to RFR at a SAR of 1.4 W/kg. Furthermore, tumor multiplicity on rats bearing tumors was inconsistent in the 1.4-W/kg RFR-exposed groups between the two studies. In the second experiment, the tumor-bearing rats in the 1.4-W/kg RFR-exposed group had fewer malignant tumors than tumor-bearing rats in the sham group, whereas no such difference was observed in the first study. The inconsistency in the results, which may be partially attributable to low animal numbers per group, led the authors to conclude that this study added no new evidence as to the co-promoting effect of RFR. More recently, Yu et al. (2006) also investigated the effects of 900 MHz GSM-modulated RFR exposure on mammary tumor promotion in female rats with DMBA-induced mammary tumors. The authors exposed rats (100 per group) to cage conditions or to RFR for 4 h/day, 5 days/week for 26 weeks at SARs of 0, 0.44, 1.33 and 4.0 W/kg. All mammary glands were palpated weekly and analyzed histopathologically. The authors reported that no statistically significant differences were observed between RFR-exposed and sham rats for latency to tumor onset, mammary tumor incidence, tumor multiplicity or tumor size on tumor-bearing rats. However, the investigators did observe a statistically significant increase in body weight and tumor incidence in the cage control group and the latency to tumor onset was also significantly shorter in these animals compared with that in the sham and RFR-exposed groups. The authors mentioned that these differences may have been due to daily food deprivation of the sham and RFR-exposed groups, which may have led to reduced tumorigenesis in these groups relative to the cage controls. In summary, when these studies using mammary tumor prone mice and chemically induced mammary tumors are assessed collectively, there is no compelling evidence of altered mammary tumor incidence or development by RFR exposure.

The possible association between RFR exposure and brain tumor risk has been of great public and scientific concern for the past decade, owing to the large and ever-increasing number of mobile-phone users worldwide who are exposed to RFR in close proximity to the head. Since 2000, several animal cancer studies have investigated whether RFR exposure may influence brain tumor initiation and/or promotion. Adey et al. (2000) exposed ethylnitrosourea (ENU) treated and normal Fischer 344 rats continuously to 836 MHz CW RF fields in utero on gestational day 18 for 3–4 days until parturition, then at 10 days after weaning the pups were exposed intermittently to RFR for the remainder of their lifespan (approximately 2 years). The brain-averaged SARs were estimated to range from 1 to 1.2 W/kg. The authors found no evidence that RFR exposure had any effect on either spontaneous or ENU-induced brain tumor incidence. In an earlier study by the same group (Adey et al. 1999), the authors employed a similar experimental design but the rats were exposed to 836 MHz pulse-modulated RFR at brain-averaged SARs of 1.0–1.6 W/kg. In their earlier study, the authors also reported no evidence of any tumorigenic effects from RFR exposure, but did note that RFR exposure caused a trend towards a reduced incidence in ENU-induced brain tumors. More recently, Shirai et al. (2005) exposed normal and ENU-treated Fischer 344 rats to 1.439 GHz pulse-modulated RFR at brain-averaged SARs of 0.67 and 2 W/kg. RFR exposures commenced at 5 weeks of age for 90 min/day, 5 days/week for 104 weeks. The authors reported that there were no differences in survival rates, incidences or numbers of brain and/or spinal cord tumors or tumor types in ENU-treated rats among the RFR-exposed groups relative to sham animals. Anderson et al. (2004) conducted a similar study, where pregnant Fischer 344 rats were exposed to a 1.6 GHz Iridium signal. Similar to the previous studies, rats were exposed in utero on gestational day 19 until weaning (about 23 days) for 2 h/day, 7 days/week, then the offspring were exposed for 2 h/day for 5 days/week under near-field conditions to 1.6 GHz RFR at brain-averaged SARs ranging from 0.16–1.6 W/kg for 2 years. No statistically significant differences in the incidence of cancer in RFR-treated normal rats relative to sham rats was observed in this 2-year study.

La Regina et al. (2003) exposed normal adult Fischer 344 rats to either 835 or 847 MHz RFR for 4 h/day, 5 days/week for 2 years at a time-averaged brain SAR of 1.3 W/kg. At the end of the study, all major organs were examined grossly and histopathologically for the incidence of spontaneous tumor formation. The authors found no evidence of altered tumor formation or tissue hyperplasia in the brain or other organs of RFR-exposed rats, relative to sham controls. Zook and Simmens (2001) also examined the effect of long-term exposure to 860 MHz pulse-modulated or CW RFR on tumor initiation and promotion in the brains of rats during a 2-year bioassay. ENU-treated and untreated rats were exposed to RFR for 6 h/day, 5 days/week from 2 up to 24 months of age. The brain-averaged SAR was calculated to be approximately 1 W/kg. At the end of the exposure period, all but two rats (898 in total) were necropsied and all major tissues were studied histopathologically. The authors reported no statistically significant differences in tumor incidence, volume, multiplicity, histological type, malignancy or fatality due to brain tumors from RFR exposure. Furthermore, there was no evidence of RFR-induced neoplasia in any other tissue examined. However, animals exposed to 860 MHz pulse-modulated RFR in the highest-dosed ENU group tended to develop fatal brain tumors at a shortened latency compared with the sham group. Although the trend was not statistically significant, the authors further examined this possible RFR-promoting effect in a more recent study (Zook and Simmens 2006). In this study, a similar experimental design was employed, however an equal number of RFR-exposed and control animals were sacrificed at 30-day intervals between 171 and 325 days of exposure. As with their previous study, histopathological analysis was performed on all rats (1,080 in total). The authors again found no statistically significant differences in tumor incidence, volume, multiplicity, histological type, malignancy or fatality associated with any type of neurogenic tumor. Furthermore, Zook and Simmens (2006) found no evidence of RFR-induced shortening of brain tumor latency in this study. The authors concluded that their studies found no evidence that RFR could induce a tumor-promoting effect.

Heikkinen et al. (2001) examined the tumor-promoting ability of RFR on tumors initiated with ionizing radiation. The authors exposed 200 female CBA/S mice to 4-Gy ionizing radiation (given in weekly 1.33-Gy fractions) followed by 902 MHz GSM-modulated or CW RFR at SARs ranging from 0.35–1.5 W/kg for 1.5 h/day, 5 days per week for 78 weeks. No significant differences were reported in the incidence of any neoplastic lesions in mice exposed to RFR, relative to the sham control group. In a later study, Heikkinen et al. (2003) examined the ability of RFR to influence the promotion of ultraviolet (UV) light induced skin tumorigenesis in transgenic ornithine decarboxylase overexpressing mice and in normal mice. In this study, mice were exposed to UV radiation in the presence/absence of 849 or 902 MHz pulse-modulated RFR for 1.5 h/day, 5 days/week for 52 weeks. The authors observed no statistically significant changes due to RFR exposure in any of the parameters examined, although a trend of accelerated tumor development was seen in both RFR groups. However, the authors had low animal numbers in all of their groups, ranging from eight in their non-transgenic cage control group to only 27 in their non-transgenic GSM-modulated RFR group, thereby reducing the statistical power of the study to detect differences between groups. Imaida et al. (2001) investigated the co-promoting ability of 1.5 GHz pulse-modulated RFR (whole body average SAR 0.08 W/kg; peak skin SAR 2 W/kg) in ICR mice initiated with DMBA. The animals were exposed to RFR for 90 min/day, 5 days/week for 19 weeks. At 20 weeks, all animals were killed and skin tumors were analyzed histopathologically. No evidence of RFR-induced skin tumor co-promotion was observed. Similarly, Huang et al. (2005) assessed skin tumor co-promotion in ICR mice exposed to either 849 or 1,763 MHz CDMA RFR (whole body average SAR 0.4 W/kg) for 2×45 min/day, 5 days/week for 19 weeks following DMBA initiation and found no evidence of RFR-induced skin tumor promotion.

Overall, the results of animal cancer studies relating to RFR exposure are overwhelmingly negative, but it is important to note that these types of animal studies have inherent limitations. Rodent strains react differently to restraint stress, food deprivation, acclimatization as well as to environmental conditions in various testing laboratories. This phenomenon is readily observable by comparing the low reproducibility in the rate of tumor incidence by known carcinogens in these animal studies, even when conducted within the same laboratory. Therefore, the limited number of studies with positive effects should not be dismissed entirely without additional replication studies.

28.7 Summary

Over the past 30 years, hundreds of studies have been conducted in laboratories around the world to evaluate the ability of non-thermalizing RFR to induce DNA strand breaks, chromosome aberrations, MN, SCE and cancer in a variety of animal and human systems. The vast majority of in vitro, in vivo and human RFR investigations conducted to date have found no consistent evidence of RFR-induced cytogenetic, genotoxic or carcinogenic effects. Of the limited number of in vitro studies reporting RFR-induced cytogenetic effects, thermal effects appear to be a possible confounding variable. Relatively few studies have assessed cytogenetic end points in animal and human tissue following in vivo RFR exposure and obvious methodological and/or statistical shortcomings are evident in many of these studies. The results of animal cancer studies relating to RFR exposure have reported a nearly uniform lack of association between RFR exposure and cancer initiation and/or promotion. However, current epidemiological studies of RFR have been hindered by the relatively recent introduction (within the last 10 years) of a wide variety of RFR-emitting devices and the long latency phase (more than 20 years) for the occurrence of most humans cancers. Therefore, despite the extensive body of experimental studies conducted to date, the issue surrounding the safety of RFR will likely remain a significant public concern for years to come. In the absence of conclusive epidemiology, future well-characterized prospective human and animal RFR cytogenetics studies would be beneficial to permit a more complete RFR health risk assessment.

References

Adey WR, Byus CV, Cain CD, Higgins RJ, Jones RA, Kean CJ, Kuster N, MacMurray A, Stagg RB, Zimmerman G, Phillips JL, Haggren W (1999) Spontaneous and nitrosourea-induced primary tumors of the central nervous system in Fischer 344 rats chronically exposed to 836 MHz modulated microwaves. Radiat Res 152:293–302

Adey WR, Byus CV, Cain CD, Higgins RJ, Jones RA, Kean CJ, Kuster N, MacMurray A, Stagg RB, Zimmerman G (2000) Spontaneous and nitrosourea-induced primary tumors of the central nervous system in Fischer 344 rats chronically exposed to frequency-modulated microwave fields. Cancer Res 60:1857–1863

Aitken RJ, Bennetts LE, Sawyer D, Wiklendt AM, King BV (2005) Impact of radio frequency electromagnetic radiation on DNA integrity in the male germline. Int J Androl 28:171–179

Anane R, Dulou P-E, Taxile M, Geffard M, Crespeau F-L, Veyret B (2003) Effects of GSM-900 microwaves on DMBA-induced mammary gland tumors in female Sprague-Dawley rats. Radiat Res 160:492–497

Anderson LE, Sheen DM, Wilson BW, Grumbein SL, Creim JA, Sasser LB (2004) Two-year chronic bioassay study of rats exposed to a 1.6 GHz radiofrequency signal. Radiat Res 162:201–210

Balcer-Kubiczek EK, Harrison GH (1985) Evidence for microwave carcinogenesis in vitro. Carcinogenesis 6:859–864

Balcer-Kubiczek EK, Harrison GH (1989) Induction of neoplastic transformation in C3H/10T1/2 cells by 2.45 GHz microwaves and phorbol ester. Radiat Res 117:531–537

Balcer-Kubiczek EK, Harrison GH (1991) Neoplastic transformation of C3H/10T1/2 cells following exposure to 120-Hz modulated 2.45 GHz microwaves and phorbol ester tumor promoter. Radiat Res 126:65–72

Bartsch H, Bartsch C, Seebald E, Deerberg F, Dietz K, Vollrath L, Mecke D (2002) Chronic exposure to a GSM-like signal (mobile phone) does not stimulate the development of DMBA-induced mammary tumors in rats: Results of three consecutive studies. Radiat Res 157:183–190

Beechey CV, Brooker D, Kowalczuk CI, Saunders RD, Searle AG (1986) Cytogenetic effects of microwave irradiation on male germ cells of the mouse. Int J Radiat Biol Phys Chem Med 50:909–918

Belyaev IY, Koch CB, Terenius O, Roxstrom-Lindquist K, Malmgren LOG, Sommer WH, Salford LG, Persson BRR (2006) Exposure of rat brain to 915 MHz GSM microwaves induces changes in gene expression but not double-stranded DNA breaks or effects on chromatin conformation. Bioelectromagnetics 27:295–306

Bisht KS, Moros EG, Straube WL, Baty JD, Roti Roti JL (2002) The effect of 835.62 MHz FDMA or 847.74 MHz CDMA modulated radiofrequency radiation on the induction of micronuclei in C3H 10T1/2 cells. Radiat Res 157:506–515

Brusick D, Albertini R, McRee D, Peterson D, Williams G, Hanawalt P, Preston J (1998) Genotoxicity of radiofrequency radiation. Environ Mol Mutagen 32:1–16

Cain CD, Thomas DL, Adey WR (1997) Focus formation of C3H10T1/2 cells and exposure to a 836.55 MHz modulated radiofrequency field. Bioelectromagnetics 18:237–243

Chou C-K, Swicord M (2003) Comment on "Exposure of human peripheral blood lymphocytes to electromagnetic fields associated with cellular phones leads to chromosomal instability, by Mashevich et al." Bioelectromagnetics 2003:24(2):82–90. Bioelectromagnetics 24:582

Chou C-K, Guy AW, Kunz LL, Johnson RB, Crowley JJ, Krupp JH (1992) Long-term, low-level microwave irradiation of rats. Bioelectromagnetics 13:469–496

d'Ambrosio G, Massa R, Scarfi MR, Zeni O (2002) Cytogenetic damage in human lymphocytes following GMSK phase modulated microwave exposure. Bioelectromagnetics 23:7–13

Diem E, Schwartz C, Adlkofer F, Jahn O, Rudiger H (2005) Non-thermal DNA breakage by mobile-phone radiation (1800 MHz) in human fibroblasts and in transformed GFSH-R17 rat granulosa cells in vitro. Mutat Res 583:178–183

Fahim MS, Fahim Z, Der R, Hall DG, Harman J (1975) Heat in male contraception (hot water 60 degrees C, infrared, microwave, and ultrasound). Contraception 11:549–562

Frei MR, Berger RE, Dusch SJ, Guel V, Jauchem JR, Merritt JH, Stedham MA (1998a) Chronic exposure of cancer-prone mice to low-level 2450 MHz radiofrequency radiation. Bioelectromagnetics 19:20–31

Frei MR, Jauchem JR, Dusch SJ, Merritt JH, Berger RE, Stedham MA (1998b) Chronic, low-level (1.0 W/kg) exposure of mice prone to mammary cancer to 2450 MHz microwaves. Radiat Res 150:568–576

Fucic A, Garaj-Vrhovac V, Skara M, Dimitrovic B (1992) X-rays, microwaves and vinyl chloride monomer: their clastogenic and aneugenic activity, using the micronucleus assay on human lymphocytes. Mutat Res 282:265–271

Gadhia PK, Shah T, Mistry A, Pithawala M, Tamakuwala D (2003) A preliminary study to assess possible chromosomal damage among users of digital mobile phones. Electromagn Biol Med 22:149–159

Garaj-Vrhovac V, Fucic A (1993) The rate of elimination of chromosomal aberrations after accidental exposure to microwave radiation. Bioelectrochem Bioenerg 30:319–325

Garson OM, McRobert TL, Campbell LJ, Hocking BA, Gordon I (1991) A chromosomal study of workers with long-term exposure to radio-frequency radiation. Med J Aust 155: 289–292

Görlitz B-D, Müller M, Ebert S, Hecker H, Kuster N, Dasenbrock C (2005) Effects of 1-week and 6-week exposure to GSM/DCS radiofrequency radiation on micronucleus formation in B6C3F1 mice. Radiat Res 164:431–439

Heikkinen P, Kosma V-M, Hongisto T, Huuskonen H, Hyysalo P, Komulainen H, Kumlin T, Lahtinen T, Lang S, Puranen L, Juutilanen J (2001) Effects of mobile phone radiation on x-ray-induced tumorigenesis in mice. Radiat Res 156:775–785

Heikkinen P, Kosma V-M, Alhonen L, Huuskonen H, Komulainen H, Kumlin T, Laitinen JT, Lang S, Puranen L, Juutilainen J (2003) Effects of mobile phone radiation on UV-induced skin tumorigenesis in ornithine decarboxylase transgenic and non-transgenic mice. Int J Radiat Biol 79:221–233

Hook GJ, Zhang P, LaGroye I, Li L, Higashikubo R, Moros EG, Straube WL, Pickard WF, Baty JD, Roti Roti JL (2004) Measurement of DNA damage and apoptosis in Molt-4 cells after in vitro exposure to radiofrequency radiation. Radiat Res 161:193–200

Huang T-Q, Lee J-S, Kim T-H, Pack J-K, Jang J-J, Seo J-S (2005) Effect of radiofrequency radiation exposure on mouse skin tumorigenesis initiated by 7, 12-dimethybenz[α]anthracene. Int J Radiat Biol 81:861–867

Imaida K, Kuzutani K, Wang J, Fujiwara O, Ogiso T, Kato K, Shirai T (2001) Lack of promotion of 7,12-dimethylbenz[a]anthracene-initiated mouse skin carcinogenesis by 1.5 GHz electromagnetic near fields. Carcinogenesis 22:1837–1841

International Commission on Non-Ionizing Radiation Protection (1998) Guidelines for limiting exposure to time-varying electric, magnetic, and electromagnetic fields (up to 300 GHz). Health Phys 74:494–522

Komatsubara Y, Hirose H, Sakurai T, Koyama S, Suzuki Y, Taki M, Miyakoshi J (2005) Effect of high-frequency electromagnetic fields with a wide range of SARs on chromosomal aberrations in murine m5S cells. Mutat Res 587:114–119

Koyama S, Nakahara T, Wake K, Taki M, Isozumi Y, Miyakoshi J (2003) Effects of high frequency electromagnetic fields on micronucleus formation in CHO-K1 cells. Mutat Res 541:81–89

LaGroye I, Anane R, Wettring BA, Moros EG, Straube WL, LaRegina M, Niehoff M, Pickard WF, Baty J, Roti Roti JL (2004a) Measurement of DNA damage after acute exposure to pulsed-wave 2450 MHz microwaves in rat brain cells by two alkaline comet assay methods. Int J Radiat Biol 80:11–20

LaGroye I, Hook GJ, Wettring BA, Baty JD, Moros EG, Straube WL, Roti Roti JL (2004b) Measurements of alkali-labile DNA damage and protein-DNA crosslinks after 2450 MHz microwave and low-dose gamma irradiation in vitro. Radiat Res 161:201–214

Lai H, Singh NP (1995) Acute low-intensity microwave exposure increases DNA single-strand breaks in rat brain cells. Bioelectromagnetics 16:207–210

Lai H, Singh NP (1996) Single- and double-strand breaks in rat brain cells after acute exposure to radiofrequency electromagnetic radiation. Int J Radiat Biol 69:513–521

Lai H, Singh NP (1997) Melatonin and a spin-trap compound block radiofrequency electromagnetic radiation-induced DNA strand breaks in rat brain cells. Bioelectromagnetics 18:446–454

Lalic H, Lekic A, Radosevic-Stasic (2001) Comparison of chromosome aberrations in peripheral blood lymphocytes from people occupationally exposed to ionizing and radiofrequency radiation. Acta Med Okayama 55:117–127

La Regina M, Moros EG, Pickard WF, Straube WL, Baty J, Roti Roti JL (2003) Effect of chronic exposure to 835.62 MHz FDMA or 847.74 MHz CDMA radiofrequency radiation on the incidence of spontaneous tumors in rats. Radiat Res 160:143–151

Li L, Bisht KS, LaGroye I, Zhang P, Straube WL, Moros EG, Roti Roti JL (2001) Measurement of DNA damage in mammalian cells exposed in vitro to radiofrequency fields at SARs of 3-5 W/kg. Radiat Res 156:328–332

Maes A, Collier M, Slaets D, Verschaeve L (1995) Cytogenetic effects of microwaves from mobile communication frequencies (954 MHz) Electro-Magnetobiol 14:91–98

Maes A, Collier M, Slaets D, Verschaeve L (1996) 954 MHz microwaves enhance the mutagenic properties of mitomycin C. Environ Mol Mutagen 28:496–501

Maes A, Collier M, Van Gorp U, Vandoninck S, Verschaeve L (1997) Cytogenetic effects of 935.2-MHz (GSM) microwaves alone and in combination with mitomycin C. Mutat Res 393:151–156

Maes A, Collier M, Verschaeve L (2001) Cytogenetic effects of 900 MHz (GSM) microwaves on human lymphocytes. Bioelectromagnetics 22:91–96

Maes A, Van Gorp U, Verschaeve L (2006) Cytogenetic investigation of subjects professionally exposed to radiofrequency radiation. Mutagenesis 21:139–142

Malyapa RS, Ahern EW, Straube WL, Moros EG, Pickard WF, Roti Roti JL (1997a) Measurement of DNA damage after exposure to 2450 MHz electromagnetic radiation. Radiat Res 148:608–617

Malyapa RS, Ahern EW, Straube WL, Moros EG, Pickard WF, Roti Roti JL (1997b) Measurement of DNA damage after exposure to electromagnetic radiation in the cellular phone communication frequency band (835.62 and 847.74 MHz). Radiat Res 148:618–627

Malyapa RS, Ahern EW, Bi C, Straube WL, LaRegina M, Pickard WF, Roti Roti JL (1998) DNA damage in rat brain cells after in vivo exposure to 2450 MHz electromagnetic radiation and various methods of euthanasia. Radiat Res 149:637–645

Manikowski-Czerska E, Czerski P, Leach WM (1985) Effects of 2.45 GHz microwaves on meiotic chromosomes of male CBA/CAY mice. J Hered 76:71–73

Mashevich M, Folkman D, Kesar A, Barbul A, Korenstein R, Jerby E, Avivi L (2003) Exposure of human peripheral blood lymphocytes to electromagnetic fields associated with cellular phones leads to chromosomal instability. Bioelectromagnetics 24:82–90

McNamee JP, Bellier PV, Gajda GB, Miller SM, Lemay EP, Lavallee BF, Marro L, Thansandote A (2002a) DNA damage and micronucleus induction in human leukocytes after acute in vitro exposure to a 1.9 GHz continuous-wave radiofrequency field. Radiat Res., 158:523-533

McNamee JP, Bellier PV, Gajda GB, Lavallee BF, Lemay EP, Marro L, Thansandote A (2002b) DNA damage in human leukocytes after acute in vitro exposure to a 1.9 GHz pulse-modulated radiofrequency field. Radiat Res 158:534–537

McNamee JP, Bellier PV, Gajda GB, Lavallee BF, Marro L, Lemay E, Thansandote A (2003) No evidence for genotoxic effects from 24 h exposure of human leukocytes to 1.9 GHz radiofrequency fields. Radiat Res 159:693–697

Meltz ML (2003) Radiofrequency exposure and mammalian cell toxicity, genotoxicity, and transformation. Bioelectromagnetics Suppl 6:S196–213

Moulder JE, Foster KR, Erdreich LS, McNamee JP (2005) Mobile phones, mobile phone base stations and cancer: a review. Int J Radiat Biol 81:189–203

Paulraj R, Behari J (2006) Single stranded DNA breaks in rat brain cells exposed to microwave radiation. Mutat Res 596:76–80

Phillips JL, Ivaschuk O, Isida-Jones T, Jones RA, Campbell-Beachler M, Haggren W (1998) DNA damage in Molt-4 T-lymphoblastoid cells exposed to cellular telephone radiofrequency fields. Bioelectrochem. Bioenergetics 45:103–110

Repacholi MH, Basten A, Gebski V, Noonan D, Finnie J, Harris AW (1997) Lymphomas in Eμ-*Pim1* transgenic mice exposed to pulsed 900 MHz electromagnetic fields. Radiat Res 147:631–640

Roti Roti JL, Malyapa RS, Bisht KS, Ahern EW, Moros EG, Pickard WF, Straube WL (2001) Neoplastic transformation in C3H10T1/2 cells after exposure to 835.62 MHz FDMA and 847.74 MHz CDMA radiations. Radiat Res 155:239–247

Royal Society of Canada (1999) A review of the potential health risks of radiofrequency fields from wireless telecommunication devices. An expert panel report prepared at the request of the Royal Society of Canada for Health Canada. http://www.rsc.ca//files/publications/expert_panels/RF//RFreport-en.pdf

Sakuma N, Komatsubara Y, Takeda H, Hirose H, Sekijima M, Nojima T, Miyakoshi J (2006) DNA strand breaks are not induced in human cells exposed to 2.1425 GHz band CW and W-CDMA modulated radiofrequency fields allocated to mobile radio base stations. Bioelectromagnetics 27:51–57

Sarker S, Ali S, Behari J (1994) Effect of low power microwave on the mouse genome: A direct DNA analysis. Mutat Res 320:141–147

Saunders RD, Kowalczuk CI, Beechey CV, Dunford R (1988) Studies of the induction of dominant lethals and translocations in male mice after chronic exposure to microwave radiation. Int J Radiat Biol Phys Chem Med 53:983–992

Shirai T, Kawabe M, Ichihara T, Fujiwara O, Taki M, Watanabe S, Wake K, Yamanaka Y, Imaida K, Asamoto M, Tamano S (2005) Chronic exposure to a 1.439 GHz electromagnetic field used for cellular phones does not promote N-ethylnitrosourea induced central nervous system tumors in F344 rats. Bioelectromagnetics 26:59–68

Sommer AM, Streckert J, Bitz AK, Hansen VW, Lerchl A (2004) No effects of GSM-modulated 900 MHz electromagnetic fields on survival rate and spontaneous development of lymphoma in female AKR/J mice. BMC Cancer 4:77

Stronati L, Testa A, Moquet J, Edwards A, Corelli E, Villani P, Marino C, Fresegna AM, Appolloni M, Lloyd D (2006) 935 MHz cellular phone radiation. An in vitro study of genotoxicity in human lymphocytes. Int J Radiat Biol 82:339–346

Sykes PJ, McCallum BD, Bangay MJ, Hooker AM, Morley AA (2001) Effect of exposure to 900 MHz radiofrequency radiation on intrachromosomal recombination in pKZ1 mice. Radiat Res 156:495–502

Tice RR, Hook GG, Donner M, McRee DI, Guy AW (2002) Genotoxicity of radiofrequency signals. I. Investigation of DNA damage and micronuclei induction in cultured human blood cells. Bioelectromagnetics 23:113–126

Toler JC, Shelton WW, Frei MR, Merritt JH, Stedham MA (1997) Long-term, low-level exposure of mice prone to mammary tumors to 435 MHz radiofrequency radiation. Radiat Res 148:227–234

Trosic I, Busljeta I, Kasuba V, Rozgaj R (2002) Micronucleus induction after whole-body microwave irradiation of rats. Mutat Res 521:73–79

Trosic I, Busljeta I, Modlic B (2004) Investigation of the genotoxic effect of microwave irradiation in rat bone marrow cells: in vivo exposure. Mutagenesis 19:361–364

Utteridge TD, Gebski V, Finnie JW, Vernon-Roberts B, Kuchel TR (2002) Long-term exposure of Eμ-*Pim1* transgenic mice to 898.4 MHz microwaves does not increase lymphoma incidence. Radiat Res 158:357–364

Verschaeve L (2005) Genetic effects of radiofrequency radiation (RFR). Toxicol Appl Pharmacol 207:S336–S341

Verschaeve L, Heikkinen P, Verheyen G, Van Gorp U, Boonen F, Vander Plaetse F, Maes A, Kumlin T, Maki-Paakkanen J, Puranen, Juutilainen J (2006) Investigation of co-genotoxic effects of radiofrequency electromagnetic fields in vivo. Radiat Res 165:598–607

Vijayalaxmi, Obe G (2004) Controversial cytogenetic observations in mammalian somatic cells exposed to radiofrequency radiation. Radiat Res 162:481–496

Vijayalaxmi, Frei MR, Dusch SJ, Guel V, Meltz ML, Jauchem JR (1997) Frequency of micronuclei in the peripheral blood and bone marrow of cancer-prone mice chronically exposed to 2450 MHz radiofrequency radiation. Radiat Res 147:495–500

Vijayalaxmi, Frei MR, Dusch SJ, Guel V, Meltz ML, Jauchem JR (1998) Correction of an error in calculation in the article "Frequency of micronuclei in the peripheral blood and bone marrow of cancer-prone mice chronically exposed to 2450 MHz radiofrequency radiation" [Radiat. Res. 147:495–500]. Radiat Res 149:308–312

Vijayalaxmi, Leal BZ, Szilagyi M, Prihoda TJ, Meltz ML (2000) Primary DNA damage in human blood lymphocytes exposed in vitro to 2450 MHz radiofrequency radiation. Radiat Res 153:479–486

Vijayalaxmi, Pickard WF, Bisht KS, Leal BZ, Meltz ML, Roti Roti JL (2001a) Cytogenetic studies in human blood lymphocytes exposed in vitro to radiofrequency radiation at a cellular telephone frequency (835.62 MHz, FDMA). Radiat Res 155:113–121

Vijayalaxmi, Bisht KS, Pickard WF, Meltz ML, Roti Roti JL, Moros EG (2001b) Chromosome damage and micronucleus formation in human blood lymphocytes exposed in vitro to radiofrequency radiation at a cellular telephone frequency (847.74 MHz, CDMA). Radiat Res 156:430–432

Vijayalaxmi, Pickard WF, Bisht HS, Prihoda TJ, Meltz ML, La Regina MC, Roti Roti JL, Straube WL, Moros EG (2001c) Micronuclei in the peripheral blood and bone marrow cells of rats exposed to 2450 MHz radiofrequency radiation. Int J Radiat Biol 77:1109–1115

Vijayalaxmi, Sasser LB, Morris JE, Wilson BW, Anderson LE (2003) Genotoxic potential of 1.6 GHz wireless communication signal: In vivo two-year bioassay. Radiat Res 159:558–564

Vijayalaxmi, McNamee JP, Scarfi MR (2006) Comments on: "DNA strand breaks" by Diem et al. [Mutat. Res. 583 (2005) 178-183] and Ivancsits et al. [Mutat. Res. 583 (2005) 184-188]. Mutat Res 603:104–106

Wang J, Sakurai T, Koyama S, Komatubara Y, Suzuki Y, Taki M, Miyakoshi J (2005a) Effects of 2450 MHz electromagnetic fields with a wide range of SARs on methylcholanthrene-induced transformation in C3H10T1/2 cells. J Radiat Res 46:351–361

Wang B, He J, Jin L, Lu D, Zheng W, Lou J, Deng H (2005b) Studying the synergistic damage effects induced by 1.8 GHz radiofrequency field radiation (RFR) with four chemical mutagens on human lymphocyte DNA using comet assay in vitro. Mutat Res 578:149–157

Wu RY, Chiang H, Shao BJ, Li NG, Fu YD (1994) Effects of 2.45 GHz microwave radiation and phorbol ester 12-O-tetradecanoylphorbol-13-acetate on dimethylhydrazine-induced colon cancer in mice. Bioelectromagnetics 15:531–538

Yu D, Shen Y, Kuster N, Fu Y, Chiang H (2006) Effects of 900 MHz GSM wireless communication signals on DMBA-induced mammary tumors in rats. Radiat Res 165:174–180

Zhang M-B, He J-L, Jin L-F, Lu D-Q (2002) Study of low-intensity 2450 MHz microwave exposure enhancing the genotoxic effects of mitomycin C using micronucleus test and comet assay in vitro. Biomed Environ Sci 15:283–290

Zeni O, Chiavoni AS, Sannino A, Antolini A, Forigo D, Bersani F, Scarfi MR (2003) Lack of genotoxic effects (micronucleus induction) in human lymphocytes exposed in vitro to 900 MHz electromagnetic fields. Radiat Res 160:152–158

Zeni O, Romano M, Perrotta A, Lioi MB, Barbieri R, d'Ambrosio G, Massa R, Scarfi MR (2005) Evaluation of genotoxic effects in human peripheral blood leukocytes following an acute in vitro exposure to 900 MHz radiofrequency fields. Bioelectromagnetics 26:258–265

Zook BC, Simmens SJ (2001) The effects of 860 MHz radiofrequency radiation on the induction or promotion of brain tumors and other neoplasms in rats. Radiat Res 155:572–583

Zook BC, Simmens SJ (2006) The effects of pulsed 860 MHz radiofrequency radiation on the promotion of neurogenic tumors in rats. Radiat Res 165:608–615

Zotti-Martelli L, Peccatori M, Scarpato R, Migliore L (2000) Induction of micronuclei in human lymphocytes exposed in vitro to microwave radiation. Mutat Res 472:51–58

Zotti-Martelli L, Peccatori M, Maggini V, Ballardin M, Barale R (2005) Individual responsiveness to induction of micronuclei in human lymphocytes after exposure in vitro to 1800-MHz microwave radiation. Mutat Res 582:42–52

29 Radiofrequency Radiation and Replication Studies

Maria Rosaria Scarfì and Ferdinando Bersani

Abstract

The induction of genotoxic effects after exposure to radiofrequency radiation represents one of the important aspects in the risk assessment of human exposure because of a close correlation between DNA damage and carcinogenesis. Most of the papers on this topic, published in peer-reviewed scientific journals, refer to experiments conducted by independent investigations who employed different experimental conditions, such as exposure characteristics, biological material examined as well as laboratory techniques. In particular, genotoxicity has been evaluated using several cytogenetic techniques, such as chromosomal aberrations, sister-chromatid exchanges, micronuclei and single-cell gel electrophoresis/comet assay. Just a few studies were "confirmatory," carried out adopting exactly the same experimental conditions. In this chapter the studies reported in the literature on the genotoxic potential of *in vivo* and *in vitro* exposure to radiofrequency radiation where several authors employed similar experimental conditions are compared and discussed.

29.1 Introduction

The aim of this chapter is to compare and discuss the studies reported in the literature where several authors employed similar experimental conditions to investigate the genotoxic potential of *in vivo* and *in vitro* exposure to radiofrequency radiation (RFR) using the classical (chromosomal aberrations, CA, sister-chromatid exchanges, SCE, and micronuclei) and more recent (single-cell gel electrophoresis/comet assay) cytogenetic techniques. Genotoxicity represents one of the important aspects in the risk assessment of human exposure to RFR because of a close correlation between DNA damage and carcinogenesis. Most of the *in vivo* and *in vitro* studies which have been published in peer-reviewed scientific journals were conducted by independent investigators. A small number of these were 'replicate' studies conducted under similar experimental conditions, while just a few of them were 'confirmatory' studies which adopted exactly the same experimental conditions. In all of these studies, several variables were included in the experimentation, viz.

Günter Obe and Vijayalaxmi (Eds.)
Chromosomal Alterations: Methods, Results and Importance in Human Health

RFR exposure characteristics, biological material examined as well as laboratory techniques. Moreover, not all researchers have provided detailed information on the experimental procedures used to enable independent investigators to replicate and confirm their observations. In some publications the experimental protocols were not adequately described. This makes it extremely difficult to evaluate whether or not the replicate and confirmatory studies were 'real' replications and confirmations.

29.2 DNA Strand Breaks

Lai and Singh applied the comet assay to evaluate DNA damage in brain cells of Sprague-Dawley rats which were exposed for 2 h to 2,450 MHz RFR in a cylindrical waveguide. The authors found no significant increase in DNA single-strand breaks when the alkaline comet assay was applied immediately after exposure to pulsed-wave (PW; 2 μs width, 500 pulses/s) RFR, while a specific absorption rate (SAR) dependent (0.6 and 1.2 W/kg) increase in single-strand breaks was observed both immediately and at 4 h after exposure to 2,450 MHz continuous wave (CW) RFR (Lai and Singh 1995). In a second paper the authors confirmed their previous results on single-strand breaks and also reported the increased DNA double-strand breaks at 4 h after exposure to both CW and PW RFR (Lai and Singh 1996). Malyapa et al. (1998) repeated the experiments of Lai and Singh by exposing male Sprague-Dawley rats to 2,450 MHz RFR for 2 h in a cylindrical waveguide system. The animals were killed immediately and at 4 h after RFR or sham exposure. Cells from cerebral cortex and hippocampus were used to perform the alkaline comet assay. The results did not indicate any significant differences in single-strand breaks between RFR-exposed and sham-exposed rats. As an extension, the latter authors also applied two methods of euthanasia, viz. CO_2 asphyxia or decapitation by guillotine. No significant increase in DNA strand breaks were observed between RFR-exposed and sham-exposed controls. Malyapa et al. (1997a) also performed in vitro experiments where cultured mammalian cells were exposed to RFR. Exponentially growing human glioblastoma U87MG cells and rodent C3H 10T1/2 cells were exposed for 2 h to 2,450 MHz RFR (CW) in a radial transmission line at a calculated SAR of 0.7 and 1.9 W/kg. The cells were examined immediately after RFR exposure as well as at 4 h after exposure. No significant differences in DNA single-strand breaks were observed between RFR-exposed and sham-exposed cells. Malyapa et al. (1997b) also investigated DNA strand breaks in the same cell types (in plateau phase or in exponential growth stage) which were exposed for 2, 4 or 24 h to 835.62 MHz frequency modulated CW (FMCW) RFR and 847.74 MHz code division multiple access (CDMA) RFR, at a SAR of 0.6 W/kg. No significant RFR exposure effect on DNA single-strand breaks was observed in both plateau phase cells and in exponentially growing cells.

It is important to consider here the differences in the comet assay protocols used by Lai and Sing (1995, 1996) and Malyapa et al. (1997a, 1998) in the above mentioned *in vivo* and *in vitro* studies. Malyapa et al. (1997a, 1998) used a protocol which did not include the treatment of the cells with proteinase K, while Lai and Singh (1995, 1996) included the proteinase K digestion of the cells which removes possible DNA–protein cross-links that hinder the movement of DNA strand breaks into the comet tail. To address this issue, Lagroye et al (2004) used both methods (with and without proteinase K treatment of the cells) and evaluated DNA strand breaks, alkali-labile lesions and DNA–protein or DNA–DNA cross-links in one cell type used by Malyapa et al. (1997b). C3H 10T1/2 cells were exposed for 2 h to 2,450 MHz CW RFR at a SAR of 1.9 W/kg. Cells treated with 1 mM cisplatin, a well-known DNA cross-link inducer, were included as positive controls. The results obtained indicated no significant differences in DNA damage between RFR-exposed and sham-exposed cells and confirmed the observations of Malyapa et al. (1997b). These findings also indicate that the slight differences in the protocols adopted to perform the comet assay were not responsible for the earlier contradictory results (Lagroye et al. 2004).

In other investigations, Phillips et al. (1998) evaluated the induction of DNA single-strand breaks in Molt-4 T-lymphoblastoid cells exposed for 2, 3 and 21 h to 813.5 MHz (Integrated Digital Enhanced Network, iDEN) and 836.55 MHz (time-division multiple access, TDMA) RFR at SAR values of 2.4 and 26 mW/kg. The exposures/sham exposures were conducted in transverse electromagnetic (TEM) cells under controlled conditions of temperature. The authors reported an increase, decrease or no significant effects in DNA damage, depending on the experimental parameters investigated (signal, SAR, exposure duration). The authors suggested that the observed effects could be due to a balance between RFR-induced damage and DNA repair. Recently, Hook et al. (2004) repeated the experiments of Phillips et al. (1998) using the same Molt-4 T-lymphoblastoid cells and the same RFR-exposure conditions in terms of frequency, signals, SAR values, exposure duration and biological assay. The induction of DNA damage was evaluated by means of the alkaline comet assay, performed immediately after RFR exposure as well as at 4 h after exposure. For all the experimental conditions tested, Hook et al. (2004) failed to confirm the observations of Phillips et al. (1998).

29.3 Micronucleus Test

Tice et al. (2002) used the micronucleus (MN) technique to determine the genotoxic potential in human blood lymphocytes (G_0, unstimulated) exposed in vitro to several RFR (837 and 1,909.8 MHz) signals, such as CW, global system for mobile communications (GSM), TDMA and CDMA. RFR exposures for 3 and 24 h were conducted in a TEM cell exposure system. No significant induction of MN was observed when the cells were exposed to RFR at a SAR below 1 W/kg; however, a consistent and reproducible increase in the incidence of MN was detected when the cells were exposed for 24 h a SAR of 5 or 10 W/kg. The increased incidence of MN was independent of the RFR signal employed. Tice et al. (2002) stated that the increased MN could be due to the localized heat production with consequent temperature elevation during RFR exposure.

Several other research groups examined the incidence of MN in human blood lymphocytes under conditions similar to those used by Tice et al (2002), and with appropriate temperature controls during RFR exposure. No significant increase in the incidence of MN (and also of CA) was observed when human blood lymphocytes were exposed for 24 h to 835.5 MHz frequency division multiple access (FDMA) RFR or 847.7 MHz (CDMA) RFR (Vijayalaxmi et al. 2001a, b). McNamee et al. (2002a, b, 2003) also reported similar 'negative' results in human blood lymphocytes exposed for 24 h to 1,900 MHz (CW, PW) at SAR values ranging from 0 to 10 W/kg. In a more recent study, Tice et al. (2004) could not confirm their original observations when some of their experimental conditions were modified, such as the exposure system (radial transmission line instead of a TEM cell), the containers for blood exposures (Petri dishes instead of tubes) and the number of donors employed in the study (three instead of one). In the latter study, Tice et al. (2004) also evaluated the effect of maintaining unstimulated whole-blood cultures at different temperatures (35, 39, 42°C) for 24 h before stimulation with phytohemagglutinin, with and without RFR exposure. No increase in the incidence of MN was observed under these experimental conditions. The latter data indicated that the increase in temperature might not be responsible for the genotoxic effects observed in their previous investigation. Bisht et al. (2002) confirmed the lack of MN induction in C3H 10T1/2 cells (exponentially growing and plateau-phase cells) exposed for 3, 8, 16 or 24 h to 835.5 MHz (FDMA) and 847.7 MHz (CDMA) RFR at SARs ranging from 3.2 to 5.1 W/kg. More recently, Scarfi et al. (2006) repeated the experiment of Tice et al (2002) using a wire patch cell with four channels and a double-blind procedure. The data indicated an absence of MN induction in human peripheral blood lymphocytes from ten healthy donors exposed for 24 h to 900 MHz (TDMA in GSM modulation) RFR at lower SAR levels (0, 1, 5 or 10 W/kg peak values).

29.4 Combination Studies – RFR and Other Mutagens

The possibility that RFR could enhance the genotoxic effect of well-known mutagens was investigated by several researchers. Kerbacher et al. (1990) evaluated the induction of CA in Chinese hamster ovary cells after 2-h exposure to 2,450 MHz RFR at a SAR of 33.8 W/kg. They also evaluated the effects of simultaneous exposures to RFR and two genotoxic chemicals, viz. mitomycin C (MMC) and adriamycin (ADR). The results indicated no significant induction of CA in cells exposed to RFR alone or in cells exposed to RFR plus MMC/ADR. Under the same experimental conditions, Ciaravino et al (1991) confirmed the absence of effects of RFR alone and in combination with ADR, by evaluating the incidence of SCE.

Maes et al. (1996) conducted combination experiments using RFR and MMC. In the first investigation, human whole-blood lymphocytes (G_0, unstimulated) were exposed for 2 h to 954 MHz RFR emitted by an antenna from a GSM base station. During the exposure samples were positioned 5 cm from the antenna and the power input was 15 W, with a calculated SAR of 1.5 W/kg. The authors reported no genotoxic effects in terms of SCE induction in RFR-exposed samples, but a clear and reproducible synergistic effect was detected when RFR-exposed samples were co-cultured with MMC. The authors hypothesized that there may exist a microwave-induced sensitization of RFR-exposed cells when they are treated with MMC. Maes et al (1997) repeated the experiments where human whole blood (G_0, unstimulated) was exposed for 2 h to 935.2 MHz RFR in a TEM transmission line that allowed a homogeneous radiation field. The power input was lower that in the previous experiments (4.5 W), with a SAR of 0.3–0.4 W/kg. The genotoxic effect of RFR was investigated by evaluating three cytogenetic end points: CA, SCEs and single-strand breaks. The combination effect of RFR was determined using SCE frequency. The authors confirmed the lack of effects due to RFR alone, while a slight increase in SCE frequency was detected in cells exposed to RFR plus MMC; the increase in SCE frequency was lower than that reported in the previous experiments (Maes et al. 1996) and the authors hypothesized that the difference in the two sets of experiments may be due to the mode of exposure (continuous vs. pseudorandom). In further experiments, Maes et al. (2001) evaluated the induction of genotoxicity in human lymphocytes exposed to 900 MHz RFR (GSM) with and without MMC treatment (SCE) and X-irradiation to determine the incidence of SCE and CA, respectively. Three modes of exposure were applied: continuous carrier wave, pseudorandom signal (corresponding to a GSM signal as obtained during conversation) and dummy burst signal (corresponding to the signal in standby mode). In the experiments devoted to evaluation of the effects of RFR alone, for each exposure condition the calculated SARs were 0.4, 2.0, 3.5, 5.5 and 10 W/kg. The possible synergistic effect with MMC or X-rays was examined at SAR values of 2.0 and 3.5 W/kg. In all the experimental conditions investigated neither RFR exposure nor co-exposures induced genotoxic effects.

More recently, Stronati et al. (2006) confirmed the absence of synergistic genotoxic effects in human peripheral blood lymphocytes (G_0, unstimulated) exposed for 24 h to 935 MHz RFR alone or coexposed with X-rays. The tests employed were alkaline comet assay, CA, MN and SCE. The experimental conditions adopted were similar but not identical to those tested by Maes et al. (2001). In particular, exposures to a 935 MHz GSM basic signal were carried out at SAR values of 1 or 2 W/kg, while 1 Gy/min was applied immediately before or after RFR.

29.5 Discussion

The effects of RFR on biological systems are generally divided into two categories, viz. thermal and nonthermal. While there is sufficient knowledge of thermal effects of RFR, the nature of alleged nonthermal effects and even their existence is still largely controversial. On the other hand, if RFR exposure has genotoxic potential, it could be considered more due to a nonthermal effect or at the edge of a thermal (athermal) effect. In this context, owing to the absence of convincing and comprehensive hypotheses about the biophysical mechanisms underlying the nonthermal effects of RFR exposure, it is difficult to conduct hypothesis-driven experiments. The choice of the physical parameters of the RFR exposures (frequency, amplitude, modulation, etc.) are dictated either by an arbitrary choice among the huge variety of combinations (we can name this approach as a 'fishing' approach) or by the physical characteristics of the RFR field signals more often used in daily-life applications, such as in wireless technology.

The possibility that RFR can affect DNA integrity has often been challenged on the physical consideration that RFR-associated photon energy is too low to break chemical bonds. Indeed, if real genotoxic effects could be demonstrated, the mechanisms underlying such effects could be due a more subtle though as yet unknown mechanism. For example, it could be due to an indirect effect of RFR on the enzymes involved in the repair of DNA damage or on other processes potentially amplified by the cell machinery. In the context of this low plausibility, the experimental evidence for a significant genotoxic effect of RFR should be sound and unquestionable. Besides, high-quality experiments together with a high level of reproducibility is required. Furthermore, independent replications of positive findings are more a cogent requirement than an academic exercise.

In general, scientists do not like to simply 'repeat' experiments conducted by other investigators, since this appears like a pure technical and uncreative activity; they prefer to extend the previous observations by the examination of other genotoxic end-points or, at least, to significantly improve some technical details. As described above, true 'replication' and 'confirmation' studies on the genotoxic effects of RFR are rare.

It is important to evaluate the replication experiments of Lai and Singh (1995, 1996) with respect to DNA strand breaks. The investigation of Malyapa et al. (1998) was a replication with slight differences in the comet assay protocol (inclusion of proteinase K) but included a methodological extension of the use of two different kinds of euthanasia. Perhaps a true replication study was carried out by Lagroye et al (2004) in which the laboratory protocols used by the earlier two research groups were compared. The data confirmed those of Malyapa et al (1998) and strongly suggest an absence of genotoxic effects of RFR exposure. The 'positive' observations of Lai and Singh (1995, 1996) could thus be due to unknown reasons and not RFR exposure. The *in vitro* studies of Malyapa (1997a, b) represent a sort of extension or a parallel confirmation of their *in vivo* experiments. Bisht et al. (2002) extended these *in vitro* observations of Malyapa et al (1998) which included RFR exposure at higher SAR levels and another genotoxic end point, viz. MN. The data from these studies again were 'negative'.

There were several 'partial replication' investigations (Vijayalaxmi et al. 2001a, b; McNamee et al. 2002a, b, 2003) which did not confirm the original 'positive' MN data in human blood lymphocytes exposed to RFR at 5 or 10 W/kg (Tice et al. 2002). This again points to 'potential differences' in RFR exposure conditions used in the study of Tice et al. (2002) or to other unknown reasons but not directly related to RFR exposure.

It is difficult to draw definitive conclusions from replication/confirmation experiments. If two series of independent experiments conducted using similar experimental conditions report contradictory results, the final conclusion remains uncertain. One of the ways to resolve this uncertainty would be to find a reasonable rationale for different outcomes, which is often difficult to achieve. At the moment, only a restricted number of experiments have been replicated. It is suggested that a meta-analysis should be performed with all of the available cytogenetic data to resolve the open question of the genotoxic potential of RFR exposure. It is also desirable to initiate an international collaborative investigation with the participation of experienced cytogeneticists.

References

Bisht KS, Moros EG, Straube WL, Baty JD, Roti Roti JL (2002) The effect of 835.62 MHz FDMA or 847.74 MHz CDMA modulated radiofrequency radiation on the induction of micronuclei in C3H 10T(1/2) cells. Radiat Res 157:506–515

Ciaravino V, Meltz ML, Erwin DN (1991) Absence of a synergistic effect between moderate-power radio-frequency electromagnetic radiation and Adriamycin on cell-cycle progression and sister-chromatid exchange. Bioelectromagnetics 12:289–298

Hook GJ, Zhang P, Lagroye I, Li L, Higashikubo R, Moros EG, Straube WL, Pickard WF, Baty JD, Roti Roti JL (2004) Measurement of DNA damage and apoptosis in Molt-4 cells after in vitro exposure to radiofrequency radiation. Radiat Res 161:193–200

Kerbacher JJ, Meltz ML, Erwin DN (1990) Influence of radiofrequency radiation on chromosome aberrations in CHO cells and its interaction with DNA-damaging agents. Radiat Res 123:311–319

Lagroye I, Hook GJ, Wettring BA, Baty JD, Moros EG, Straube WL, Roti Roti JL (2004) Measurements of alkali-labile DNA damage and protein-DNA crosslinks after 2450 MHz microwave and low-dose gamma irradiation in vitro. Radiat Res 161:201–214

Lai H, Singh NP (1995) Acute low-intensity microwave exposure increases DNA single strand breaks in rat brain cells. Bioelectromagnetics 16:207–210

Lai H, Singh NP (1996) Single- and double-strand DNA breaks in rat brain cells after acute exposure to radiofrequency electromagnetic radiation. Int J Radiat Biol 69:513–521

Maes A, Collier M, Slaets D, Verschaeve L (1996) 954 MHz microwaves enhance the mutagenic properties of mitomycin C. Environ Mol Mutagen 28:26–30

Maes A, Collier M, Van Gorp U, Vandoninck S, Verschaeve L (1997) Cytogenetic effects of 935.2-MHz (GSM) microwaves alone and in combination with mitomycin C. Mutat Res 393:151–156

Maes A, Collier M, Verschaeve L (2001) Cytogenetic effects of 900 MHz (GSM) microwaves on human lymphocytes. Bioelectromagnetics 22:91–96

Malyapa RS, Ahern EW, Straube WL, Moros EG, Pickard WF, Roti Roti JL (1997a) Measurement of DNA damage after exposure to 2450 MHz electromagnetic radiation. Radiat Res 148:608–617

Malyapa RS, Ahern EW, Straube WL, Moros EG, Pickard WF, Roti Roti JL (1997b) Measurement of DNA damage after exposure to electromagnetic radiation in the cellular phone communication frequency band (835.62 and 847.74 MHz). Radiat Res 148:618–627

Malyapa RS, Ahern EW, Bi C, Straube WL, La Regina M, Pickard WF, Roti Roti JL (1998) DNA damage in rat brain cells after in vivo exposure to 2450 MHz electromagnetic radiation and various methods of euthanasia. Radiat Res 149:637–645

McNamee JP, Bellier PV, Gajda GB, Miller SM, Lemay EP, Lavallee BF, Marro L, Thansandote A (2002a) DNA damage and micronucleus induction in human leukocytes after acute in vitro exposure to a 1.9 GHz continuous-wave radiofrequency field. Radiat Res 158:523–533

McNamee JP, Bellier PV, Gajda GB, Lavallee BF, Lemay EP, Marro L, Thansandote A (2002b) DNA damage in human leukocytes after in vitro exposure to a 1.9 GHz pulse-modulated radiofrequency field. Radiat Res 158:534–537

McNamee JP, Bellier PV, Gajda GB, Lavallee BF, Marro L, Lemay EP, Thansandote A (2003) No evidence for genotoxic effects from 24 h exposure of human leukocytes to 1.9 GHz radiofrequency fields. Radiat Res 159:693–697

Phillips JL, Ivaschuk O, Ishida-Jones T, Jones RA, Campbell-Beachler M (1998) DNA damage in Molt-4 T-lymphoblastoid cells exposed to cellular telephone radiofrequency fields in vitro. Bioelectrochem Bioenerg 45:103–110

Scarfì MR, Fresegna AM, Villani P, Pinto R, Marino C, Sarti M, Sannino A, Altavista P, Lovisolo GA (2006) Exposure to radiofrequency radiation (900 MHz, GSM signal) does not affect micronucleus frequency and cell proliferation in human peripheral blood lymphocytes. Radiat Res 165:655–663

Stronati L, Testa A, Moquet J, Edwards A, Cordelli E, Villani P, Marino C, Fresegna AM, Appolloni M, Lloyd D (2006) 935 MHz cellular phone radiation. An in vitro study of genotoxicity in human lymphocytes. Int J Radiat Biol 82:339–346

Tice RR, Hook GG, Donner M, McRee DI, Guy AW (2002) Genotoxicity of radiofrequency signals. I. Investigation of DNA damage and micronuclei induction in cultured human blood cells. Bioelectromagnetics 23:113–126

Tice RR, Baldetti C, Kehl M, Levesque B, Winters J, Roti Roti J, Moros E, Mueller J, Cha B (2004) Effect of exposure to radiofrequency radiation (837 MHz RF) emitted by a TDMA cellular telephone on the frequency of micronuclei in cultured human blood lymphocytes. Tutorial session at the 26th annual meeting of the Bioelectromagnetic Society on research related to the effect of radiofrequency energy on micronucleus formation, Washington, 25 June 2004

Vijayalaxmi, Pickard WF, Bisht KS, Leal BZ, Meltz ML, Roti Roti JL, Straube WL, Moros EG (2001a) Cytogenetic studies in human blood lymphocytes exposed in vitro to radiofrequency radiation at a cellular telephone frequency (835.62 MHz, FDMA). Radiat Res 155:113–121

Vijayalaxmi, Bisht KS, Pickard WF, Meltz ML, Roti Roti JL, Moros EG (2001b) Chromosome damage and micronucleus formation in human blood lymphocytes exposed in vitro to radiofrequency radiation at a cellular telephone frequency (847.74 MHz, CDMA). Radiat Res 156:430–432

30 Chromosome Analysis in Cancer Patients: Applications and Limitations

SIDDHARTH G. ADHVARYU

Abstract

Since altered genetic mechanisms lead to the development of cancer, chromosome analysis plays a very important role in diagnosis and treatment-monitoring of patients with various types of cancer. The precise methodology involves several steps. Short-term cell culture, arrest of cell division in metaphase, hypotonic treatment, and cell harvest help in getting a cell suspension rich in metaphase cells. Slide preparation and staining protocols (G-banding) allow microscopic analysis and precise identification of the chromosome complement. Malignant growth involving the hematopoietic system is categorized as myeloid or lymphoid, depending upon the origin of the cells. The myeloid neoplasms are further subgrouped into myeloproliferative disorders, myelodysplastic syndromes, and acute myeloid leukemia (AML). In the leukemic cells of a chronic myeloid leukemia (CML) patient, the presence of t(9;22)(q34;q11.2) determines the diagnosis. Just like CML, very specific chromosome abnormalities help in diagnosis of different types of myeloid leukemias. The lymphoid neoplasms include acute lymphoblastic leukemia (ALL), chronic lymphoproliferative disorders, and lymphoma. The detection of t(12;21)(p13;q22) or a chromosome count of more than 50 chromosomes is considered a good prognosis in children with ALL. Similarly, the presence of a specific chromosome abnormality helps in the classification of and treatment for lymphoma. The information on chromosome abnormalities in solid tumors has also been gaining importance in recent years. To address the limitation of resolution and to bridge the gap between classical cytogenetic analysis and molecular diagnosis, techniques such as fluorescence in situ hybridization (FISH), comparative genomic hybridization, multiplex FISH, and spectral karyotyping are now available and are discussed briefly in this chapter.

30.1 Introduction

Cancer develops by the clonal expansion of a cell having altered genetic mechanisms. These genetic defects are restricted to the tumor tissue and are not present in other cells of the body. The defects may be very subtle at gene

Günter Obe and Vijayalaxmi (Eds.)
Chromosomal Alterations: Methods, Results and Importance in Human Health

level or they may be easily detectable as chromosomal changes. Both benign and malignant tumors have been shown to exhibit chromosome abnormalities. The chromosomal changes can be numerical or structural (Mitelman et al. 2006) and have been classified into two main categories: primary and secondary abnormalities. Primary abnormalities are specific or causative for a specific tumor type, may be the only cytogenetic abnormality, and may play an important role in the earliest stages of tumor initiation. Secondary abnormalities are also nonrandom, but less disease specific, and are postulated to be later events contributing to the process of tumor progression. In the clinical practice, cytogenetic analysis is an invaluable tool for diagnosis, staging, prognosis, and treatment monitoring of hematological malignancies and many solid tumors.

A deleted G group chromosome described as Philadelphia (Ph) chromosome by Nowell and Hungerford (1960) in chronic myeloid leukemia (CML) was the first chromosome abnormality in any cancer. The rapid advances, however, started following the discovery of banding techniques by Casperson et al. (1969) and Seabright (1971) and with the availability of high-resolution banding protocols described by Yunis (1974). Since then a new era in cancer cytogenetics and a parallel progress in the application of molecular genetic techniques have not only increased our understanding of the pathogenetic mechanisms underlying the neoplastic process, but have also led to development of targeted therapies for CML (Druker 2001).

30.2 Methods

The basic requirement for chromosome analysis in any cancer tissue is the need to have actively dividing cells. If sufficient mitotic activity is present in the specimen collected (e.g., bone marrow, peripheral blood with circulating blasts, or fresh tumor tissue) cells can be harvested after a 1-h colcemid treatment or overnight incubation in the culture medium containing colcemid followed by cell harvest can be performed. If sufficient mitotic activity is not present, a short-term or long-term cell culture procedure is employed for propagating the cells. In clinical laboratories various protocols are used to obtain the needed cytogenetic information. These protocols mainly depend upon the tissue to be processed. When the cells are actively dividing, the cell division is blocked at metaphase stage by adding spindle poisons (e.g., colchicine, colcemid, vinblastine). The cells are subjected to hypotonic treatment so as to swell the cell and make the cell membrane easy to rupture when the cell suspension is dropped on a glass slide and to the fixation with acetic acid–alcohol (1:3) mixture to stop the subsequent enzyme degradation. The cell pellet obtained is optimally diluted and the suspension is dropped on cleaned slides. The optimal quality depends upon the humidity and temperature and many laboratories use a mechanical dropping chamber for this purpose. Then the slides are subjected to a combination of time and temperature treatment for aging before subjecting them to one of the banding protocols. The most commonly used is G-banding, which includes trypsin treatment followed by one of the nuclear stains (Giemsa, Wright, or Leishman) to reveal the characteristic banding patterns (Seabright 1971; Sumner 1971). Usually 15–20 metaphase cells are completely analyzed preferably from more than one culture source. This analysis is the most important step and the technologist performing the microscope analysis must have sufficient technical experience in identifying various abnormalities and in selection of cells for analysis. In many cancers the abnormal cells are not well banded and care must be exercised in not excluding them from analysis. It is desirable to have the involvement of more than one technologist in finishing analysis on any case. The technologist finishing the microscope analysis decides after close review if there are any clonal abnormalities. In recent years, multiple metaphase cells have been imaged using computerized imaging systems. Use of a darkroom has become almost obsolete since the late 1990s. Two karyotypes of the main clone and one of each side clone are prepared. If the analysis yields a lack of any clonal abnormalities, a careful review of clinical information is done by the qualified cytogeneticist to determine if further tests are needed. A final report is submitted to the referring physician explaining the details of findings and the clinical significance of these findings. It is impossible to encompass the methodology and applications of chromosome alterations in cancer patients in a few pages. Those interested in learning the details will find useful protocols in Barch et al. (1997), Keagle and Gersen (1999), and Rooney (2001).

Owing to technical reasons of easy tissue sampling and to the availability of repeat specimens, most cytogenetic knowledge initially came from hematological disorders. In terms of human morbidity and mortality, solid tumors are very important and our knowledge of cytogenetic changes in solid tumors has increased in recent years. In the following sections the cytogenetic aberrations have been grouped under headings of hematological disorders and solid tumors.

30.3 Hematologic Cancers

The French–American–British (FAB) classification had been used most extensively till the recently adopted WHO system (Jaffe et al. 2001) became available. According to the origin of the cells, hematologic neoplasms are described as myeloid or lymphoid neoplasms. The myeloid neoplasms have four major subgroups: (1) chronic myeloproliferative disorders (CMPDs), (2) myeloproliferative/myelodysplastic disorders, (3) myelodysplastic syndromes (MDS), and (4) acute myeloid leukemia (AML). The data on the lymphoid neoplasms (incidence of acute leukemia in different age groups) which are summarized

Table 30.1 Age distribution of acute myeloid leukemia (*AML*) and acute lymphocytic leukemia (*ALL*)

	Age groups	AML	ALL
Acute leukemia	Children	15%	85%
	Adults	80%	20%
Annual incidence	Children	1.2:100,000	9.8:100,000
	Age 50	3.5:100,000	12.8:100,000
	Age 70	15:100,000	
	Age 90	35:100,000	

in Table 30.1 indicate that acute lymphocytic leukemia (ALL) is predominantly a pediatric disease, whereas AML is more common in adults.

30.3.1 Myeloid Neoplasms

CMPDs are clonal stem cell disorders resulting in elevated white cell counts in the blood. CMPDs occur primarily in adults during their 50s and 70s. The classic CMPD is CML. This was the first malignancy reported to have a chromosomal marker in the form of t(9;22)(q34;q11.2). This rearrangement results in the formation of BCR/ABL chimeric gene and is present in 100% of CML cases. In about 2–5% of these CML cases, the translocation may not be visible at the chromosome level. However, the BCR/ABL rearrangement is detectable at the molecular level by polymerase chain reaction (PCR) or by fluorescence in situ hybridization (FISH). Juxtaposition of ABL to BCR results in enhanced tyrosine kinase activity and this knowledge led to the effective treatment for CML using the drug imanitib, which is a very effective inhibitor of the hybrid BCR/ABL protein product (Druker 2001). During the accelerated phase or blast crisis of CML, additional abnormalities such as +8, +Ph, iso(17q), or +19 are detected either alone or in combination (Heim and Mitelman 1995). The detection of chromosome abnormality in the case of CML establishes a marker for monitoring the treatment effectiveness, presence of minimal residual disease, and early prediction of relapse or progression to the accelerated phase. For treatment-monitoring, molecular techniques such as reverse transcription PCR have proven to be much more sensitive. Nonetheless, chromosome analysis is still the preferred technique for detecting the acute phase of CML or blast crisis of CLM where additional cytogenetic markers [+8, +19, iso(17q), or +Ph] are detected well before the clinical manifestation. Unlike CML, there are no specific chromosomal changes associated with the other myeloproliferative disorders. A variety of nonspecific changes are indeed observed with some frequency. These include deletions of the long arms of chromosomes 13 and 20, monosomy 7, and trisomies 8 and 9. Information about chromosome abnormalities in CMPDs is presented in Table 30.2.

Table 30.2 Chromosomal rearrangements in myeloproliferative disorders

	CML	CNL	CEL	PV	CIMF	ET
Specific chromosome abnormality	t(9;22)	None	None	None	None	None
Other common nonspecific abnormalities	+8[a], +Ph, iso(17q), +19	+8, +9, del(20q), del(11q)	+8, t(5;12)	+8, +9, del(20q), del(13q)	+8, del(20q),-7,del(7q), del(11q), del(13q)	+8, del(13q)

CEL chronic eosinophilic leukemia, *CIMF* chronic idiopathic myelofibrosis, *CNL* chronic neutrophilic leukemia, *ET* essential thrombocythemia, *PV* polycythemia vera
[a]Only during the acute phase/blast crisis of CML

The MDS are a heterogeneous group of disorders showing clonal stem cell abnormalities leading to dysplasia and disturbed hematopoesis in one or more myeloid cell lineages. This may result in the presence of some myeloblast (below the 20% threshold limit). Primary MDS usually occurs in older individuals; however, secondary MDS can occur at an earlier age as a result of effective chemotherapies for various other types of cancers. Among these are refractory anemia (RA), RA with ringed sideroblasts, RA with multilineage dysplasia, RA with excess blasts, unclassified MDS, and MDS due to 5q deletion. The latter, commonly identified as "5q– syndrome" is more common in women. As the name suggests the deletion of 5q is the sole abnormality and is associated with a longer survival rate. Additional cytogenetic abnormalities usually indicate progression to AML. According to the cytogenetic abnormalities detected, MDS is grouped as (1) normal cytogenetics, isolated del(5q), del(20q), and –Y as "good risk," (2) chromosome 7 abnormalities or more than three recurring abnormalities as "poor risk," and (3) all other chromosome abnormalities as "intermediate risk."

When the number of myeloblasts exceeds 20%, the condition is labeled as AML. Under WHO classification, AMLs have been grouped as (1) AML with recurrent genetic abnormality, (2) AML with multilineage dysplasia, (3) therapy-related or secondary AML/MDS, and (4) AML not otherwise categorized. This WHO system is based on morphology, immunophenotypic, genetic, and clinical features; however, the reader will find the FAB classification (AML M0–M7) being commonly used in laboratory settings. There is no scientific conflict between the two systems and those working solely in the laboratory setting will be better off knowing both systems. Different classes of myeloid neoplasms indicating the characteristic chromosome abnormalities and related gene rearrangements are summarized in Table 30.3.

AML with translocation involving 11q23 (MLL gene rearrangement) have been considered indicative of an adverse outcome, whereas t(8;21), t(15;17), and inv(16) in AML is considered good prognosis. During the disease

Table 30.3 Chromosomal rearrangements in myeloproliferative disorders

Chromosomal abnormality	Gene/s involved	Neoplasm
t(8;21)(q22;q22)	*AML1-ETO*	AML-M2, other AML
inv(16),t(16;16)	*CBFB-MYH11*	AML-M4Eo, other AML
t(15;17)(q24;q21)	*PML-RARa*	AML-M3 (APL)
11q23 translocation	*MLL*	MLL rearranged AML

progression in about 75–80% of the patients, secondary chromosome abnormalities develop. A loss of chromosome 7 or 7q deletion, loss of chromosome 5 or 5q deletion, 11q23 abnormalities, or 12p deletions are often detected as additional abnormalities during progression of myeloid neoplasms. Some of these abnormalities have been related to the exposure to the chemotherapeutic agents. In general, the presence of multiple clonal abnormalities is indicative of a more aggressive disease.

Some specific chromosome abnormalities in myeloid malignancies are presented in Fig. 30.1.

30.3.2 Lymphoid Neoplasms

These include ALLs (when 30% or more lymphoblasts are detected in the bone marrow), chronic lymphoproliferative disorders (CLPD) and lymphoma. Depending on the involvement of the abnormal cell type, lymphoid neoplasms are of B-cell or T-cell origin. The most common recurrent genetic abnormalities in ALL are indicated in Table 30.4.

ALL is a malignant proliferation of immature lymphoid cells in the bone marrow and is usually present in the peripheral circulation. Peripheral blood studies reveal anemia, thrombocytopenia, and usually increased numbers of lymphoblasts. Chromosomes from ALL cases often exhibit poor quality and

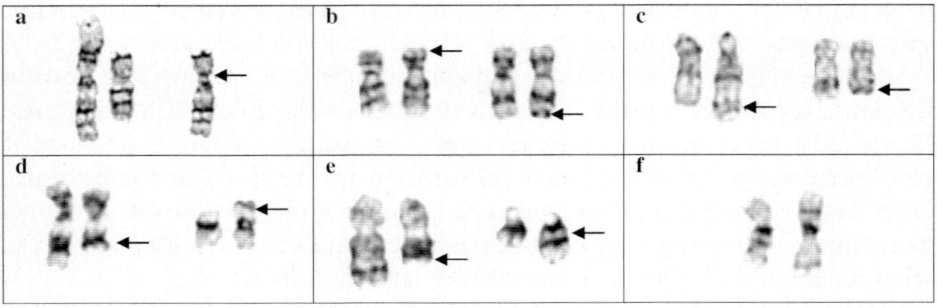

Fig. 30.1 Common abnormalities associated with acute myeloid leukemia. **a** Translocation (1;7)(q10;q10). **b** Translocation (9;11)(p21;q23). **c** Translocation (15;17)(q22;q11). **d** Translocation (11;19)(q23;p13). **e** Translocation (8;21)(q22;q22). **f** Inversion (16)(p13q22)

Table 30.4 Chromosomal rearrangements and genes involved in ALL

t(9;22)	*BCR-ABL1*	ALL
t(1;19)	*E2A-PBX1*	Pre-B ALL
t(12;21)	*TEL(ETV6)-AML1*	Common B ALL
t(4;11), 11(q23)	*MLL*	Infant ALL
t(8;14)	*c-MYC*	Burkitt/ALL-L3

poor staining, making them diagnostically challenging. Nevertheless, about two thirds of cases of ALL demonstrate recognizable cytogenetic changes, and the cytogenetic findings often have prognostic significance. Massive hyperdiploidy (50+ chromosomes), and especially with extra copies of chromosomes 4, 10, and 17 without structural anomalies, carries a good prognosis. Near-haploidy (a chromosome number of less than 30 and usually 23–28) is generally associated with a poor outcome. It is important not to overlook apparently broken metaphase spreads when performing the cytogenetic evaluation, as they could represent near-haploid cells. The translocation t(12;21)(p13;q22) is considered a good prognosis. This abnormality cannot be visualized cytogenetically; however, it can be easily confirmed with FISH probes. Another abnormality t(1;19)(q23;p13.3) is considered good prognosis with certain therapy regimens. Translocations involving a 14q11 breakpoint are seen with some frequency in T-cell ALL. Some examples are t(10;14)(q24;q11), t(11;14)(p13-15;q11), and t(8;14)(q24;q11). The T-cell receptor alpha (TCRA) locus resides at 14q11 and it appears that juxtaposition of it with a variety of oncogenes results in their activation. In general, T-cell ALL patients have a poorer prognosis than patients with B-cell ALL. The translocation t(9;22) (q34.1;q11) detected in ALL looks cytogenetically similar to that observed in CML; however, the breakpoint on chromosome 22 is different at the molecular level. This translocation is associated with an extremely poor prognosis and is seen more frequently as patient age increases. Any rearrangement involving 11q23 (MLL locus) seen in infants with ALL also carries a poor prognosis (Harrison 2001; Pui et al. 2004). When seen in adults, it is usually secondary to the treatment of their cancer with topoisomerase II inhibitors.

Among the CLPD, chronic lymphocytic leukemia (CLL) and multiple myeloma account for almost one third of all leukemias in the USA and Europe. CLL is a disease of middle-aged and elderly adults. The peripheral blood shows increased numbers of white blood cells, with mature lymphocytes predominating. Over 95% of CLLs are of B-cell origin; hence, the use of a B-cell mitogen is generally recommended for cell cultures. B-cell CLL has a good prognosis, whereas T-cell CLL carries a poor prognosis. CLL can transform, becoming an acute leukemia, most often ALL. Trisomy 12 and the translocation t(11;14)(q13;q32) are the most common abnormalities detected in B-cell CLL. Deletions involving the long arm of chromosome 13 and the short arm of chromosome 17 are fairly common and may require FISH analysis for detection.

The inversion, inv(14)(q11q32) is commonly seen in T-cell CLL as are other rearrangements involving q11 and q32 breakpoints on chromosome 14.

Multiple myeloma is a B-cell tumor in which there is a malignant proliferation of plasma cells that impairs bone marrow function. Multiple myeloma is also characterized by lytic bone lesions and impaired renal function. Plasma cells divide slowly and obviously care must be taken to improve the likelihood that the appropriate cells are being detected. FISH can be helpful by allowing large number of cells to be screened. Thirty to fifty percent of cases show cytogenetic changes including many numerical abnormalities, t(11;14)(q13;q32), 13q deletions, 6q deletions, or abnormalities of chromosome 1 (Harousseau et al. 2004).

Lymphomas can be broadly classifies as Hodgkin lymphoma and non-Hodgkin lymphoma. Hodgkin lymphoma is the most common lymphoma. Non-Hodgkin lymphoma is a complex group of neoplasms that over the years have been classified in numerous ways. Cytogenetic abnormalities are seen in 90% of cases, far greater than in any leukemia. The cytogenetics show highly specific and consistent cytogenetic changes; however, Burkitt lymphoma is an immunoglobulin-producing B-cell tumor. The Epstein–Barr virus might play a role in its development. Eighty percent of cases of Burkitt lymphoma show a specific translocation, t(8;14)(q24;q32), whereas the remaining 20% show involvement of 2p12 or 22q11.2, with 8q24 as variant translocations. These translocations juxtapose c-*myc* oncogene with the immunoglobulin gene promoter region, resulting in MYC activation. Secondary changes include abnormalities of the short arm of chromosome 1, seen in 30% of cases, and abnormalities involving a 13q34 breakpoint, seen in 15% of cases.

Some specific chromosome abnormalities in lymphoid malignancies are presented in Fig. 30.2.

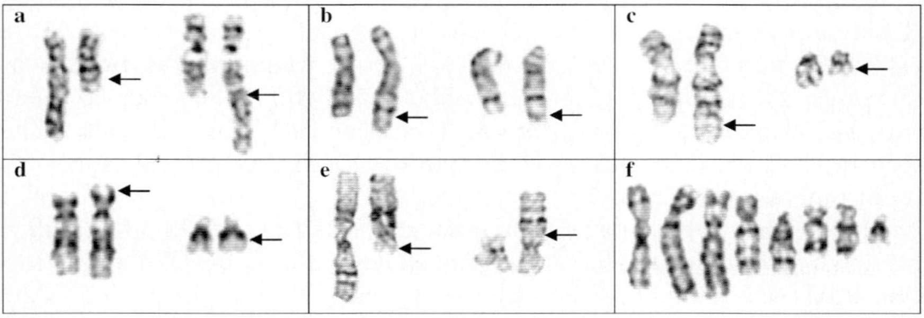

Fig. 30.2 Common abnormalities associated with lymphoma or lymphocytic leukemia. **a** Translocation (4;11)(q21;q23). **b** Translocation (8;14)(q24;q11). **c** Translocation (9;22)(q34;q11.2). **d** *Cryptic* translocation (12;21). The chromosomes pictured are from a case reported as normal in routine chromosome analysis but that was 95% positive for TEL/AML1 rearrangement by fluorescence in situ hybridization. This picture illustrates the difficulty in identifying this translocation in routine chromosome analysis. **e** Translocation (1;19)(q23;p13). **f** Common trisomies in hyperdiploidy (chromosomes X, 4, 6, 10, 14, 17, 18, and 21)

30.4 Solid Tumors

Compared with the details reported on chromosome analysis in hematological neoplasms, similar information on solid tumors is very limited. This is mainly due to reasons such as specimen collection requiring an operative procedure, symptoms only being noticeable when malignancy has already progressed, and repeated specimens not being available in most cases. Secondly, each solid tumor needs to be processed differently and may need different cell culture media, etc. In general, simple chromosome alterations are found in solid tumors of benign or initial malignant stages, whereas highly malignant tumors have more complex changes sometimes involving multiple abnormal cell lines. Important chromosomal aberrations, possible genetic alterations, and estimated prevalence in different solid tumor types are presented in Table 30.5.

Some specific chromosome abnormalities in solid tumors are presented in Fig. 30.3.

Cytogenetic analysis has become an integral part of the tumor diagnosis and can provide vital evidence for differential diagnosis. The (X;18)(p11;q11) is pathognomonic for synovial sarcoma and with the availability of FISH probes can help in making a diagnosis in the absence of dividing cells. The t(12;22)(q13;q12) is found as a characteristic abnormality in clear-cell sarcoma as well as in a minor subset of myxoid liposarcoma. In spite of recent advances in cytogenetic analysis of solid tumors, the impact on deciding prognosis is significantly less than for hematological neoplasms.

Table 30.5 Important chromosomal rearrangements in solid tumors

Tumor type	Translocation	Fusion transcript	Estimated prevalence (%)
Ewing sarcoma/peripheral neuroectodermal tumor	t(11;22)(q24;q12)	EWS-FL11	95
	t(21;22)(q22;q12)	EWS-ERG	5
Desmoplastic small round cell tumor	t(11;22)(p13;q12)	EWS-WT1	99
Myxoid/round-cell liposarcoma	t(12;16)(q13;p11)	TLS-CHOP	>95
	t(12;22)(q13;p12)	EWS-CHOP	<5
Extraskeletal myxoid chondrosarcoma	t(9;22)(q22;q12)	EWS-CHN	75
	t(9;17)(q22;q11)	TAF2N-CHN	25
Malignant melanoma of soft parts/ clear cell sarcoma	t(12;22)(q13;p12)	EWS-ATF1	>90
Synovial sarcoma	t(X;18)(p11.23;q11)	SYT-SSX1	65
	t(X;18)(p11.21;q11)	SYT-SSX2	35
Alveolar rhabdomyosarcoma	t(2;13)(q35;q14)	PAX3-FKHR	72
	t(1;13)(p36;q14)	PAX7-FKHR	9
Dermatofibrosarcoma protuberans	t(17;22)(q22;q13)	COL1A-PDGFB	>99
Congenital fibrosarcoma	t(12;15)(p13;q25)	ETV6-NTRK3	>99
Alveolar soft part sarcoma	t(X;17)(p11;q25)	ASPL-TFE3	>99

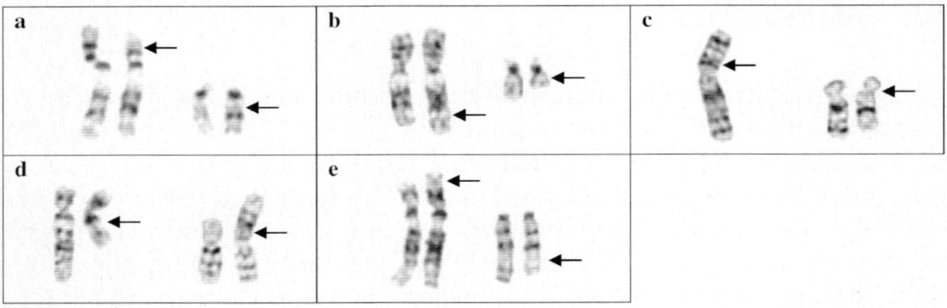

Fig. 30.3 Important chromosome abnormalities in solid tumors. **a** Translocation (11;22)(p13;q12) in desmoplastic round cell tumor. **b** Translocation (11;22)(q24;q12) in Ewing sarcoma. **c** Translocation (X;18)(p11.23;q11) in synovial sarcoma. **d** Translocation (12;16)(q13;p11) in liposarcoma. **e** Translocation (12;15)(p13;q25) in congenital fibrosarcoma and congenital mesoblastic nephroma

There are two major limitations of classical cytogenetic analysis: (1) it requires the presence of mitotic cells; (2) even in very good quality banding preparations, the resolution is very limited and it may not be possible to detect translocations such as t(12;21) in case of pre-B-cell ALL. At the 550-band level each band represents approximately 5×10^6–10×10^6 base pairs. The FISH technique provides a very useful advancement in overcoming resolution limitations of G-banded chromosome analysis. In recent years, techniques such as comparative genomic hybridization (CGH), multiplex FISH, and spectral karyotyping have become available and are briefly discussed next.

30.5 Molecular Cytogenetics

When conventional analysis fails to detect chromosomal defects with the use of banded chromosome spreads or fails to provide any mitotic activity, alternative techniques may reveal the presence of cytogenetic abnormalities or gene alterations commonly associated with them. These include molecular cytogenetics in the form of FISH (Pinkel et al. 1986) and the PCR, which, although plagued by their own limitations, circumvented some technical problems associated with routine cytogenetic analysis. These recent techniques have tremendously increased the sensitivity level for detection of genetic defects in cancer specimens (Fan 2002).

The FISH technique involves the hybridization of single-stranded DNA probe(s) to hybridize with homologous single-stranded sequence(s) in chromosomes of metaphase or interphase nuclei. Various types of FISH probes are available, such as whole-chromosome-specific or paint probes, centromere-specific or locus-specific. These probes are now available in different forms, such as dual color break apart probes, dual color single fusion, dual color dual fusion, and so on. Each type of probe has a specific ability and provides very specific answers about deletions or translocations or both (Fig. 30.4).

Probe Type	*Application*	*Signals*
LSI: Locus-Specific Identifier Dual-Color Break-Apart Rearrangement Probes	Useful for detecting multiple translocations involving a specific genetic breakpoint. Normal cells display a yellow fusion signal. A rearrangement will split the signal into a separate red and green signal	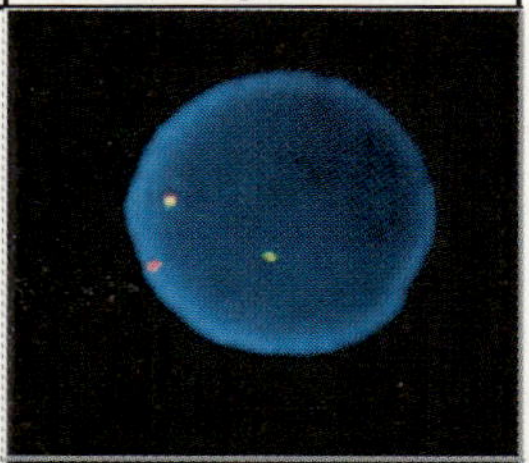
LSI: Locus-Specific Identifier Dual-Color, Single Fusion Probes.	Detects high percentages of cells possessing specific chromosomal translocations. Normal cells will display 2 red and 2 green signals. Translocations show as a single fusion signal.	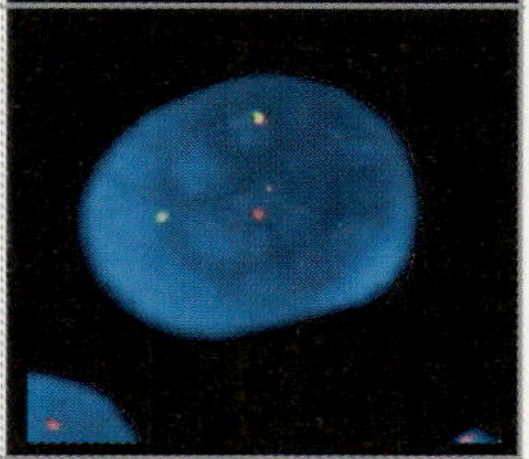
LSI: Locus-Specific Identifier Dual-Color, Dual Fusion Probes.	Detects low percentages of cells possessing specific chromosomal translocations. The dual fusion design greatly reduces false positive signals due to random co-localization of chromosomes. Translocations appear as 2 fusion signals. One on each derivative chromosome.	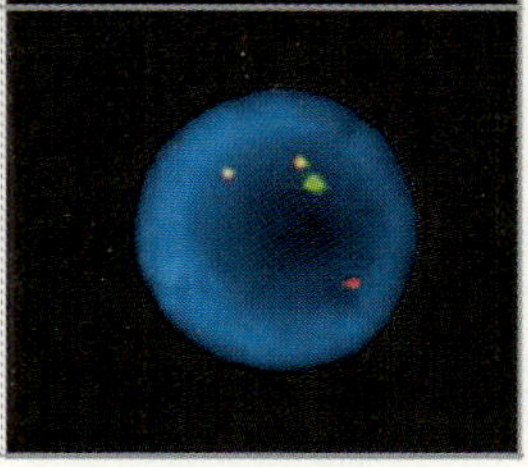

Probe Type	*Application*	*Signals*
CEP: Chromosome Enumeration Probe	Chromosome-specific probes which hybridize to satellite DNA sequences. Useful for enumeration of specific chromosomes in both interphase and metaphase cells.	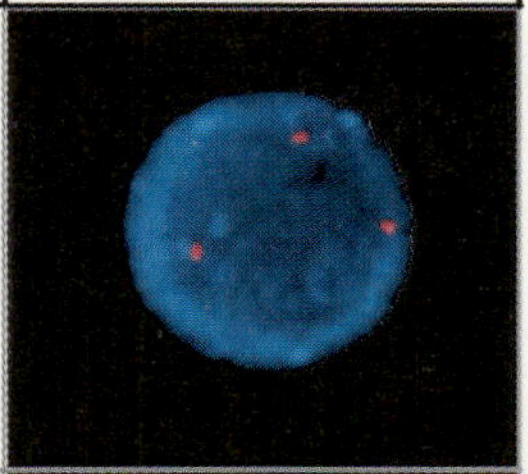
WCP: Whole Chromosome Paint Probes	Labels entire chromosome. The cell shows presence of t(9;22), t(9;11) and deletion of 17p in same cell with help of WCPs for 9, 11, 17 and 22. Useful for detection of complex or unknown translocations as well as identification of marker chromosomes.	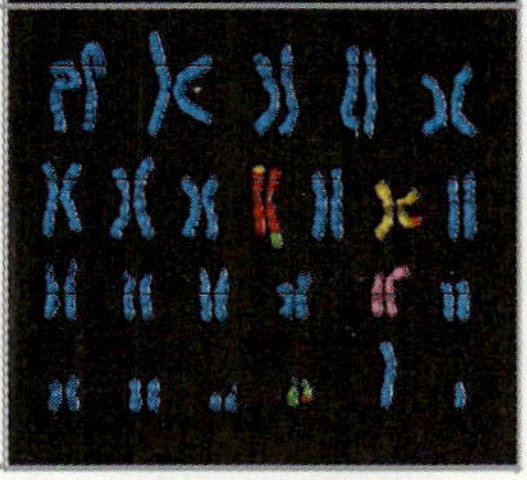

Fig. 30.4 Types of probes available and commonly used in molecular cytogenetics. Dual-color probes are labeled *red* (one locus) and *green* (second locus). When a translocation occurs the red and green signals come very close to each other on the derivative chromosome. This proximity (fusion) of red and green appears as a "*yellow*" signal under the fluorescence microscope. Normal is two red + two green, abnormal is one red + one green + one or two fusion signals on an interphase nucleus. In the break-apart probe the 3′ end of a locus is labeled green and the 5′ end of same locus is labeled red. So normal condition is a fusion (yellow) signal. When the locus is involved in translocation the 3′ and 5′ regions (green and red) separate and the signals on an abnormal interphase nucleus are one fusion (yellow), one green (3′), and one red (5′), separated owing to translocation. These probes are useful when the same locus is involved in translocation with several different chromosomes in causing clinically identical cancer

The molecular probes are fluorescently tagged either directly or by various "sandwich" methods and are detected with a fluorescence microscope. In this way, genetic alterations can be detected at the single-cell level, in a dividing or a nondividing cell, and correlated with morphology and immunophenotypic features of that cell. FISH is particularly advantageous in patients with acute leukemia in clinical remission. In this situation, cells carrying a patient's diagnostic cytogenetic aberration can be readily enumerated by counting fluorescence signals or hybrid (merged) signals when using translocation-specific probes. The major limitation of the FISH technique is that only those aberrations that are targeted by the probes used will be detected. In other words one can get an answer only specific to the question, whereas routine chromosome analysis provides visualization of all the abnormalities in the cell under the microscope. FISH is therefore not recommended for establishing an aberrant karyotype at the time of diagnosis, unless the clinical presentation is suggestive of a specific cytogenetic abnormality [e.g., abnormal eosinophilia suggesting inv(16)]. Although excellent for monitoring the t(9;22) in CML, for instance, when monitoring disappearance of the translocation under imatinib mesylate therapy, FISH will not detect the various cytogenetic abnormalities (+8, iso17q, +19, or +Ph) associated with blast crisis of CML.

The use of FISH technology involves tedious validation, quality assurance, and quality control issues as well as the availability of a fluorescence microscope and the imaging system. However, in addition to aiding in detecting a known abnormality on an interphase nucleus, the FISH technique helps very

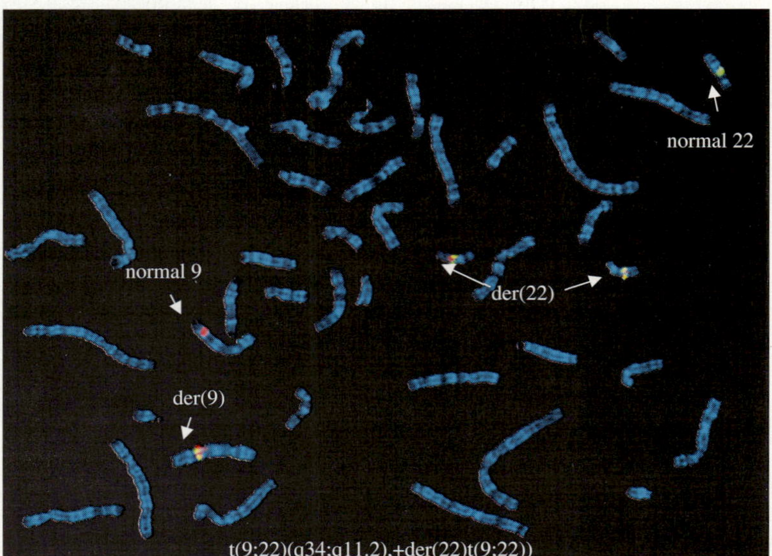

Fig. 30.5 Metaphase fluorescence in situ hybridization. A translocation (9;22) with an extra copy of the Philadelphia chromosome [*der(22)*]

effectively in characterization of complex abnormalities and in identification of the marker chromosomes. It has also proven very useful in the detection of submicroscopic deletions (RB1, p53), amplifications (e.g., MLL, MYC, HER2), or masked rearrangements (Fig. 30.5).

In addition to the FISH protocols briefly described here, other FISH-based protocols are being used in various situations. These include metaphase CGH, which can identify chromosomal losses or gains in tumor cells (Kallioniemi et al. 1992), multiplex FISH (Speicher et al. 1996), spectra karyotyping (Schrock et al. 1996), or array CGH (Schoch et al. 2004). Similarly

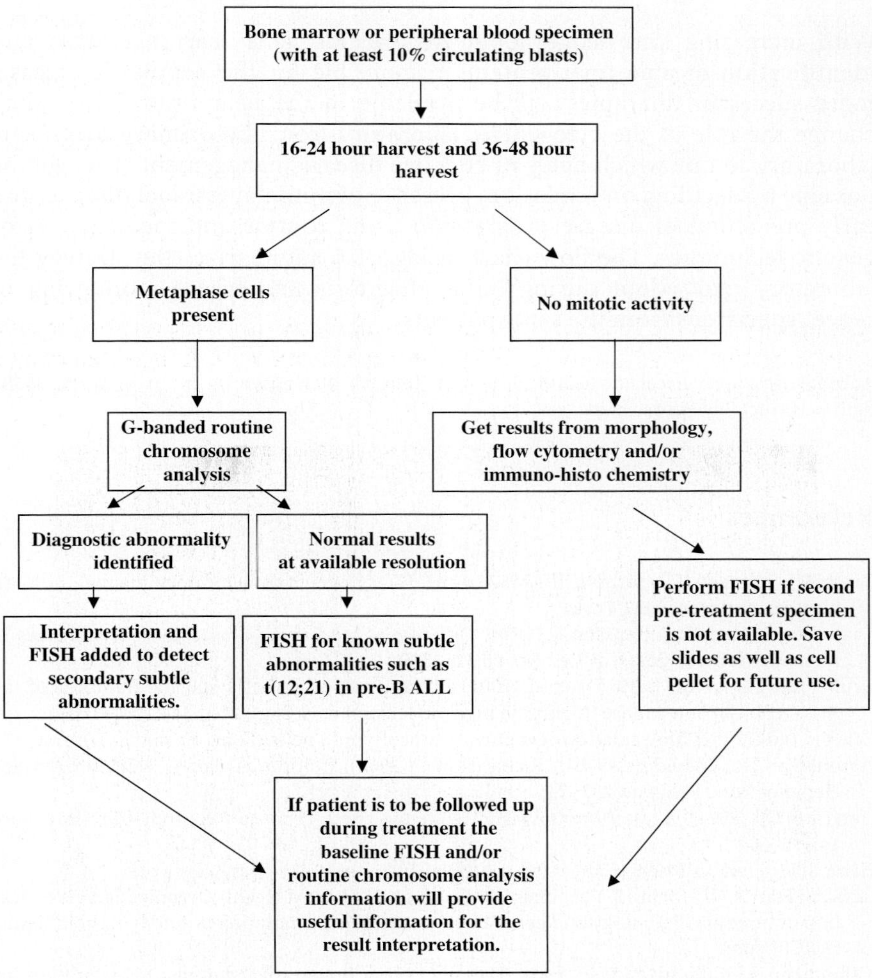

Fig. 30.6 Summary of useful strategies during initial diagnosis and subsequent treatment-monitoring of a patient. *FISH* fluorescence in situ hybridization, *ALL* acute lymphocytic leukemia

evolving newer techniques are coming into the clinical arena very rapidly. An emerging technology that addresses the limitations of metaphase CGH is microarray CGH. This technology offers multiple platforms and has revolutionized the field of conventional cytogenetics. Those readers interested in learning more of these newer technologies should refer to Fan (2002) and other recent review articles on these topics.

30.6 Future for Cytogenetics in Cancer Investigations

With increasing knowledge about the specific gene rearrangements and identification of abnormal proteins responsible for the neoplastic disease, more successful therapies will be available in the near future. This shall change the role of the cytogenetic laboratory from the primary diagnostic laboratory to one which helps in effective disease management. This will be possible by deciding on remission, presence of minimal residual disease, and early prediction of disease progression using routine and molecular cytogenetic techniques. The flow chart in Fig. 30.6 summarizes the strategy the laboratory may adopt during initial diagnosis and future monitoring of repeat specimens from the same patients.

Acknowledgement: Excellent technical help of Kenneth A. Hughes in the preparation of the figures is sincerely appreciated.

References

Barch MJ, Knutsen T, Spurbeck JL (1997) The AGT cytogenetics laboratory manual, 3rd edn Lippincott-Raven, New York

Casperson T, Zech L, Johansson C (1970) Differential binding of alkylating fluorochromes in human chromosomes. Exp Cell Res 60:315–319

Druker BJ, Talpaz M, Resta DJ et al (2001) Efficacy and safety of a specific inhibitor of the BCR-ABL tyrosine kinase in chronic myeloid leukemia. N Eng J Med 344:1031–1037

Fan Y-S (ed) (2002) Molecular cytogenetics, protocols and applications. Humana, Totowa

Harousseau J-L, Shaughnessy J Jr, Richardson P (2004) Multiple myeloma. Hematol Am Soc Hematol Educ Program 237–256

Harrison CJ (2001) Acute lymphoblastic leukemia. Best Practice Research Clin Haematol 14:593–607

Heim S, Mitelman F (1995) Cancer cytogenetics, 2nd edn. Wiley–Liss, New York

Jaffe ES, Harris NL, Stein H, Vardiman JW (eds) (2001) World Health Organization classification of tumours. Pathology and genetics of tumours of haematopoietic and lymphoid tissue. IARC, Lyons

Kallioniemi A, Kallioniemi O-P, Sudar D et al (1992) Comparative genomic hybridization for molecular cytogenetic analysis of solid tumors. Science 258:818–821

Keagle MB, Gersen SL (1999) Basic laboratory procedures. In: Gersen SL, Keagle MB (eds) The principles of clinical cytogenetics. Humana, Totowa, pp 71–89

31 Chromosomal Aberration in Peripheral Blood Lymphocytes of Healthy Subjects and Risk of Cancer

Stefano Bonassi

Abstract

The pioneering papers of Theodor Boveri opened the way to extensive research on the mechanism linking chromosomal abnormalities to the pathogenesis of cancer. As a result of this effort, robust theoretical and empirical evidence accumulated, and an increased cancer risk was postulated in healthy subjects with high levels of chromosomal aberrations (CA). The first cohort study which directly evaluated the presence of this association was carried out within the framework of a collaborative study in northern Europe in 1994. Subjects classified as having high frequency of CA showed incidence rates that were more twice those of the general population. The results of this first study were replicated a few years later in a parallel research initiative carried out in Italy, and the subsequent pooled analysis of these two cohorts published in 1998 refined the quantitative estimate of the CA–cancer association. More recent results referring to cohorts in the Czech Republic (2005) and in five European countries (2007) contributed further evidence, suggesting a stronger association with chromosome-type aberrations. A pooled analysis of all cohort studies published involving more that 22,000 subjects from 11 countries has been carried out within the framework of the European project Cytogenetic Biomarkers and Human Cancer Risk, and preliminary results are discussed. The leading priorities to be addressed in the field are the application of existing evidence to the control of occupational and environmental exposure to genotoxic agents, and the evaluation of individual cancer risk assessment procedures based on markers of chromosome damage.

31.1 Biological Rationale

The first consistent theory postulating the importance of chromosomes for cancer was developed at the turn of the twentieth century by Theodore Boveri (1914), who elaborated the idea that loss or gain of chromosomes by aberrant mitoses may lead to cancer. This pioneering view was formalized nearly 50 years later, in the early 1960s, with the description of the first specific chromosome abnormality in a human primary tumor, i.e., the Philadelphia

Günter Obe and Vijayalaxmi (Eds.)
Chromosomal Alterations: Methods, Results and Importance in Human Health

chromosome (Novell and Hungerford 1960), and this was substantiated in the 1970s with the introduction of banding techniques.

During the 1980s, a number of new cancer-specific chromosomal rearrangements (many of them translocations) mostly occurring in hematological tumors were found. These alterations allowed a deeper insight into mechanistic aspects; in particular it was clear that many genes affected by chromosomal rearrangements were involved in some critical stages in cell growth, development, or survival. These results led to a better understanding of the mechanisms of formation of chromosomal alterations and their role in cancer development (Rowley 1998).

The model linking chromosome breakage to the activation/deactivation of genes relevant to early stages of carcinogenesis was further refined in the 1990s by the use of the fluorescence in situ hybridization technique. These studies contributed to the understanding of the role of DNA double-strand breaks (DSBs) in the formation of chromosomal aberrations (CA). A further refinement in the understanding of the origin of CA was provided by a number of papers published at the turn of the millennium, which revealed the major role of homology-dependent and homology-independent pathways of DSB repair (Richardson and Jasin 2000; Pfeiffer et al. 2000).

The clastogenic action of many environmental agents has well been documented and their role in turning on the process leading from DSB to altered function of cancer genes is generally recognized. It is also interesting to note that many DSBs are formed endogenously as a consequence of physiologic processes, such as replication, meiosis, V(D)J recombination, and DNA repair (Pfeiffer et al. 2000; Vilenchik and Knudson 2003). Although endogenous DSBs are repaired with high fidelity, errors in DSB repair are considered to contribute significantly to human cancer rate (Vilenchik and Knudson 2003).

The role of chromosomal alterations in the early stages of carcinogenesis provides support to observations, from both laboratory and human studies, linking the clastogenic effect of many compounds to their carcinogenic properties (Bonassi 1999). Furthermore, the availability of a reliable model of causality supports the biological plausibility of studies linking chromosomal damage to the risk of cancer, and is the driving force that justifies the use of the CA assay for cancer-prevention policies.

31.2 The Epidemiologic Approach

The identification of an accurate mechanistic model corroborates the presence of association between the frequency of CA in healthy subjects and the risk of cancer; however, the role of chromosome damage in carcinogenesis is substantiated by more evidence. The high frequencies of CA in patients affected by cancer, the parallel increase of CA and risk of cancer in subjects affected by genetic diseases interfering with DNA repair, and the already mentioned association between clastogenic and carcinogenic properties are the most convincing.

All these observations, despite their importance, provide only circumstantial evidence for a causal relationship and cannot provide any real data about the strength of the association between CA and cancer in the general population, which would be required from a public health perspective. The availability of risk estimates based on the frequency of a biomarker in healthy individuals is a prerequisite to validate the use of CA and other potential surrogate markers for cancer-prevention purposes. This approach is particularly appropriate when the biomarker studied is possibly affected by the presence of the disease, i.e., the so-called *reverse causality*; in this case prospective cohort studies are preferable for the biomarker validation (Rothman 1986).

Major problems in planning a cohort study based on biomarker frequency are the small size of most biomonitoring studies, the heterogeneity of technical protocols among laboratories and the differences in methods over the years, the high number of subjects without demographic data, and the easy misclassification due to the use of only a single (or a few) measure(s) in their life. A cohort approach is particularly problematic for those biomarkers, such as cytogenetic biomarkers, which need fresh tissues and therefore cannot be easily measured in frozen samples.

Most published cohort studies linking CA frequency to the risk of cancer adopted the methodology of the historical cohort (Rothman 1986).

31.3 Literature Review of Cohort Studies Linking CA Frequency to the Risk of Cancer

The first cohort study on healthy subjects screened for CA (but also for sister-chromatid exchanges, SCE, and micronuclei frequency) which had enough statistical power to evaluate the presence of a causal link was published in 1994 as a result of a collaborative initiative carried out in northern Europe (Hagmar et al. 1994). A group of 3,182 subjects from Finland, Sweden, Norway, and Denmark, free of cancer at the time of cytogenetic analysis, were followed up from 1970 to 1991. Eighty-five cancer cases were found. In order to standardize for the interlaboratory variation, the results were trichotomized for each laboratory into three strata: low (1–33 percentile), medium (34–66 percentile), or high (67–100 percentile). Subjects classified as having a high frequency of CA showed incidence rates that were more than twice those of the general population (standardized incidence ratio, SIR 2.1, 95% confidence interval, CI, 1.5–2.8), and a highly significant trend of SIRs by the frequency of CA was observed ($p = 0.0009$).

These findings were confirmed 1 year later in an independent cohort of 1,455 subjects gathered from 17 Italian laboratories and followed up from 1969 to 1994 for a total of 16,190 person-years. A significant increase in mortality rate for all cancers was found in subjects with high and medium frequency of CA (standardized mortality ratio, SMR, 1.79, 95% CI 1.02–2.90; and SMR 1.82, 95% CI 1.11–2.81, respectively) (Bonassi et al. 1995). In 1998 a first collaborative European study, pooling the cohorts from Nordic countries and Italy, produced a joint data analysis and a further extension of the follow-up. The results strengthened the confidence in the previous findings, while the survival analysis improved the understanding of the CA–cancer relationship, showing highly different patterns of survival in the three CA frequency categories (Hagmar et al. 1998). To investigate the possible effect of smoking habit and occupational exposures to carcinogens as confounders or effect modifiers of the relationship evaluated, a case-control study nested within the cohort was performed. The results showed that CA frequencies could predict cancer independently of exposure to major carcinogens, making the point that the predictivity of high rates of CA does not require such exposures (Bonassi et al. 2000). The first non-European report showing an association between CA and cancer risk was published in 1999 by a Taiwanese group (Liou et al. 1999). A case-control study nested in a small cohort of 686 residents in a blackfoot endemic area showed that a frequency of chromosome-type CA over the median is highly predictive of cancer risk (odds ratio 12.0, 95% CI 1.56–92.29). No association was found with chromatid-type CA and SCE.

A new independent cohort study was carried out in the Czech Republic, based on data from a large national program of occupational safety surveillance ongoing since 1975. Various groups of workers with heterogeneous exposures had been investigated. Overall, 11,834 subjects were followed up for cancer incidence and mortality from 1975 to 2000, accounting for a total of 113,967 person-years (Rossner et al. 2005). A significant association was found between the overall cancer incidence and the presence of chromosome-type aberrations (chromosome-type breaks, ring chromosomes, marker chromosomes, and dicentrics) (relative risk, RR, for high vs. low CA frequency 1.24, 95% CI 1.03–1.50), but not chromatid-type aberrations (chromatid-type breaks and chromatid exchanges). Stomach cancer showed a strong association with the frequency of total CA (RR 7.79, 95% CI 1.01–60.0). The predictivity of CA observed in subjects exposed to various classes of carcinogens did not significantly differ from that for the group of nonexposed subjects.

The issue whether subclasses of CA are more effective in predicting cancer incidence was specifically addressed in a paper of the Nordic–Italian cohorts, showing no evidence of a different predictivity (Hagmar et al. 2004). This result was contradicted by the latest European cohort study, which evaluated a cohort of 6,430 individuals from nine laboratories in Croatia, Hungary, Lithuania, Poland, and Slovakia, and included CA surveys performed during 1978–2002. In this study 224 cancer cases were observed. The RR of cancer was 1.57 (95% CI 1.08, 2.28) for a medium CA level and 1.75 (95% CI

1.20, 2.55) for a high level. The RR for the group with a high frequency of chromosome-type CA was 1.47 (95% CI 1.11, 1.93), while that for chromatid-type CA was 1.09 (95% CI 0.82, 1.45). The analyses of risk of specific cancers were limited by small numbers, but the association was stronger for stomach cancer (Boffetta et al. 2007).

The contrasting findings concerning different predictivity of subclasses of CA are possibly due to intrinsic limitations of data. From a mechanistic point of view, chromosome-type aberrations occur directly from DSB in G0 lymphocytes, or during repair of lesions other than DSB (e.g., adducts), and therefore seem to be more causally related to early stages of carcinogenesis than chromatid-type aberrations, mostly arising during replication of damaged DNA in culture. The interlaboratory difference in the criteria used for CA classification, and the measurement of specific chromosome damage in a surrogate tissue, made the historical cohort a poor study design for addressing this issue.

A summary of the major findings from cohort studies on CA frequency and risk of cancer is reported in Table 31.1.

31.4 Pooled Analysis of Published Data

The experience from clinical trials indicates that a summary of the published studies (meta-analysis) could be useful even in observational epidemiology and molecular genetic research to obtain organized information from the existing results and to address future studies. Another way to summarize data in observational studies is to pool individual records and reanalyze the data. This approach, despite the highest costs, has several advantages over the classic meta-analysis of published data, mainly the possibility to perform interaction tests and subgroup analysis including dose–response curves (Taioli and Bonassi 2003). The pooled approach has been widely used in those cohort studies described above to validate predictive biomarkers. In particular, a European effort to collect all available information about population studies using cytogenetic biomarkers has been conducted within the framework of the research project Cytogenetic Biomarkers and Human Cancer Risk (CRB). This European collaborative project was established in 2001 to further study cancer risk prediction by cytogenetic biomarkers. The project had a number of scientific objectives, which included the quantitative assessment of risk prediction, the role of CA subclasses, the association with specific cancer sites, and the role of genetic susceptibility. The most challenging aim of the project was to create a pooled cohort of all cohort studies published so far. This achievement required 5 years of work, and in Table 31.2 a general description of the pooled database is reported that will be analyzed to address all items formulated in the rationale of the CRB project.

Table 31.1 Overview of published cohort studies evaluating chromosomal aberrations as cancer-risk biomarkers (all studies categorized subjects in three groups of chromosomal aberration frequency, i.e., low, medium, and high, and relative risks in the table are reported accordingly)

Study	Country	Subjects	Person-years	Follow-up period	Cancer cases (deaths)	SIR (95% CI)	SMR (95% CI)	RR (95% CI) incidence	RR (95% CI) mortality
The Nordic Study Group on the Health Risk of Chromosome Damage (1990)	Sweden, Finland, Norway	1,787	7,547	1970–1985	26	0.9 (0.3–1.9) 0.9 (0.3–2.0) 1.8 (0.9–3.0)			
Hagmar et al. (1994)	Sweden, Finland, Norway, Denmark	1,979	17,666	1970–1988	66	0.9 (0.5–1.4) 0.7 (0.3–1.2) 2.1 (1.5–2.8)			
Bonassi et al. (1995)	Italy	1,455	16,190	1969–1994	(44)		0.8 (0.7–1.6) 1.8 (1.0–2.9) 1.8 (1.1–2.8)		
Hagmar et al. (1998)	Sweden, Finland, Norway, Denmark, Italy	3,541	43,827	1970–1995	91 (64)	0.8 (0.5–1.2) 0.8 (0.5–1.3) 1.5 (1.1–2.0)	0.8 (0.5–1.8) 1.2 (0.7–1.8) 2.0 (1.4–2.9)	1.0 (–) 1.3 (0.8–2.2) **2.1 (1.3–3.5)**	1.0 (–) 1.3 (0.7–2.6) **2.5 (1.3–4.7)**
Rossner et al. (2005)	Czech Republic	11,834	113,967	1975–2000	485			1.0 (–) 1.2 (0.9–1.5) 1.1 (0.9–1.4) 1.0 (–)[a] **1.2 (1.03–1.50)**	
Boffetta et al. (2007)	Croatia, Hungary, Poland, Slovakia, Lithuania	6,430	54,655	1978–2002	224			1.0 (–) **1.8 (1.29–2.7)** **1.8 (1.2–2.7)**	

Relative risk (*RR*) values in *bold* correspond to $p<0.05$.
SIR standardized incidence ratio, *CI* confidence interval, *SMR* standardized mortality ratio
[a]Chromosome-type chromosomal aberration (high vs. low)

Table 31.2 Selected characteristics of the study population included in the pooled European study on chromosomal aberration (*CA*) frequency and risk of cancer

Country	Subjects	Cancer cases	Period of test	Mean CA frequency (SD)	Mean age (SD)	Males (%)	Exposed (%)	Current smokers (%)
Croatia	1,320	24	1982–2000	3.5 (2.3)	44.4 (9.7)	55.8	100.0	44.8
Hungary	840	46	1978–2001	1.9 (2.1)	45.3 (12.5)	64.8	25.5	40.1
Lithuania	812	24	1981–2002	3.3 (2.0)	48.7 (9.4)	78.4	91.6	48.4
Slovakia	2,994	55	1985–2001	2.1 (1.7)	47.8 (10.5)	57.8	76.4	42.5
Poland	456	22	1981–2001	1.5 (1.8)	55.1 (14.5)	84.2	57.5	55.3
Czech Republic	11,991	363	1975–1999	2.4 (1.8)	47.5 (11.7)	55.4	84.5	39.2
Sweden	756	48	1970–1987	1.8 (1.8)	54.6 (11.7)	72.9	–	–
Norway	471	48	1970–1988	1.5 (1.3)	57.6 (14.4)	78.0	65.8	38.2
Denmark	197	6	1987	1.8 (1.5)	50.7 (10.8)	100.0	66.0	43.1
Finland	557	39	1974–1988	2.3 (2.1)	54.6 (9.6)	69.5	–	–
Total incidence	20,394	675	1970–2002	2.4 (1.9)	48.2 (11.7)	60.0	75.5	40.9
Italy (mortality)	1,964	99	1965–1995	3.6 (4.2)	56.3 (12.9)	72.4	62.1	29.8
Total	22,358	774	1965–2002	2.5 (2.2)	48.2 (11.7)	60.8	74.2	39.8

SD standard deviation

31.5 Preliminary Results and Open Issues

As evident from Table 31.2, the pooled cohort was largely unbalanced, mostly owing to the large size of the Czech cohort that represented by itself about 50% of the whole database. Despite this limitation and the obvious heterogeneity due to discrepancy in methods and different life-styles in various countries, reliable estimates of the association between CA frequency and the risk of cancer could be provided. This was done by comparing cancer incidence rates for the medium and high levels versus the low level, after adjusting for the confounding effects of age, gender, smoking habit, and occupational exposure to mutagens or carcinogens. A *random effect* term was then included in the models to adjust for the differences in cancer rates occurring among different countries. The presence of effect modification was tested by computing the log-likelihood ratio test for two hierarchical models, the first with and the second without interaction terms involving CA (Lindsey 1995).

Preliminary results showed a significant increase of cancer risk in subjects with medium and high frequencies of CA when compared with those in the lowest tertile of the CA distribution. The increase of RRs ranged between 30 and 46%, with the highest risks found in the Hungarian and Norwegian cohorts, 4.24 and 2.21, respectively. The higher risk described in the Czech (Rossner et al. 2005) and in the new central and eastern European countries (Boffetta et al. 2007) cohorts for stomach cancer was confirmed by the pooled reanalysis which in subjects in the highest tertile found a RR for this tumor 3 times higher than those in the lowest tertile ($p<0.01$). Finally, the evidence obtained from these two cohorts that chromosome-type CA are a better predictor for cancer risk than chromatid type CA was largely confirmed by the pooled analysis, which reported the maximum risk in subjects with medium and high frequency of this CA type (1.32 and 1.44, respectively, $p<0.01$).

The large amount of data available from the pooled analysis, however, cannot address all open issues. Among these, establishing the role of individual genetic variability and, as a consequence, the reliability of individual cancer risk assessment are the most urgent. This item was partially addressed by a work package of the CRB project which evaluated, in a subgroup of the Italian, Norwegian, and Danish cohorts, the potential modification effect induced by most common genetic polymorphisms. Preliminary results from this ongoing study revealed that the role of genetic polymorphisms – at least those of the glutathione *S*-transferase family – is marginal, and its influence on the association between CA frequency and risk of cancer is negligible. This study, nested within the main European cohort, revealed an unexpected finding concerning the role of technical variability. In the frame of the nested study it was shown that the strength of the association between CA and cancer in the subgroup evaluated in Pisa – where frozen slides were homogeneously restained and rescored – was much more evident than in any other previous study, with an odds ratio for subjects with high CA frequency of 13.9 ($p<0.001$).

31.6 Conclusion

After 25 years of research aimed at demonstrating that the frequency of CA in healthy subjects can be used to predict cancer some general conclusions have been reached. All cohort studies measuring the link between CA frequency and risk of cancer found positive results. In some cases these results were unequivocal, as in the Nordic/Italian experience, in other cases the association was not so strong, such as in the Czech and Slovak cohorts, which nevertheless showed a significant association of cancer risk with chromosome-type CA. These findings, supported by the strong theoretical background linking chromosomal rearrangements to early stages of carcinogenesis, provide a significant contribution to the understanding of early stages of carcinogenesis, as shown by the high number of citations (about 500 overall) of the cohort studies described above, and the dramatically increasing trends of publications on chromosome damage as biomarkers for cancer (Bonassi et al. 2005).

Despite many issues remaining to be defined, especially from the mechanistic point of view, the current priorities to be addressed in the field are the application of existing evidence to regulation in the management of occupational exposure to genotoxic carcinogens, and the evaluation of an individual cancer risk assessment based on the use of markers of chromosome damage.

Future studies in this direction, corroborated by similar evidence from studies performed on the micronucleus assay (Bonassi et al. 2007) (but not on SCE, Norppa et al. 2006) are necessary to validate the use of biomarkers of chromosomal alterations in comprehensive cancer-prevention programs.

References

Boffetta P, van der Hel O, Norppa H, Fabianova E, Fucic A, Gundy S, Lazutka J, Cebulska-Wasilewska A, Puskailerova D, Znaor A, Kelecsenyi Z, Kurtinaitis J, Rachtan J, Forni A, Vermeulen R, Bonassi S (2007) Chromosomal aberrations and cancer risk: results of a cohort study from Central Europe. Am J Epidemiol 165:36–43

Bonassi S (1999) Combining environmental exposure and genetic effect measurements in health outcome assessment. Mutat Res 428:177–185

Bonassi S, Abbondandolo A, Camurri L, Dal Pra' L, De Ferrari M, Degrassi F, Forni A, Lamberti L, Lando C, Padovani P, Sbrana I, Vecchio D, Puntoni R (1995) Are chromosome aberrations in circulating lymphocytes predictive of a future cancer onset in humans? Preliminary results of an Italian cohort study. Cancer Genet Cytogenet 79:133–135

Bonassi S, Hagmar L, Strömberg U, Huici Montagud A, Tinnerberg H, Forni A, Heikkilä P, Wanders S, Wilhardt P, Hansteen I-L, Knudsen LE, Norppa H, for the European Study Group on Cytogenetic Biomarkers and Health (ESCH) (2000) Chromosomal aberrations in lymphocytes predict human cancer independently of exposure to carcinogens. Cancer Res 60:1619–1625

Bonassi S, Ugolini D, Kirsch-Volders M, Strömberg U, Vermeulen R, Tucker JD (2005) Human population studies with cytogenetic biomarkers: review of the literature and future prospectives. Environ Mol Mutagen 54:258–270

Bonassi S, Znaor A, Ceppi C, Lando L, Chang WP, Holland N, Kirsch-Volders M, Zeiger E, Ban S, Barale R, Bigatti MP, Bolognesi C, Cebulska-Wasilewska A, Fabianova E, Fucic A, Hagmar

L, Joksic G, Martelli A, Migliore L, Mirkova E, Scarfi' MR, Zijno A, Norppa H, Fenech M (2007) An increased micronucleus frequency in peripheral blood lymphocytes predicts the risk of cancer in humans. Carcinogenesis (in press). doi:10.1093/carcin/bgl177

Boveri T (1914) Zur Frage der Entstehung maligner Tumoren. Fischer, Jena Hagmar L, Brøgger A, Hansteen I-L, Heim S, Högsted B, Knudsen L, Lambert B, Linnainmaa K, Mitelman F, Nordenson I, Reuterwall C, Salomaa S, Skerfving S, Sorsa M (1994) Cancer risk in humans predicted by increased levels of chromosomal aberrations in lymphocytes: Nordic study group on the health risk of chromosomal damage. Cancer Res 54:2919–2922

Hagmar L, Bonassi S, Strömberg U, Brøgger A, Knudsen L, Norppa H, Reuterwall C (1998) Chromosomal aberrations in lymphocytes predict human cancer – A report from the European Study Group on Cytogenetic Biomarkers and Health (ESCH). Cancer Res 58:4117–4121

Hagmar L, Brøgger A, Hansteen I-L, Heim S, Högsted B, Knudsen L, Lambert B, Linnainmaa K, Mitelman F, Nordenson I, Reuterwall C, Salomaa S, Skerfving S, Sorsa M (1994) Cancer risk in humans predicted by increased levels of chromosomal aberrations in lymphocytes: Nordic study group on the health risk of chromosomal damage. Cancer Res 54:2919–2922

Hagmar L, Strömberg U, Bonassi S, Hansteen I-L, Knudsen LE, Lindholm C, Norppa H (2004) Impact of types of lymphocyte chromosomal aberrations of human cancer risk: Results from Nordic and Italian cohorts. Cancer Res 64:2258–2263

Lindsey JK (1995) Modelling frequency and count data. Oxford University Press, Oxford

Liou S-H, Lung J-C, Chen Y-H, Yang T, Hsieh L-L, Chen C-J, Wu T-N (1999) Increased chromosome-type chromosome aberration frequencies as biomarkers of cancer risk in a blackfoot endemic area. Cancer Res 59:1481–1484

Nordic Study Group on the Health Risk of Chromosome Damage (1990) An inter-Nordic prospective study on cytogenetic endpoints and cancer risk. Cancer Genet Cytogenet 45:85–92

Norppa H, Bonassi S, Hansteen I-L, Hagmar L, Strömberg U, Rössner P, Boffetta P, Lindholm C, Gundy S, Lazutka J, Cebulska-Wasilewska A, Fabiánová E, Šrám RJ, Knudsen LE, Barale R, Fucic A (2006) Chromosomal aberrations and SCEs as biomarkers of cancer risk. Mutat Res 600:37–45

Nowell P, Hungerford D (1960) A minute chromosome in human granulocytic leukemia. Science 132:1497

Pfeiffer P, Goedecke W, Obe G (2000) Mechanisms of DNA double-strand break repair and their potential to induce chromosomal aberrations. Mutagenesis 15:289–302

Richardson C, Jasin M (2000) Frequent chromosomal translocations induced by DNA double-strand breaks. Nature 405:697

Rossner P, Boffetta P, Ceppi M, Bonassi S, Smerhovsky Z, Landa K, Juzova D, Sram RJ (2005) Chromosomal aberration in lymphocytes of healthy subjects and risk of cancer. Environ Health Perspect 113:517–520

Rothman KJ (1986) Modern epidemiology. Little, Brown, Boston

Rowley JD (1998) The critical role of chromosome translocations in human leukemias. Annu Rev Genet 32:495–519

Taioli E, Bonassi S (2002) Methodological issues in pooled analysis of biomarker studies. Mutat Res 512:85–92

Vilenchik MM, Knudson AG (2003) Endogenous DNA double-strand breaks: Production, fidelity of repair, and induction of cancer. Proc Natl Acad Sci USA 100:12871–12876

Index

J

K

L

M

Printing: Krips bv, Meppel
Binding: Stürtz, Würzburg